Analysis and Control of Production Systems and Operations and Production Management

Analysis and Control of Production Systems and Operations and Production Management

Rajagopal Kurnool PhD MTech BTech

Professor and Head
Department of Mechanical Engineering
KSRM College of Engineering
Kadapa 516003
Andhra Pradesh

CBS

CBS Publishers & Distributors Pvt Ltd

New Delhi • Bengaluru • Chennai • Kochi • Mumbai • Pune
Hyderabad • Kolkata • Nagpur • Patna • Vijayawada

Analysis and Control of Production Systems and Operations and Production Management

ISBN: 978-81-239-2469-4

Copyright © Author and Publisher

First Edition: 2015

Published by Satish Kumar Jain for

CBS Publishers & Distributors Pvt Ltd
4819/XI Prahlad Street, 24 Ansari Road, Daryaganj, New Delhi 110 002, India.
Ph: 23289259, 23266861, 23266867   Website: www.cbspd.com
Fax: 011-23243014   e-mail: delhi@cbspd.com; cbspubs@airtelmail.in.
Corporate Office: 204 FIE, Industrial Area, Patparganj, Delhi 110 092

Ph: 4934 4934   Fax: 4934 4935   e-mail: publishing@cbspd.com; publicity@cbspd.com

Branches

- **Bengaluru:** Seema House 2975, 17th Cross, K.R. Road, Banasankari 2nd Stage, Bengaluru 560 070, Karnataka
  Ph: +91-80-26771678/79   Fax: +91-80-26771680   e-mail: bangalore@cbspd.com
- **Chennai:** 20, West Park Road, Shenoy Nagar, Chennai 600 030, Tamil Nadu
  Ph: +91-44-26260666, 26208620   Fax: +91-44-42032115   e-mail: chennai@cbspd.com
- **Kochi:** 36/14 Kalluvilakam, Lissie Hospital Road, Kochi 682 018, Kerala
  Ph: +91-484-4059061-65   Fax: +91-484-4059065   e-mail: kochi@cbspd.com
- **Mumbai:** 83-C, Dr E Moses Road, Worli, Mumbai-400018, Maharashtra
  Ph: +91-22-24902340/41   Fax: +91-22-24902342   e-mail: mumbai@cbspd.com
- **Pune:** Bhuruk Prestige, Sr. No. 52/12/2+1+3/2 Narhe, Haveli (Near Katraj-Dehu Road Bypass), Pune 411 041, Maharashtra
  Ph: +91-20-64704058, 64704059, 32392277   Fax: +91-20-24300160   e-mail: pune@cbspd.com

Representatives

- **Hyderabad**   0-9885175004
- **Nagpur**   0-9021734563
- **Vijayawada**   0-9000660880
- **Kolkata**   0-9831437309, 0-9051152362
- **Patna**   0-9334159340

Printed at Goyal Offset Printers

Preface

This book covers various topics of production systems and production and operations management with special emphasis on latest techniques. This book with its 35 chapters covers almost the complete syllabus of subjects such as industrial engineering, management science, quality control and inspection, production and operations management, production planning and control, CAD/CAM and operations research being taught in various universities, management institutes and engineering institutes. Emphasis is on applications and solved problems. Solved problems have been used to clarify quantitative approaches and techniques. This book provides students with basic understanding of the issues facing today's operations and production managers.

This book is organized into 10 main parts, divided into 35 chapters.

Production is a process whereby raw material is converted into finished products. Production function encompasses the activities of procurement, allocation and utilization of resources. Part I is intended to provide number of topics related to production. These topics include basic production concepts, production costs involved in manufacturing operations, basic concepts of production and operations management, and various types of production systems.

Product design is concerned with development of a new product or new model of an existing product. Part II is intended to provide a number of topics related to product design. These topics include product lifecycle, product design and concurrent engineering and reliability measurement techniques.

Plant location refers to the choice of region and the selection of a particular site for setting up a business or factory. Facility layout is a plan of an optimum arrangement of facilities. Part III deals with facility location and layout. Methods of evaluating location alternatives, facility layout techniques, computerized facility layout techniques and assembly line balancing techniques.

Computer-aided manufacturing (CAM) refers to the use of computers in converting engineering designs into finished products. Part IV is intended to provide a number of topics related to computer-aided manufacturing (CAM). These topics include group technology, CAM software, industrial robots and numerical control and manufacturing systems.

To convert the product design into physical entity process plan is needed. Part V covers a number of topics related to process planning. These topics include process selection, different approaches for solving speed and feed problem and different approaches in computer-aided process planning.

Production planning includes aggregate production planning, material requirement planning and capacity planning. Production control attempts to take corrective action when actual production is not progressing as per plan. Part VI covers a number of topics related to production planning and control. These topics include forecasting techniques, aggregate production planning methods, capacity planning, MRP I, MRP II, ERP, supply chain management, line of balance, loading and scheduling techniques.

Companies which want to remain competitive have to provide quality product. Part VII is intended to provide a number of topics related to quality. These topics include the dimensions in product quality, quality costs, statistical quality control techniques, Taguchi methods, six sigma and total quality management. The use of computers for quality control of the product is called computer-aided quality control (CAQC). The two major parts of computer-aided quality control are computer-aided inspection (CAI) and computer-aided testing (CAT); CAI and CAT are presented in Chapter 35.

Network helps in designing, planning, coordinating, controlling and decision-making to accomplish the project economically in the minimum available time with the limited available resources. Part VIII is intended to provide a number of topics related to project management. These topics include PERT and CPM, crashing of network, resource leveling, resource aggregation and heuristic methods of project scheduling under multiple resource constraints.

Maintenance encompasses all those activities that relate to keeping facilities and equipment in good working order and making necessary repairs when breakdowns occur, so that the system can perform as intended. Part IX is intended to provide a number of topics related to maintenance. These topics include maintenance strategies, maintenance functions and equipment replacement.

Operations research models are also useful for making decisions. That is why these models are also called decision support system models. Operations research models consist of mathematical structure and these models are solved by using mathematical techniques. Part X is intended to provide a number of topics related to operations research. These topics include linear programming, queuing theory and inventory.

While writing this book I have referred to several books, papers and articles. I take this opportunity to thank the authors of books, papers and articles. In the course of writing this book I have received much encouragement from my management, director, principal and colleagues. I express my sincere gratitude to them. I wish to acknowledge my sincere thanks to my PhD guide Dr A Ramakrishna Rao for his inspiration and encouragement in this project. I wish to thank CBS Publishers & Distributors, in particular their editorial and production teams, for the meticulous and speedy processing of the manuscript. Any constructive criticism and suggestions will be generally appreciated.

Rajagopal Kurnool
PhD MTech BTech

Contents

Preface   *v*

Part I: Production Concepts, Productivity and Production Systems

1. Production Concepts and Productivity   1–13
   1.1   Introduction 1
   1.2   Production Concepts Using Mathematical Models 1
       1.2.1 Solved Problems 3
   1.3   Production System 4
       1.3.1 Production System Facilities 4
       1.3.2 Manufacturing Support System 4
   1.4   Production Function 4
   1.5   Production and Productivity 5
   1.6   Ways to Improve Productivity 6
       1.6.1 Benefits of Productivity Measurement in Organizations 6
   1.7   Productivity Improvement Techniques 6
   1.8   Production and Operation Management-based Productivity
       Improvement Techniques 6
   1.9   Solved Problems 7
   1.10 Manufacturing Operations 11
   1.11 Questions and Problems 12

2. Costs of Manufacturing Operations and Breakeven Analysis   14–29
   2.1   Elements of Cost 14
   2.2   Determining Total Cost 15
   2.3   Determining Selling Price 15
   2.4   Solved Problems 16
   2.5   Methods of Allocating Overhead Costs 18
   2.6   Solved Problems 19
   2.7   Depreciation 20
   2.8   Methods of Calculating Depreciation 20
   2.9   Solved Problems 21
   2.10 Nature of Costs 23
   2.11 Total Revenue 23
   2.12 Breakeven Point 23
   2.13 Solved Problems 25
   2.14 Questions and Problems 27

3. Production and Operations Management   30–36
   3.1   Introduction 30
   3.2   Production and Operations Management 30
   3.3   Objectives of Production and Operations Management 30
   3.4   Scope of Production and Operations Management 31
   3.5   Production Management v/s Operations Management 32
   3.6   Modern Approaches to Production and Operations Management 33
   3.7   Benefits of Production and Operations Management 35
   3.8   Questions 36

4. Production Systems or Manufacturing Systems   37–49
  4.1   Definition 37
  4.2   Objectives of Manufacturing System 38
  4.3   Components of Production System 38
  4.4   Types of Production Systems 39
         4.4.1   Continuous Production System 40
                 4.4.1.1   Mass Production System 40
                 4.4.1.2   Flow Production System 40
         4.4.2   Intermittent Production System 41
                 4.4.2.1   Batch Production System 41
                 4.4.2.2   Job Shop Production System 42
                 4.4.2.3   Project Type Production System 42
         4.4.3   Cellular Manufacturing Systems 43
         4.4.4   Automated Manufacturing Systems 43
         4.4.5   Flexible Manufacturing System 44
         4.4.6   Lean Production 44
         4.4.7   Agile Manufacturing 46
         4.4.8   Just in Time Production System (JIT) 47
         4.4.9   Kanban System 48
  4.5   Questions 49

Part II: Product Design

5. The Product Life Cycle, Product Design and Concurrent Engineering   50–59
  5.1   The Product Life Cycle 50
         5.1.1   Stages or Phases of Product Life Cycle 50
  5.2   Product Design 52
         5.2.1   Objectives of Product Design 52
         5.2.2   Factors Influencing Product Design 53
         5.2.3   Stages of Product Design 53
  5.3   Concurrent Engineering 55
         5.3.1   Ten Commandments of Concurrent Engineering 56
         5.3.2   Benefits of Concurrent Engineering 57
  5.4   Designing for Manufacturability (DFM) and Design for
         Assembly (DFA) 57
         5.4.1   DFM and DFA General Guidelines 58
         5.4.2   DFM Benefits 58
  5.5   Questions 59

6. Computer-aided Design   60–67
  6.1   Introduction 60
  6.2   Computers in Design and Manufacturing 60
  6.3   Product Cycle 60
  6.4   Computer-aided Design (CAD) 62
  6.5   Classification of Geometric Models 62
  6.6   Three-dimensional Geometric Construction Methods 63
  6.7   CAD Software 66
  6.8   Hardware Required 67
  6.9   Benefits of Computer-aided Design Software 67
  6.10   Questions 67

7. Reliability   68–76
  7.1   Definition and Objectives of Reliability 68
  7.2   Failure Rates and Mean Time Between Failures (MTBF) 68
         7.2.1   Solved Problems 69

7.3 Determination of Reliability of a System for a Period of Time 't'   70
    7.3.1 Series System   70
    7.3.2 Solved Problems   70
    7.3.3 Parallel System   71
    7.3.4 Solved Problems   72
7.4 Ways to Improve Reliability   74
7.5 Reliability and Quality   74
7.6 Cost of Reliability   75
7.7 Questions and Problems   76

Part III: Facility Layout and Material Handling

8. Facility Location   77–89
8.1 Definition   77
8.2 Factors Influencing Location   77
8.3 Methods of Evaluating Location Alternatives   78
    8.3.1 Cost Analysis   78
    8.3.2 Profit Analysis   78
    8.3.3 Return on Investment   79
    8.3.4 Factor Rating System   79
    8.3.5 Centre of Gravity (Grid) Method   79
    8.3.6 Solved Problems   80
8.4 Multiplant Locations   84
    8.4.1 Methods of Evaluating Multiplant Locations   85
        8.4.1.1 Cost-Volume Analysis   86
        8.4.1.2 Profit-Volume Analysis   86
        8.4.1.3 Linear Transportation Model   86
8.5 Questions and Problems   88

9. Facility Layout   90–114
9.1 Definition   90
9.2 Factors Influencing Plant Layout   90
9.3 Material Flow Patterns   90
9.4 Objectives of Plant Layout   92
9.5 Types of Layout   92
    9.5.1 Product Layout or Line Layout   93
    9.5.2 Functional Layout or Process Layout   93
    9.5.3 Combination Layout   94
    9.5.4 Fixed Position Layout   94
    9.5.5 Group Technology Layout or Cellular Layout   95
9.6 Tools and Techniques of Plant Layout   96
    9.6.1 Process Charts   96
    9.6.2 Travel Charts/From-To Charts   97
    9.6.3 Solved Problems   97
    9.6.4 Relationship Chart   102
    9.6.5 Solved Problems   103
9.7 Computer Packages for Layout Analysis   105
    9.7.1 Computerized Relative Allocation of
           Facility Technique (CRAFT)   105
    9.7.2 Automated Layout Design Program (ALDEP)   106
    9.7.3 Solved Problems   106
    9.7.4 Computerized Relationship Layout Planning (CORELAP)   111
    9.7.5 Solved Problems   111
9.8 Questions and Problems   113

10. Material Handling 115–124

10.1 Definition 115

10.2 Objectives of Material Handling 115

10.3 Determination of Cost of Handling 115

10.4 Manufacturing Cycle Time 116

10.5 Material Handling Principles 116

10.6 Need to Design a Material Handling System 117

10.7 Factors for Consideration of Material Handling System Design 117

10.8 Material Handling Equipment 117

10.9 Types of Material Handling Equipment 118

    10.9.1 Conventional Material Handling Equipment 118

    10.9.2 Computer-aided/Automated Material Handling Equipment 120

    10.9.3 Unit Load Principle 121

    10.9.4 Characteristics, Classification, Handling and Storage of Materials 122

    10.9.5 Factors Affecting the Selection of Material Handling Equipment 123

    10.9.6 Advantages of Material Handling System 123

    10.9.7 Disadvantages of Material Handling System 124

    10.9.8 Benefits of Automated Handling System (AHS) 124

10.10 Questions 124

11. Assembly Line Balancing 125–146

11.1 Introduction 125

    11.1.1 Solved Problem 126

11.2 Terms Used in Assembly Line 126

    11.2.1 Solved Problems 127

11.3 Line Balancing Algorithms 127

    11.3.1 Ranked Positional Weight Technique (RPW) 127

    11.3.2 Solved Problems 128

    11.3.3 COMSOAL (Computer Method for Sequencing Operations for Assembly Lines) 134

    11.3.4 Solved Problem 134

    11.3.5 Largest Candidate Rule 140

    11.3.6 Solved Problem 140

    11.3.7 Kilbridge-Wester Method 142

    11.3.8 Solved Problem 142

11.4 Objectives of Line Balancing 144

11.5 Questions and Problems 144

Part IV: Manufacturing

12. Group Technology and Cellular Manufacturing 147–157

12.1 Introduction 147

12.2 Benefits of Group Technology 147

12.3 Part Family 148

    12.3.1 Group Technology Layout or Cellular Layout 148

12.4 Methods of Grouping Parts into Part Family 149

    12.4.1 The Visual Inspection Method 149

    12.4.2 Part Classification and Coding System 149

        12.4.2.1 MICLASS Coding and Classification System 150

        12.4.2.2 Opitz Classification System 150

    12.4.3 Production Flow Analysis 152

    12.4.4 Solved Problem 155

12.5 Composite Part 156

12.6 Limitations of Group Technology or Cellular Manufacturing 157

12.7 Application of Group Technology in CAPP 157
12.8 Questions and Problem 157

13. **Computer-aided Manufacturing (CAM)** 158–167
13.1 Definition 158
13.2 Computers in Manufacturing 158
13.3 Applications of CAM system 164
13.4 CAM Software 165
13.5 CAM Solutions for Machining Processes 165
13.6 Benefits of CAM 166
13.7 Computer-aided Engineering (CAE) 167
13.8 Questions 167

14. **Industrial Robots** 168–177
14.1 Introduction 168
14.2 Characteristics of Robot 168
14.3 Industrial Robots 168
14.4 Anatomy of Robot or Elements of Robot System 169
14.5 Degrees of Freedom (DOF) 172
14.6 Kinematics 172
14.7 Work Envelope or Work Volume 172
14.8 Operating and Performance Parameters 172
14.9 Types of Robots 172
14.10 Industrial Robot Applications 174
14.11 Robot Programing 175
14.12 Advantages of Robots 175
14.13 Disadvantages of Robots 175
14.14 Robot Accuracy and Repeatability 176
14.15 Questions 176

15. **Numerical Control** 178–202
15.1 Introduction 178
15.2 Numerical Control Modes/CNC Motion Control Systems 178
15.3 Interpolation 181
15.4 Applications of CNC 182
15.5 Numerical Control Elements or Basic Components of NC System 183
　　15.5.1 Machine Control Unit 183
　　15.5.2 Part Program 184
　　　　15.5.2.1 Manual Part Programing 184
　　　　15.5.2.2 Solved Problems 188
　　　　15.5.2.3 Computer-aided Part Programing 193
　　　　15.5.2.4 Part Programing Using CAD/CAM Systems/
　　　　　　　　Automatic Part Programing 195
　　　　15.5.2.5 Manual Data Input 196
　　15.5.3 Machine Tool/Machining Center 196
　　　　15.5.3.1 Structure of CNC Machine Tools 196
　　　　15.5.3.2 Axes of Machine Tool 196
　　　　15.5.3.3 Drives 197
　　　　15.5.3.4 Automatic Tool Changer 198
　　　　15.5.3.5 Machine Tool 199
　　　　15.5.3.6 Cutting Tool 200
15.6 Advantages of CNC 200
15.7 Questions 201
15.8 G- and M-codes 201

16. **Flexible Manufacturing System** 203–211
16.1 Introduction 203

16.2 Types of Flexibility 203
16.3 FMS Components 204
    16.3.1 Workstations 204
    16.3.2 Material Handling and Storage System 204
    16.3.3 FMS Computer Control System 207
    16.3.4 Human Resources or Human Labor in Flexible System 208
16.4 Types of FMS Layout 208
    16.4.1 Progressive or Line Type Layout 209
    16.4.2 Loop Type Layout 209
    16.4.3 Ladder Type Layout 209
    16.4.4 Open Field Type Layout 210
    16.4.5 Robot-centered Type Layout 210
16.5 FMS Applications 210
16.6 FMS Benefits 211
16.7 FMS Disadvantages and Limitations 211
16.8 Questions 211

Part V: Process Planning

17. Process Selection 212–216
17.1 Introduction 212
17.2 Factors that Affect Selection of Manufacturing Process 212
17.3 Manufacturing Processes for Metals 213
17.4 Material Removal Process or Machining Process 213
17.5 Cutting Parameters 213
17.6 Different Approaches for Solving Speed/Feed Selection Problem 213
17.7 Elements of Cost in Machining Operation 214
17.8 Optimization Model to Predict Optimum Cutting Speed 214
17.9 Breakeven Analysis in Selection of Process 215
17.10 Solved Problems 215
17.11 Questions and Problems 216

18. Process Planning 217–228
18.1 Introduction 217
18.2 Information Required for Process Planning System 217
18.3 Steps in Process Planning 217
18.4 Route Sheet 218
18.5 Approaches to Process Planning 219
    18.5.1 Manual Approach 220
    18.5.2 Computer-aided Process Planning (CAPP) or
            Automated Process Planning 220
        18.5.2.1 Retrieval CAPP System 221
        18.5.2.2 Generative CAPP System 223
                18.5.2.2.1 Elements in a Generative CAPP System 223
        18.5.2.3 Hybrid Approach 227
18.6 CAPP Applications 227
18.7 Facts about CAPP Technology 227
18.8 Limitations of CAPP 227
18.9 Questions 227

Part VI: Production Planning and Control

19. Forecasting 229–251
19.1 Introduction 229
19.2 Types of Forecasting 229

19.3 Forecasting Techniques 230

    19.3.1 Qualitative Techniques 230

    19.3.2 Quantitative Techniques 230

        19.3.2.1 Regression Analysis 233

        19.3.2.2 Solved Problems 234

        19.3.2.3 Smoothing Methods 240

            19.3.2.3.1 Moving Average Forecasting 240

            19.3.2.3.2 Solved Problem 241

            19.3.2.3.3 Weighted Moving Average 242

            19.3.2.3.4 Solved Problem 242

            19.3.2.3.5 Exponential Smoothing Method 243

            19.3.2.3.6 Solved Problem 243

            19.3.2.3.7 Forecast for Seasonal Variations 245

            19.3.2.3.8 Solved Problem 246

        19.3.2.4 Measures of Forecast Accuracy or Forecast Error 247

            19.3.2.4.1 Mean absolute deviation 248

            19.3.2.4.2 BIAS 248

            19.3.2.4.3 Mean Squared Error (MSE) 248

            19.3.2.4.4 Standard Deviation of Error 248

            19.3.2.4.5 Tracking Signal 248

            19.3.2.4.6 Solved Problems 248

19.4 Questions and Problems 249

20. Production Planning and Control Systems   **252–272**

20.1 Introduction 252

20.2 Activities in Production Planning 252

    20.2.1 Aggregate Production Planning 253

        20.2.1.1 Aggregate Production Planning Methods 254

        20.2.1.2 Solved Problem 254

    20.2.2 Master Production Schedule 259

    20.2.3 Material Requirements Planning (MRP I) 259

    20.2.4 Capacity Planning 261

    20.2.5 Production Control 262

    20.2.6 Solved Problems 263

20.3 Line of Balance 266

    20.3.1 Solved Problem 267

20.4 Questions and Problems 271

21. MRP II, ERP and Supply Chain Management   **273–279**

21.1 Manufacturing Resource Planning (MRP II) 273

    21.1.1 Basic Modules in MRP II System 273

    21.1.2 Benefits of MRP II Systems 274

    21.1.3 Disadvantages of MRP II Systems 274

21.2 Enterprise Resource Planning (ERP) 275

    21.2.1 ERP and Internet 275

    21.2.2 Characteristics of ERP Systems 275

    21.2.3 ERP Modules 275

    21.2.4 ERP Implementation 276

    21.2.5 Benefits of ERP 276

    21.2.6 Disadvantages of ERP 276

21.3 Supply Chain Management (SCM) 276

    21.3.1 Components of Supply Chain Management 277

    21.3.2 Reverse Supply Chain 278

    21.3.3 Supply Chain Structure of Industry 278

    21.3.4 Application of Supply Chain Management 278

    21.3.5 Benefits of Supply Chain Management 279

21.4   A Comparison of ERP and Supply Chain Management 279
21.5   Questions 279

22. Loading and Scheduling   280–287
22.1   Terms Used in Scheduling 280
22.2   Scheduling 280
22.3   Factors Affecting Scheduling 281
22.4   Objectives of Production Scheduling 281
22.5   Methods Used in Scheduling 281
    22.5.1   Forward Scheduling 281
    22.5.2   Backward Scheduling 282
    22.5.3   Solved Problem 282
    22.5.4   Gantt Chart 283
    22.5.5   Solved Problem 283
    22.5.6   Johnson's Algorithm 284
    22.5.7   Index Method 284
    22.5.8   Solved Problem 285
    22.5.9   Critical Path Analysis 286
    22.5.10   Critical Ratio Scheduling 286
    22.5.11   Solved Problem 286
22.6   Questions and Problems 287

23. Sequencing   288–312
23.1   Priority Sequencing Rules 288
23.2   Parameters for Comparing Performance of Various
    Sequencing Rules 288
23.3   Solved Problems 289
23.4   Job Shop and Flow Shop Sequencing Problems 292
23.5   Gantt Chart 293
23.6   Model 1: Processing N Jobs Through Two Machines 293
23.7   Johnson's Algorithm 293
23.8   Solved Problems 294
23.9   Model 2: Processing of N jobs Through Three Machines 296
23.10   Solved Problems 297
23.11   Model 3: Processing of N jobs Through M Machines 298
23.12   Solved Problems 299
23.13   Model 4: Two Jobs Through M Machines 301
23.14   Solved Problems 301
23.15   Model 5: Traveling Salesman Problem 304
23.16   Solved Problems 304
23.17   Questions and Problems 309

Part VII: Quality Management

24. Quality and Taguchi Methods   313–324
24.1   Definition of Quality 313
24.2   Factors Affecting Product Quality 313
24.3   Product Quality 313
24.4   Dimensions in Product Quality 313
24.5   Quality Costs 314
24.6   Quality Assurance and Quality Control 314
24.7   Taguchi Methods 315
    24.7.1   Offline Quality Control 315
        24.7.1.1 Robust Product Design 316
    24.7.2   Online Quality Control 316

24.8 Taguchi's Loss Function 317
24.8.1 Quality Loss Functions for Various Quality Characteristics 318
24.9 Solved Problems 319
24.10 Specifying Tolerances for a Process 321
24.11 Uses of Quality Loss Function 322
24.12 Eight Steps in Taguchi Methodology 322
24.13 Questions and Problems 323

25. **Inspection and Statistical Quality Control (SQC)** 325–350
25.1 Inspection 325
25.2 Statistical Quality Control (SQC) 325
25.3 Sources of Variation 326
25.4 Descriptive Statistics 326
25.4.1 Mean 326
25.4.2 Range and Standard Deviation 326
25.4.3 Distribution of Data 327
25.4.4 Solved Problems 327
25.5 Control Charts 329
25.5.1 Control Chart for Variables 329
25.5.1.1 $\bar{X}$ Charts 330
25.5.1.2 Range (R) Charts 330
25.5.1.3 Process Capability 331
25.5.1.4 Solved Problems 333
25.6 Control Charts for Attributes 338
25.6.1 p Charts 339
25.6.2 c Charts 339
25.6.3 Solved Problems 339
25.7 Sampling Plans 343
25.7.1 Single Sampling Plan 344
25.7.2 Double Sampling Plan 344
25.7.3 Multiple Sampling Plans 345
25.8 Operating Characteristic (OC) Curves 345
25.9 Average Outgoing Quality (AOQ) 347
25.10 Comparison between Attribute Charts and Variable Charts 348
25.11 Questions and Problems 348

26. **Normal Distribution and Six Sigma** 351–363
26.1 Introduction 351
26.2 Standard Deviations 352
26.3 Solved Problems 353
26.4 Six Sigma 356
26.5 Comparison of ± 3 Sigma with ± 6 Sigma 357
26.6 Statistical Representation of Six Sigma 357
26.7 Six Sigma Methodologies 359
26.8 Benefits that an Organization Derives as the Result of Implementing a Six Sigma Project 360
26.9 Questions and Problems 361
26.10 Normal Distribution Tables 362

27. **Deming's 14 Points and Total Quality Management** 364–371
27.1 The Deming Cycle (or Shewhart Cycle) 364
27.2 Deming's 14 Points 364
27.3 Total Quality Management 366
27.4 The Importance of Customer-Supplier Relationships (Quality Chains) 366

27.5   Main Principles of TQM   366
27.6   Introducing TQM into a Business   367
27.7   Benefits of TQM   370
27.8   Quality in Manufacturing and Service Organizations   370
27.9   Quality cost   371
27.10 Questions   371

Part VIII: Project Management

28.   PERT and CPM   372–396
28.1   Introduction   372
28.2   Network Techniques   372
28.3   Terms Used in PERT and CPM   372
28.4   Critical Path Method (CPM)   374
28.5   Steps Required for Determining Various Parameters of a Project
       Using CPM   374
28.6   Rules for Drawing Network Diagram   374
28.7   Solved Problems   374
28.8   Program Evaluation Review Technique (PERT)   381
28.9   Solved Problems   382
28.10 Differences Between PERT and CPM   385
28.11 Crashing of Network   385
28.12 Solved Problems   386
28.13 Resource Management   391
28.14 Resource Allocation   391
28.15 Resource Aggregation (Loading)   391
28.16 Resource leveling (Smoothing)   391
28.17 Applications of PERT and CPM   394
28.18 Questions and Problems   394

29.   Project Scheduling under Resource Constraints   397–406
29.1   Introduction   397
29.2   Heuristic Methods of Project Scheduling Under
       Resource Constraints   397
29.3   Solved Problem   398
29.4   Questions and Problems   406

Part IX: Maintenance Management

30.   Maintenance   407–416
30.1   Introduction   407
30.2   Equipment Failure Rate Over Time   407
30.3   Maintenance Objectives   408
30.4   Consequences of Breakdowns   408
30.5   Maintenance Functions   408
30.6   Maintenance Strategies   409
       30.6.1   Reactive Maintenance (Breakdown Maintenance)   409
       30.6.2   Preventive Maintenance   410
       30.6.3   Predictive Maintenance (Condition-based Maintenance)   410
       30.6.4   Reliability Centered Maintenance (RCM)   411
       30.6.5   Total Productive Maintenance   412
       30.6.6   Determining Optimum Period Between
                Periodic Maintenance   412
30.7   Solved Problems   413
30.8   Questions and Problems   415

31. Replacement and Capital Investment Decisions   **417–429**
31.1 Introduction 417
31.2 Decision to be Made in Replacement Problems 417
31.3 Replacement Models 417
31.4 Model 1 417
31.5 Solved Problems 418
31.6 Model 2 419
31.7 Solved Problems 419
31.8 Model 3 422
31.9 Solved Problems 422
31.10 Capital Investment Decision Using Net Present Value (NPV) 425
31.11 Solved Problems 426
31.12 Questions and Problems 428

Part X: Operations Research Models

32. Introduction to Operations Research Models and
    Linear Programing   **430–458**
32.1 Definitions and Explanations of Operations Research 430
32.2 Overview of Operations Research Models or Various
    Operations Research Models 430
32.3 Types of Operations Research Models or Classification of Operations
    Research Models or Nature of Operations Research Models 432
32.4 Decision Making 433
32.5 Steps in Decision Making 433
32.6 Definition of Linear Equation 434
32.7 Applications of Linear Programing Models 434
32.8 Methods of Solving Linear Programing Problem 435
32.9 Graphical Method 435
    32.9.1 Solved Problems 435
32.10 Simplex Method 439
    32.10.1 Solved Problems 441
32.11 Transportation Problem 444
    32.11.1 Solved Problems 446
32.12 The Assignment Model 449
    32.12.1 Differences Between Transportation and Assignment Models 450
    32.12.2 Solved Problems 450
32.13 Questions and Problems 453

33. Queuing Theory (Waiting Lines)   **459–477**
33.1 Introduction 459
33.2 Elements or Parameters of Queuing System 459
33.3 Kendall's Notation for Representing Queuing Models 461
33.4 Model 1: Single Server Model (M/M/1/∞/∞/FCFS) 461
33.5 Solved Problems 462
33.6 Model 2: M/M/1/∞/N/FCFS 466
33.7 Solved Problems 467
33.8 Model 3: Multiserver Model M/M/S/∞/∞/FCFS 470
33.9 Solved Problems 471
33.10 Machine Servicing Model (M/M/R/k/k/FCFS) 473
33.11 Solved Problems 474
33.12 Applications of Queuing Models 475
33.13 Questions and Problems 476

34.   Inventory   **478–513**
34.1   Introduction   478
34.2   Costs Involved in Inventory   479
34.3   Classification of Inventory Models or Various Inventory Models   480
34.4   Model 1a: Uniform Demand, Instantaneous Replenishment
         and Shortages Are not Allowed (Simple EOQ Model)   481
34.5   Solved Problems   482
34.6   Model 1b: Uniform Demand, Instantaneous Replenishment
         with Shortages   484
34.7   Solved Problems   485
34.8   Model 1c: Uniform Demand, Finite Production Rate Without Shortages   487
34.9   Solved Problems   487
34.10 Model 1d: Finite Production Rate, Uniform Demand with Shortages   489
34.11 Solved Problem   489
34.12 Models with Price Breaks   490
34.13 Model with Single Price Break   491
34.14 Solved Problems   491
34.15 Model with Number of Price Breaks   494
34.16 Solved Problems   494
34.17 Probabilistic Models or Stochastic Models   497
34.18 Model 1: Instantaneous Demand (Discrete Demand)
         No Setup Cost   498
34.19 Solved Problems   498
34.20 Model 2: Instantaneous Demand (Continuous Demand)
         No Setup Cost   500
34.21 Solved Problems   501
34.22 Replenishment Systems (Inventory Control Systems)   502
34.23 Solved Problems   505
34.24 Empirical Formulae for Safety Stock   507
34.25 Solved Problems   507
34.26 Selective Control   509
34.27 Solved Problem   510
34.28 Questions and Problems   511

35.   Computer-aided Quality Control   **514–522**
35.1   Introduction   514
35.2   Advantages of Computer-aided Quality Control or CAQC   514
35.3   Characteristics of Measuring Instruments   515
35.4   Inspection Techniques   516
         35.4.1   Contact Inspection Techniques   516
                  35.4.1.1   Coordinate Measuring Machines   516
         35.4.2   Non-contact Inspection Techniques   518
                  35.4.2.1   Machine Vision   518
                  35.4.2.2   3D Laser Scanning   521
                  35.4.2.3   Non-contact Non-optical Inspection Techniques   521
                  35.4.2.4   Advantages of Non-contact Inspection Techniques   521
35.5   Integration of CAQC with CAD/CAM   521
35.6   Questions   522

Index   **523–525**

Part I

Production Concepts, Productivity and Production Systems

Production Concepts and Productivity

1.1 INTRODUCTION

Production is a process whereby raw material is converted into finished products. Production function encompasses the activities of procurement, allocation and utilization of resources. The main objective of production function is to produce the goods and services demanded by the customers in the most efficient and economical way. Therefore, efficient management of the production function is of utmost importance in order to achieve this objective.

Production can broadly categorize into following based on technique:

- **Production through separation:** Desired output is achieved through separation or extraction from raw materials. A classic example of separation or extraction is oil into various fuel products.
- **Production by modification or improvement:** It involves change in chemical and mechanical parameters of the raw material without altering physical attributes of the raw material. Production of engine shaft from raw material using machining process is an example of Production. Annealing process (heating at high temperatures and then cooling) is an example of production by modification or improvement.
- **Production by assembly:** Car assembly and computer assembly are examples of production by assembly.

1.2 PRODUCTION CONCEPTS USING MATHEMATICAL MODELS

A number of production concepts are quantitative. They require mathematical approach to measure them.

a. **Production rate:** The production rate for individual processing or assembly operation is expressed as an hourly rate. The number of parts produced per unit time is called production rate.

$$\text{Production rate} = \frac{\text{Number of units produced in time '}t\text{'}}{\text{Time '}t\text{'}}$$

b. **Production capacity:** Maximum rate of output by production facility.

c. **Utilization:** It is ratio of actual output to the capacity of the production facility

$$\text{Utilization} = \frac{\text{Actual output}}{\text{Capacity}}$$

d. **Cycle time:** Time between two successive units coming out of processing station or assembly workstation.

$$\text{Cycle time} = \frac{\text{Production time}}{\text{Number of units produced}}$$

Typical cycle time for a production operation:

$$T_c = T_o + T_h + T_{th}$$

Where

T_c = Cycle time

T_o = Processing time for the operation

T_h = Handling time (e.g. loading and unloading the job on production machine)

T_{th} = Tool handling time (e.g. time to change tools)

e. **Availability:** Availability is the common measure of reliability for equipment

$$\text{Availability} = \frac{\text{MTBF} - \text{MTTR}}{\text{MTBF}}$$

Where

MTBF = Mean time between failures, and

MTTR = Mean time required to repair

f. **Manufacturing lead time:** Manufacturing lead time is the total time required to process a given part or product through the plant.

g. **Work-in-progress (WIP):** Work-in-progress is the quantity of parts or products currently in the factory between processing operations or in semi-finished state.

h. **Production quantity:** Production quantity refers to the number of units of a given part or product produced annually by plant.

i. **Product variety:** Product variety refers designs or types that are produced in a plant. Soft product variety means there is only small difference between the varieties of products produced by the company. Hard product variety means there is substantial difference between the varieties of products produced by the company. High product variety requires highly skilled labor, general purpose machines, detailed production planning and control system. On the other hand, low product variety (i.e. one or few products produced in large volumes) enables the use of low skilled labor, highly automated mass production processes using special purpose machines and simple production planning and control systems.

j. **Process types:** The projected sales volume is a major influencing factor in determining whether the firm should go for intermittent or continuous process. Fixed costs are high for continuous process and low for intermittent process while variable costs are more for the intermittent process and less for continuous process. Intermittent process therefore, will be cheaper to install and operate at low volumes and continuous process will be economical to use at high volume. Process types are given below:

Job Shops: Small lots, low volume, general equipment, skilled workers and high-variety.

Examples: Tool and die shop

Batch Processing: Moderate volume and variety

Example: Paint production

Repetitive/Assembly: Semi-continuous, high volume of standardized items and limited variety.

Examples: Auto plants and cafeteria

Continuous Processing: Very high volume and no variety.

Examples: Steel mill and chemical plants

Projects: No routine jobs

Characteristics of various process types are given in Table 1.1.

Table 1.1 Product-process matrix

Dimension	Job shop	Batch	Repetitive	Continuous
Job variety	Very high	Moderate	Low	Very low
Process flexibility	Very high	Moderate	Low	Very low
Unit cost	Very high	Moderate	Low	Very low
Volume of output	Very low	Low	High	Very high

1.2.1 Solved Problems

Problem 1.1: The CNC lathe section has seven machines, all devoted to the production of the same part. The section operates 10 shifts/ Wk. The number of hours per shift is 8. Production rate of each machine is 17 units per hour. Determine weekly production capacity of a CNC lathe section.

Solution:

Production capacity $= 7 \times 10 \times 8 \times 17 = 9,520$ parts/week

Problem 1.2: A production machine operates for 80 hours/week (two shifts, 5 days) at full capacity. It production rate is 20 units per hour. During certain week the machine produced 1000 parts and remaining time was idle.

  i. Determine production capacity of the machine
  ii. What is the utilization of machine during the week under consideration?

Solution:

  i. Production capacity $= 80 \times 20 = 1,600$ parts/week

  ii. Utilization $= \dfrac{\text{Actual output}}{\text{Capacity}} = \dfrac{1,000}{1,600} = 0.625$

Problem 1.3: A certain part is produced in a batch size of 100 units. Five operations are required to produce each part. Average setup time is 3 hours/operation. Average operation time is 6 minutes. Average nonoperation time due to handling, delays and inspections are 7 hours for each operation. Determine how many hours to complete the batch assuming the plant runs 8 hours shift/day.

Solution:

Number of units to be produced $=100$

Number of operations are required to produce each part $= 5$

Number of operations required to produce 100 parts $= 5 \times 100 = 500$

Average setup time $= 3$ hours/operation

Total setup time for 500 operations $= 500 \times 3 = 1,500$ hours

Average nonoperation time $= 7$ hours for each operation

Total nonoperation time for 500 operations $= 500 \times 7 = 3,500$ hours

Total time spent by 500 jobs or each batch $= 1,500$ hours $+ 3,500$ hours $= 5,000$ hours

Total time spent by each job $=$ Manufacturing lead time $= \dfrac{5,000}{500} = 100$ hours

1.3 PRODUCTION SYSTEM

Production system is a collection of people, equipment, and procedures organized to accomplish the manufacturing operations of a company (or other organization). Production systems can be divided into facilities and manufacturing support systems as indicated in Fig. 1.1.

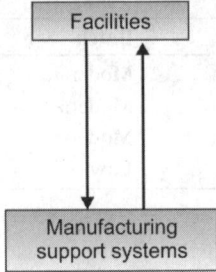

Fig. 1.1 Levels of production system

1.3.1 Production System Facilities

The facilities in the production system are the factory, production machines and tooling, material handling equipment, inspection equipment, and the computer systems that control the manufacturing operations. Facilities also include the facility layout. Facility layout is the way the equipment is physically arranged in the factory. Facilities also include People. In general, direct labor people (blue color workers) are responsible for operating the facilities, and professional staff people (white color workers) are responsible for the manufacturing support systems.

1.3.2 Manufacturing Support System

This is the set of procedures used by the company to manage production and to solve the technical and logistics problems encountered in ordering materials, moving work through the factory, and ensuring that products meet quality standards. Product design and certain business functions are included among the manufacturing support systems. In modern manufacturing operations, portions of the production system are automated and/or computerized.

1.4 PRODUCTION FUNCTION

Production is an organized activity of transforming raw materials into finished products. It is an intentional act of producing something useful. In production systems we have different resources as inputs. The inputs are processed in a series of operations. The sequence, number, and type of operations (mechanical, chemical, electrical, assembly, inspection, transportation, etc.) are specified for each input. The output of the system will be the complete parts, products, chemicals, etc. Production function shows the relationship between the input and output of an organization. The objective of studying production function is to maximize output with given inputs, or say resources. The production function can be represented by the simple mathematical equation which relates the outputs as the function of inputs, that is

$$Y = f(X_1, X_2, \ldots, X_n)$$

Where

Y = Output, which is the function of the inputs

X_i = Input 'i', $i = 1, 2, \ldots, n$

n = Number of inputs

1.5 PRODUCTION AND PRODUCTIVITY

The concept of productivity and production are totally different. Production is concerned with the activity of producing goods and services. Production directly refers to the number of units produced. Productivity is concerned with the efficient utilization of inputs in producing outputs. Productivity is a relative term. Productivity is the ratio of output to input.

Measurements of productivity:
- Partial productivity
- Total factor productivity
- Total productivity

Partial productivity: Partial productivity is the ratio of output to one class of input. Partial productivity measure reflects the impact of each input in producing the output.

Labor productivity: It reflects the effectiveness and efficiency of labor in the generation of output.

$$\text{Labor productivity} = \frac{\text{Output}}{\text{Human input}}$$

Material productivity: It reflects the effectiveness and efficiency of material in the generation of output.

$$\text{Material productivity} = \frac{\text{Output}}{\text{Material input}}$$

Capital productivity: Capital productivity measures the effectiveness and efficiency of capital in the generation of output.

$$\text{Capital productivity} = \frac{\text{Output}}{\text{Capital input}}$$

Energy productivity: Energy productivity measures the utilization of energy in the generation of output.

$$\text{Energy productivity} = \frac{\text{Output}}{\text{Energy input}}$$

Other expenses productivity:

$$\text{Other expenses productivity} = \frac{\text{Output}}{\text{Other expenses input}}$$

Total factor productivity: Total factor productivity is the ratio of output to the sum of labor and capital inputs. Total factor productivity measure reflects the joint impact of capital and labor inputs in producing the output.

$$\text{Total factor productivity} = \frac{\text{Output}}{\text{Labor input} + \text{Capital input}}$$

For calculation, convert both input and output into same unit (usually currency, i.e. rupee).

Total productivity: Total productivity is the ratio of total output to the sum of all input factors. Total productivity measure reflects the joint impact of all the inputs in producing the output.

$$\text{Total productivity} = \frac{\text{Output}}{\text{Total input}}$$

For calculation, convert both, i.e. input and output into same unit (usually currency, i.e. rupee).

Efficiency: Efficiency is the ratio of actual output attained to standard output expected. How well the resources are utilized to accomplish the results refer to the efficiency.

Effectiveness: Effectiveness is the degree of accomplishment of objectives.

Productivity is a combination of both effectiveness and efficiency, since effectiveness is related to performance while efficiency is related to resource utilization.

$$\text{Productivity index} = \frac{\text{Effectiveness}}{\text{Efficiency}}$$

$$\text{Productivity index of period `i'} = \frac{\text{Productivity of period `i'}}{\text{Productivity of base period}}$$

where i means during the time "i". For example, productivity index for the month of January.

All input costs and output prices are calculated based on base period costs and prices.

Base period means any time period that is a reference period for computing the productivity values and indices. For example, if a company institutes a productivity measurement for the first time in April 2010, then April 2010 can be the base period provided that there have been no strikes, lockouts, etc.

1.6 WAYS TO IMPROVE PRODUCTIVITY

The following are the ways to improve productivity:
- Increase output by using the same or a lesser amount of input.
- Reduce amount of inputs used while keeping output constant or increasing it.
- Proportionate increase of output is more than proportionate increase of input.

1.6.1 Benefits of Productivity Measurement in Organizations

Benefits of productivity measurement in organizations are given below:
- Judging the performance of an entire industry
- Productivity measurement helps in comparing the productivity levels between organizations within a particular category.
- Helpful for planning resources
- Product produced will be cheaper
- High productivity improves standards of living
- The organization can assess the efficiency of conversion of its resources to finished products.
- Improvement in living standard of workers

1.7 PRODUCTIVITY IMPROVEMENT TECHNIQUES

The following are productivity improvement techniques:
- Increasing output
- Improving methods
- Reducing overheads
- Minimizing waste
- Minimizing absenteeism
- Install modern equipment
- Train and motivate employees
- Equal distribution of wok to all workstations

1.8 PRODUCTION AND OPERATION MANAGEMENT-BASED PRODUCTIVITY IMPROVEMENT TECHNIQUES

Production and operation management-based productivity improvement techniques are as follows:
- Group technology
- Inventory control

- Material requirements planning
- Aggregate production planning
- Assembly/production line balancing
- Forecasting techniques
- PERT and CPM
- Priority scheduling
- Facility layout techniques
- Lean production
- Agile production
- JIT production system

1.9 SOLVED PROBLEMS

Problem 1.4: The output of an operator is 120 pieces per hour, while the standard rate is 180 pieces per hour. What is the operator's efficiency?

Solution:

$$\text{Operator's efficiency} = \frac{\text{Output of an operator}}{\text{Standard rate}} = \frac{120}{180} = 0.6667 = 66.67\%$$

Problem 1.5: A company is produced 20,000 calculators by employing 50 people at 8 hours/day for 25 days. Calculate labor productivity.

Solution:

$$\text{Labor productivity} = \frac{\text{Output}}{\text{Human input}} = \frac{20,000}{50 \times 8 \times 25} = 2 \text{ calculators/man-hour}$$

Problem 1.6: An accounting firm generates service valued ₹ 8,00,00,000 per day and total costs of service is ₹ 5,00,00,000. What is total productivity?

Solution:

$$\text{Total productivity} = \frac{\text{Output}}{\text{Total input}} = \frac{8,00,00,000}{5,00,00,000} = 1.6$$

Problem 1.7: In a saving company there are 5 tellers. During past week (5 working days/ week), they served a total of 3,000 customers. Business hour is 10 a.m to 4 p.m. Compute productivity of tellers.

Solution:

$$\text{Labor productivity} = \frac{\text{Output}}{\text{Human input}} = \frac{3,000}{5 \times 6 \times 5} = 20 \text{ calculators/man-hour}$$

Problem 1.8: The input and output data of xyz company for a particular time period is given below.

$$\text{Output} = ₹ 1,00,00,000$$
$$\text{Human input} = ₹ 1,00,000$$
$$\text{Material input} = ₹ 25,00,000$$
$$\text{Capital input} = ₹ 15,00,000$$
$$\text{Energy input} = ₹ 50,000$$
$$\text{Other expenses input} = ₹ 40,000$$

Determine partial, total factor and total productivity.

Solution:

$$\text{Labor productivity} = \frac{\text{Output}}{\text{Human input}} = \frac{1,00,00,000}{1,00,000} = 100$$

$$\text{Material productivity} = \frac{\text{Output}}{\text{Material input}} = \frac{1,00,00,000}{25,00,000} = 4$$

$$\text{Capital productivity} = \frac{\text{Output}}{\text{Capital input}} = \frac{1,00,00,000}{15,00,000} = 6.66$$

$$\text{Energy productivity} = \frac{\text{Output}}{\text{Energy input}} = \frac{1,00,00,000}{50,000} = 200$$

$$\text{Other expenses productivity} = \frac{\text{Output}}{\text{Other expenses input}} = \frac{1,00,00,000}{40,000} = 250$$

$$\text{Total factor productivity} = \frac{\text{Output}}{\text{Labor input} + \text{Capital input}} = \frac{1,00,00,000}{1,00,000 + 15,00,000} = 6.25$$

$$\text{Total productivity} = \frac{\text{Output}}{\text{Total input}} = \frac{1,00,00,000}{1,00,000 + 25,00,000 + 15,00,000 + 50,000 + 40,000}$$

$$= 2.38$$

Problem 1.9: Compute the productivity per machine hour with the data given in Table 1.2. Also draw interpretation.

Table 1.2 Productivity input data

Months	Number of machines employed	Working hours	Number of units produced
March	400	225	99,000
April	500	200	1,00,000
May	600	250	1,35,000

Solution: Productivity calculations are given in Table 1.3.

Table 1.3 Productivity calculations

Months	Number of machines employed (a)	Working hours (b)	Total number of machine hours (a) × (b)	Number of units produced	Productivity = $\dfrac{\text{Number of units produced}}{\text{Total number of machine hours}}$ (Units/machine-hour)
March	400	225	90,000	99,000	$\dfrac{99,000}{90,000} = 1.1$
April	500	200	1,00,000	1,00,000	$\dfrac{1,00,000}{1,00,000} = 1$
May	600	250	1,50,000	1,35,000	$\dfrac{1,35,000}{1,50,000} = 0.9$

Interpretation: Though total production in number of units is increasing, the productivity is declining.

Problem 1.10: An electric equipment manufacturing company manufactures AC motors, DC motors and transformers. During the month of December, the production of three items in rupees has been ₹ 140 million, ₹ 250 million and ₹ 90 million respectively. The inputs of human resources, capital and materials and power have been given in Table 1.4.

Table 1.4 Various inputs

Inputs ↓	Products		
	AC motors	DC motors	Transformers
Human (₹ in millions)	14	23	12
Capital (₹ in millions)	28	81	14
Material (₹ in millions)	72	108	24
Power (₹ in millions)	09	20	10

a. What is the partial productivity of each of the inputs?
b. What is the total productivity of each of the three products?
c. What is the total factor productivity of a company?

Solution:

Total output = 140 + 250 + 90 = ₹ 480 millions

Partial productivity computations of each of the inputs are given in Table 1.5.

Table 1.5 Partial productivity of each of the inputs

Inputs ↓	Products			Total inputs	Partial productivity
	AC motors	DC motors	Trans-formers		
Human (₹ in millions)	14	23	12	49	Labor productivity = $\dfrac{\text{Output}}{\text{Human input}}$ $= \dfrac{480}{49} = 9.79$
Capital (₹ in millions)	28	81	14	123	Capital productivity = $\dfrac{\text{Output}}{\text{Capital input}}$ $= \dfrac{480}{123} = 3.90$
Material (₹ in millions)	72	108	24	204	Material productivity = $\dfrac{\text{Output}}{\text{Material input}}$ $= \dfrac{480}{204} = 2.35$
Power (₹ in millions)	09	20	10	39	Power productivity = $\dfrac{\text{Output}}{\text{Power input}}$ $= \dfrac{480}{39} = 12.30$

Total productivity computations of each of the products are given in Table 1.6.

Table 1.6 Total productivity of each of the three products

Inputs ↓	Products		
	AC motors	DC motors	Transformers
Human	14	23	12
Capital	28	81	14
Material	72	108	24
Power	09	20	10
Total input	123	232	60
Output	140	250	90
Total productivity = $\dfrac{\text{Output}}{\text{Total input}}$	1.13	1.077	1.5

Total factor productivity of a company is given in Table 1.7.

Table 1.7 Total factor productivity of a company

Inputs ↓	Products			Total input	(Human input + Capital input) of a company	Total factor productivity of a company $= \dfrac{\text{Output}}{\text{Human input} + \text{Capital input}}$
	AC motors	DC motors	Trans-formers			
Human	14	23	12	49	49 + 123 = 172	$\dfrac{480}{172} = 2.79$
Capital	28	81	14	123		

Problem 1.11: There are two factories manufacturing same type of plugs. The standard time per piece is 1.5 minutes. The outputs of two factories are 300 and 200 respectively per shift of 8 hours.

    a. What is the productivity of each factory/shift of 8 hours?

    b. What is the production of each factory per week (6 days) on the basis of double shift?

Solution:

    Given data:

    The standard time per piece = 1.5 minutes

    The outputs of two factories are 300 and 200 respectively per shift of 8 hours

    a. Actual time per piece for factory $1 = \dfrac{8 \times 60}{300} = 1.6$ minutes

        Actual time per piece for factory $2 = \dfrac{8 \times 60}{200} = 2.4$ minutes

        $\text{Productivity} = \dfrac{\text{Standard time}}{\text{Actual time}}$

        Productivity of factory $1 = \dfrac{1.5}{1.6} = 0.9375$

        Productivity of factory $2 = \dfrac{1.5}{2.4} = 0.625$

    b. Production of factory 1/week = $300 \times 2 \times 6 = 3{,}600$

        Production of factory 2/week = $200 \times 2 \times 6 = 2{,}400$

Problem 1.12: Consider the data for a company shown in Table 1.8. Calculate partial productivities, total productivity and productivity indexes.

Table 1.8 Data for a company

		2011 (Costs and prices are in rupees)		2012 (Costs and prices are in rupees)	
		Quantity	Price	Quantity	Price
Output:	Product 1	1,000	30	1,100	35
	Product 2	100	190	80	200
Inputs:	Labor				
	Category 1	3,000	5	2,500	7
	Category 2	600	6	500	8
Material:	Material 1	6,000	1	7,000	1.3
	Material 2	200	6	150	7.5
	Material 3	300	2	300	3
Energy:	Type 1	10,000	0.15	8,000	0.20
	Type 2	200	1.00	250	1.10
Capital:	Depreciation	1,00,000	0.10	1,00,000	0.10
	Others	1,50,000	0.073	1,80,000	0.073

Solution: Calculation of partial productivities, total productivity and productivity indexes are given in Table 1.9. All input costs and output prices for the period 2012 are calculated based on base period costs and prices.

Table 1.9 Calculation of productivity indexes

	2011 (Costs and prices are in rupees)				2012 (Costs and prices are in rupees)				Producti-vity index
	Quantity	Price	Quantity × Price	Total	Quantity	Price	Quantity × Price	Total	
Output:									
Product 1	1,000	30	30,000		1,100	30	33,000		
Product 2	100	190	19,000	49,000	80	190	15,200	48,200	
Total output									
Inputs:									
Labor									
Category 1	3,000	5	15,000		2,500	5	12,500		
Category 2	600	6	3,600	18,600	500	6	3,000	15,500	
Material:									
Material 1	6,000	1	6,000		7,000	1	7,000		
Material 2	200	6	1,200		150	6	900		
Material 3	300	2	600	7,800	300	2	600	8,500	
Energy:									
Type 1	10,000	0.15	1,500		8,000	0.15	1,200		
Type 2	200	1.00	200	1,700	250	1.00	250	1,450	
Capital:									
Depreciation	1,00,000	0.100	10,000	20,950	1,00,000	0.100	10,000		
Others	1,50,000	0.073	10,950		1,80,000	0.073	13,140	23,140	
Total input				49,050				48,590	
Labor productivity		2.634				3.110			1.181
Material productivity		6.282				5.671			0.903
Energy productivity		28.823				33.241			1.153
Capital productivity		2.338				2.083			0.889
Total productivity		1.00				0.992			0.992

1.10 MANUFACTURING OPERATIONS

Many companies use manufacturing operations to produce goods for consumers. Before selling, the company has to produce. It means that the first task of any business is to manufacture or assemble goods and services. Manufacturing is a process where raw material is converted to finished product with the help of machinery and skilled labor. Manufacturing provides millions of jobs to people across all the fields.

Example of producing a good:

Let's suppose that your business idea is to produce plate steels from sponge iron or salvaged materials. To produce plate steel the following operations are needed:

Operation 1: Collect and store the raw material sponge iron

Operation 2: Melting Sponge iron in an electric arc furnace

Operation 3: Refining the melting metal. It means that you separate the chemical elements in order to get the specific steel you need.

Operation 4: Casting the liquid steel into products such as billets

Operation 5 (Rolling): The billets are heated at 1,200°C and then rolled in order to get the plates.

The manufacturing process is a very technical matter and you should need to have specific knowledge or to partner with a good engineer.

There are certain basic activities that must be carried out in a factory to convert raw material into finished products.

Important manufacturing operations are as follows:

1. *Processing and assembly operations*: Processing and assembly operations alter the geometry, properties and appearance of the work unit. Processing operation uses energy to alter work part shape or physical properties. In assembly operation, two or more separate parts are joined to form new entity.

2. *Material handling*: The product must be moved from one operation to the next in the manufacturing sequence. In most manufacturing plants material spend more time being moved and stored than being processed.

3. *Inspection and testing*: The raw material or semi-finished product or finished product must be inspected to insure high quality. Inspection and testing are quality control activities. The purpose of inspection is to determine whether the manufactured product meet the established design standards and specifications.

4. *Coordination and control*: Control at the plant level includes effective use of labor, maintenance of equipment, shipping product of good quality on schedule and keeping plant operating costs at a minimum possible level.

Problem 1.13: A company product line produces 50 different models. Annual production of each model is 1,000 units. Each product is the assembly of 400 components. 6 processing steps are required to produce each component. Each step takes 1.0 minute. All processing operations are performed at workstations. Each workstation requires floor space of 250 ft², one production machine and is operated by one worker. Factory operates 2,000 hours/year. Determine:

    a. How many production operations are there?

    b. How many workers will required in the plant?

    c. How much floor space is required?

Solution:

    a. Number of production operations $= 50 \times 1,000 \times 400 \times 6 = 12,00,00,000$

        Production time required $= 12,00,00,000 \times 1 = 12,00,00,000$ minutes

    b. Number of workstations required

$$= \frac{\text{Production time required}}{\text{Factory operation time}} = \frac{12,00,00,000}{2,000 \times 60} = 1,000 \text{ workstations}$$

       $\therefore$ Number of workers required $= 1,000$

    c. Floor space required $= 1,000 \times 250 = 2,50,000$ ft^2

1.11 QUESTIONS AND PROBLEMS

1. What is the difference between the effectiveness and efficiency?

2. Describe the term productivity and how it is used to measure in goods/service industry with suitable examples.

3. Write notes on productivity index.
4. Compute productivity per machine hour with the following data:

Months	Number of machines employed	Working hours	Number of units produced
March	400	225	99,000
April	500	200	1,00,000
May	600	250	1,35,000

Answer: Productivity for March = 1.1; productivity for April = 1; productivity for May = 0.9

5. Differentiate between production and productivity.
6. Explain the following with respect to production:
   a. Production rate
   b. Production capacity
   c. Utilization
   d. Cycle time
   e. Availability
   f. Manufacture lead time
   g. Work progress
7. Describe production system facilities.
8. Explain manufacturing support systems.
9. Explain the importance of productivity.
10. Describe the various ways to improve productivity.
11. Explain partial productivity, total factor productivity and total productivity.
12. An automobile equipment supplier wishes to install a sufficient number of ovens to produce 4,00,000 good castings per year. The baking operation takes 2.0 minutes per casting, but the oven output is typically about 6 percent defective. How many ovens will be required, if each one is available 1800 hours per year?

   Hint: Required system capacity $= \dfrac{\text{Actual good output}}{0.94} = \dfrac{4,00,000}{0.94} = 4,25,532$ units/year

   Number of ovens required $= \dfrac{4,25,532 \times 2}{1,800 \times 60} = 7.88 \approx 8$ ovens

13. A company is capable of producing 2,00,000 good parts per year. 12% of the production will have to be scrapped because of defects. What is the required system capacity in parts per year?

   Answer: Required system capacity $= \dfrac{\text{Actual good output}}{0.88} = \dfrac{2,00,000}{0.88}$

   $= 2,27,273$ units/year

14. In a transport company, design capacity = 50 trucks/day; effective capacity = 40 trucks/day and actual output = 36 units/day. Determine the efficiency and utilization.

   Answer: Efficiency $= \dfrac{\text{Actual capacity}}{\text{Standard output}} = \dfrac{36}{40} = 90\%$

   Utilization $= \dfrac{\text{Actual output}}{\text{Capacity}} = \dfrac{36}{50} = 72\%$

2

Costs of Manufacturing Operations and Breakeven Analysis

2.1 ELEMENTS OF COST

Product cost may be defined as the amount of expenditure incurred to produce and sell a product.

The product cost is divided into three principle elements:
1. Material cost
2. Labor cost
3. Overhead or Indirect expenses

1. **Material cost:** Material cost is divided into two types:
  a. Direct material cost
  b. Indirect material cost

  a. Direct materials are the raw materials that become part of the product. For example, bricks and bath tubs would be the direct materials when building a house. Paper would be a direct material when making grocery bags. Steel is the direct material when making a shaft.
  b. Indirect materials also include other materials used in the production process such as oil for machines, welding rods for repairs, sand paper for making tables, and any other small miscellaneous materials.

2. **Labor cost:** Labor cost is divided into two types:
  a. Direct labor cost
  b. Indirect labor cost

  a. Direct labor represents the people who do the core work of the business. For example, if the business is a construction company, direct labor would be the people actually constructing the building. They would be the people with hammers and saws in their hands. In a retail store, direct labor would be the people helping on the sales floor doing the basic work that takes place serving the customers. In a grocery bag factory, direct labor is the people running the machines actually making the bags. Direct labor is the people who make or build the product.

  b. Indirect labors are maintenance men, helpers, machine setters, supervisors and foreman, etc.

3. **Overhead:** The definition for overhead is easy. If a cost is not direct labor or direct materials, the cost is overhead.

Examples of overhead:
- Rent
- Depreciation
- Audit and legal
- Administrative staff
- Equipment rental
- Fuel (it can be a direct cost in transportation industry)
- Maintenance
- Generator
- Security
- Telephone bill expense
- Business travel expenditure

Overhead costs are broadly divided into four types. They are:
- Manufacturing overhead costs
- Selling overhead costs
- Distribution overhead costs
- Administrative overhead costs

Factory overhead costs: Factory overhead costs are all costs except direct labor cost, direct material cost which are incurred in the factory.

Selling overhead costs: Selling overhead costs include cost of advertising, cost of marketing, etc.

Distribution overhead costs: Distribution overhead costs cover all expenses connected with transporting products to the customers.

Administrative overhead costs: Administrative overhead costs consist of expenses incurred for the purpose of administration.

2.2 DETERMINING TOTAL COST

Prime cost: Prime cost is the direct cost incurred in manufacturing division.

$$\text{Prime cost} = \text{Direct material cost} + \text{Direct labor cost} + \text{Direct expenses}$$

Factory cost: Factory cost is the cost incurred in manufacturing division.

$$\text{Factory cost} = \text{Prime cost} + \text{Factory overhead costs}$$

Total cost: Total cost incurred in manufacturing, administration and selling

$$\text{Total cost} = \text{Factory cost} + \text{Selling overhead costs} + \text{Distribution overhead costs}$$
$$+ \text{Administrative overhead costs}$$

2.3 DETERMINING SELLING PRICE

Sales revenue = Total cost ± Profit or Loss

Selling price × Number of units produced = Total cost ± Profit or Loss

Businesses must make a profit to survive. To make a profit; sales revenue must be higher than total costs.

2.4 SOLVED PROBLEMS

Problem 2.1: A factory producing 150 electric bulbs a day, involves direct material cost ₹ 250, direct labor cost ₹ 200 and factory overheads of ₹ 225. Assuming a profit of 10% of the sales revenue and selling overhead is 30% of the factory cost, calculate the selling price of one electric bulb.

Solution:

Production rate = 150 electric bulbs a day

Direct material cost = ₹ 250

Direct labor cost = ₹ 200

Factory overheads = ₹ 225

Prime cost = Direct material + Direct labor = 250 + 200 = ₹ 450

Factory cost = Prime cost + Factory overhead costs = 450 + 225 = ₹ 675

Selling overhead = 30% of the factory cost

Total cost = Factory cost + Selling overhead costs = 675 + 202.5 = ₹ 877.5

Sales revenue = Total cost ± Profit or loss

Profit = 10% of the Sales revenue

150 × Selling price = Total cost + Profit or loss = 877.5 + 0.1 × 150 × Selling price

0.9 × 150 × Selling price = 877.5

Selling price = ₹ 6.5

Problem 2.2: A cast iron foundry employs 30 persons. It consumes material worth ₹ 25,000, pays workers at the rate of ₹ 10 per hour and incurs total overheads of ₹ 10,000. In a particular month (25 days) workers had an overtime of 150 hours and were paid double on their normal rate. Find:

    a. The total cost

    b. The man-hour rate of overhead.

Assume an eight hours working day.

Solution:

a. Total cost = Direct material + Direct labor + Overhead expenses

$$= 25,000 + 25 \times 8 \times 10 \times 30 + 150 \times 20 + 10,000 = 98,000$$

b. Man-hour rate of overhead $= \dfrac{\text{Total overhead costs}}{\text{Total number of man-hours}}$

Man-hour rate of overhead $= \dfrac{10,000}{25 \times 8 \times 30 + 150} = ₹\,1.626$

Problem 2.3: Two molders can cast 25 gears in a day. Each gear weighs 3 kg and the gear material cost ₹ 12.5 per kg. If the overhead expenses are 150% of direct labor cost and two molders are paid ₹ 70 per day, calculate the cost of producing a gear.

Solution: Factory cost for 25 gears = Direct material + Direct labor + Direct expenses

$$= 25 \times 3 \times 12.5 + 70 = ₹\,1,007.5$$

Overhead = 150% of the Direct labor cost

Total cost for 25 gears = Prime cost + Factory overhead costs = 1,007.5 + 1.5 × 70 = ₹ 1,112.5

Cost of gear $= \dfrac{1,112.5}{25} = ₹\,44.5$

Problem 2.4: Calculate the selling price of a fountain pen from the data given below:

    Number of fountain pens produced = 135

Labor cost = ₹ 200
Material cost = ₹ 160
Factory overheads = 35% of Prime cost
Administrative and selling overheads = 20% of Factory cost
Profit = 10% of Total cost

Solution: Factory cost = Direct material + Direct labor + Factory overhead expenses

Factory cost for 135 pens = 200 + 160 + 0.35 (360) = ₹ 486

Total cost = Factory cost + Selling overhead

Total cost for 135 pens = 486 + 0.2 (486) = ₹ 583.2

Selling price × Number of units produced = Total cost ± Profit or loss

135 × Selling price = 583.2 + 0.1 (583.2) = 641.52

$$\text{Selling price} = \frac{641.52}{135} = ₹ 4.752$$

Problem 2.5: A factory is making a pipefitting by (a) Casting and (b) Forging. The cost data is shown in Table 2.1.

Table 2.1 Cost data

	Casting	Forging
Material cost/piece	₹ 2	₹ 2
Labor rate	₹ 0.8 per hour	₹ 0.8 per hour
Time required to make one fitting	3 hours	48 minutes
Overheads	25% of labor cost	150% of labor cost

Calculate and compare the total cost of each fitting in two cases.

Solution:

Calculation of total cost of pipefitting by casting and forging is given in Table 2.2.

Table 2.2 Calculating total cost of pipefitting

	Casting	Forging
Material cost/piece	₹ 2	₹ 2
Labor rate	₹ 0.8 per hour	₹ 0.8 per hour
Time required to make one fitting	3 hours	48 minutes
Overheads	25% of labor cost	150% of labor cost
Prime cost	2 + 0.8 (3) = 2 + 2.4 = ₹ 4.4	2 + 0.8 (48/60) = 2 + 0.64 = ₹ 2.64
Total cost	4.4 + 0.25 (2.4) = ₹ 5	2.64 + 1.5 (0.64) = ₹ 3.60

Problem 2.6: The catalogue price of drilling machine is ₹ 6,100 and discount allowed to distributor is 20%. The administrative and selling expenses are 50% of the factory cost and the material cost, labor cost and factory overheads are in the ratio of 1:3:2. If the cost of labor on the manufacture of machine is ₹ 1,200, determine profit on each machine.

Solution: The catalogue price of drilling machine = ₹ 6,100

Discount allowed to distributor = 20%

The administrative and selling expenses = 50% of the factory cost

The material cost, labor cost and factory overheads are in the ratio of 1:3:2

Labor cost = ₹ 1,200

$$\frac{3}{6} \text{ of Factory cost} = \text{Labor cost} = ₹ 1,200$$

$$\text{Factory cost} = \frac{1,200 \times 6}{3} = ₹\,2,400$$

Total cost = Factory cost + Administrative and Selling expenses = $2,400 + 0.5 \times 2,400$

Total cost = ₹ 3,600

$$\text{Profit} = \text{Selling price} - \text{Total cost} = 4,800 - 3,600 = ₹\,1,200$$

2.5 METHODS OF ALLOCATING OVERHEAD COSTS

In the same factory different jobs are done or different products are produced then total overhead charges are distributed among these products. There are number of methods of allocation of overhead costs. The choice of a particular method depends upon the nature of work, type of organization and type of machines used.

The different methods of allocation of overhead costs are:

- Percentage on prime cost
- Percentage on direct labor cost
- Percentage on direct material cost
- Man-hour rate
- Machine hour rate
- Unit rate method

Percentage on prime cost: This method is generally used when the expenses on direct labor and direct material constitute the main cost in determining total cost.

Steps:

1. Determine total overhead costs.
2. Determine percentage for the product 'i'.

$$\% \text{ for the product '}i\text{'} = \frac{\text{Total overhead costs}}{\text{Prime cost for the product '}i\text{'}} \times 100$$

3. Overhead cost for product 'i' = Percentage × Prime cost for the product 'i'
4. Repeat the steps 2 and 3 for all products.

Percentage on direct labor cost: This method is generally used when the expenses on direct labor constitute the main cost in determining total cost.

Steps:

1. Determine total overhead costs.
2. Determine percentage for the product 'i'.

$$\% \text{ for the product '}i\text{'} = \frac{\text{Total overhead costs}}{\text{Direct labor cost for the product '}i\text{'}} \times 100$$

3. Overhead cost for product 'i' = Percentage × Direct labor cost for the product 'i'
4. Repeat the steps 2 and 3 for all products.

Percentage on direct material cost: This method is generally used when the expenses on direct material constitute the main cost in determining total cost.

Steps:

1. Determine total overhead costs.
2. Determine percentage for the product 'i'.

$$\% \text{ for the product '}i\text{'} = \frac{\text{Total overhead costs}}{\text{Direct material cost for the product '}i\text{'}} \times 100$$

3. Overhead cost for product 'i' = Percentage × Direct material cost for the product 'i'
4. Repeat the steps 2 and 3 for all products.

Man-hour rate: In this method the overhead cost is allocated on the total hours spent by the direct labor.

Steps:

1. Determine total overhead cost.
2. Determine man-hour rate.

$$\text{Man-hour rate} = \frac{\text{Total overhead cost}}{\text{Total man-hour}}$$

3. Overhead cost for product 'i' = Man-hour rate × Man-hours taken by the product 'i'
4. Repeat the step 3 for all products.

Machine-hour rate: This method is generally used where the work is done mostly by the machines.

Steps:

1. Determine total overhead cost.
2. Determine machine-hour rate.

$$\text{Machine-hour rate} = \frac{\text{Total overhead cost}}{\text{Total machine-hour}}$$

3. Overhead cost for product 'i' = Machine-hour rate × Machine-hours taken by the product 'i'
4. Repeat the step 3 for all products.

Unit rate method: In this method the total overhead cost is divided by total output. Overhead costs are determined on the basis of number of units produced. This method is applicable for the factories which manufacture one type of product only.

2.6 SOLVED PROBLEMS

Problem 2.7: A sugar mill had its overhead cost of ₹ 60,000 while it purchased sugarcane worth ₹ 2,40,000. Find the percentage overhead using the percentage direct material cost method. If a particular batch had a direct material cost of ₹ 30,000 determine it overheads.

Solution:

Total overhead cost = ₹ 60,000

Direct material or sugarcane cost = ₹ 2,40,000

$$\% \text{ for the batch} = \frac{\text{Total overhead cost}}{\text{Direct material cost for the batch '}i\text{'}} \times 100 = \frac{60,000}{2,40,000} \times 100 = 25\%$$

Direct material cost for a particular batch = ₹ 30,000

Overhead cost for a particular batch = 0.25 (30,000) = ₹ 7,500

Problem 2.8: A fabrication assembly shop had its total overhead of ₹ 10,000. It used direct material worth ₹ 10,000 and paid ₹ 15,000 as direct labor charges. Calculate % overhead. If one product has its prime cost as ₹ 5,000 determine overhead related to it.

Solution: Total overhead = ₹ 10,000

Prime cost = Direct material cost + Direct labor cost = 10,000 + 15,000 = ₹ 25,000

$$\% \text{ for the product} = \frac{\text{Total overhead cost}}{\text{Prime cost for the product '}i\text{'}} \times 100 = \frac{10,000}{25,000} \times 100 = 40\%$$

Overhead for a particular product = 0.4 (25,000) = ₹ 10,000

Problem 2.9: A fitting and assembly shop had its factory overheads of ₹ 1,20,000 and the production for the period in terms of direct labor was 24,000 hours. Find the rate per direct labor-hour. If a particular job takes 20 labor-hours, calculate the overhead applied.

Solution:

$$\text{Man-hour rate} = \frac{\text{Total overhead cost}}{\text{Total man-hour}} = \frac{1,20,000}{24,000} = ₹ \text{ 5/labor-hour}$$

Time taken by a particular job = 20 labor-hours

Overhead cost for a particular job = 20 (5) = ₹ 100

2.7 DEPRECIATION

Depreciation is a term that describes the decline in value of an asset over time. Depreciation is listed as an expense on an income statement because of the loss of value that occurs each year. Assets are depreciated over time due to use.

Depreciable assets: Depreciable assets are any assets that have a useful life of more than one year, normally capital assets. Examples of depreciable assets include machinery, equipment, and breeding livestock. Items that appreciate in value, such as land, cannot be depreciated.

Salvage value: Salvage value is the remaining value of an asset at the end of its useful life

2.8 METHODS OF CALCULATING DEPRECIATION

Depreciation expense is calculated utilizing either a straight line depreciation method or an accelerated depreciation method. The straight line method calculates depreciation by spreading the cost evenly over the life of the fixed asset. Accelerated depreciation methods such as declining balance and sum of year's digits calculate depreciation by expensing a large part of the cost at the beginning of the life of the fixed asset.

The required variables for calculating depreciation are the cost and the expected life of the fixed asset. Salvage value may also be considered. Examples of depreciation calculations for both straight line and accelerated methods are provided below.

Straight line depreciation method: The straight line method of depreciation is calculated by taking the original cost of the item minus the salvage value, divided by useful life in years. It is the easiest and most commonly used method of depreciation.

The same amount of depreciation is taken each year of the item's useful life.

$$\text{Annual depreciation charges} = \frac{C - S}{N}$$

Where,

C = Initial cost of the machine

S = Scrap value in rupees

N = Estimated life of the machine

Diminishing balance method: The machine or equipment depreciates rapidly in the early years and later on slowly. Depreciation is calculated based on current value of asset.

$$X = 1 - \left(\frac{S}{C}\right)^{\frac{1}{N}}$$

Where,

X = Fixed percentage for calculating yearly depreciation

C = Initial cost of the machine

S = Scrap value

N = Estimated life of the machine

Depreciation charge for the year 'i' = Current value of the item for year 'i' $\times X$

Sum of the digits method: Sum of the digits method is a form of depreciation that uses the sum of the years of useful life, original cost, and the salvage value of an asset. The percentage of depreciation declines each year of useful life of the asset. To calculate the sum of the digits method of depreciation, the following variables are required:

- Useful life of an item
- Salvage value
- Original cost

Steps:

The first step is to sum the digits or numbers starting with the life and going back to one. For example, an asset with a life of 5 would have a sum of digits as follows: $5 + 4 + 3 + 2 + 1 = 15$.

Find the percentage for each year and divide the year's digit by the sum. In the example above the percentage would be calculated as shown in Table 2.3.

Calculate depreciation expense for all the years

Depreciation charge for the year 'i' = Initial cost of the item × percentage for the year 'i'

Table 2.3 Percentage for each year

Year	Percentage
1	$\dfrac{5}{15} = 33.34\%$
2	$\dfrac{4}{15} = 26.67\%$
3	$\dfrac{3}{15} = 20.00\%$
4	$\dfrac{2}{15} = 13.33\%$
5	$\dfrac{1}{15} = 6.67\%$

Double declining-balance method: Double declining-balance method is an accelerated form of depreciation that takes into account the original cost, salvage value, and useful life of an asset while also considering the fact that an asset will lose most of its values during the first few years of its useful life.

To calculate the double declining-balance method of depreciation, the following variables are required:

- Current value of the asset
- Useful life of the item

$$\text{Depreciation expense} = \text{Current value} \times \frac{2}{\text{Expected life}}$$

2.9 SOLVED PROBLEMS

Problem 2.10: A machine costing ₹ 24,000 was purchased on 2005. The installation and erecting charges were ₹ 1,000 and useful life is expected to be 10 years. The scrap value of the machine at the end of the useful life is ₹ 5,000. Calculate the yearly depreciation by straight line method.

Solution:

C = Initial cost of the machine = 24,000 + 1,000 = ₹ 25,000

S = Scrap value in rupees = ₹ 5,000

N = Estimated life of the machine = 10 years

$$\text{Annual depreciation charges} = \frac{C - S}{N} = \frac{24,000 + 1,000 - 5,000}{10} = ₹\ 2,000$$

Problem 2.11: An engine lathe was purchased for ₹ 20,000. Its useful life was estimated 10 years and the salvage value is ₹ 5,000. Using the diminishing balance method, calculate the depreciation ratio. Also estimate depreciation fund at the end of two years.

Solution:

C = Initial cost of the machine = ₹ 20,000

S = Scrap value in rupees = ₹ 5,000

N = Estimated life of the machine = 10 years

$$X = 1 - \left(\frac{S}{C}\right)^{\frac{1}{N}}$$

$$X = 1 - \left(\frac{5,000}{20,000}\right)^{\frac{1}{N}} = 1 - \left(\frac{5,000}{20,000}\right)^{0.1} = 0.1294$$

The reduction in the value of lathe after 1 year = CX = 20,000 (0.1294) = ₹ 2,588

Value of lathe after 1 year = 20,000 – 2,588 = ₹ 17,412

The reduction in the value of lathe after 2nd year = CX = 17,412 (0.1294) = ₹ 2,253

Depreciation in first and second year = 2,588 + 2,253 = ₹ 4,841

Problem 2.12: A desk is purchased for $487.65. The expected life is 5 years. Calculate the annual depreciation.

Solution:

C = Initial cost of the desk = $487.65

S = Scrap value in rupees = 0

N = Estimated life of the machine = 5 years

$$\text{Annual depreciation charges} = \frac{C - S}{N} = \frac{487.65}{5} = \$97.53$$

Problem 2.13: A conference table is purchased for ₹ 1,467.89. The expected life is 5 years. Calculate depreciation expenses using sum of the digits method.

Solution: Since this is a 5 year asset the yearly factors have been calculated as shown in Table 2.4.

Table 2.4 Calculating depreciation expenses

Year	Percentage	Depreciation expense (Initial cost of item × Percentage)
1	$\frac{5}{15} = 33.34\%$	1,467.89 × 0.3334 = 489.40
2	$\frac{4}{15} = 26.67\%$	1,467.89 × 0.2667 = 391.49
3	$\frac{3}{15} = 20.00\%$	1,467.89 × 0.2000 = 293.58
4	$\frac{2}{15} = 13.33\%$	1,467.89 × 0.1333 = 195.67
5	$\frac{1}{15} = 6.67\%$	1,467.89 × 0.0667 = 97.91

Problem 2.14: A copy machine is purchased for ₹ 3,217.89. The expected life is 6 years. Calculate depreciation expenses using double declining balance.

Solution: Purchasing cost of a copy machine = ₹ 3,217.89

The expected life = 6 years

$$\text{Depreciation expense} = \text{Current value} \times \frac{2}{\text{Expected life}}$$

$$\text{Depreciation expense} = \text{Current value} \times \frac{2}{6}$$

$$\text{Depreciation expense} = \text{Current value} \times 0.33$$

Calculation of depreciation expenses is given in Table 2.5.

Table 2.5 Calculating depreciation expenses

Year	Current value	Depreciation expense (Current value × 0.33)	Accumulated depreciation
1	3,217.89	3,217.89 × 0.33 = 1,061.9	1,061.9
2	2,155.9	2,156.9 × 0.33 = 711.5	1,773.4
3	1,444.32	1,444.32 × 0.33 = 476.6	2,250.0
4	967.77	967.77 × 0.33 = 319.3	2,569.3
5	648.53	648.53 × 0.33 = 214.0	2,783.3
6	434.48	434.48 × 0.33 = 143.3	2,926.6

2.10 NATURE OF COSTS

There are two types of costs.

- Fixed costs
- Variable costs

Fixed costs remain same regardless of sales/output levels. Fixed costs have to be paid even if no products are sold. Examples include: Rent, insurance and wages.

Variable costs: Variable costs are costs that change with changes in production levels or sales. Examples include: Costs of materials, cost of labor and cost of energy used in the production of the goods.

The total cost: The Total cost is calculated by first multiplying the Variable costs per unit by the Volume. This product is then added to the Fixed costs to determine the Total costs.

Total cost = Variable cost/unit × Volume of production + Fixed costs

2.11 TOTAL REVENUE

The Total revenue is calculated by multiplying the Price per unit by the Volume. The Volume is the number of units that will be sold.

The Total revenue = Price per unit × Volume of production

2.12 BREAKEVEN POINT

The "breakeven" point is the point at which total revenue is equal to total cost. Breakeven point is the point at which a company makes neither a profit nor a loss. Companies tend to look at the breakeven point in terms of sales volume. For example, "Company A will breakeven after selling 500 units. Breakeven analysis is also known as C-V-P analysis (Cost Volume Profit Analysis). Representation of revenue and cost with respect to volume of production is shown in Fig. 2.1.

Notations used for calculating Breakeven formula:

F = Total fixed cost

P = Unit sale price

V = Unit variable cost.

Assumptions made in Breakeven analysis:

- All fixed and variable costs can be identified
- Variable costs are assumed to vary directly with output
- Fixed costs will remain constant
- Selling prices are assumed to remain constant for all levels of output
- Breakeven charts cannot handle multi-product situations
- It is assumed that all production will be sold
- The volume of activity is the only relevant factor which will affect costs

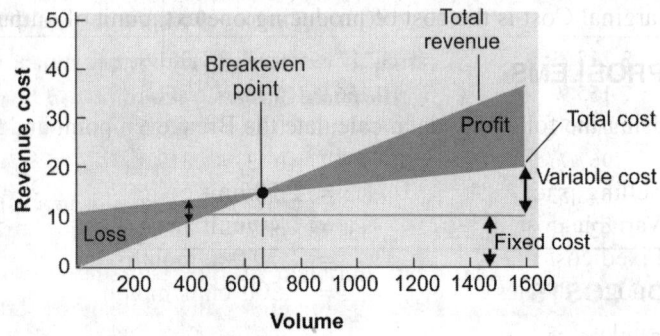

Fig. 2.1 Representation of revenue and cost with respect to volume of production

At Breakeven point (Fig. 2.1):

Total revenue = Total cost

(Price/unit × Volume) = (Variable cost/unit × Volume + Fixed cost)

$$\text{Breakeven volume} = \frac{\text{Fixed cost}}{\text{Price} - \text{Variable cost per unit}}$$

$$\text{Breakeven volume} = \frac{F}{P - V}$$

Applications of Breakeven analysis:

- Breakeven analysis helps managers to find the combination of costs, output and selling price that permits the firm to breakeven with no profits and losses
- Breakeven analysis can also be used to calculate profit (or loss) at a given level of output
- Selection of plant location
- Make or buy decision

Effect of changes in the costs and price on Breakeven volume:

- If variable cost per unit increases breakeven volume increases
- If variable cost per unit decreases breakeven volume decreases
- If fixed cost increases breakeven volume increases
- If fixed cost decreases breakeven volume decreases
- If selling price increases breakeven volume decreases
- If selling price decreases breakeven volume increases

Limitations of Breakeven analysis:

- In breakeven analysis costs are assumed either fixed or variable.
- In breakeven analysis production is assumed to equal to sales.
- Selling price is assumed remains constant
- Volume is the only factor affecting costs
- Costs and revenues are not always represented by a straight line.
- Not all variable costs increase directly with output.
- Not all costs can be categorized into fixed or variable costs

Contribution per unit: The sales price/unit minus the variable cost per unit.

Margin of safety: Margin of safety represents the strength of the business.

$$\text{Margin of safety} = \text{Current output} - \text{Breakeven output}$$

Marginal cost: Marginal Cost is the cost of producing one extra unit of output.

2.13 SOLVED PROBLEMS

Problem 2.15: Using the following data, calculate the Breakeven point and margin of safety in units.

Selling price	= ₹ 50/unit
Variable cost	= ₹ 40/unit
Fixed cost	= ₹ 70,000/month
Budgeted sales	= 7,500 units/month

Solution: Given data:

Selling price = P	= ₹ 50/unit
Variable cost per unit = V	= ₹ 40/unit
Fixed cost = F	= ₹ 70,000/month
Budgeted sales	= 7,500 units/month
Contribution/unit = 50 – 40	= ₹ 10 per unit

$$\text{Breakeven volume} = \frac{F}{P-V} = \frac{70,000}{50-40} = 7,000 \text{ units/month}$$

Margin of safety = (Current output – Breakeven output) = 7,500 – 7,000 = 500 units

Problem 2.16: Using the following data, calculate the level of sales required to generate a profit of ₹ 10,000/year

Selling price	= ₹ 35/unit
Variable cost	= ₹ 20/unit
Fixed cost	= ₹ 50,000/year

Solution: Given data:

Selling price = P	= ₹ 35/unit
Variable cost = V	= ₹ 20/unit
Fixed cost = F	= ₹ 50,000/year

Level of sales required to generate profit of ₹ 10,000/year:

$$\text{Profit} = (\text{Price/unit} \times \text{Volume}) - (\text{Variable cost/unit} \times \text{Volume} + \text{Fixed cost})$$
$$10,000 = 35 \times \text{Volume} - (20 \times \text{Volume} + 50,000)$$
$$10,000 = 15 \times \text{Volume} - 50,000$$
$$\text{Volume} = \frac{60,000}{15} = 4,000 \text{ units/year}$$

Problem 2.17: A plant produces 15,000 units per month. Find breakeven level, if fixed cost is ₹ 75,000/month, selling price is ₹ 8/unit and variable cost is ₹ 2.50/unit. Determine expected monthly profit or loss.

Solution:

Given data:

Selling price   = ₹ 8/unit

Variable cost per unit = V = ₹ 2.5

Fixed cost = F = ₹ 75,000/month

Budgeted sales   = 15,000 units/month

$$\text{Breakeven volume} = \frac{F}{P-V} = \frac{75,000}{8-2.5} = 13,636 \text{ units/month}$$

Profit = (Price/unit × Volume) – (Variable cost/unit × Volume + Fixed cost)

  = 8 × 15,000 – (2.5 × 15,000 + 75,000) = ₹ 7,500/month

Problem 2.18: Using the following data, calculate Breakeven volume.

Fixed cost = $2,00,000/year

Contribution per unit = $50

What is the Breakeven level of output?

Solution:

$$\text{Breakeven volume} = \frac{F}{P-V} = \frac{2,00,000}{50} = 4,000 \text{ units/year}$$

Problem 2.19: The fixed cost for the year 2012–2013 is ₹ 80,000. The estimated sales for the period are valued at ₹ 2,00,000. The variable cost per unit for the single product is ₹ 4. Each unit sells at ₹ 20.

 I. Construct the Breakeven chart.

 II. Determine the Breakeven point.

 III. Above how many units the company should produce in order to seek profit?

 IV. Determine the profit earned at a turnover of ₹ 1,60,000.

 V. Determine margin of safety.

Solution: Given data:

Selling price   = ₹ 20/unit

Variable cost per unit = V = ₹ 4

Fixed cost = F = ₹ 80,000/year

Budgeted sales   = ₹ 2,00,000/year

$$\text{Actual volume of production} = \frac{2,00,000}{20} = 10,000 \text{ units/year}$$

 I. Fixed cost, sales revenue and variable cost at various volumes of production are given in Table 2.6.

Table 2.6 Fixed cost, sales revenue and variable cost at various volumes of production

Volume of production	Fixed costs	Sales revenue	Total costs
0	80,000	0	80,000
1,000	80,000	20,000	84,000
2,000	80,000	40,000	88,000
3,000	80,000	60,000	92,000
8,000	80,000	1,60,000	1,12,000

Using the data in Table 2.6, Breakeven chart is drawn and is shown in Fig. 2.2.

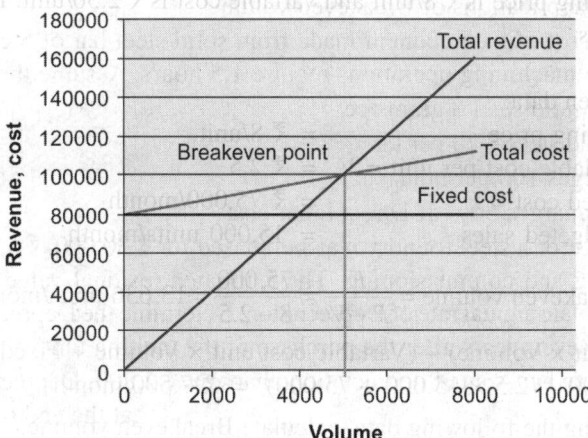

Fig. 2.2 Breakeven volume chart

II. Breakeven volume $= \dfrac{F}{P-V} = \dfrac{80,000}{20-4} = 5,000$ units/year

III. Company has to produce more than 5,000 units per year to seek the profit

IV. Volume of production at ₹ 1,60,000 $= \dfrac{1,60,000}{20} = 8,000$ units/year

Profit = (Price/unit × Volume of production) – (Variable cost/unit × Volume of production + Fixed cost)

$= 20 \times 8,000 - (4 \times 8,000 + 80,000) = ₹ 48,000$/year

V. Margin of safety = Actual production – Breakeven volume

$= 10,000 - 5,000 = 5,000$ units

∴ Margin of safety when represented as % $= \dfrac{\text{Margin of safety} \times 100}{\text{Total sales}} = \dfrac{5,000}{10,000} \times 100 = 50\%$

2.14 QUESTIONS AND PROBLEMS

1. Describe the various elements of cost and explain how the selling price of the product is determined.
2. Differentiate between Prime cost and Overhead cost with suitable examples.
3. Differentiate between Fixed and Variable costs.
4. State the relationship between various elements of cost.
5. What is depreciation, why assets are depreciated, and what are some examples of depreciable assets?
6. How do you calculate depreciation using the Straight line method?
7. How do you calculate depreciation using the Sum of the Digits method?
8. State the assumptions made in Breakeven analysis.
9. State the formulae for calculating:
   a. Breakeven point
   b. Margin of safety
10. Describe the effect of increase or decrease in Fixed cost and Variable cost on Breakeven point.

11. Describe the effect of increase or decrease in sales price on Breakeven point.
12. What are the applications of Breakeven volume?
13. Find factory cost of a component made from solid steel bar of 5 cm long and 2 cm in diameter. The machining operations require 1.5 hours. Assume the following data.
    - Density of mild steel 8 grams/cc
    - Cost of mild steel ₹ 20 per kg
    - Labor charges 100 per hour
    - Overhead charges 50% of Direct labor cost
14. A melting unit for a steel foundry was purchased for ₹ 30,000. ₹ 5,000 more was spent on its erecting and commissioning. The estimated residual value after ten years was ₹ 7,000. Calculate annual rate of Depreciation. Determine the Depreciation fund collected at the end of seven years after the purchase of the melting unit.
    Answer: Annual depreciation charge = ₹ 2,800, Deprecation fund collected at the end of 7 years = ₹ 19,600
15. An old car was purchased for ₹ 32,000. Its life was estimated as ten years and the scrap value as ₹ 8,000. Using Reducing Balance method, calculate Depreciation rate. Estimate Depreciation fund at the end of two years.   **Answer:** Depreciation rate =12.94%, Deprecation fund collected at the end of 2 years = ₹ 7,746
16. A copy machine is purchased for $3,217.89. The expected life is 4 years. Calculate depreciation expenses using double declining balance.
    Answer: Depreciation expenses: year 1 = 1,608, year 2 = 804, year 3 = 402 and year 4 = 201
17. A factory is producing 6,000 components per month. The fixed overhead for the month is ₹ 9,000 and the variable cost of each component is ₹ 10. If the selling price of each component is ₹ 12, estimate the minimum monthly production that may not cause any loss to the owner.   **Answer:** 4,500 units
18. A drill press costs ₹ 6,000. A discount of 25% of the price is given to the distributor. If Labor cost, Material cost and Factory Overhead are in the ratio of 4:1:2, and selling expenses are 25% of factory cost, calculate the profit of the factory for one drill press. Assume factory overheads of ₹ 800.   **Answer:** ₹ 1,000
19. A company is considering three different locations (A, B and C) for a new facility that will have annual costs ₹ 27,500, ₹ 25,400 and ₹ 29,000 respectively. The annual revenues are ₹ 40,000 for location A, ₹ 36,000 for location B and ₹ 42,000 for location C. Which location will maximize the net return per year?
    Answer: Net return per year = Annual revenue – Annual costs. C is the optimal location.
20. Annual Fixed cost at a small textile shop is ₹ 46,000 and Variable costs are estimated at 50% of the ₹ 40 per unit selling price.
    a. Find BEP.
    b. What profit would result from a volume of 3,000 units?
    Answer: a. BEP = 2,300 units b. Profit = ₹ 14,000
21. Price of the product is ₹ 650 per unit, Fixed cost is equal to ₹ 82,000 per year and Variable cost is ₹ 240 per unit.
    a. What is the Breakeven point?
    b. What volume is needed to generate a profit of ₹ 10,250?
    Answer: a. 200 units b. 225 units

22. Fixed cost is ₹ 40,000 per year, Variable cost is ₹ 50 per unit and the Selling price is ₹ 90 each. Find the BEP.　　　　　　　　**Answer:** 1,000 units

23. If the Sales price of a product is ₹ 8 and the Variable cost is ₹ 2. What is the contribution?
　　　　　　　　　　　　　　　　　　　　Answer: ₹ 6

24. A travel agency has an excursion package that sells for ₹ 125. Fixed cost is ₹ 80,000 and at the present volume of 1,000 customers, Variable cost is ₹ 25,000 and profit is ₹ 20,000. What is the Breakeven point volume?

25. Assuming the fixed costs remain constant, how many additional customers will be required for the agency to increase profit by ₹ 1,000.
　　　　　　　　　　　　　　　　Answer: a. 800 customers b. 10 customers

Production and Operations Management

3.1 INTRODUCTION

A production system takes inputs, raw material, personnel, machines, buildings, technology, cash, and other resources and converts them into products and services. This conversion process is called production and is the predominant activity of a production system. Operations managers manage the production system; their primary concern is with the activities of the conversion process.

3.2 PRODUCTION AND OPERATIONS MANAGEMENT

Production and operations management can be defined as the management of all activities directly related to the creation of goods and/or services through the transformation of inputs into output. We use the term "Production management" when the result is a physical product such as electronics, appliances or automobiles. Manufacturing operations produce tangible goods, which are physical products that can be held and seen. We use the term "operations management" when the output is a "service". Service operations provide certain intangible services such as banking, hospitality, advertising and consultancy that may not be easily identifiable.

3.3 OBJECTIVES OF PRODUCTION AND OPERATIONS MANAGEMENT

The overall objective of production and operations management is to deliver only the necessary parts in their right quality at the right time and at right place while using the minimum of facilities.

Objectives of production and operations management are given below:

- **Production of right quality:** The quality of product is established based upon the customer's needs. The right quality is not necessarily best quality. It is determined by the cost of the product and the technical characteristics as suited to the specific requirements.

- **Production of right quantity:** The manufacturing organization should produce the products in right number. If they are produced in excess of demand, the capital will block up in the form of inventory and if the quantity is produced in short of demand, leads to shortage of products.

- **Production at right time:** Timeliness of delivery is one of the important parameters to judge the effectiveness of production department. So, the production department has to make the optimal utilization of input resources to achieve its objective.
- **Minimizing cost:** Manufacturing costs are established before the product is actually manufactured. Hence, all attempts should be made to produce the products at pre-established cost, so as to reduce the variation between actual and the standard (pre-established) costs.

3.4 SCOPE OF PRODUCTION AND OPERATIONS MANAGEMENT

Fig. 3.1 Scope of production and operations management

The activities under production and operations management functions are given in Fig. 3.1. Following are the activities which are listed under production and operations management functions:

Forecasting: Success in business depends greatly on the ability to estimate future values of sales, output, price, etc. Forecasting is a tool used for predicting future sales, output, price, etc. using qualitative and quantitative techniques.

Product design: Product design is the process of creating a new product to be sold by a business to its customers. The product designer's role is to combine art, science, and technology to create new products that other people can use. Their evolving role has been facilitated by digital tools that allow designers to communicate, visualize, analyze and actually produce tangible ideas in a way that would have taken greater manpower in the past.

Process selection: Process selection refers to the strategic decisions of selecting the kind of production process to have in a manufacturing plant. Manufacturing process is the process of converting raw materials into products.

Process planning: To convert the product design into physical entity, manufacturing plan is needed. The activity of developing such plan is called process planning.

Facility location: Facility location refers to the choice of region and the selection of a particular site for setting up a business or factory. But the choice is made only after considering cost and benefits of different alternative sites.

Facility layout: Facility layout is a plan of an optimum arrangement of facilities including personnel, operating equipment, storage space, material handling equipment and all other supporting services along with the design of the best structure to contain all these facilities. The overall objective of facility layout or plant layout is to design a physical arrangement that meets the required output quantity and quality in most economical way.

Line balancing: Objective of the line balancing problem is to group the tasks at workstations so that the total time required at each workstation is approximately same. In a perfectly balanced assembly line, total time taken by all operations assigned to each workstation is equal.

Production planning: Production planning is concerned with the following activities:
- Aggregate production planning: This involves planning of the production output levels for major product lines produced by the firm.
- Master production schedule: Aggregate production plan must be converted into a master production schedule which is a specific plan of the quantities produced individual models within product line.
- Material requirements planning: Material requirements planning is a computational technique that converts master production schedule for the end product into detail purchasing and manufacturing schedule of raw materials and components used in end product.
- Capacity planning: Capacity planning is concerned with determining labor and equipment capacity required to meet the current master production schedule as well as the long-term future production needs of the firm.

Shop floor control: The shop floor control is concerned with the following:
- Releasing production orders to the production shop
- Scheduling of production order
- Monitoring and controlling progress of the production order
- Acquiring current information on the status of the production

Quality control: Quality control (QC) is the collection of methods and techniques for ensuring that a product or service is produced and delivered according to given requirements. This includes the development of specifications and standards, performance measures, and tracking procedures, and corrective actions to maintain control.

Maintenance management: In modern industry, equipment and machinery are very important parts of the total productive effort. Therefore, their idleness or downtime becomes very expensive. Hence, it is very important that the plant machinery should be properly maintained.

Determining Total cost: Total cost incurred in manufacturing, administration and selling.

Determining Selling price: Businesses must make a profit to survive. To make a profit; sales revenue must be higher than Total costs.

Determining Breakeven point: The "Breakeven" point is the point at which total revenues equal to total costs. Breakeven point is the point at which a company makes neither a profit nor a loss. Companies tend to look at the Breakeven point in terms of sales volume.

All the above activities related to production and operations management are discussed in subsequent chapters.

3.5 PRODUCTION MANAGEMENT V/S OPERATIONS MANAGEMENT

Comparison of production and operations management can be done on following characteristics:
- Output: Production management deals with manufacturing of products (like computer, car, air conditioner, cement bags, televisions, soaps, etc.) while operations management covers both products and services.

- Usage of output: Products like computer, car, air conditioner, cement bags, televisions, soaps, etc. are utilized over a period of time whereas services (service after sales, hospitals, insurance companies, banks, etc.) need to be consumed immediately.

- Classification of work: To produce products like computer, car, air conditioner, cement, televisions, soaps, etc., more of capital equipment and less labor are required while services require more labor and lesser capital equipment.

- Customer contact: There is no participation of customer during production whereas for services a constant contact with customer is required.

Comparison of production and operations management is given in Table 3.1.

Table 3.1 Comparison of production and operations management

Production management	Operations management
1. It is concerned with manufacturing	1. It is concerned with services
2. Output is tangible	2. Output is intangible
3. In this job, less labor and more equipment are used.	3. In this job, more labor and less equipment are used.
4. There is no customer participation	4. Frequent customer participation

Production management and operations management both are very essential in meeting objective of an organization.

3.6 MODERN APPROACHES TO PRODUCTION AND OPERATIONS MANAGEMENT

Modern production concepts can be considered as an essential field of economics now-a-days. They help to give valuable insights and thus provide important competitive advantages. There is a broad variety of new approaches to production and operations management. These approaches are:

Concurrent engineering: Concurrent Engineering brings together multidisciplinary teams, in which product developers from different functions work together in parallel from the start of a project with the intention of getting things right as quickly as possible, and as early as possible. In concurrent engineering, different tasks are tackled at the same time, and not necessarily in the usual order.

Lean production: Lean production can be defined as adaptation of mass production in which workers and work cells are more flexible and efficient by adopting methods that reduce waste in all forms.

Agile manufacturing: Agile manufacturing can be defined as: (i) An enterprise level manufacturing strategy of introducing new product in rapidly changing markets (ii) An organizational ability to thrive in a competitive environment characterized by continuous and sometimes unforeseen changes.

Just in time production system (JIT): JIT is viewed as production methodology which aims to improve overall productivity through elimination of waste and which leads to improve quality. JIT provides for the cost efficient production in an organization and delivery of only the necessary parts in their right quality at the right time and at right place while using the minimum of facilities.

Computer-aided design (CAD): Computer-aided design can be defined as the use of computer systems to assist in the creation, modification, analysis and optimization of design. All CAD/CAM systems offer almost the same set of features.

Computer-aided manufacturing (CAM): CAM refers to the use of computers in converting engineering designs into finished products. In CAM, computers assist managers, manufacturing engineers, and production workers by automating many production tasks. CAM is often used with computer-aided design (CAD). A direct link between product design and manufacturing can be established using CAD/CAM. Product engineers use a CAD system to establish the part geometry, dimensions, and tolerances. This design data can be transferred to the CAM system where the part programmers develop the CNC program to machine the part.

Computer-aided process planning (CAPP) or Automated process planning: Use of computers in process planning is usually called computer-aided process planning (CAPP) or automated process planning. Computer-aided process planning is a means to automatically develop the process plan from geometric image of the component. CAPP is usually considered to be part of computer-aided manufacturing (CAM).

Group technology and cellular manufacturing: Group technology is a manufacturing philosophy in which similar parts are identified and grouped together to take advantage of their similarities in design and production. Similar parts are arranged into part families, where each part family possesses similar design and/or manufacturing characteristics.

Numerical control: Numerical control of machine tools is defined as a method of automation in which various functions of machine tools are controlled by letters, numbers and symbols. Basically a NC machine runs on a part program. The part program consists of instructions written in numerical codes. These instructions are entered into the input medium. The controller translates these numerical codes into the machine's actual details, which are used to control machine functions such as feed, speed, tool change, movement of axes, etc. Each of the machines axes are connected to the servo motor or a stepper motor. The movement of the cutting tool with respect to the work piece is given in terms of coordinates.

Automated guided vehicles: Vehicles operate independently and are driven by electric motors that pick up power from batteries and moves generally in a fixed path. Automated guided vehicles are used to move work parts between machine tools. Imbedded guide wires in the floor emit electromagnetic signal that the vehicles follow.

Robot: Robot is a system that contains sensors, control systems, manipulators, power supplies and software all working together to perform a task. Designing, building, programing and testing a robot is a combination of physics, mechanical engineering, electrical engineering, structural engineering, mathematics and computing. In some cases biology, medicine and chemistry might also be involved. Robot is basically a computerized mechanical device that performs a series of complicated and complex tasks automatically. The robot can be programed to perform sequence of mechanical motions, and it can repeat that motion sequence over and over until the program to perform some other job.

Flexible manufacturing system (FMS): FMS is "A highly automated GT machine cell, consisting of a group of processing stations (usually CNC machine tools), interconnected by an automated material handling and storage system, and controlled by an integrated computer system". The FMS relies on the principles of GT. An FMS is capable of producing a single part family or a limited range of part families.

Computer packages for layout analysis: Several computerized approaches are available for developing and analyzing process layouts.

COMSOAL (Computer Method for Sequencing Operations for Assembly Lines): For the large scale problems the optimal solutions can be generated by computerizing the line balancing techniques. The basic methodology of COMSOAL is based on the generation of fairly large number of feasible solutions to the line balancing problem by sampling.

Material requirements planning I (MRP I): Material requirements planning is a computer package that converts master production schedule for the end product into detail purchasing and manufacturing schedule of raw materials and components used in end product.

MRP II: MRP II stands for Manufacturing Resource Planning II. MRP II is a computer modeling technique for planning, analyzing and controlling manufacturing operations. MRP II includes software for many areas of manufacturing such as purchasing, inventory, material requirements planning, shop floor scheduling, capacity planning, customer order entry and accounting. Advantages are standardization and automation of business processes leading to improvements in cost control and revenue. The goal of MRP II is to provide consistent data to all players in the manufacturing process as the product moves through the production line.

Enterprise resource planning (ERP): ERP systems were originally extensions of MRP II systems. An ERP system integrates areas such as planning, purchasing, inventory, sales, marketing, finance, human resources, etc. ERP facilitates information sharing, business planning, and decision making on an enterprise.

Total quality management (TQM): TQM is a comprehensive system approach that works horizontally across an organization, involving all departments and employees and extending backward and forward to include both suppliers and clients/customers. Total quality management measures its success on basis of customer satisfaction with regard to all aspects of the product (i.e. quality, price and availability). Like most quality management concepts, TQM views "quality" entirely from the point of view of "the customer".

Six sigma: It combines elements of statistical quality control, breakthrough thinking and management science. Six sigma means a measure of quality for near perfection. Six sigma is a disciplined, data-driven approach and methodology for eliminating defects in any process from manufacturing to service.

All these modern approaches are presented in detail in different chapters.

3.7 BENEFITS OF PRODUCTION AND OPERATIONS MANAGEMENT

- **Accomplishment of firm's objectives:** Production management helps the business firm to achieve all its objectives. It produces products, which satisfy the customers' needs and wants. So, the firm will increase its sales. This will help it to achieve its objectives.
- **Reputation, goodwill and image:** Production management helps the firm to satisfy its customers. This increases the firm's reputation, goodwill and image. A good image helps the firm to expand and grow.
- **Introducing new products in the market:** Production management helps to introduce new products in the market. It conducts Research and development (R & D). This helps the firm to develop newer and better quality products.
- **Support to other functional areas in an organization:** Production management supports other functional areas in an organization, such as marketing, finance, and personnel. The marketing department will find it easier to sell good quality products, and the finance department will get more funds due to increase in sales. It will also get more loans and share capital for expansion and modernization. The personnel department will be able to manage the human resources effectively due to the better performance of the production department.
- **To face competition in the market:** Production management helps the firm to face competition in the market. This is because production management produces products of right quantity, right quality, and right price at the right time. These products are delivered to the customers as per their requirements.

- **Optimum utilization of resources:** Production management facilitates optimum utilization of resources such as manpower, machines, etc. So, the firm can meet its capacity utilization objective.
- **Minimize the cost of production:** Production management helps to minimize the cost of production. It tries to maximize the output and minimize the inputs.
- **Expansion of the firm:** The Production management helps the firm to expand and grow. This is because it tries to improve quality and reduce costs. This helps the firm to earn higher profits. These profits help the firm to expand and grow.
- **Higher standard of living:** Production management conducts continuous research and development (R & D). So they produce new and better varieties of products. People use these products and enjoy a higher standard of living.
- **Improves job opportunities:** Production activities create many different job opportunities in the country, either directly or indirectly. Direct employment is generated in the production area, and indirect employment is generated in the supporting areas such as marketing, finance, customer support, etc. Because of production, other sectors also expand. Companies making spare parts will expand. The service sector such as banking, transport, communication, insurance, BPO, etc. also expand. This spread effect offers more job opportunities and boosts economy.
- **Boosts economy:** Production management ensures optimum utilization of resources and effective production of goods and services. This leads to speedy economic growth and well-being of the nation.

3.8 QUESTIONS

1. What is production and operations management?
2. What is production?
3. Explain in brief the objectives of production and operations management.
4. Distinguish between production management and operations management.
5. Explain the scope of production and operations management.
6. Explain modern approaches to production and operations management.
7. Explain benefits of production and operations management.

Production Systems or Manufacturing Systems

4.1 DEFINITION

Manufacturing system consists of two words manufacturing and system, both are having distinct meanings. Manufacturing means to convert raw materials into useful products. System is a group of interacting elements which are operated together in order to achieve efficiency. The integration of these two words results into a manufacturing system. Basically, manufacturing system is the best example of a system where the combination of men, machines, material flow along with the flow of information. Complete manufacturing system can be divided based on input materials, output products, production process, handling devices and plant services.

Manufacturing system based on input materials

- **Agro industries:** Industries based on the agriculture products.
  Examples: oil mills, cotton mills, rice mills, food processing industries, etc.
- **Mineral-based industries:** Industries based on minerals.
  Examples: Oil refineries, steel industries, cement industries, etc.

Industries based on output products

Examples: Automobiles, engineering industries and food processing industries

Industries based on production process

Examples: Welding, foundries, machining processes, etc.

Industries based on material handling devices

Examples: Conveyors, cranes, automatic guided vehicles, trucks, etc.

Selection of manufacturing technology

Selection of manufacturing technology depends on:
- Quantity/Unit time
- Quality level

Quality of output depends upon quality of input. Normal inputs may include:
- Engineering information through drawings and instructions
- Product design and process design

- Materials
- Maintenance policies

4.2 OBJECTIVES OF MANUFACTURING SYSTEM

- Earnings and profits to the share holders and owners
- Acceptable quality product to the customers at acceptable price
- Economic gains such as salaries, education, health, etc. to the employees

4.3 COMPONENTS OF PRODUCTION SYSTEM

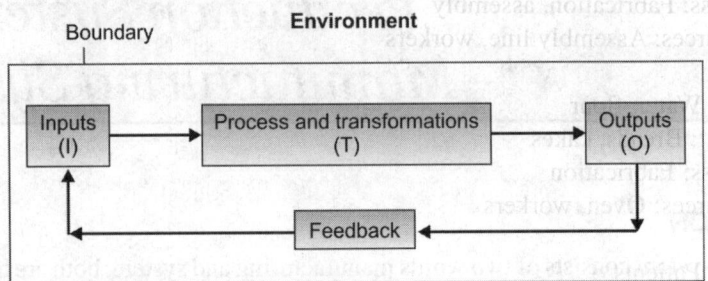

Fig. 4.1 Components of a system

A novel way to study a complex system such as manufacturing is to treat it as an Input-Output model. Components of production system are shown in Fig. 4.1. The components of production system are:

1. Input
2. Output
3. Transformations
4. Feedback
5. Environment

Inputs: Inputs are defined as the energizing or startup component on which the system operates. The quality of input affects the quality of outputs. In a manufacturing system, the input comprises raw material, capital, information, men, etc.

Output: Output is defined as the result of an operation. Anything exported from the system into the environment is an output. Some of the outputs such as waste matter and pollution are undesirable. Products and services are desirable outputs.

Transformation: Transformation transforms inputs into output. Efficiency of process is the ratio of output to inputs. Production process must be technically and economically efficient. Technical efficiency relates to the greater output for a given inputs. Economic efficiency relates to value added output per given capital investment. Engineering, operations, personnel and marketing activities change the raw material into end products. The transformation is responsible for adding value to the raw material. Activities in transformation of inputs to output are:

i. Engineering: R & D, product design and process design
ii. Marketing: Sales, advertising and forecasting
iii. Operation: Production and inspection control
iv. Accounting
v. Personnel

Feedback: Feedback is the step at which outputs are compared to desire standards or goals. If the outputs fail to meet these standards or goals, remedial action can then be taken. Feedback indicates that some aspect of the system varies from a standard or goal and requires attention.

Environment: A system is often affected by changes occurring outside the system. Such changes occurring outside the system are said to be occurring in the system environment. Social factors, government, consumer behavior and market trends are the examples of environment.

Examples of Production Systems:

1. *Automobile Factory*
   - Input: Raw material
   - Output: Completed cars
   - Process: Fabrication, assembly
   - Resources: Assembly line, workers
2. *Bakery*
   - Input: Water, flour
   - Output: Breads, cakes
   - Process: Fabrication
   - Resources: Oven, workers
3. *Hospital*
   - Input: Patients
   - Output: Healthy individuals
   - Process: Health care
   - Resources: Medical doctors, nurses, medical supplies, equipment
4. *University*
   - Input: Students
   - Output: Educated individuals
   - Process: Imparting knowledge and skills
   - Resources: Teachers, books, classrooms

A manufacturing system can be studied to solve the following problems:

  i. *Analysis*: Identify the contents in inputs, outputs and transformation
 ii. *Operation*: Finding output for a given input and specific transformation function
iii. *Synthesis*: Determining a suitable transformation for a given input and output
 iv. *Inversion*: Finding input for a given output and transformation function
  v. *Optimization*: Selecting or determining value of variables to maximize profit or minimize cost.

4.4 TYPES OF PRODUCTION SYSTEMS

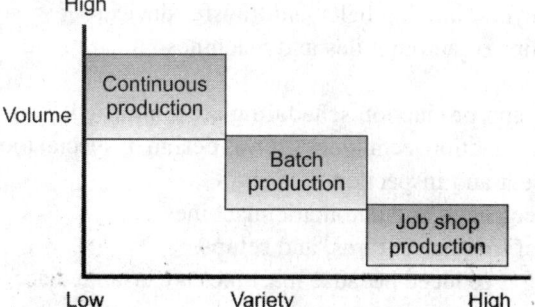

Fig. 4.2 Types of production systems

Traditionally, the manufacturing systems can be classified on the basis of production volume and variety of products. Types of production systems on the basis of production volume and variety of products are given in Fig. 4.2. The different production systems are:
1. Continuous production system
   - Mass production system
   - Flow production system
2. Intermittent production system
   - Batch production system
   - Job shop production system
   - Project production system

4.4.1 Continuous Production System

Continuous production is usually associated with large quantity with high rate of demand in the market.

Features of continuous production systems:
- Flow of production is continuous and not intermittent
- Products are standardized
- Products are produced as per quality standards
- Products are produced in anticipated demand
- Standardized routing sheets and schedules are prepared

Following are examples of continuous production system:
- The production system of a food industry is purely based on the demand forecast. Here large scale production of food takes place. It is also continuous.
- Similarly the production and processing system of oil refineries is also based on demand forecast. Petrol, diesel, etc. are produced from crude oil continuously on a large scale.

Types of continuous production systems:
  i. Mass production
  ii. Flow production

4.4.1.1 Mass production system

In mass production, large quantities of identical parts are produced. Examples: Manufacturing of products such as pumps, motors, etc. Process industries like cement, oil refineries, rice mills, etc.

4.4.1.2 Flow production system

In flow production, items are made to flow through a sequence of operations by some handling devices such as conveyors, moving belts and transfer devices.

Example: Assembly line of automobiles and machines

Advantages:
- Process planning and production scheduling are simplified.
- Improvement in production techniques such as design of special tools, jigs, fixtures, material handling equipment and inspection devices
- Use of highly specialized and automatic machines
- Standardization of tooling, fixtures, and setups
- Material handling is reduced because machines are arranged according to the sequence of operations.
- Automatic material handling equipment can be used.

- Work-in-process is reduced.
- Improvement in worker satisfaction
- Use of inspection on quality control methods
- Higher quality work is accomplished
- Productivity is high
- Lowest cost per unit

Disadvantages:
- Both plant and equipment are less flexible.
- It involves high initial expenditure.
- High degree of risk because of production of large number of similar products.

4.4.2 Intermittent Production System

In the intermittent production system, goods are produced based on customer's order. These goods are produced on a small scale. The flow of production is intermittent. In other words, the flow of production is not continuous. In this system, the large variety of products are produced. Features of intermittent production system:
- Flow of production is not continuous
- Variety of products are produced
- Volume of production is small
- General purpose machines are used
- Sequence of operations change as per design. Production depends upon the customer's orders.

Figure 4.3 highlights the concept of intermittent production system.

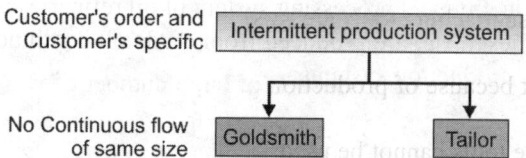

Fig. 4.3 Intermittent production system

Types of intermittent production systems:
- Batch production system
- Job shop production system
- Project production system

4.4.2.1 Batch production system

To manufacture medium variety of products, the volume of production is a medium for each product. So periodically production run is undertaken to produce a batch that will satisfy demand. Demand is, therefore, met from inventory which is replenished periodically.

Advantages:
- Higher degree of flexibility
- The general purpose machine tools are used
- Initial cost is less
- Less degree of risk because of production of medium number of variety products

Disadvantages:
- Production planning is difficult
- Production scheduling is difficult

- Automatic machine tools cannot be used
- Specialized machines cannot be used
- Automatic material handling equipment cannot be used
- Standardization of jigs and fixtures is not possible
- Lot of unproductive time

Some of the constituents of unproductive time in a manufacture cycle time:
  i. Setup time: Time consumed in mounting tools at each time to start the required machining operation.
  ii. Part loading and unloading: Time consumed in loading the work into jig or fixture and also unloading the job from the machine.
  iii. Work transport: Time consumed in transporting the material from one machine tool to another machine tool.
  iv. Time consumed in loading and unloading parts into/from material handling equipment
  v. Inspection time: Time consumed in the inspection activity at different stages of production.

4.4.2.2 Job shop production system

It is the manufacture of the product to meet customer requirements where quantity involved is very less. Large engines, boilers, material handling devices, ship building, etc. are the activities belong to this group. Examples: BHEL, CNC machine tool manufacturing industries, HMT tools, repair and servicing shops, etc.

Advantages:
- Higher degree of flexibility
- The general purpose machine tools are used
- Initial cost is less
- Less degree of risk because of production of large number of variety products

Disadvantages:
- Automatic machine tools cannot be used
- Specialized machines cannot be used
- Automatic material handling equipment cannot be used
- Standardization of jigs and fixtures is not possible
- Lot of unproductive time

Some of the constituents of unproductive time in a manufacture cycle time:
  i. Setup time: Time consumed in mounting tools at each time to start the required machining operation.
  ii. Part loading and unloading: Time consumed in loading the work into jig or fixture and also unloading the job from the machine.
  iii. Work transport: Time consumed in transporting the material from one machine tool to another machine tool.
  iv. Time consumed in loading and unloading parts into/from material handling equipment.
  v. Inspection time: Time consumed in the inspection activity at different stages of production.

4.4.2.3 Project type production system

Project consists of number of activities. All the activities are completed in order to complete the project. CPM and PERT techniques are used for project scheduling and controlling various activities of the project. Men, material, machines and money are required to complete the project.

Examples of project:

- Construction of buildings
- Construction of dam
- Starting of industry
- Overhauling of boilers, turbines, etc.

Other types of production systems are given below:

4.4.3 Cellular Manufacturing Systems

Group technology is a manufacturing philosophy in which similar parts are identified and grouped together to take advantage of their similarities in design and production. Similar parts are arranged into part families, where each part family possesses similar design and/or manufacturing characteristics. For example, a plant producing 10,000 different part numbers may be able to group the vast majority of these parts into 30–40 distinct families. It is reasonable to believe that the processing of each member of a given family is similar, and this should result in manufacturing efficiencies. The efficiencies are generally achieved by arranging the production equipment into machine groups, or cells, to facilitate workflow. Grouping the production equipment into machine cells, where each cell specializes in the production of a part family, is called cellular manufacturing. There are two major tasks that a company must undertake when it implements group technology. These two tasks represent significant obstacles to the application of GT.

1. Identifying the part families: If the plant makes 10,000 different parts, reviewing all of the part drawings and grouping the parts into families is a substantial task that consumes a significant amount of time.

2. Rearranging production machines into machine cells: It is time consuming and costly to plan and accomplish this rearrangement and the machines are not producing during the changeover.

Benefits of group technology:

The benefits include:

- GT promotes standardization of tooling, fixtures, and setups.
- Material handling is reduced because parts are moved within a machine cell rather than within the entire factory.
- Process planning and production scheduling are simplified. Setup times are reduced, resulting in lower manufacturing lead times.
- Work-in-process is reduced.
- Worker satisfaction usually improves.
- Higher quality work is accomplished using group technology.

4.4.4 Automated Manufacturing Systems

The level of automation deployed is an important characteristic of the manufacturing system. Workstation machines may be manually-operated, semi-automated, or fully-automated. There are production systems in which there is no human interference. Automatic production systems consist of automatic machines (usually CNC machine tools), interconnected by an automated material handling and storage system, and controlled by an integrated computer system". Flexible manufacturing system is an example of automatic system.

4.4.5 Flexible Manufacturing System

Definition: "A highly automated GT machine cell, consisting of a group of processing stations (usually CNC machine tools), interconnected by an automated material handling and storage system, and controlled by an integrated computer system". The FMS relies on the principles of GT. An FMS is capable of producing a single part family or a limited range of part families.

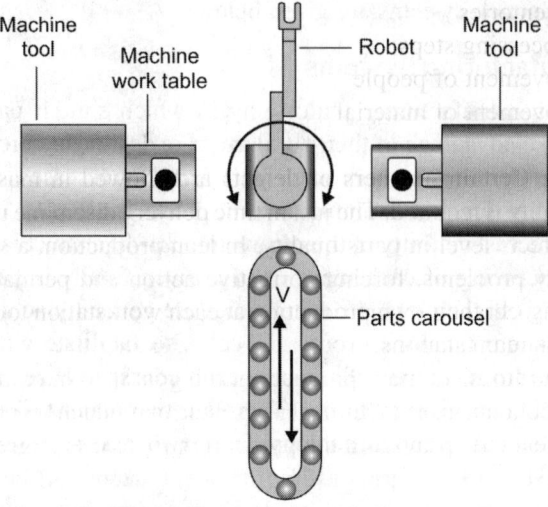

Fig. 4.4 Automated manufacturing cell with two machine tools and robot

Figure 4.4 is a flexible manufacturing system consisting of two CNC machine tools that are loaded and unloaded by an industrial robot from a parts carousel. Periodically a worker must unload completed parts from carousel and replace them with new parts.

Type of flexibilities:
- Machine flexibility: Capability of machine to perform variety of operations.
- Product flexibility: It processes different part styles.
- Schedule flexibility: It accepts changes in production schedule.
- Routing flexibility: It accepts the parts through alternative workstations' sequences in response to equipment malfunctions and breakdowns.
- Expansion flexibility: It accommodates introduction of new part designs, adding of work-stations, etc.
- Volume flexibility: It accepts change in volume of each part produced.

FMS components: There are several basic components of FMS:
1. Workstations
2. Material handling and storage system
3. Computer control system
4. Human labor

4.4.6 Lean Production

Lean production can be defined as adaptation of mass production in which workers and work cells are more flexible and efficient by adopting methods that reduce waste in all forms. Lean production is based on four principles:
- Minimize waste
- Perfect first time quality

- Flexible production lines
- Continuous improvement

All four principles are derived from the first principle, i.e. minimize waste.

Minimize waste: Waste forms can be listed as follows:

- Production of defective parts
- Production of more than the number of items needed
- Unnecessary inventories
- Unnecessary processing steps
- Unnecessary movement of people
- Unnecessary movement of material
- Workers waiting

First time quality: Certain numbers of defects are allowed in mass production. In lean production, perfect quality is required. The just in time delivery discipline used in lean production necessitates a zero defects level in parts quality. In lean production, a single defective draws attention to the quality problems, forcing corrective action and permanent solution. In lean production, workers inspect their own production at each workstation to minimize the delivery of defects to the downstream stations.

Flexible production lines: In mass production, the goal is to maximize efficiency. This is achieved using long production run of identical parts. In lean production, procedures are designed to speed the changeovers thus providing the production system with greater flexibility.

Continuous improvement: Continuous improvement means constantly searching ways to reduce cost, improve quality and increase productivity, setup time reduction, manufacturing lead time and improvement in product design.

Lean production approach can be applied to all aspects of business from design to distribution of finished product to the markets.

An example of lean production in a car manufacturing company:

- A car manufacturing company did not store car seats in the factory. Instead they order car seats through phone. They receive the stock after 5 hours when the car was ready for the seats to be put in. Thus, they eliminate space required to store car seats and also they minimize inventories.
- Better quality control: Instead of doing a comprehensive quality control at the end of the production, the car manufacturing company adopted the method of checking each item at all stages of production. This saves money because items are checked at all stages of production to prevent useless machining.

Comparison of mass production and lean production is given in Table 4.1.

Table 4.1 Comparison of mass production and lean production

Mass production	Lean production
• Mass production techniques like specialized machines, automatic material handling equipment, product layout, etc. are used.	• Lean production techniques are used in which machines are flexible.
• Acceptable quality level concept is used.	• Zero defective concept is used.
• Fraction of defective items are allowed.	• Zero defectives are allowed.
• Inspection is done at the end of the production line.	• Workers inspect their own production at each workstation to minimize the delivery of defects to the downstream.

(Contd...)

(Contd...)

Mass production	Lean production
• Buffers and inventories are allowed.	• Minimization of inventories.
• System is not flexible. Therefore, continuous improvement is not possible.	• Flexible production lines are used. Therefore, continuous improvement is possible.

4.4.7 Agile Manufacturing

Agile manufacturing can be defined as: (i) An enterprise level manufacturing strategy of introducing new product in rapidly changing markets (ii) An organizational ability to thrive in a competitive environment characterized by continuous and sometimes unforeseen changes.

Necessity of agile manufacturing:

- Global competition is intensifying.
- Mass markets are fragmenting into niche markets.
- Cooperation among companies is becoming necessary, including companies who are in direct competition with each other.
- Very short product life cycles, development time, and production lead times are required.

Manufacturing companies that are agile tends to exhibit four principles or characteristics. The four principles are:

- Organize to master change
- Leverage the impact of people and information
- Cooperate to enhance competitiveness
- Enrich the customer

Organize to master change: If the company is agile, the human and physical resources can be rapidly reconfigured to adapt the changing environment and market opportunities.

Leverage the impact of people and information: In an agile company knowledge has value. Innovations are rewarded. Authority is distributed to the appropriate levels of organization. Management provides the resources that personnel need. There is a climate of mutual responsibility for joint success.

Cooperate to enhance competitiveness: Partnership with other companies to form virtual enterprises. Required resources and competencies are used wherever they exist to bring the product to the market as rapidly as possible. Suppose, knowledge is available with one company and physical resources are available with other company. Knowledge of one company and physical facilities of other company are used to bring new product as early as possible into the market.

Enrich the customer: An agile company is perceived by the customer. This is possible only if the company gives complete information about the product to the customer. The products of the company are perceived as solution to the customer problems. Price of the product can be based on the value of the solution to the customer rather than manufacturing cost.

Agile methodology is one of the most popular methodologies currently being used for software development. Many software companies in India are employing agile approaches to software development because they deliver value to organizations and end users faster and with higher quality. It has helped to successfully release numerous products into the market. This approach has resulted in better end products that can meet clients overall expectations. Comparison of mass production and agile manufacturing is given in Table 4.2. Comparison of lean production and agile production is given in Table 4.3.

Table 4.2 Comparison of mass production and agile manufacturing

Mass production	Agile manufacturing
• Company produces large quantity of identical parts	• Large quantity of products having features that have been specified by their respective customers.
• Long market life is expected	• Short market life is expected
• Produce to forecast	• Produce to order
• Low information content	• High information content
• Single time sales	• Continuous relationship
• Pricing by the production cost	• Pricing by customer value.

Table 4.3 Comparison of lean production and agile production

Lean production	Agile production
• Minimize waste	• Organize to master change
• Perfect first time quality	• Leverage the impact of people and information
• Flexible production lines	• Cooperate to enhance competitiveness
• Continuous improvement	• Enrich the customer

4.4.8 Just in Time Production System (JIT)

According to Voss, JIT is viewed as production methodology which aims to improve overall productivity through elimination of waste and which leads to improve quality. JIT provides for the cost efficient production in an organization and delivery of only the necessary parts in their right quality at the right time and at right place while using the minimum of facilities.

Seven wastes:

- **Waste of production time:** Eliminate waste of production time by reducing setup times. Synchronizing quantities and timing between the processes, eliminating traffic jams, make only what is needed now.

- **Waste of waiting:** Waiting time of material is reduced by eliminating bottlenecks and balancing uneven loads by flexible workforce and equipment.

- **Waste of transportation:** Eliminate transportation and handling if possible, by using flexible manufacturing system to perform more than one operation. Reduce the transportation time by designing optimal layout and material handling equipment.

- **Waste of processing itself:** Eliminate unnecessary features which add neither functional value nor appearance value. This reduces unnecessary features and eliminates unnecessary operations.

- **Waste of stock:** Reduces all other wastes reduce stock.

- **Waste of motions:** Eliminate unnecessary motions by motion study. Economy of motions improves productivity and consistency improves quality.

- **Waste of making defective parts:** Develop the production process to prevent production of defective parts. A quality process always yield quality product.

Benefits of JIT:

- Production cost is greatly reduced due to the reduction of manufacturing cycle time, reduction of waste and inventories and elimination of non-value added operations.

- Quality is improved because of continuous quality improvement program.

- Due to fast response to engineering change, alternative designs can be quickly brought on the shop floor.

- Higher production system flexibility
- Administrative ease and flexibility

JIT manufacturing includes three supportive components such as:

1. People participation and involvement
2. Total quality control
3. Just in flow

1. **People participation and involvement:** The success of JIT depends upon how the companies train their human resource to have an appropriate skill, responsibility and coordinate and motivate people. JIT fully utilizes the creative talents of employees, suppliers, subcontractors and others who contribute the company's improvement.

2. **Total quality control:** Total quality control refers to the achievement and improvement in quality of a JIT company, which involves every department and each employee in the company. All employees should find the ways to serve the final customer better so that the company can remain competitive.

3. **JIT flow:** The major objective of JIT is to have only the right item at the right time. This practice reduces work-in-process inventory and capital requirements but also the floor space and flow time.

4.4.9 Kanban System

Kanban system is a kind of production system which operates based on the information contained in cards called "Kanbans". JIT objectives are met by using pull-based production planning and control systems. A "pull" system is one in which you produce only when you receive a signal from the user. Kanban system is based on a pull system of production control in which the order to make and deliver parts at each workstation in the production sequence comes from the downstream station that uses those parts. When supply of parts at a given workstation '$(i+1)$' is about to be exhausted, that station '$(i+1)$' orders the upstream station 'i' to produce the parts. Order to produce the parts in station 'i' and transport of parts from station 'i' comes from station '$(i+1)$'. One way to implement pull system is to use Kanbans. The word Kanban means card in Japanese. The Kanban system of production control was developed by Japanese automobile company. There are two types of Kanbans: (1) Production Kanbans (2) Transport Kanbans. A production Kanban (P-Kanban) authorizes the upstream station to produce a batch of parts. As they are produced the parts are placed in the containers, so the batch quantity is just sufficient to fill the container. Production of more than this quantity is not allowed in the Kanban system. A transport Kanban (T-Kanban) authorizes transport of the containers of parts to the downstream station.

Let us describe the operation of Kanban system with reference to Fig. 4.5. The flow of work is from station 'i' to station '$(i+1)$'.

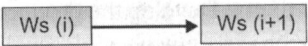

Fig. 4.5 Operation of Kanban system between workstations

The sequence of steps in Kanban system is as follows:

I. Station '$(i+1)$' authorizes using P-Kanban to the station 'i' to process parts.

II. As they are produced these parts are placed in the container.

III. A transport Kanban (T-Kanban) authorizes the workstation 'i' to transport the parts to the station '$(i+1)$'.

Advantages of Kanban system:

- A simple and understandable process
- Minimize inventories
- Avoids over production
- Delegates' responsibility to line workers
- Push Systems pushes the parts to downstream station irrespective of the pats needed by the downstream station. Push system results large number of parts in front of the machines.

4.5 QUESTIONS

1. Define production system. Explain various components of the production system.
2. Explain the different types of production systems stating the advantages and limitations each.
3. What is production system? Explain different types of production systems.
4. Explain agile manufacturing. What are its advantages and disadvantages?
5. What is lean manufacturing? Explain special features of lean manufacturing.
6. What are the basic elements of JIT? Explain.
7. Explain flexible manufacturing system.
8. Differentiate between job shop, batch and mass production systems.
9. Classify various production systems and compare their characteristics.
10. Differentiate lean and agile manufacturing.
11. Explain briefly Kanban system.
12. Differentiate lean and mass productions.

Product Design

5

The Product Life Cycle, Product Design and Concurrent Engineering

A product is an article obtained by transformation of raw material and is sold by manufacturer. A product can be developed by:

Imitation: Making another product similar to one in the market.

Adaptation: Developing an improved product for an already existing in the product.

Examples: CNC machines in the place of Turret lathe, LCD televisions in the place of CRT televisions.

Invention: Making new products.

Examples: Diesel engines in the place of petrol engines, cell phones production.

5.1 THE PRODUCT LIFE CYCLE

Like human beings, products also have their own life cycle. From birth to death human beings pass through various stages, e.g. birth, growth, maturity, decline and death. A similar life cycle is seen in the case of products. And just like us, these products have a life cycle. Older, long-established products eventually become less popular, while in contrast, the demand for new, more modern goods usually increase quite rapidly after they are launched. The product life cycle is an important concept in marketing. It describes the stages a product goes through from when it was first thought of until it finally is removed from the market. Not all products reach this final stage. Some continue to grow and others rise and fall. About 95 percent of all newly introduced products fail each year, according to a March 2010 article by Forbes.com. And even the few that succeed have certain life spans called product life cycles.

5.1.1 Stages or Phases of Product Life Cycle

The product life cycle is shown in Fig. 5.1. Product life cycle is broken down into four phases. These phases are:

1. Introduction
2. Growth
3. Maturity
4. Decline

Above are four phases to each product's life cycle, not including the innovation and development stage. Companies use various marketing strategies in each stage to prolong the life cycles of their products. Most strategies are implemented to counter key moves and strategies of competitive companies.

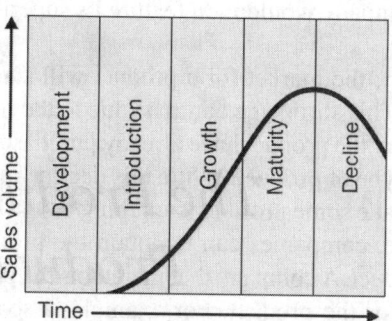

Fig. 5.1 Product life cycle

Introduction phase: This stage of the cycle could be the most expensive for a company launching a new product. The size of the market for the product is small, which means sales are low, although they will be increasing. On the other hand, the cost of things like research and development, consumer testing, and the marketing needed to launch the product can be very high, especially if it's a competitive sector. The product appears in stores for the first time, and people start seeing print and television ads. During this phase, a company may choose one of two pricing strategies. They may set prices high to recoup initial expenses that went into producing the product. For example, a cell phone manufacturer with new technology may introduce cell phones 10 percent to 20 percent above the prices of most premium cell phones. They may price their phones higher because of the hype and anticipation of the new technology. The company also knows enough people will pay the extra 10 to 20 percent for it to earn substantial profits. Contrarily, the same cell phone company may introduce a cell phone with basic features at reduced prices in hopes of gaining lots of new customers.

Growth phase: The growth stage is typically characterized by a strong growth in sales and profits, and because the company can start to benefit from economies of scale in production, the profit margins, as well as the overall amount of profit, will increase. This makes it possible for businesses to invest more money in the promotional activity to maximize the potential of this growth stage. Product quality is also maintained. However, a company will usually expand its product distribution during the growth stage. For example, a consumer products company might start selling its product in new markets, based on positive marketing research from consumers. Eventually, the product will start appearing in stores across the country. Company marketers usually increase advertising during the growth phase.

Maturity stage: During the maturity stage, the product is established and the aim for the manufacturer is now to maintain the market share they have built up. This is probably the most competitive time for most products and businesses need to invest wisely in any marketing they undertake. They also need to consider any product modifications or improvements to the production process which might give them a competitive advantage. Success leads to increased competition. Other companies eventually will start introducing similar products, especially if the initial product is highly successful. Consequently, the demand for the product and its competitors will peak at some point. Sales growth will start to decline. Some companies may lower prices to capture additional market share or new customers. At this point, a company may need to develop new product features or services to differentiate its products from the competitions. For example, the company that first introduced the product may enhance its customer service department to establish itself as the service leader in the industry. The company's stellar customer service may be the distinguishing element that spurs additional sales and customers. The company would then feature its superior customer service in most of its advertisings.

Decline stage: Eventually, the market for a product will start to shrink, and this is what's known as the decline stage. This shrinkage could be due to the market becoming saturated (i.e. all the customers who will buy the product have already purchased it), or because the consumers are switching to a different type of product. While this decline may be inevitable, it may still be possible for companies to make some profit by switching to less expensive production methods and cheaper markets. Hence, companies can maintain the product, sell it at heavily reduced prices or discontinue the product. A company that maintains the product may continue increasing sales by finding new uses for the product. For example, a soap manufacturer may discover through marketing research that restaurants and industrial companies like the cleaning properties of its soap. Subsequently, the company would start selling its soap to both consumers and businesses. This strategy could help extend the life of the product. Innovation rate in the mobile industry is very high. That is why product life is less for mobile phone models. Even highly successful Nokia mobile company is not able to maintain its market share. The Ambassador has been relegated by the new brigade of Japanese, Korean, European and American cars. However, the key to successful manufacturing does not just understand this life cycle, but also proactively managing products throughout their lifetime, applying the appropriate resources and sales and marketing strategies, depending on what stage products are at in the cycle.

5.2 PRODUCT DESIGN

Product design is the process of creating a new product to be sold by a business to its customers. This is usually accomplished by adventurous people. The product designer's role is to combine art, science, and technology to create new products that other people can use. Their evolving role has been facilitated by digital tools that allow designers to communicate, visualize, analyze and actually produce tangible ideas in a way that would have taken greater manpower in the past.

The idea for new or improved products comes from many sources such as:

- Customer's suggestions and complaints
- Research and development department
- Other competitor products in the market
- Marketing department
- Design department
- Production department

5.2.1 Objectives of Product Design

- The overall objective is profit generation in the long-run.
- To achieve the desired product quality
- To reduce the cost of the product
- To ensure manufacturability
- Adequate profit to the manufacturer
- To achieve customer satisfaction

5.2.2 Factors Influencing Product Design

- Customer requirements
- Function
- Appearance
- Materials used
- Equipment used
- Layout
- Tolerance
- Surface finish required
- Cost
- Product quality
- Process capability
- Maintenance

5.2.3 Stages of Product Design

Design is basically a problem solving exercise. The design of a new product consists of the following stages:

- Design brief
- Product design specifications
- Concept design
- Testing
- Detail design
- Engineering analysis
- Construction of prototype
- Evaluate the final product

In practice, the stages could be defined in different ways and the sequence may vary.

Design brief: The design brief is typically a statement of intent.

Example: "We will design and make a Formula One racing car".

Product design specification (PDS): Before producing a 'solution', there is a true understanding of the actual problem. The PDS is a document listing the problems in detail. To produce the PDS, you will have to research the problem and analyze competing products and all important points and discoveries should be included in the PDS. Using the PDS as the basis, the designer attempts to produce an outline of a solution.

Concept design: A conceptual design is to determine key components of the product and their arrangement. For example, a concept design for a car might consist of a sketch showing a car with four wheels and the engine mounted at the front of the car. The exact details of the components such as the diameter of the wheels, or the size of the engine are determined at the detail design stage. However, the degree of detail generated at the conceptual design stage will vary depending on the product being designed. When designing a product you also consider the activities such as manufacturing, quality, assembly, maintenance, sales, transportation, etc. By considering these stages early, you can eliminate problems that may occur at these stages. This stage of the design involves drawing up a number of different viable concept designs which satisfy the requirements of the product outlined in the PDS and then evaluating them to decide on the most suitable to develop further.

Concept evaluation: Once a suitable number of concepts have been generated, it is necessary to choose the design most suitable to fulfill the requirements set out in the PDS. The product design specification should be used as the basis of any decision being made. Ideally a multifunction design team should perform this task so that each concept can be evaluated from a number of angles or perspectives. The chosen concept will be developed in detail.

Detail design: In this stage of the design process, the chosen concept design is designed in detailed with all the dimensions and specifications necessary to make the design specified on a detailed drawing of the design. Creating a final product is the culmination of the product design process.

Engineering analysis: Basically, it is the breaking down of an object, system or problem, into its fundamental parts to understand their relationships to each other and to outside elements. Analysis of a design prior to implementation leads to increased safety and efficiency in using the product. Finite element analysis (FEA) is a type of computer program that uses the finite element method to analyze a material or object and find how applied stresses will affect the material or design. FEA can help to determine any points of weakness in a design before it is manufactured. FEA programs are more widely available with the spread of more powerful computers, but are still mostly used in aerospace and other high-stress applications. The analysis is done by creating a mesh of points in the shape of the object that contains information about the material and the object at each point for analysis. In addition to determining the reaction to stress upon an object, FEA can also analyze the effect of vibrations, fatigue, and heat transfer. The finite element method is a computational scheme to solve field problems in engineering and science. The technique has very wide application, and has been used on problems involving stress analysis, fluid mechanics, heat transfer, diffusion, vibrations, electrical and magnetic fields, etc. The fundamental concept involves dividing the body under study into a finite number of pieces (sub domains) called elements (see Fig. 5.2). Particular assumptions are then made on the variation of the unknown dependent variable(s) across each element using so-called interpolation or approximation functions. This approximated variation is quantified in terms of solution values at special element locations called nodes. Through this discretization process, the method sets up an algebraic system of equations for unknown nodal values which approximate the continuous solution. Because element's size, shape and approximating scheme can be varied to suit the problem, the method can accurately simulate solutions to problems of complex geometry and loading and thus this technique has become a very useful and practical tool.

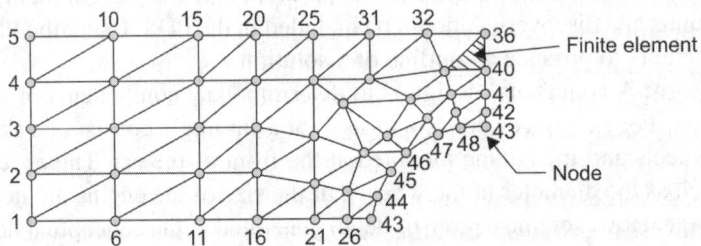

Fig. 5.2 Dividing the body under study into a finite number of elements

Common types of elements

One-dimensional elements: Rods, beams, trusses, frames, etc.

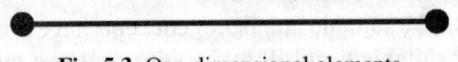

Fig. 5.3 One-dimensional elements

Two-dimensional elements: Triangular, quadrilateral plates, shells, 2-D continua, etc.

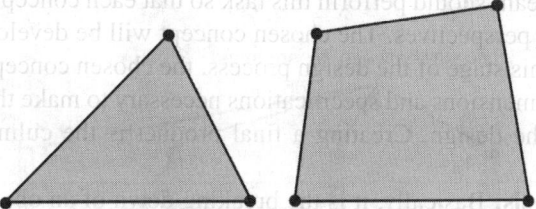

Fig. 5.4 Two-dimensional elements

Three-dimensional elements: Tetrahedral, rectangular prism (brick), etc.

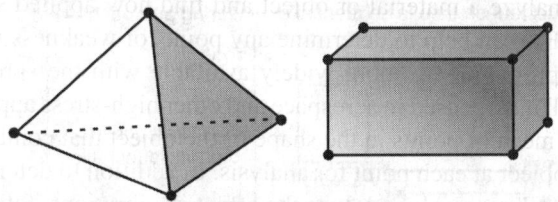

Fig. 5.5 Three-dimensional elements

Basic steps in the Finite Element method:
- Select element's type (Shape and Approximation).
- Derive element's equations (Variation and Energy Methods).
- Assemble element's equations to form global system.

$$[K]\{U\} = \{F\}$$

where

   $[K]$ = Stiffness or Property matrix
   $\{U\}$ = Nodal displacement vector
   $\{F\}$ = Nodal force vector

- Incorporate boundary and initial conditions.
- Solve assembled system of equations for unknown nodal displacements and secondary unknowns of stress and strain values.

Construct prototype: A prototype is a working model of a product that is used for testing, before it is manufactured. Prototypes help designers learn about the manufacturing process of a product, how people will use the product, and how the product could fail or break. Designers often iterate many times before determining the final solution to a problem. Once a successful prototype has been developed, the engineering team can use it as a mock-up for full scale manufacturing. A successful prototype is used as the basis for creating a final product.

Evaluate the final product: Creating a final product is the culmination of the product design process.

5.3 CONCURRENT ENGINEERING

Sequential engineering or traditional engineering is the term used to describe the method of production in a linear format. The different steps are done one after another (Fig. 5.6). In traditional serial development, the product is first completely defined by the design engineering department, after which the manufacturing process is defined by the manufacturing engineering department, etc. Usually this is a slow, costly and low quality approach, leading to a lot of engineering changes, production problems, product introduction delays, and a product that is less competitive than desired.

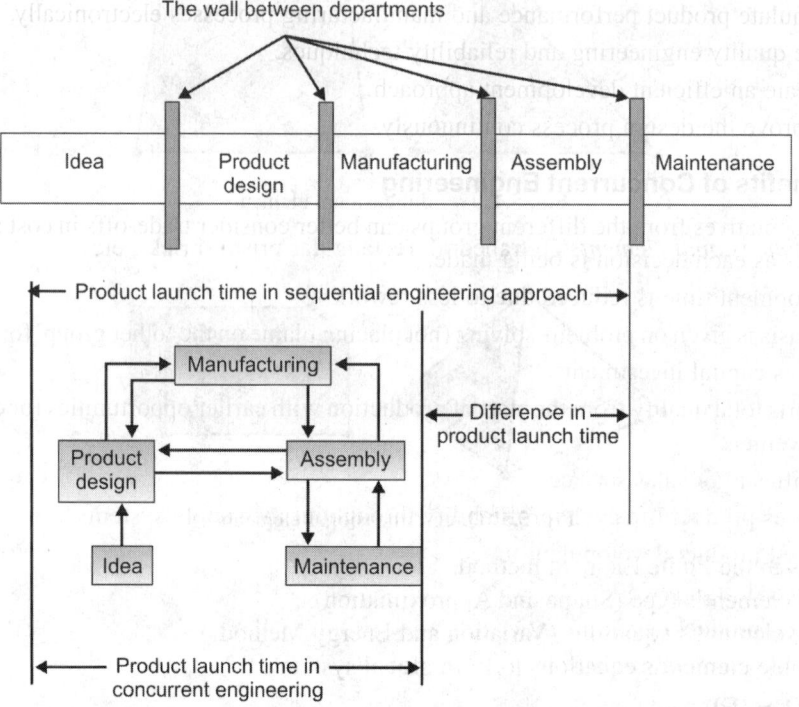

Fig. 5.6 Comparison of sequential approach and concurrent engineering

Concurrent engineering (Fig. 5.6) brings together multidisciplinary teams, in which product developers from different functions work together in parallel from the start of a project with the intention of getting things right as quickly as possible, and as early as possible. In concurrent engineering, different tasks are tackled at the same time, and not necessarily in the usual order.

The concurrent engineering is a product design approach during which all phases of manufacturing operate at the same time or simultaneously. Decision making involves full team participation and involvement. The team often consists of product design engineers, manufacturing engineers, marketing personnel, purchasing, finance, and suppliers. The overall time taken to design and manufacture a new product can be substantially reduced, if all the activities are carried out together rather than in series. The reductions in design cycle time that result from concurrent engineering invariably reduce total product cost. Sometimes, only design engineers and manufacturing engineers are involved in concurrent engineering. In other cases, the cross-functional teams include representatives from purchasing, marketing, production, quality assurance and other functional groups. Sometimes customers and suppliers are also included in the team.

5.3.1 Ten Commandments of Concurrent Engineering

1. Understand your customer.
2. Use product development teams.
3. Integrate process design.

4. Involve suppliers and subcontractors early.
5. Use digital product models.
6. Integrate CAE, CAD, and CAM tools.
7. Simulate product performance and manufacturing processes electronically.
8. Use quality engineering and reliability techniques.
9. Create an efficient development approach.
10. Improve the design process continuously.

5.3.2 Benefits of Concurrent Engineering

- Representatives from the different groups can better consider trade-offs in cost and design choices as each decision is being made.
- Development time is reduced due to less rework.
- Emphasis is given on problem-solving (not placing blame on the 'other group' for mistakes).
- Reduces capital investment
- Supports total quality from the start of production with earlier opportunities for continuous improvement
- Simplifies after sales service
- Increases product life cycle profitability throughout the supply system
- Reduced product development time
- Reduced design rework
- Reduced product development cost
- Improved communications
- Increases in overall quality
- Reductions in design changes after release

Concurrent engineering has become the standard method for product design in most progressive companies. To make concurrent engineering a real success, all the necessary information concerning products, parts and processes, have to be available at the right time.

5.4 DESIGNING FOR MANUFACTURABILITY (DFM) AND DESIGN FOR ASSEMBLY (DFA)

Understanding design for manufacturability is paramount especially when design requirements exceed technology or manufacturing capability. More and more companies who are outsourcing design are facing this problem, where engineering designs have to be modified to match technology or process capability. Companies are left with the dilemma of how to communicate their capabilities to designers. One approach in communicating manufacturing capabilities to designers is to develop DFM and DFA handbooks.

Design for Manufacturability (DFM) is a proven design methodology that works for any level company. Designing for manufacturability is a principle that engineers use to design parts that can be easily manufactured. DFM is usually a principle that concentrates on reducing the cost of part production. Designing for manufacturability should start at the beginning of product development and should be improved upon throughout the design process. Engineers should work with manufacturing and other functional groups to design parts that can be easily manufactured. By thinking about the manufacturability of parts from the beginning of a project, engineers and manufacturers can work together to create cost effective parts that satisfy both groups. Parts that are designed for manufacturability and assembly generally cost less to produce than parts that are not designed with these considerations. Since cost reduction is one of the main goals of DFM, it is a principle known and used at most companies.

Examples of designing for manufacturability:

Using the same screws throughout a product reduces the number of tools on the assembly line, the number of tools change when machining and the number of different screws that are held in inventory.

Creating access holes for assembly workers allows for easy installation of screws in tight spaces.

Reducing the number of parts within an assembly so there is less time spent in assembling.

5.4.1 DFM and DFA General Guidelines

It is vital for the designers to understand the significance of designing for manufacturability. A brief summary of these guidelines is given below:

- Simplify the design and reduce the number of parts.
- Standardize and use common parts and materials.
- Design for ease of fabrication
- Design for ease of assembly
- Design for efficient joining and fastening
- Design for ease of service
- Select processes compatible with the materials and production volumes
- Select materials compatible with production processes and that minimize processing time while meeting functional requirements.
- Avoid unnecessary part features because they involve extra processing effort.
- For higher volume parts, consider castings or stampings to reduce machining.
- Use near net shapes for molded and forged parts to minimize machining and processing efforts.
- Design for ease of fixtures by providing large solid mounting surface and parallel clamping surfaces.
- Minimize flexible parts and interconnections.
- Avoid designs requiring sharp corners or points in cutting tools, since they break easier.
- Avoid thin walls, thin webs and deep holes to withstand clamping and machining without distortion.
- Avoid tapers and contours, as much as possible, in favor of rectangular shapes.
- Avoid undercuts that require special operations and tools.
- Avoid hardened or difficult machined materials, unless essential to requirements.
- Put machined surfaces on same plane or with same diameter to minimize number of operations.
- Design workpieces to use standard cutters, drill bit sizes, or other tools.
- Avoid small holes (drill bit breakage greater) and length to diameter ratio greater than 3 (chip clearance and straightness deviation).

5.4.2 DFM Benefits

Companies that have applied DFM have realized substantial benefits. Costs and time to market are often cut in half with significant improvements in quality, reliability, serviceability, product line breadth, delivery, customer acceptance, and, in general, competitive posture.

5.5 QUESTIONS

1. What is product life cycle? Explain different stages of product life cycle.
2. What are the benefits of concurrent engineering?
3. Differentiate between sequential engineering and concurrent engineering.
4. What are DFM and DFA guidelines?
5. Explain concurrent engineering.
6. Explain design for manufacturability and design for assembly.
7. What are DFM benefits?
8. What are the factors influencing product design?
9. Explain finite element analysis.

6

Computer-aided Design

6.1 INTRODUCTION

Computer-aided design (CAD), Computer-aided manufacturing (CAM) and Flexible manu-facturing system (FMS) reflect the trend towards fully automated manufacturing facilities that are designed to integrate product design and manufacturing activities.

6.2 COMPUTERS IN DESIGN AND MANUFACTURING

The role of computers in design and manufacturing may be broadly classified into following groups:

- **Computer-aided design (CAD):** The use of computer methods to develop the geometric model of the product in three dimensional forms, such that geometric and manufacturing requirements can be examined.
- **Computer-aided design and drafting (CADD):** Combining the CAD function with drafting to generate the production drawing of the part for the purpose of downstream processing.
- **Computer-aided engineering (CAE):** Generally refers to the computer software used to develop the computer's numerical part programs for machining and other processing applications.
- **Computer-aided manufacturing (CAM):** The use of computers to generate the process plan for the complete manufacture of products and parts.
- **Computer-aided tool design (CATD):** Computer assistance to be used for developing the tools for manufacture such as jigs and fixtures, dies and moulds.
- **Computer-aided planning (CAP):** The use of computers for many of the planning functions such as computer-aided scheduling, material requirements planning, etc.

6.3 PRODUCT CYCLE

The product cycle includes all the activities starting from identification of product to deliver the finished product to the customer. Figure 6.1 shows various steps in the product cycle.

The product cycle starts from customers and markets which need for a new product. The basic work for product development such as synthesis, analysis, evaluation and documentation are carried out by the design engineering. In some place, even the prototype testing of the product is carried before going for actual production. The detailed design of the product is

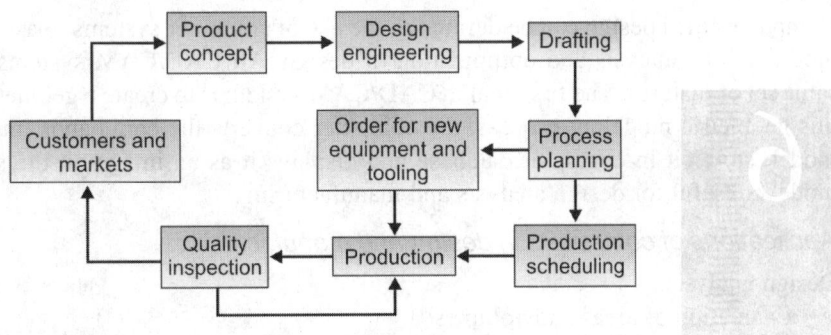

Fig. 6.1 Product cycle (design and manufacturing)

drafted and given to the process planning department. A process plan is formulated which specifies the sequence of production operations to be carried out to produce the new product. Sometimes, for the manufacturing of new product, new equipments and tools may be required which will be ordered at this stage. Based on the process plan, production schedule will be prepared. This scheduling provides a plan to the company that a certain quantities of the product should be manufactured within the specified time period. The production is followed by quality testing and delivery to the customer. All these activities are described in subsequent chapters.

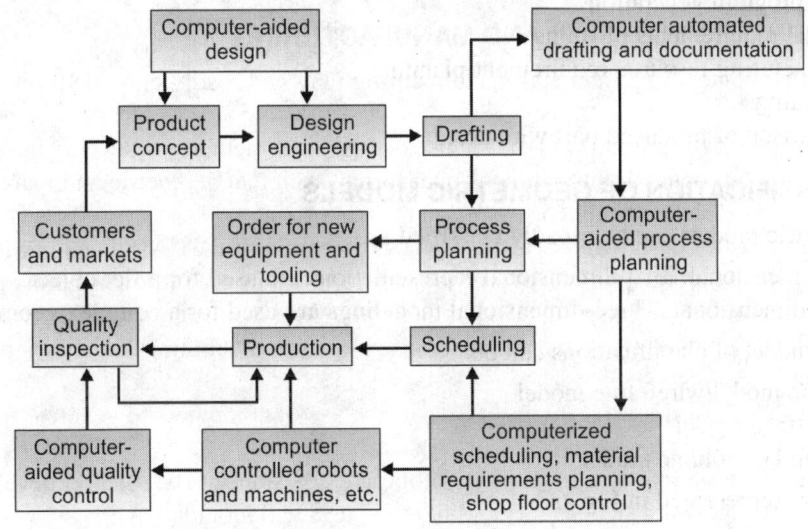

Fig. 6.2 Computer-aided product cycle

From Fig. 6.2, it can be understood that the computer aids each and every activities of the product cycle. Computer-aided design and automated drafting are utilized for design engineering works and for detail drawing of the product. The computers are also used in process planning and scheduling works to perform their functions efficiently. Even the product monitoring and control of shop floor activities, computers are used. Computer controlled robots, CNC machines, etc. are used in production activities. In quality control, the computers are used for quality checking/inspection and performance testing on the products and its components. In the modern manufacturing process, computers have become very useful and essential tools to carry out the design and production activities effectively and efficiently.

6.4 COMPUTER-AIDED DESIGN (CAD)

Computer-aided design can be defined as the use of computer systems to assist in the creation, modification, analysis and optimization of design. All CAD/CAM systems offer almost the same set of features. The first goal in CAD/CAM system is to create a geometry model. During this geometric modeling process, the computer converts the commands into a mathematical model, stores it in computer database and displays it as an image on the screen. Geometry model is useful for design analysis and manufacturing.

Applications of computers in design and manufacturing:

Design analysis:
- Evaluation of areas and volumes
- Evaluation of mass and inertia properties
- Interference checking in assemblies
- Analysis of tolerances in assemblies
- Analysis of kinematics
- Automatic mesh generation for finite element analysis

Manufacturing:
- Parts classification
- Process planning
- Numerical control program generation
- Robot program generation
- Material requirements planning
- Manufacturing resource requirement planning
- Scheduling
- Comparison of produced part with design

6.5 CLASSIFICATION OF GEOMETRIC MODELS

The geometric models can be broadly classified as:

- Two-dimensional: Two-dimensional representations are used for a flat object.
- Three-dimensional: Three-dimensional modelings are used for a complex geometry.

The other model of classifications can be:

- The line model/wireframe model
- The surface model
- The solid or volume model

These are represented in Fig. 6.3.

The line model/wireframe model: In line model, the complete object is represented by a number of lines with their end point coordinates (X, Y, Z) and their connectivity relationship. The line model is in adequate for representing more complex objects. Figure 6.3a is a geometric model represented in wireframe model.

The surface model: The surface model is constructed from surfaces such as planes (Fig. 6.3b), rotated curved surfaces and even very complex surfaces. There are number of techniques available for handling these surfaces such as Bezier and B-splines. By using this model calculation of properties such as mass and inertia would be difficult.

Solid/Volume model: The most advanced method of geometric modeling is solid modeling in three dimensions. This method is illustrated in Fig. 6.3c, typically uses solid geometric shapes called primitives to construct the object.

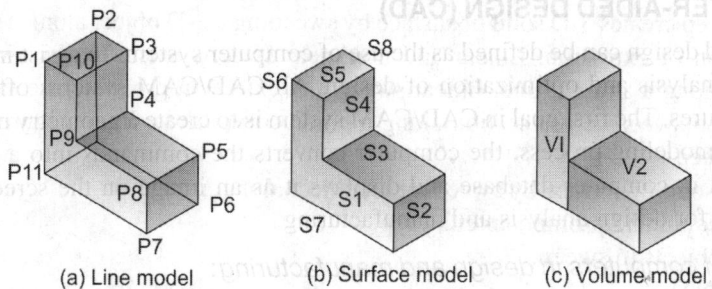

(a) Line model   (b) Surface model   (c) Volume model

Fig. 6.3 3-D geometric representation techniques

6.6 THREE-DIMENSIONAL GEOMETRIC CONSTRUCTION METHODS

The three-dimensional geometric construction methods normally used are:

Extruded feature: Creates 3-D solid or surface by extruding a 2-D object. If you extrude a closed object, the result is 3-D solid. If you extrude an open object, the result is a surface. The extrusion direction is always perpendicular to sketch plane of the profile. Figures 6.4 and 6.5 show extrusions.

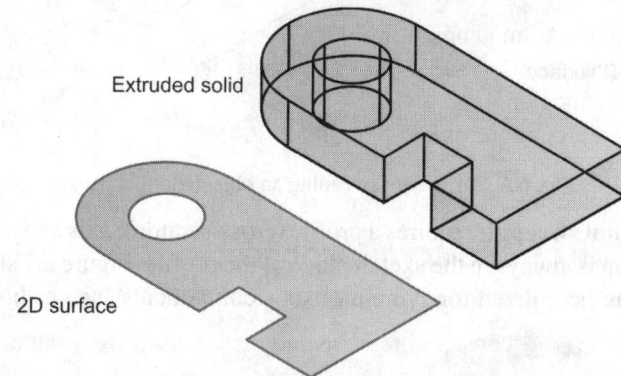

Extruded solid

2D surface

Fig. 6.4 Extruded feature

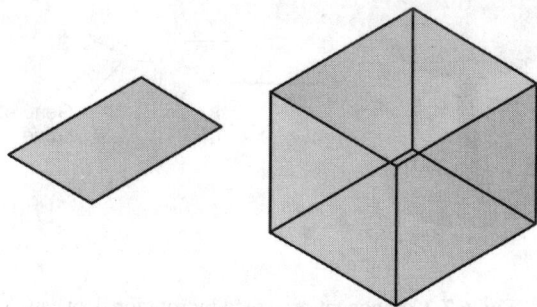

Fig. 6.5 Extruded feature

To extrude an object:
1. At the command prompt, enter extrude.
2. Select the object(s) to extrude, and then press ENTER.
3. Specify the height.

Sweep: Sweep creates 3-D solid or surface by sweeping a 2-D object along a path. Sweep is a generalization of extrusion. Linear and nonlinear sweeps exist. Linear sweep sweeps a cross-section along a straight line. Linear sweep and extrusion produces the same feature. A nonlinear sweep sweeps cross-section along a guide curve (not a line). Figure 6.6 shows an example of sweep.

To create a solid or surface by sweeping an object along a path:

1. At the command prompt, enter sweep.
2. Select objects to sweep.
3. Press ENTER.
4. Select a sweep path.

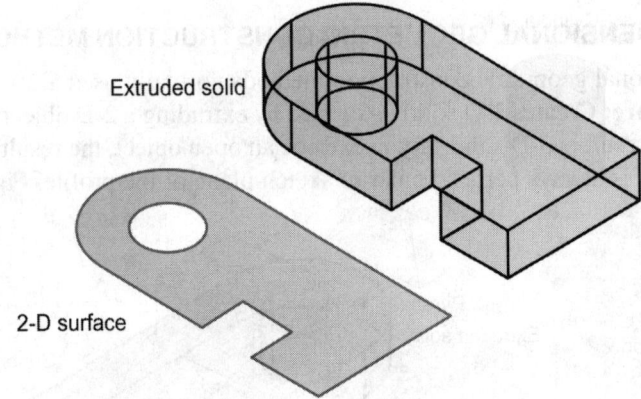

Fig. 6.6 Surface by sweeping an object along a path

Revolve/Rotational sweep: It requires a profile (cross-section), axis and angle of revolution. The axis of revolution is always in the sketch plane of the profile. Figure 6.7 shows a revolution. Rotational sweep can be utilized for symmetric axis components such as bottles.

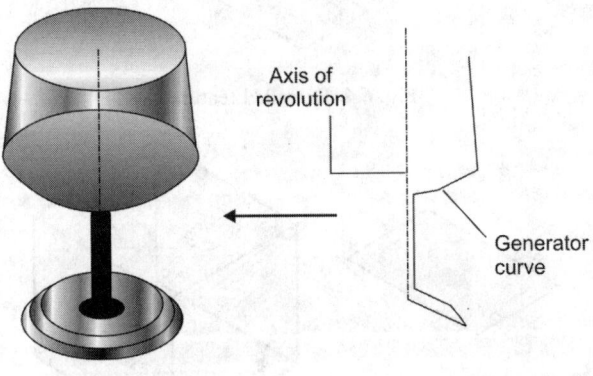

Fig. 6.7 Component produced by rotational sweep

To revolve objects about an axis:

1. At the command prompt, enter revolve.
2. Select the object(s) to revolve.
3. Specify the start point and end point of the axis of revolution.
4. Specify the angle of revolution.

Constructive solid geometry (CSG) or Solid modeling with primitives:

The above methods are not suited for the purpose of inputting geometry. The best method for three-dimensional solid modeling is constructive solid Geometry. In this method, number of solid models are provided as primitives. Some typical primitives are shown in Fig. 6.8. From these solid primitives, the complex objects are created by adding or subtracting the primitives (Fig. 6.9). The basic Boolean operators like union, intersection and difference are used for combining the primitives to form the complex solid. Figure 6.10 is another example of the CSG tree. However, the non-analytical surfaces such as Bezier surfaces cannot be modeled using the CSG representation.

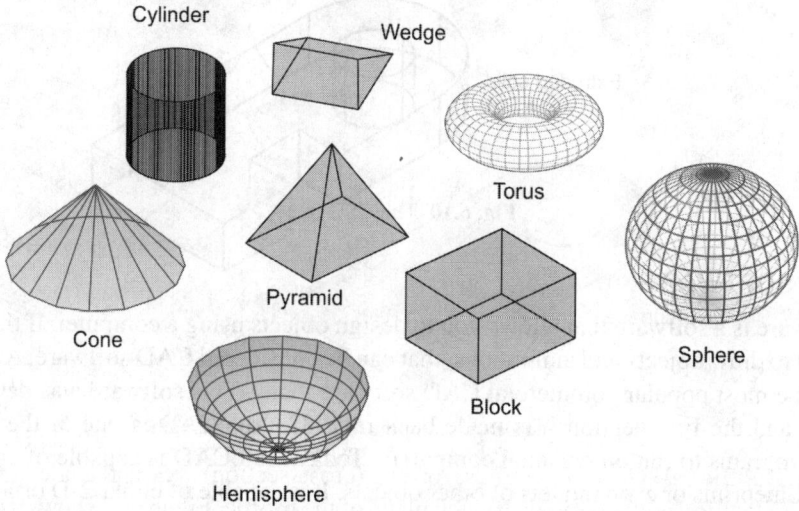

Fig. 6.8 Various solid modeling primitives

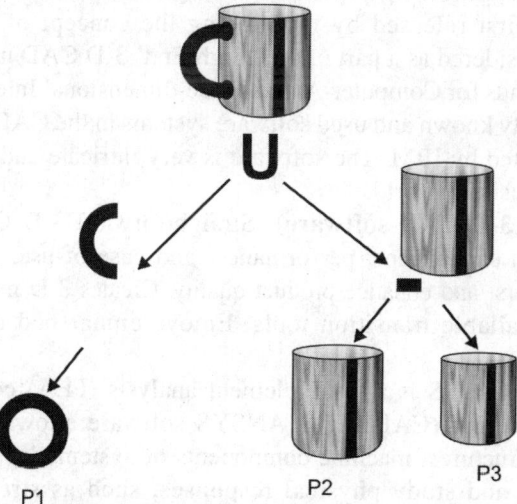

Fig. 6.9 The CSG tree (P1, P2, and P3 are primitives)

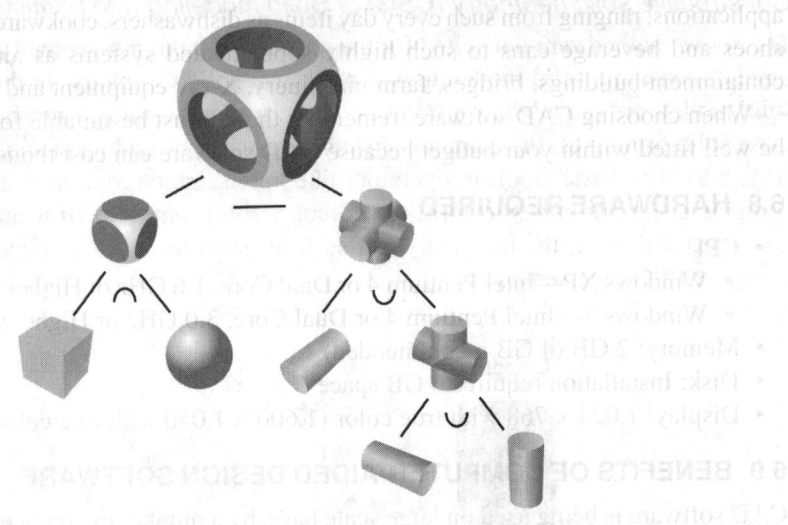

Fig. 6.10 The CSG tree

6.7 CAD SOFTWARE

CAD software is a software that allows you to design objects using a computer. If the software allows you to draw objects and animations, that can be considered CAD software. AutoCAD is probably the most popular commercial CAD software today. This software was developed by Autodesk, and the first version was made back in 1982. AutoCAD is one of the first CAD software programs to run on personal computers. Today, AutoCAD is capable of creating full plans like blueprints or even models of other objects. It is capable of either 2-D or 3-D designs that are very detailed.

Pro/ENGINEER: Pro/ENGINEER is a mechanical engineering and design CAD tool capable of creating complex 3-D models, assemblies, and 2-D measured drawings; it does not support architectural or civil engineering practices. It originally caused a major change in the CAD industry when first released by introducing the concept of *Parametric Modeling*. Pro/ENGINEER is considered as a part of the 'High End' 3-D CAD modeling packages.

CATIA: CATIA stands for Computer-Aided Three-dimensional Interactive Application and is one of the most widely known and used software systems in the CAD world that is marketed and technically supported by IBM. The software is very intricate and is used by some of the biggest names in the business world.

SolidWorks 2013 (3-D CAD software): Straightforward 3-D CAD software, offering unmatched 3-D design capabilities, performance, and ease-of-use. Evaluates more design alternatives, reduce errors, and enhance product quality. Creates 3-D models from existing 2-D data with the best available transition tools. Enjoys unmatched design communication capabilities.

ANSYS software: ANSYS is a finite element analysis (FEA) code widely used in the computer-aided engineering (CAE) field. ANSYS software allows engineers to construct computer models of structures, machine components or systems; apply operating loads and other design criteria; and study physical responses, such as stress levels, temperature distributions, pressure, etc. It permits an evaluation of a design without having to build and destroy multiple prototypes in testing. The ANSYS program has a variety of design analysis applications, ranging from such every day items as dishwashers, cookware, automobiles, running shoes and beverage cans to such highly sophisticated systems as aircraft, nuclear reactor containment buildings, bridges, farm machinery, X-ray equipment and orbiting satellites.

When choosing CAD software, remember that it must be suitable for your needs and must be well fitted within your budget because CAD software can cost thousands of dollars.

6.8 HARDWARE REQUIRED

- CPU
  - Windows XP—Intel Pentium 4 or Dual Core, 1.6 GHz or Higher with SSE2 technology
  - Windows 7—Intel Pentium 4 or Dual Core, 3.0 GHz or Higher with SSE2 technology
- Memory: 2 GB (4 GB recommended)
- Disk: Installation requires 6 GB space
- Display: 1,024 × 768 with true color (1,600 × 1,050 with true color recommended)*

6.9 BENEFITS OF COMPUTER-AIDED DESIGN SOFTWARE

CAD software is being used on large scale basis by a number of engineering professionals and firms for various applications. The most common application of CAD software is designing and drafting. Here are some of the benefits of implementing CAD systems in the companies:

1. **Increase in the productivity of the designer:** The CAD software helps designer in visualizing the final product that is to be made, it subassemblies the constituent parts. The product can also be given in animation and see how the actual product will work, thus helping the designer to immediately make the modifications, if required. CAD software helps designer in synthesizing, analyzing, and documenting the design. All these factors help in drastically improving the productivity of the designer that translates into fast designing, lower designing cost and shorter project completion times.

2. **Improve the quality of the design:** With the CAD software the designing professionals are offered large number of tools that help in carrying out thorough engineering analysis of the proposed design. The tools also help designers to consider large number of investigations. Since the CAD systems offer greater accuracy, the errors are reduced drastically in the designed product leading to better design. Eventually, better design helps in carrying out manufacturing faster and reducing the wastages that could have occurred because of the faulty design.

6.10 QUESTIONS

1. Explain the role of computers in design and manufacturing.
2. Explain computer-aided product cycle.
3. What is CAD?
4. What are the uses of geometry model in design analysis and manufacturing?
5. How are geometric models classified?
6. What are geometric models? Explain various methods of constructing them.
7. What are the benefits of computer-aided design software?
8. Explain CAD software.

Reliability

7.1 DEFINITION AND OBJECTIVES OF RELIABILITY

Product reliability is the mathematical probability that an item can perform its intended function for a specified interval under stated conditions.

- Reliability is a probability-based concept. The numerical value of the reliability is between 0 and 1.
- It implies successful operation over a certain period of time.
- Operating on environmental conditions under which product use takes place are specified.
- Reliability = 1 – Probability of failure

Objectives of reliability: During the design phase of any product/system, it is desired that the said system should meet the performance standards as expected within the specified constraints. These constraints include cost of equipment/product, environmental conditions and availability of materials/parts, etc. A system/equipment normally comprises of many units/components and their interconnection and integration should result in satisfactory performance. The number of components and units make a system complex and therefore, system is dependent on complexity of the functioning of the units. It is further more difficult to achieve satisfactory performance from such system/equipment. Therefore, the objectives of reliability are many folds and include the following:

- Trouble free running of system/equipment
- The adequate performance for a stated period of time
- The equipment/system should work under the specified environmental conditions
- Minimization of downtime of equipment/system and
- Maintainability of device/components

7.2 FAILURE RATES AND MEAN TIME BETWEEN FAILURES (MTBF)

A failure is an event that changes a product from operational to nonoperational.

$$\text{Failure rate} = \frac{\text{Number of failures}}{\text{Operating time}}$$

$$\text{MTBF} = \frac{\text{Operating time}}{\text{Number of failures}}$$

Generally, failure rate of a system follows the bathtub curve as shown in Fig. 7.1.

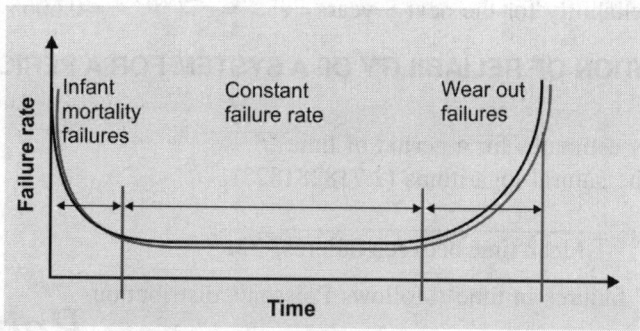

Fig. 7.1 Bathtub curve

The human body is an excellent example of a system that follows the bathtub curve. People tend to suffer a high failure rate (mortality) during their first years of life, particularly the first few years, but the rate decreases as the child grows older. Assuming a person reaches his or her teenage years; his or her mortality rate becomes fairly constant. After completion of teenage years, his or her mortality rate increases (wear out failures). Numerous influences affect mortality rates, including prenatal care and mother's nutrition, quality and availability of medical care, environment and nutrition, lifestyle choices and, of course, genetic predisposition. These factors can be compared to factors that influence machine life. Early failures may tend to follow negative exponential distribution.

7.2.1 Solved Problems

Problem 7.1: Fifty artificial heart valves were tested for 10,000 hours at a medical research center and three valves failed during the test. What is the failure rate and MTBF? On the basis of these data, how many failures could be expected during a year from installation of these valves in 100 patients?

Solution:

Nonoperating time for 3 valves $= \dfrac{10,000 \times 3}{2} = 15,000$ hours

$$\text{Failure rate} = \frac{\text{Number of failures}}{\text{Operating time}}$$

$$= \frac{3}{10,000 \times 50 - 15,000} = \frac{3}{4,85,000} = 6.18 \times 10^{-6} \text{ failures/unit-hour}$$

For 100 units,

$6.18 \times 10^{-6} \times 100 = 6.18 \times 10^{-4}$ failures/hour $= 6.18 \times 10^{-4} \times 24 \times 365 = 5.42$ failures/year

$$\text{MTBF} = \frac{\text{Operating time}}{\text{Number of failures}} = \frac{10,000 \times 50 - 15,000}{3} = \frac{4,85,000}{3} = 16,166.6667 \text{ hours}$$

Problem 7.2: Safety valves used in an oil refinery have a constant failure rate with MTBF of 16 years. What is the probability that newly installed valve will function without failure for the next 8 years?

Solution:

$$\lambda = \frac{1}{\text{MTBF}} = \frac{1}{16} = 0.0625 \text{ failures/year}$$

Probability that newly installed valve will function without failure for the next 8 years is nothing but reliability for the next 8 years.

$$\text{Reliability for the next 8 years} = e^{-\lambda t} = e^{-0.0625 \times 8} = 0.6065$$

7.3 DETERMINATION OF RELIABILITY OF A SYSTEM FOR A PERIOD OF TIME 't'

Notations used:

$R_{(t)}$ = Reliability estimates for a period of time 't'.

e = Base of the natural logarithms (2.718281828)

$$\lambda = \text{Failure rate} = \frac{1}{\text{Mean time between failures}} = \frac{1}{M}$$

Probability that 'n' failures in time 't' follows Poisson's distribution.

$$\text{Probability that '}n\text{' failures in time '}t\text{'} = P_{(n,\,t)} = \frac{e^{-\lambda t}(\lambda t)^n}{n!} \quad n = 0, 1, 2, \ldots$$

$$\text{Probability that '}0\text{' failure in time '}t\text{'} = P_{(0,\,t)} = \frac{e^{-\lambda t}(\lambda t)^0}{0!}$$

$$\text{Probability that '}0\text{' failure in time '}t\text{'} = P_{(0,\,t)} = R_{(t)} = e^{-\lambda t}$$

"If the failure rate follows Poisson's distribution then reliability follows negative exponential distribution".

The exponential distribution is the most basic and widely used reliability prediction formula.

7.3.1 Series System

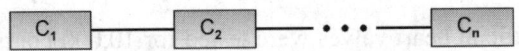

Fig. 7.2 'n' components in series

Figure 7.2 shows a system consists of 'n' components in series. If one component fails, the system fails.

Reliability of a system which consists of 'n' components in series:

The basic equation for calculating the system reliability of a simple series system (Fig. 7.2) which consists of 'n' components is:

$$R_{s,\,t} = R_{1,\,t} \times R_{2,\,t} \times R_{3,\,t} \times \ldots \times R_{n,\,t}$$

$$= e^{-\lambda_1 t} \times e^{-\lambda_2 t} \times e^{-\lambda_3 t} \times \ldots \times e^{-\lambda_n t}$$

$$= e^{-(\lambda_1 + \lambda_2 + \lambda_3 + \ldots + \lambda_n)t}$$

Hence, the system reliability is the product of component reliabilities.

Where

$R_{s,\,t}$ = System reliability for given time t

$R_{n,\,t}$ = Reliability of the component 'n' for given time t, where $n = 1, 2, 3, \ldots$

$\lambda_1, \lambda_2, \lambda_3, \ldots, \lambda_n$ are failure rates of components 1, 2, 3, $\ldots$, n respectively.

7.3.2 Solved Problems

Problem 7.3: Determine Reliability of a system which consists of three components in series (each component having a reliability of 0.90) as shown in Fig. 7.3.

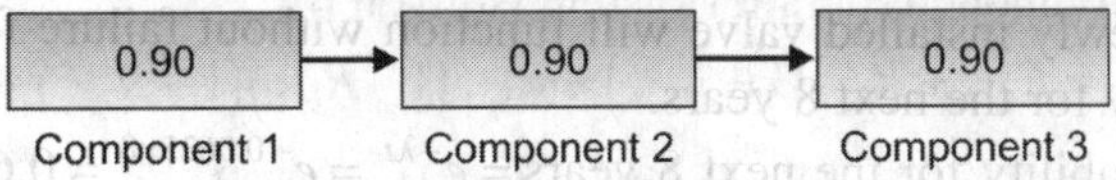

0.90	0.90	0.90
Component 1	Component 2	Component 3

Fig. 7.3 Simple serial system

Solution:

$$R_{s,t} = R_{1,t} \times R_{2,t} \times R_{3,t} = 0.9 \times 0.9 \times 0.9 = 0.729$$

7.3.3 Parallel System

A parallel system (Fig. 7.4) consists of n components, i.e. $c_1, c_2, \ldots, c_n$ arranged in such a way that the system works if and only if at least one of the n components functions properly. Often, design engineers will incorporate redundancy into critical machines. Reliability engineers call these parallel systems. These systems may be designed as active parallel systems or standby parallel systems.

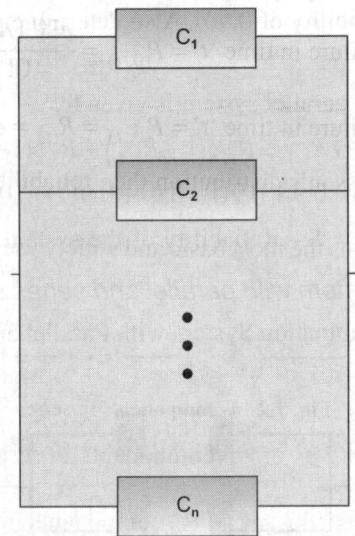

Fig. 7.4 Parallel system of n components

The block diagram for a simple two-component parallel system is shown in Fig. 7.5.

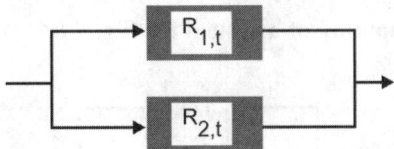

Fig. 7.5 Simple parallel system with two parallel components

Reliability of two-component parallel system is given by

$$R_{s,t} = R_{1,t} + R_{2,t}\left(1 - R_{1,t}\right)$$

Reliability of three-component parallel system is given by

$$R_{s,t} = R_{1,t} + R_{2,t}\left(1 - R_{1,t}\right) + R_{3,t}\left(1 - R_{1,t}\right)\left(1 - R_{2,t}\right)$$

Reliability of 'n'-component parallel system is given by

$$R_{s,t} = R_{1,t} + R_{2,t}\left(1 - R_{1,t}\right) + R_{3,t}\left(1 - R_{1,t}\right)\left(1 - R_{2,t}\right) +$$
$$\ldots + R_{n,t}\left(1 - R_{1,t}\right)\left(1 - R_{2,t}\right)\ldots\left(1 - R_{(n-1),t}\right)$$

7.3.4 Solved Problems

Problem 7.4: Determine reliability of a system which consists of two components in parallel (each component having a reliability of 0.90). The block diagram for two-component parallel system is shown in Fig. 7.6.

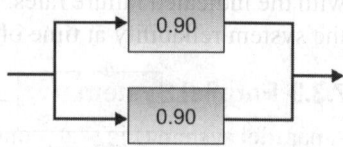

Solution:

Fig. 7.6 The block diagram

$$R_{s,t} = R_{1,t} + R_{2,t}\left(1 - R_{1,t}\right) = 0.9 + 0.9(1 - 0.9) = 0.99$$

Problem 7.5: Determine reliability of a system which consists of three components in series (each component having a reliability of 0.90). Also determine probability of failure.

Solution:

Reliability of three-component parallel system is given by

$$R_{s,t} = R_{1,t} + R_{2,t}\left(1 - R_{1,t}\right) + R_{3,t}\left(1 - R_{1,t}\right)\left(1 - R_{2,t}\right)$$
$$= 0.9 + 0.9(1 - 0.9) + 0.9(1 - 0.9)(1 - 0.9) = 0.999$$

Probability of failure = 1 – Reliability of the system = 1 – 0.999 = 0.001

Reliability of combination system with parallel and series elements:

The block diagram for the Combination System with Parallel and Series Elements is shown in Fig. 7.7.

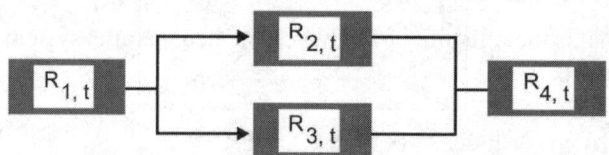

Fig. 7.7 Combination system with parallel and series elements

$$R_{s,t} = R_{1,t} \times \left(R_{2,t} + R_{2,t}\left(1 - R_{3,t}\right)\right) \times R_{4,t}$$

Problem 7.6: Determine reliability of a system for Fig. 7.8

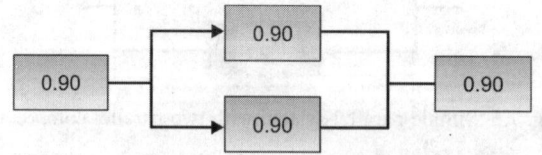

Fig. 7.8 Combination system with parallel and series elements

Solution:

$$R_{s,t} = R_{1,t} \times \left(R_{2,t} + R_{2,t}\left(1 - R_{3,t}\right)\right) \times R_{4,t} = 0.90 \times (0.9 + 0.9(1 - 0.9)) \times 0.9 = 0.8019$$

Problem 7.7: Three components each with a reliability of 0.9 are placed in series. What is the reliability of the system?

Solution: The system reliability is the product of the component reliabilities.

$$R_s = R_1 \, R_2 \, R_3 = 0.9 \times 0.9 \times 0.9 = 0.729$$

Problem 7.8: The components in the system as shown in Fig. 7.9 are exponentially distributed with the indicated failure rates. Develop an expression for the reliability of the system. What is the system reliability at time of 100 hours?

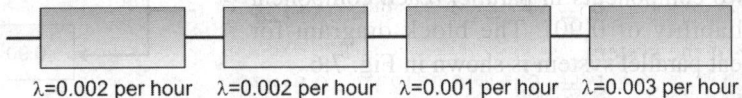

λ=0.002 per hour   λ=0.002 per hour   λ=0.001 per hour   λ=0.003 per hour

Fig. 7.9 Failure rates of each component in series

Solution:

$$R_{s,\,t} = R_{1,\,t} \times R_{2,\,t} \times R_{3,\,t} \times R_{4,\,t}$$
$$= e^{-\lambda_1 t} \times e^{-\lambda_2 t} \times e^{-\lambda_3 t} \times e^{-\lambda_4 t}$$
$$R_{s,\,t} = (e^{-0.002t})(e^{-0.002t})(e^{-0.001t})(e^{-0.003t}) = e^{-0.008t}$$

At time = 100 hours, the reliability is

$$R_{s,\,t} = e^{-0.008(100)} = 0.4493$$

Problem 7.9: Given components with the following reliabilities:

$$A = 0.99, \ B = 0.98, \ C = 0.97$$

What is the reliability of the system as shown in the Fig. 7.10?

Solution:

$$R_{s,\,t} = R_A + R_B \left(1 - R_A\right) + R_C \left(1 - R_A\right)\left(1 - R_B\right)$$
$$= 0.99 + 0.98(1 - 0.99) + 0.97(1 - 0.99)(1 - 0.98)$$
$$= 0.999994$$

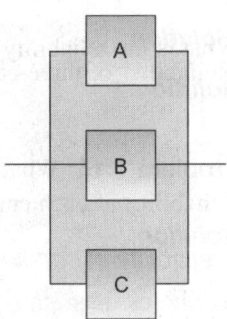

Fig. 7.10 Parallel system

Problem 7.10: What is the reliability of a five-component parallel system when the reliability of each component is 0.70?

Solution:

$$R_{s,\,t} = (0.70)^5 = 0.99757$$

Problem 7.11: What is the reliability of Fig. 7.11?

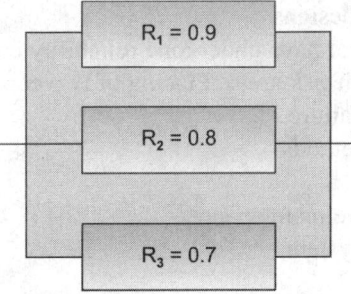

$R_1 = 0.9$

$R_2 = 0.8$

$R_3 = 0.7$

Fig. 7.11 Parallel system

Solution:

$$R_s = R_1 + R_2 \, (1 - R_1) + R_3 \, (1 - R_1) \, (1 - R_2)$$
$$= 0.9 + 0.8 \, (1 - 0.9) + 0.7 \, (1 - 0.9)(1 - 0.8) = 0.994$$

Problem 7.12: Determine the reliability expression for three independent components in series with component reliability R.

Solution: $R_s = R_1 R_2 R_3 = R^3$

As the reliabilities of these three components in series are same.

Problem 7.13: A system is made up of four independent components in series each having a failure rate of 0.005 failures per hour. Determine the reliability of the system at 10 hours.

Solution: Reliability of each component for 10 hours

$$R_t = e^{-\lambda t} = e^{-0.005 \times 10} = 0.95123$$

Reliability of the system

$$R_s = (R_t)^4 = (0.95123)^4 = 0.8187$$

Problem 7.14: Given components with the reliabilities as shown in Fig. 7.12:

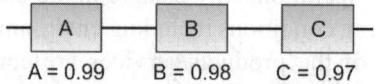

$$A = 0.99 \qquad B = 0.98 \qquad C = 0.97$$

Fig. 7.12 Series system

What is the reliability of the system?

Solution:

$$R_s = R_A R_B R_C = 0.99 \times 0.98 \times 0.97 = 0.9411$$

Problem 7.15: What is the predicted system reliability for the system shown in Fig. 7.13? Reliability of each component is 0.90.

Solution:

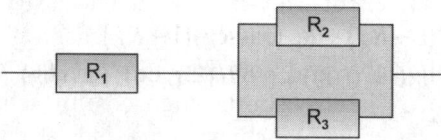

Fig. 7.13 Combination system

$$R_s = R_1 (R_2 + R_2 (1 - R_3)) = 0.90 \times (0.90 + 0.9(1 - 0.9)) = 0.891$$

7.4 WAYS TO IMPROVE RELIABILITY

- Improve the design of components
- Use the simplest possible designs
- Use proven components that have undergone reliability component testing
- Use redundant parts in high risk areas. Placing of two components in parallel will reduce the overall probability of failure.
- Improve manufacturing methods
- Install parallel system
- Perform periodic preventive maintenance
- Test the components and system
- Improve quality control

7.5 RELIABILITY AND QUALITY

The definition of reliability has been discussed in the earlier paragraphs. However, its relationship with quality is very old. Quality and reliability go hand-in-hand and are complementary of each other. Without quality there is no reliability. For a product or service to have higher reliability, it must have good quality as well. The quality of a product/service is the degree of conformance to applicable specifications and standards. It is not concerned with the elements of time and environment. Equipment which has undergone all quality tests may not necessarily be more reliable. Quality is associated with the manufacture whereas; reliability is basically associated with design/material. In a way reliability of the product is the ability of the unit to maintain its quality under specified conditions for a specified time period. The "intended function" of the product/service is related to the quality. For example, an electric generator is expected to provide an output of 5 kW at 220 V under certain conditions which is the intended function of the generator. Any deviation in any of the two parameters will be termed as the failure of the generator. This may happen today or any time in future. Environmental conditions such as temperature, humidity, vibrations, etc., affect the quality and in turn cause the failure of the equipment/service under operation. Any change in the operating parameters may also cause failure. Therefore, operating conditions including environment play an important role so far the quality and reliability of the products/services are concerned. Another important difference between quality and reliability is that one can build a reliable complex system using less reliable components but it is impossible to construct a 'good' quality system from 'poor' quality components. This is mainly due to the reason that reliability can be improved using redundant component/system. But the quality of the product cannot be improved after it has been manufactured. It is only possible through modification of the product.

7.6 COST OF RELIABILITY

It is well known that the cost of product/service increases depending upon its reliability requirements. Therefore, a cost-benefit analysis is essential to design and manufacture product at an optimum cost. Figure 7.14 shows the total cost when reliability concept is incorporated in a particular product/system. It is observed from Fig. 7.14 that quality/reliability cost increases with higher requirements of reliability. Initial high cost of achieving reliability can be compensated by minimizing the failures. Achieving reliable designs and products requires a totally integrated approach, including design, training, test, production, as well as the reliability activities.

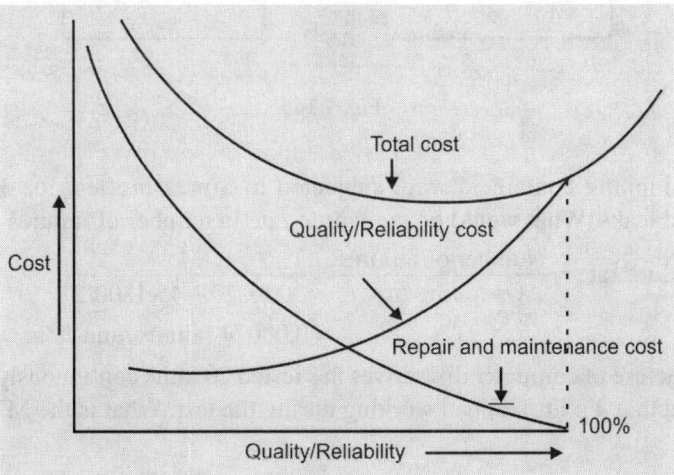

Fig. 7.14 Reliability/Quality costs

7.7 QUESTIONS AND PROBLEMS

1. What is product reliability?
2. Explain failure rates.
3. Explain mean time between failures.
4. Describe the various methods of improving reliability.
5. What is redundancy? Describe various redundancies.
6. Explain parallel configurations.
7. Determine the reliability of the system when components are in series and components are in parallel.
8. Explain the relationship between cost and reliability.
9. Compare reliability and quality.
10. An acid control system has three components in series with individual reliabilities as shown in Fig 7.15 (i) and (ii).

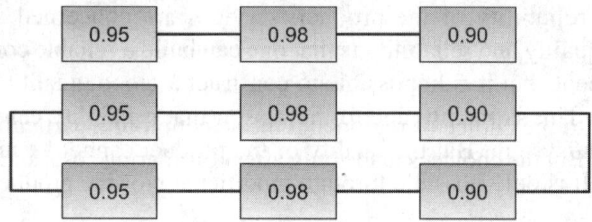

Fig. 7.15

   a. Find the reliability of the system.
   b. What would be the reliability of the system, if a parallel circuit is added?

   Answer: a. 0.84 b. 0.97

11. The reliability of a system's components are shown in Fig 7.16. What is the reliability of the system?

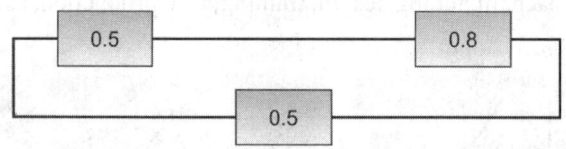

Fig. 7.16

Answer: 0.70

12. Twenty shipping containers were subjected to a pressure test for 300 hours and 4 developed leaks. What would be the failure rate in number of failures per unit hour?

   Hint: Failure rate $= \dfrac{\text{Number of failures}}{\text{Operating time}} = \dfrac{4}{300 \times 20 - 4 \times (300/2)}$

   $= 0.00074 \text{ failures/unit-hour}$

13. A manufacture of computer disk drives has tested 20 units continuously for 4,000 hours and found that 4 units stopped working during the test. What is the MTBF for the disk drives?

   Hint: $\text{MTBF} = \dfrac{\text{Operating time}}{\text{Number of failures}} = \dfrac{4,000 \times 20 - 4 \times (4,000/2)}{4} = 18,000 \text{ hours}$

Part III

Facility Layout and Material Handling

8

Facility Location

8.1 DEFINITION

Plant location refers to the choice of region and the selection of a particular site for setting up a business or factory. But the choice is made only after considering costs and benefits of different alternative sites.

8.2 FACTORS INFLUENCING LOCATION

The following are the factors influencing the location:

Proximate to the market: Locating nearer to the market is preferred: (*i*) If the products are delicate and susceptible to spoilage. (*ii*) After sales service are promptly required. (*iii*) Transportation cost is high. (*iv*) Life of the product is low.

Availability of raw material: It is essential for the organization to get raw material in right qualities at right time in order to have an uninterrupted production. Nearness to source of raw material is important when the material is bulky and transportation cost is high. Nearness to source raw material is also important, if the materials are perishable. Nearness to raw material is important in case of industries such as sugar, cement and jute and cotton textiles.

Transport facilities: Transport facilities ensure supply of raw material to the company and finished goods to the customers. There are five basic modes of physical transportation. They are air, road, rail, water and pipelines. The choice of transportation method depends on cost, convenience and suitability. Plants that are mainly intended for exports are located nearer to port or airport.

Infrastructure availability: The infrastructure facilities like power, water and waste disposal, etc. are prominent factors in deciding location. Certain type of industries like aluminum, steel chemical, etc. are located nearer to power stations. Process industries are located nearer to water for waste disposal.

Law and taxation: The policies of the state government concerning labor laws, building codes, exemption from a sales tax and excise duties, soft loans from financial institutions, subsidy in electricity charges and investment subsidy are important factors for selection of plant location.

Suitability of climate: Climate greatly influences human efficiency and behavior. Textile mills require humidity.

Supporting industries and services: Many companies will not make all the components within the company. Some of the components are purchased from the supporting industries. Availability of supporting or ancillary industries is important factor for the plant location.

Social infrastructure: Availability of facilities like housing, recreation facilities, educational institutions, medical facilities, etc. are considered for locating plant.

Availability of labor: Generally, software companies are located in urban area due to the availability of software professionals. If the company requires unskilled labor then industry can be located in rural areas. If the skilled labor is required then industries can be located in semi-urban/urban areas. Prevailing wage pattern, cost of living and trade union problems are important factors for locating plant.

8.3 METHODS OF EVALUATING LOCATION ALTERNATIVES

There are number of methods for evaluating location alternatives. The important methods are:
1. Cost analysis
2. Profit analysis
3. Returns on investment
4. Factor rating system
5. Center of gravity (Grid) method

8.3.1 Cost Analysis

Procedure for determining optimal location using cost analysis:
- Determine fixed and variable costs for each location.
- Determine total cost for all considered locations.
  Total cost = (Variable cost/unit × Volume + Fixed cost)
- Draw Total cost line with respect to volume as shown in Fig. 8.1.
- Lowest total cost location for expected output (volume) is the optimal location.

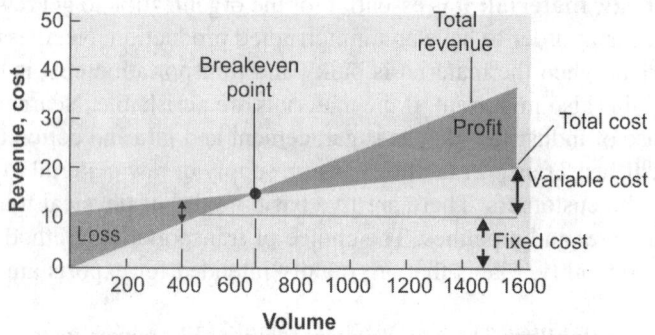

Fig. 8.1 Total cost and revenue with respect to volume

- At different volumes, other alternatives may be preferable. This analysis can be done by equating total costs of two locations.
  That is total cost of location 1 = Total cost of location 2

8.3.2 Profit Analysis

Procedure for determining optimal location using total profit technique:
- Determine fixed cost, variable cost and sales revenue for each location.
- Determine total costs for all considered locations.
  Total cost = (Variable cost/unit × Volume + Fixed cost)

- Draw Total cost line with respect to volume (Fig. 8.1).
- Determine sales revenue for each location.
  Sales revenue = (Price/unit × Volume)
- Draw sales revenue line with respect to volume (Fig. 8.1).
- Determine profit for each location.
  Profit = (Price/unit × Volume) – (Variable cost/unit × Volume + Fixed cost)
- Highest profit location for expected output (volume) is the optimal location.

8.3.3 Return on Investment

Procedure for determining optimal location using return on investment technique:
- Determine total investment, total cost (expenses) and sales revenue for each location.
- Determine return on investment for each location.

$$\text{Return on investment} = \frac{\text{Total sale} - \text{Total expense}}{\text{Total investment}} \times 100$$

- Select the location which gives highest rate of return.

8.3.4 Factor Rating System

Factor rating system is a means of assigning quantitative values to all factors related to each decision alternative and deriving a composite score that can be used for comparison. This method accommodates both quantitative and qualitative factors.

Steps to determine plant location using factor rating system:
- Develop a list of factors.
- Assign weight to each factor.
- Determine composite score for each location.

$$\text{Composite score} = \sum w_i s_i$$

Where
  w_i = Weight for factor 'i'
  s_i = Score for factor 'i'
- Choose the location with the maximum composite score.

8.3.5 Centre of Gravity (Grid) Method

This method is used to identify the location of a production center or warehouse that will minimize the costs of distributing specified volumes of a product to surrounding markets or locations.

Assumptions:
- Straight line distances are used.
- Volumes required by each location are used as weights.

Steps to identify the location of a production center or warehouse:
- Multiply X-coordinate distance of each location 'i' by corresponding volume of goods to be transported and sum all the products $(\Sigma V_i X_i)$.
- Divide sum of these $(\Sigma V_i X_i)$ by the total volume shipped to yield volume weighted X coordinate of the center of gravity (X_c) for the system.
- Volume weighted X-coordinate of the centre of gravity (X_c) for the system $= X_c = \dfrac{\Sigma V_i X_i}{\Sigma V_i}$
- Multiply Y-coordinate distance of each location 'i' by their respective volume of goods to be transported and sum all the products $(\Sigma V_i Y_i)$.

- Divide sum of these $(\Sigma V_i Y_i)$ by the total volume shipped to yield volume weighted Y-coordinate of the centre of gravity (Y_c) for the system.

- Volume weighted Y-coordinate of the centre of gravity (Y_c) for the system $= Y_c = \dfrac{\Sigma V_i Y_i}{\Sigma V_i}$

- Locate the warehouse near the coordinates (X_c, Y_c).

8.3.6 Solved Problems

Problem 8.1: Potential location in Argentina, Brazil and Chile has the cost structure shown in Table 8.1 for a product expected to sell for $130.

a. Find the most economical location for an expected volume of 6,000 units per year.
b. What is the expected profit, if the selected site is used?
c. For what output ranges each location is the best?

Table 8.1 Cost structure

Potential locations	Fixed costs per year	Variable costs/unit
Argentina	$1,50,000	$75
Brazil	$2,00,000	$50
Chile	$4,00,000	$25

Solution:

$$\text{Total cost} = \text{Fixed cost} + \text{Variable cost/unit} \times \text{Volume of production}$$

a. Volume of production = 6,000 units per year

Total cost for 6,000 units at Argentina $= \$1,50,000 + 6,000 \times \$75 = \$6,00,000$

Total cost for 6,000 units at Brazil $\quad = \$2,00,000 + 6,000 \times \$50 = \$5,00,000$

Total cost for 6,000 units at Chile $\quad = \$4,00,000 + 6,000 \times \$25 = \$5,50,000$

Total cost is less at Brazil. So Brazil is preferred for the production of volume 6,000 units/year

Expected profit at Brazil = Sales revenue – Total cost $= 130 \times 6,000 - 5,00,000 = \$2,80,000$

b. To determine preference of locations:

Comparing Argentina and Brazil:

V = Volume of production

$\$1,50,000 + V \times \$75 = \$2,00,000 + V \times \$50$

$V = 2,000$

If Volume of production < 2,000 units, Argentina is preferred.

If Volume of production > 2,000 units, Brazil is preferred.

Comparing Brazil and Chile:

V = Volume of production

$\$4,00,000 + V \times \$25 = \$2,00,000 + V \times \$50$

$V = 8,000$

If Volume of production < 2,000 units, Argentina is preferred.

If Volume of production > 2,000 units and < 8,000 units, Brazil is preferred.

If Volume of production > 8,000 units, Chile is preferred.

Problem 8.2: From the data given in Table 8.2 select the most advantageous location or setup a plant for making electronic component.

Table 8.2 Given data

	Site 'X' (₹)	Site 'Y' (₹)	Site 'Z' (₹)
Total initial investment	2,00,000	2,00,000	2,00,000
Total expected sales for the period	2,50,000	3,00,000	2,50,000
Distribution expenses	40,000	40,000	75,000
Raw material expenses	70,000	80,000	90,000
Power and water supply expenses	40,000	30,000	20,000
Wages and salaries	20,000	25,000	20,000
Other expenses	25,000	40,000	30,000
Community attitude	Indifferent	Want business	Indifferent
Employees housing facilities	Poor	Excellent	Good

Solution:

Calculation of return on investment is shown in Table 8.3.

Table 8.3 Return on investment

	Site 'X' (₹)	Site 'Y' (₹)	Site 'Z' (₹)
Total initial investment	2,00,000	2,00,000	2,00,000
Total expected sales for the period	2,50,000	3,00,000	2,50,000
Distribution expenses	40,000	40,000	75,000
Raw material expenses	70,000	80,000	90,000
Power and water supply expenses	40,000	30,000	20,000
Wages and salaries	20,000	25,000	20,000
Other expenses	25,000	40,000	30,000
Total expenses	1,95,000	2,15,000	2,35,000
Return on investment $= \dfrac{\text{Total sale} - \text{Total expense}}{\text{Total investment}} \times 100$	$\dfrac{2,50,000 - 1,95,000}{2,00,000} \times 100$ $= 27.5\%$	$\dfrac{3,00,000 - 2,15,000}{2,00,000} \times 100$ $= 42.5\%$	$\dfrac{2,50,000 - 2,35,000}{2,00,000} \times 100$ $= 7.5\%$

Site Y is the most advantageous because:

- It associates with the high rate of return, i.e. 42.5%.
- Community wants business.
- Housing facilities are excellent.
- Initial investment is same for all sites.

Problem 8.3: A company is evaluating four locations for a new plant and has weighted the relevant factors as shown in Table 8.4. Develop a quantitative factor comparison for the four locations.

Table 8.4 Weighted relevant factors

Relevant factor	Assigned weight	Location 1 score	Location 2 score	Location 3 score	Location 4 score
Production cost	0.33	50	40	35	30
Raw material supply	0.25	70	80	75	80
Labor availability	0.20	55	70	60	45
Cost of living	0.05	80	70	40	50
Environment	0.02	60	60	60	90
Market	0.15	80	90	85	50

Solution:

Calculation of maximum weighted score is given in Table 8.5.

Table 8.5 Calculation of maximum weighted score

Relevant factor	Assigned weight	Location 1		Location 2		Location 3		Location 4	
		Score	Weighted score	Score	Weighted score	Score	Weighted score	Score	Weighted score
Production cost	0.33	50	16.50	40	13.20	35	11.55	30	9.90
Raw material supply	0.25	70	17.50	80	20.00	75	18.75	80	20.00
Labor availability	0.20	55	11.00	70	14.00	60	12.00	45	9.00
Cost of living	0.05	80	4.00	70	3.50	40	2.00	50	2.50
Environment	0.02	60	1.20	60	1.20	60	1.20	90	1.80
Market	0.15	80	12.00	90	13.50	85	12.75	50	7.50
Total	1.00		62.00		65.40		58.25		50.70

Location 2 has maximum weighted score. So location 2 is preferred.

Problem 8.4: A consultant visited each potential location and rated them according to each factor, as given in Table 8.6. Each location factor is associated with certain weight. Given that all sites basically have the same leasing, labor and operation cost. Recommend a location based on the rating factor.

Table 8.6 Weighted relevant factors

Location factor	Weight	L 1	L 2	L 3	L 4
College proximity	0.30	40	60	90	60
Income	0.25	75	80	65	90
Vehicle traffic	0.25	60	90	70	85
Land quality and size	0.10	90	100	80	90
Proximity of other shopping	0.10	80	30	50	70

Solution: Calculation of maximum weighted score is given in Table 8.7.

Table 8.7 Calculation of maximum weighted score

Location factor	Weight	L 1	Weighted score	L 2	Weighted score	L 3	Weighted score	L 4	Weighted score
College proximity	0.30	40	12	60	18	90	27	60	18
Income	0.25	75	18.75	80	20	65	16.25	90	22.5
Vehicle traffic	0.25	60	15	90	22.5	70	17.5	85	21.25
Land quality and size	0.10	90	9	100	10	80	8	90	9
Proximity of other shopping	0.10	80	8	30	3	50	5	70	7
			62.75		73.5		73.75		77.75

Location 4 has maximum weighted score. So location 4 is preferred.

Problem 8.5: The grid of Fig. 8.2 shows eight market locations to which a company has to ship its products. The shipment volumes and X- and Y-coordinates of the locations are shown in Table 8.8. Using the centre gravity method:

   a. Find the volume weighted X_c- and Y_c-coordinates.

   b. Suggest a possible distribution center.

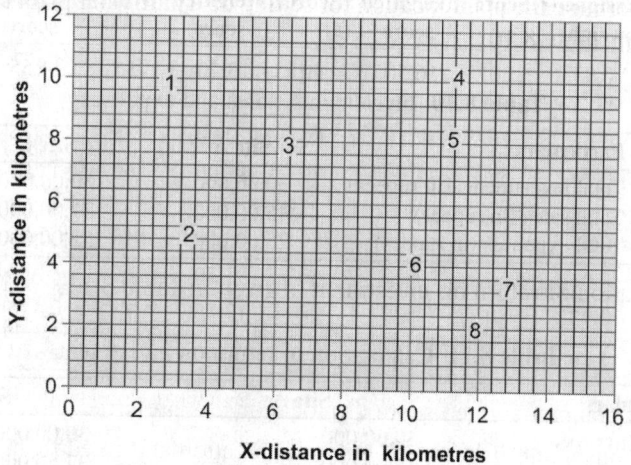

Fig. 8.2 The grid showing eight market locations

Table 8.8 X- and Y-coordinates of market locations

Market location	Volume	X	Y
1	8	2.5	10
2	20	3	5
3	12	6.5	8
4	10	11	10
5	30	11	8
6	20	10	4
7	40	13	3.5
8	30	12	2

Solution: A calculation of determining the centre of gravity of X- and Y-coordinates is shown in Table 8.9.

Table 8.9 Determining the centre of gravity of two coordinates

Market location	Volume	X_i	Y_i	V_iX_i	V_iY_i
1	8	2.5	10	20	80
2	20	3	5	60	100
3	12	6.5	8	78	96
4	10	11	10	110	100
5	30	11	8	330	240
6	20	10	4	200	80
7	40	13	3.5	520	140
8	30	12	2	360	60
	$\Sigma V = 170$			$\Sigma V_i X_i = 1{,}678$	$\Sigma V_i Y_i = 896$

The centre of gravity of two coordinates are:

$$X_c = \frac{\Sigma V_i X_i}{\Sigma V_i} = \frac{1{,}678}{170} = 9.9 \qquad Y_c = \frac{\Sigma V_i Y_i}{\Sigma V_i} = \frac{896}{170} = 5.3$$

Referring back to Fig. 8.2, the coordinates lie very close to location 8. It may be worthy of consideration for distribution centre.

Problem 8.6: Determine the plant location for setting up a particular plant using rate of return. The data is given in Table 8.10.

Table 8.10 Investments, sales and expenses

Particulars	Site A	Site B
Total investments (in rupees)	30,00,000	30,00,000
Total sales (in rupees)	45,00,000	37,50,000
Total expenses (in rupees)	28,00,000	32,00,000

Solution: Calculation of return on investment is given in Table 8.11.

Table 8.11 Calculation of return on investment

Particulars	Site A	Site B
Total investments (in rupees)	30,00,000	30,00,000
Total sales (in rupees)	45,00,000	37,50,000
Total expenses (in rupees)	28,00,000	32,00,000
Return on investment		
$= \dfrac{\text{Total sales} - \text{Total expense}}{\text{Total investment}} \times 100$	$\dfrac{45{,}00{,}000 - 28{,}00{,}000}{30{,}00{,}000} \times 100$	$\dfrac{37{,}50{,}000 - 32{,}00{,}000}{30{,}00{,}000} \times 100$
	$= 56.66\%$	$= 18.33\%$

∴ The Site A is most advantageous as it gives high rate of return.

8.4 MULTIPLANT LOCATIONS

Many firms operate several plants. The problem of the choice of an optimal pattern of locations for these plants is important. Locations which maximize profit or minimize the total cost of production and transportation are optimal plant locations. Profit/Costs are interrelated with price, demand, scale of production and transportation cost.

Factors which affect multiplant location: Multiplant location is influenced by existing location as well as all the factors are considered in single plant location. Other factors are:

- Professional managers who may live in different regions
- Increasing or decreasing of demand for the existing locations
- Shutdown or closure of existing plants
- Availability of resources
- Proximity to the markets

Objectives of Multiplant Enterprise:

- Locating plants at more than one location.
- Locating plants at a low cost production input such as raw material or human resources. This may bring considerable cost reduction for enterprise.
- Locating plants in geographically dispersed locations which are close proximity to new or special market.

Multiplant enterprises follow different types of plant arrangements to achieve these objectives.

In general, a multisite production setup can be arranged in parallel (each producing end products and supply to the market) or series (some plants producing intermediate products supplying other plants, which convert them into finished product) as shown in Fig. 8.3.

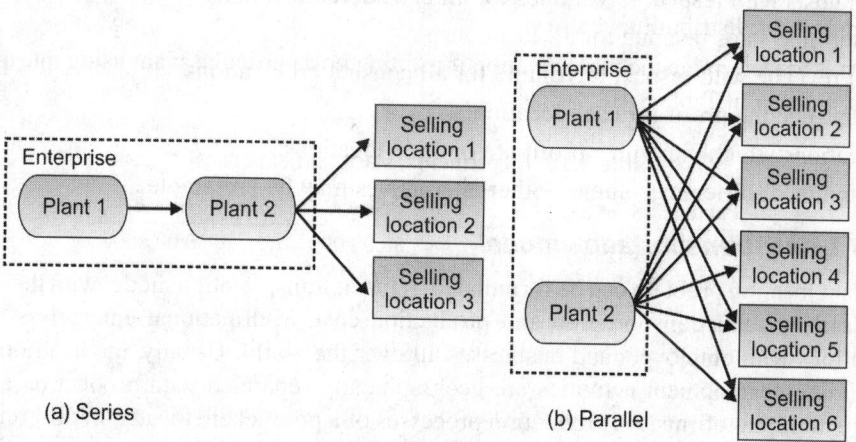

Fig. 8.3 Multiplant setups

Efficient operations of these geographically dispersed plants depend on the effective coordination of all the local plants and make their objective align to that of the enterprise. But effective coordination is a challenge that needs to overcome the different limitations of the existing networked relation in the enterprise. This limitation arises due to the complex inter-dependencies of the decision making process in managing the enterprise network.

Multiple plant strategies:

- Organize by product
- Organize by market
- Organize by process (or component)
- Combination

8.4.1 Methods of Evaluating Multiplant Locations

- Cost-Volume Analysis
- Profit-Volume Analysis
- Linear Transportation Model

8.4.1.1 Cost-Volume analysis

Steps to identify optimal location using Cost-Volume Analysis:
- Identify finite number of sites for location of plants.
- Determine demand and production costs for each location.
- Analyze the distribution network of the output for each location.
- Determine sum of the production and transportation costs for each location.
- Plot costs with respect to volumes for all considered locations.
- Locations which minimize sum of the production and transportation costs are optimal location.

8.4.1.2 Profit-Volume analysis

Steps to identify optimal location using Profit-Volume Analysis:
- Identify finite number of sites for location of plants.
- Determine demand, price/unit and production costs for each location.
- Analyze the distribution network of the output for each location.
- Determine sum of the fixed and variable costs for each location.
- Plot costs with respect to volumes for all considered locations.
- Determine sales revenue for all locations.
- Plot revenue with respect to volume for all considered locations.
- Determine the profit for each location.
- Locations which maximize profit are optimal locations.
- Note that at different volumes, other alternatives may be preferable.

8.4.1.3 Linear transportation model

Multiplant location problem can be formulated as linear transportation model with the objective of minimizing total transportation and production cost. Multinational enterprises are large corporations with employers and businesses all over the world. Usually, the head offices and research and development activities are geographically separated with production activities. They are multiplant firms. As production processes of a product are located in different plants, it is known as multipoint production. If the plants are located in different countries, it is known as multinational industry. To decide the location of new firms, it is usually dependent on professional managers who may live in different regions.

Problem 8.7: A furniture company has plants in cities A, B and C which ship to four demand locations 1, 2, 3, 4 with transporting costs (in hundred rupees) as shown in Table 8.12. Determine minimum transportation cost.

Table 8.12 Transportation costs

Supply plants	Demand locations				Capacities
	1	**2**	**3**	**4**	
A	3	5	7	4	50
B	6	8	5	2	50
C	1	9	7	3	50
Requirement	20	60	30	40	

Solution:

Step 1: The first step is balancing the given transportation problem, i.e. making Total capacity = Total requirement. In the given problem, Total capacity = Total requirement = 150. So the given problem is a balanced transportation problem.

Step 2: The second step is obtaining initial basic feasible solution. There are three important methods for obtaining initial basic feasible solution. We obtain initial basic feasible solution by using Vogel's approximation method. Initial basic feasible solution using Vogel's approximation method is given in Table 8.13. Number of positive cells in the solution is equal to $m + n - 1$ that is $3 + 4 - 1 = 6$. So the initial solution is a basic feasible solution.

Initial total cost of transportation = $50 \times 5 + 10 \times 5 + 40 \times 2 + 20 \times 1 + 10 \times 9 + 20 \times 7 = 630$.

Step 3: Third step is testing the optimality of initial basic feasible solution. There are two important methods for testing the optimality of initial basic feasible solution. Here Modified distribution method is used for obtaining optimal solution. Calculation of row numbers and column numbers for the initial basic feasible solution are shown in Table 8.13.

Table 8.13 Initial solution by Vogel's approximation method

Supply plants	Demand locations				Capacity	Row number (u_i)
	1	2	3	4		
A	3	5 ⑤⓪	7	4	50	0
B	6	8	5 ⑩ +	2 ④⓪ −	50	2
C	1 ②⓪	9 ⑩	7 ②⓪ −	3 +	50	4
Requirements	20	60	30	40	150	
Column number (v_j)	−3	5	3	0		

Opportunity costs of empty cells are determined for the initial solution and the results are given in Table 8.14.

Table 8.14 Opportunity costs of empty cells

Empty cell	A1	A3	A4	B1	B2	C4
Opportunity cost = $c_{ij} - (u_i + v_j)$	0	4	4	7	1	−1

Empty cell C4 has negative opportunity cost of ₹ −1. So we have to revise the solution by adding quantity to the cell C4. Form the closed loop with cell C4 (Table 8.13). Starting with empty cell C4, alternatively put +, − for the remaining occupied cells of closed loop. Determine the minimum quantity in − sign cell of closed loop. Minimum quantity in − sign cell of closed loop is 20. Add minimum quantity 20 to the quantity wherever + sign, i.e. to the cells B3 and C4 and subtract from the quantity wherever − sign, i.e. from the cells B4 and C3. The revised solution is given in Table 8.15.

Table 8.15 Revised solution

Supply plants	Demand locations				Capacity	Row number (u_i)
	1	**2**	**3**	**4**		
A	3	5 (50)	7	4	50	0
B	6	8	5 (30)	2 (20)	50	3
C	1 (20)	9 (10)	7	3 (20)	50	4
Requirements	20	60	30	40	150	
Column number (v_j)	–3	5	2	–1		

Number of positive cells in the solution is equal to $m + n - 1$ that is $3 + 4 - 1 = 6$. So the revised solution is a basic feasible solution. Calculation of row numbers and column numbers for the revised solution are shown in Table 8.15. Opportunity costs of empty cells are determined for the revised solution and the results are given in Table 8.16.

Table 8.16 Opportunity costs of empty cells for the revised solution

Empty cell	A1	A3	A4	B1	B2	C3
Opportunity cost = $c_{ij} - (u_i + v_j)$	6	5	5	6	0	1

Opportunity costs of all empty cells are positive. So the solution is an optimal solution.
Minimum cost of transportation $= 50 \times 5 + 30 \times 5 + 20 \times 2 + 20 \times 1 + 10 \times 9 + 20 \times 3 = 610$.

8.5 QUESTIONS AND PROBLEMS

1. The fixed and variable costs for three potential plant sites for manufacturing an item is shown below:

Site	Fixed cost/year (₹)	Variable cost/unit (₹)
A	75,000	5
B	1,00,000	4

Graph the total cost line for 2 sites.
a. What range of annual volume is preferred for each location?
b. If the expected volume is 5,000 units, which location would you recommend?
    Answer: a. Volume $\leq 25,000$, Site A is preferred; Volume $> 25,000$, Site B is preferred b. Site A is preferred

2. The individual workstation in a toy production line layout have design capacities (units/day) as shown in the figure below. If the actual output of the system is 80 toys/day. What is the system efficiency?

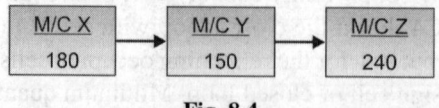

Fig. 8.4

Answer: Efficiency $= \dfrac{\text{Actual output}}{\text{Planned output}} = \dfrac{80}{150} = 53\%$

3. Explain the importance of plant location.
4. Describe the various factors to be considered in selecting location.
5. Why some industries are located near the source of raw material, whereas some near the market for their finished goods?
6. What do you understand by multiplant location problem? Explain how it differs from conventional single plant location problem?
7. A new entrepreneur wants to setup a plant. There are three possible sites with different advantages. The initial investment is ₹ 2,00,000 for all sites. Calculate the rate of return of the three cities and choose the optimal location.

Factor	Site A	Site B	Site C
Expected sales	2,50,000	2,50,000	3,00,000
Distribution expenses	40,000	40,000	75,000
Cartage	90,000	80,000	70,000
Power and water supply charges	20,000	25,000	35,000
Wages and salaries	25,000	30,000	30,000
Other Expenses	15,000	30,000	20,000

Answer: Site A

8. A company is considering three different locations (A, B and C) for a new facility that will have annual costs ₹ 27,500, ₹ 25,400 and ₹ 29,000 respectively. The annual revenues are ₹ 40,000 for location A, ₹ 36,000 for location B and ₹ 42,000 for location C. Which location will maximize the net return per year?

Answer: Net return per year = Annual revenue − Annual cost.
C is the optimal location.

9. A company has three mines A, B and C and five factories F1, F2, F3, F4 and F5. The mines can supply 80,100 and 140 tons of ore daily and the requirements of the factories are 40, 50, 70 and 80 respectively. The following Table gives the unit transportation cost of ore.

Mines	Factories				
	F1	F2	F3	F4	F5
A	4	2	3	2	6
B	5	4	5	2	1
C	6	5	4	7	3

Give a distribution plan to minimize the total cost of transportation.

Answer: Total minimum cost = 920

10. Solve the following transportation problem to maximize the total profit. Here entries are profits.

Origin	A	B	C	D	Capacity
S1	40	25	22	33	100
S2	44	35	30	30	30
S3	38	38	28	30	70
Requirement	40	20	60	30	

Hint: In general, transportation problem is minimizing to algorithm. To convert maximization problem into minimization problem subtract all the entries from highest entry. Remaining procedure is same as transportation algorithm.

9

Facility Layout

9.1 DEFINITION

Facility layout is a plan of an optimum arrangement of facilities including personnel, operating equipment, storage space, material handling equipment and all other supporting services along with the design of the best structure to contain all these facilities. The overall objective of plant layout is to design a physical arrangement that meets the required output quantity and quality in most economical way.

9.2 FACTORS INFLUENCING PLANT LAYOUT

- Types of plant: Engineering industry, assembly unit, process industry, etc.
- Types of production system: Job shop production system, batch production system, mass production system and project type production system
- Scale of production: Low volume, medium volume and high volume
- Availability of area
- Types of the material handling system
- Types of building
- Future expansion plan
- Types of production facilities
- The pattern of material flow is an important consideration in the plant layout design because good layout aims at minimizing amount of material handling.

9.3 MATERIAL FLOW PATTERNS

The pattern of material flow is an important consideration in the plant layout design because good layout aims at minimizing flow of materials. The various flow patterns are shown in Fig. 9.1.

Characteristics of different flow patterns:

Straight line:

- Shortest route and must have roads on both sides
- Plant areas has long length and narrow width
- Unsuitable for long production lines

U type:
  • Less difficulty in returning empty containers
  • Suitable for longer production lines
  • Requires square like floor area
  • One side road will be required
Serpentine:
  • Required roads on both sides
  • Suitable for longer production lines
  • Difficulty in returning empty containers
  • Requires square like floor area

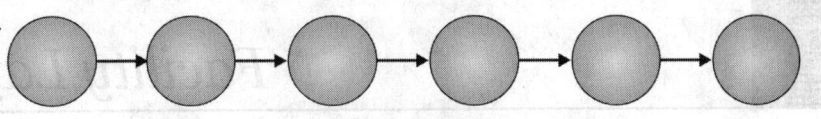

(a) Straight line

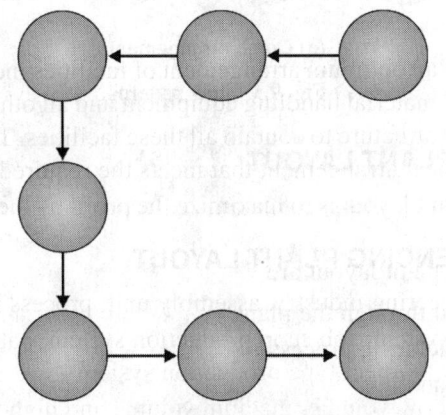

(b) U shape

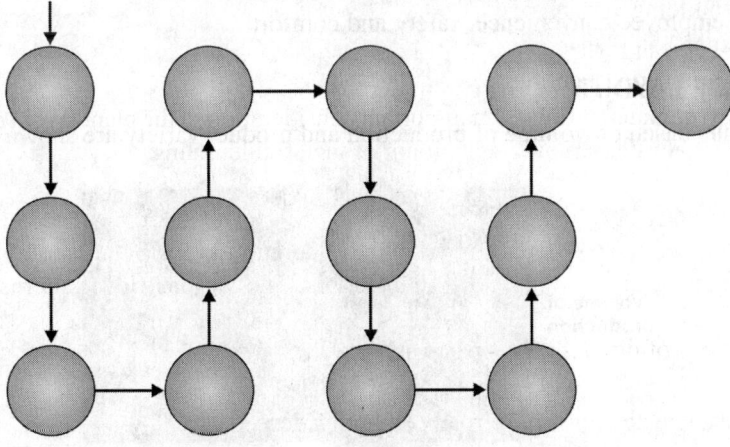

(c) Convoluted or Serpentine shape

Fig. 9.1 Flow patterns (*Contd...*)

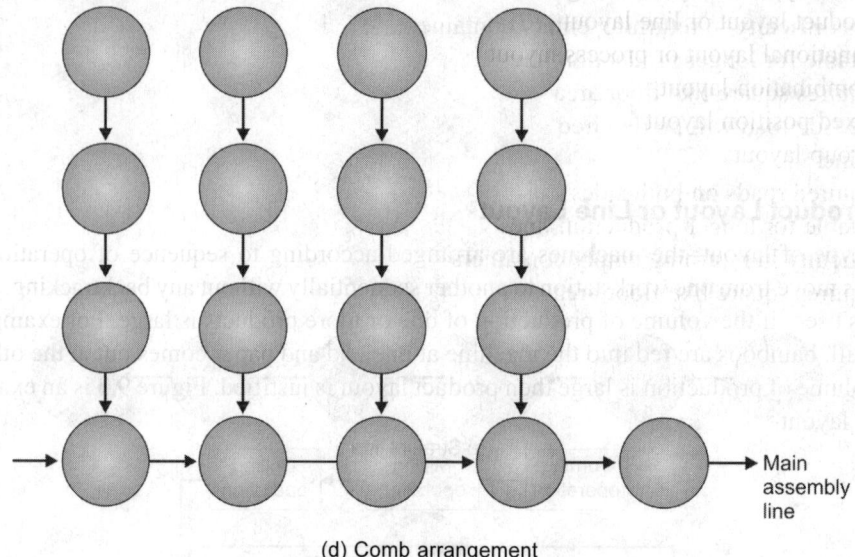

(d) Comb arrangement

Fig. 9.1 Flow patterns

9.4 OBJECTIVES OF PLANT LAYOUT

The primary goal of the plant layout is to maximize the profit by the optimal arrangement of all facilities.

The other objectives of plant layout are:

- Easy flow of material through the plant
- Facilitate the manufacturing process
- Minimize material handling
- Reduce in process inventory
- Effective utilization of men, material, space, machines and material handling equipment
- Flexibility of manufacturing operations and arrangement
- Provide employee convenience, safety and comfort.

9.5 TYPES OF LAYOUT

Types of layouts based on volume of production and product variety are shown in Fig. 9.2.

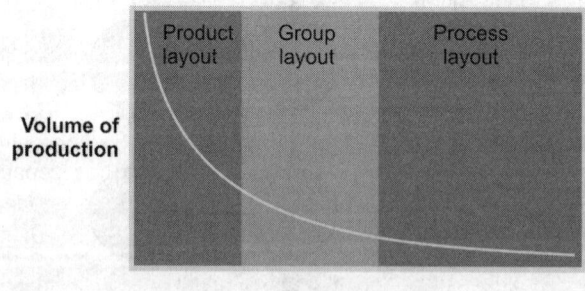

Fig. 9.2 Types of layouts based on volume of production and product variety

Different types of layouts are given below.
- Product layout or line layout
- Functional layout or process layout
- Combination layout
- Fixed position layout
- Group layout

9.5.1 Product Layout or Line Layout

In this type of layout, the machines are arranged according to sequence of operations. The materials move from one workstation to another sequentially without any backtracking. Product layout is used, if the volume of production of one or more products is large. For example, in a paper mill, bamboos are fed into the machine at one end and paper comes out at the other end. If the volume of production is large then product layout is justified. Figure 9.3 is an example of Product layout.

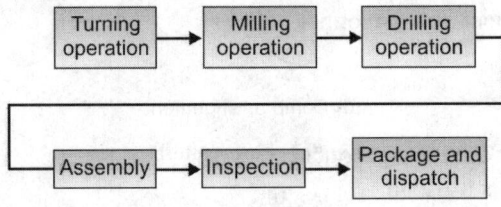

Fig. 9.3 Product layout

Advantages:

- Automatic machines can be used
- Reduced material handling
- Automatic material handling equipment can be used
- Simplified production planning and control
- Perfect line balance which eliminates bottlenecks and idle capacity
- Productivity is more
- Smaller manufacturing lead times
- Less in process inventory

Disadvantages:

- Lack of flexibility: A change in product may require the layout modification
- High capital investment
- Breakdown of any machine may result complete stoppage of production

9.5.2 Functional Layout or Process Layout

This layout (Fig. 9.4) is recommended for batch production. All the machines performing similar type of operations are grouped at one location in the process layout. For example, machines performing drilling operations are arranged in the drilling department and machines performing casting operations are grouped in the casting department. Usually paths are long and there will be possibility of back tracking.

Advantages:

- General purpose machines are used

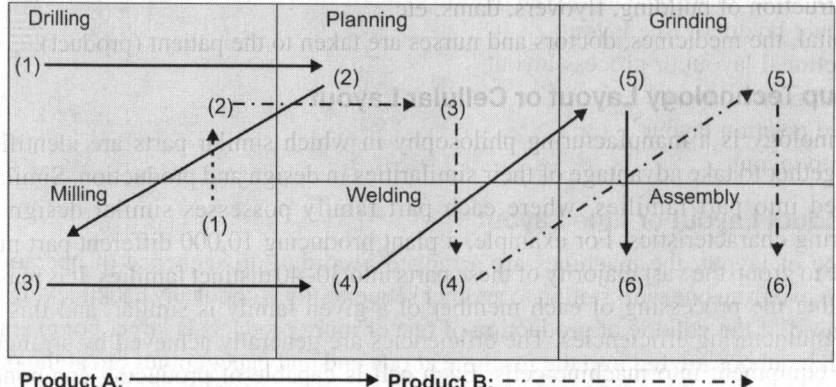

Fig. 9.4 Process layout

- Flexibility of equipment and personnel
- Variety of jobs

Disadvantages:

- Route is different for different products
- Production planning is difficult
- Lower productivity
- Automatic material handling system cannot be used
- Increase of material handling

9.5.3 Combination Layout

This is also called hybrid or mixed type layout. Usually a process layout is combined with the product layout. For example, refrigerator manufacturing unit uses a combination layout. The process layouts are used to produce various operations like stamping, welding, heat treatment. The final assembly of the product is done in a product type layout. Thus, for manufacturing various components process layout is used and assembly product layout is used.

9.5.4 Fixed Position Layout

This is also called project type layout. In this type of layout (Fig. 9.5), the material or major components remain in a fixed location and other materials are brought to this location.

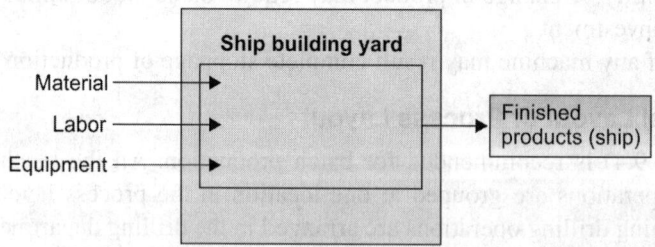

Fig. 9.5 Fixed position layout

The fixed position layout is followed in following conditions:

- Manufacture of bulky and heavy products such as locomotives, ships, boilers, generators, wagon building, aircraft manufacturing, etc.

- Construction of building, flyovers, dams, etc.
- Hospital, the medicines, doctors and nurses are taken to the patient (product).

9.5.5 Group Technology Layout or Cellular Layout

Group technology is a manufacturing philosophy in which similar parts are identified and grouped together to take advantage of their similarities in design and production. Similar parts are arranged into part families, where each part family possesses similar design and/or manufacturing characteristics. For example, a plant producing 10,000 different part numbers may be able to group the vast majority of these parts into 30–40 distinct families. It is reasonable to believe that the processing of each member of a given family is similar, and this should result in manufacturing efficiencies. The efficiencies are generally achieved by arranging the production equipment into machine cells. Each cell is capable of producing components of one part family. Process layout used to produce three different parts is shown in Fig. 9.6. Conversion of process layout into cellular layout to produce these three different parts is shown in Fig. 9.7.

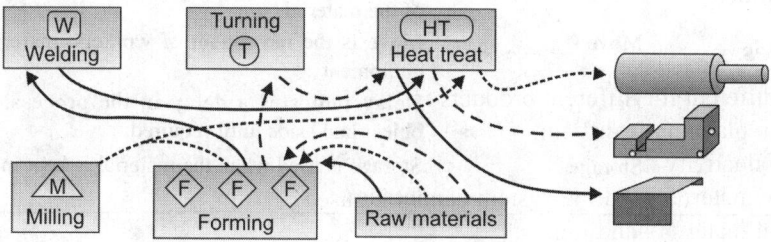

Fig. 9.6 Process layout before the use of GT cells

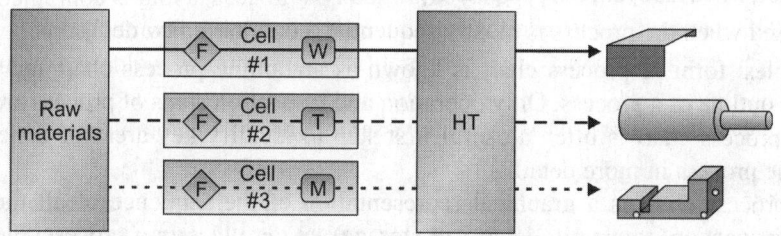

Fig. 9.7 Group technology layout or Cellular layout

Comparison of Process and Cellular Layouts is given in Table 9.1.

Table 9.1 Process layout vs Cellular layout

Dimension	Process	Cellular
Number of moves between departments	Many	Few
Travel distances	Longer	Shorter
Travel paths	Variable	Fixed
Job waiting times	Greater	Shorter
Amount of work-in-process	Higher	Lower
Supervision difficulty	Higher	Lower
Scheduling complexity	Higher	Lower
Equipment utilization	Lower	Higher

9.6 TOOLS AND TECHNIQUES OF PLANT LAYOUT

- Process charts
- Travel charts
- Relationship charts

9.6.1 Process Charts

The Process chart provides a visual representation of the steps involved in a process. Flowcharts are also referred to as Flow Diagrams. There are many symbols used to construct a flowchart; the more common symbols are shown in Table 9.2:

Table 9.2 Symbols used to construct a flowchart

Symbols	Use	Descriptions
O	Operation	Operation is a main step, where the part, material or product is usually modified or changed.
□	Inspection	Inspection is to check the quality or the quantity of the material.
→	Move	Move is the movement of workers, materials or equipment
D	Delay	Delay indicates a delay in the process, or an object laid aside until required.
∇	Storage	Storage is used when the material is kept in a safe location.

Uses of process charts:

- It is used to record a physical process.
- It is used when analyzing the steps in a process, and helps to identify and eliminate waste.
- It is used when the process is mostly sequential, containing few decisions.

The simplest form of process chart is known as an outline process chart and records an overview or outline of a process. Only operation and inspection steps of process are recorded. An outline process chart is often a useful first step to identify key areas of concern before recording the process in more detail.

A flow process chart is a graphical representation of the sequence of all the activities (operation, inspection, transport, delay and storage) taking place in a process. Flow process chart utilizes all five process chart symbols. Process charts are used for developing layout according to sequence of operations. Figure 9.8 is an example of process chart.

Subject: Request tool purchase			
Dist (ft)	Time (min)	Symbol	Description
		●⇨□D∇	Write order
		O⇨□◆∇	On desk
75		O→□D∇	To buyer
		O⇨■D∇	Examine

O = Operation; ⇨ = Transport; □ = Inspect;
D = Delay; ∇ = Storage

Fig. 9.8 Process chart

9.6.2 Travel Charts/From-To Charts

The flow of material between functional areas/departments in the plant is recorded on a travel chart. It records the distance and frequency of martial movements between various pairs of departments. This chart is used to determine the degree of closeness between the departments. It forms the basis for layout design and seeks to minimize the total material handling.

Advantages of travel charts are:
- It is a useful tool for movement analysis.
- It is useful for comparing alternative layouts.
- It is an input for developing computerized layouts.

Procedure to draw the travel chart:
- Department or work centers are listed both rowwise and columnwise in the same sequence.
- Each cell is used to record data of number of trips and distance from one department to another.
- The entries of the chart are scattered on both sides of the diagonal.
- A typical travel chart is shown in Fig. 9.9.
- The entries below the diagonal represent backtracking.
- The entries above the diagonal represent forward tracking.

To→ From↓	A	B	C	D	E	F
A		M_{AB}	M_{AC}	M_{AD}	M_{AE}	M_{AF}
B	M_{BA}		M_{BC}	M_{BD}	M_{BE}	M_{BF}
C	M_{CA}	M_{CB}		M_{CD}	M_{CE}	M_{CF}
D	M_{DA}	M_{DB}	M_{DC}		M_{DE}	M_{DF}
E	M_{EA}	M_{EB}	M_{EC}	M_{ED}		M_{EF}
F	M_{FA}	M_{FB}	M_{FC}	M_{FD}	M_{FE}	

* M_{ij} = Material handling from the department i to department j

Fig. 9.9 Travel chart

9.6.3 Solved Problems

Problem 9.1: A company is manufacturing three products P, Q and R using the same manufacturing facilities arranged in six departments A, B, C, D, E and F. The material handling is done by a forklift truck. The containers can carry 300, 400 and 600 pieces of the products P, Q and R respectively. The annual demand for each product is 1,20,000 units. Sequence operations of product movement are given below:

Product	Movement
P	A-E-B-D-C-F
Q	A-B-C-D-E-F
R	C-B-A-E-D-F

Construct travel chart.

Solution: Annual demand for any product = 12,000 units/annum

Number of trips of forklift truck between departments/year

$$\text{Product } P = \frac{12,000}{300} = 40 \text{ trips/year}$$

$$\text{Product } Q = \frac{12,000}{400} = 30 \text{ trips/year}$$

$$\text{Product } R = \frac{12,000}{600} = 20 \text{ trips/year}$$

Number of trips of forklift truck between departments/year is given in Table 9.3.

Table 9.3 Travel chart to record number of trips of forklift truck between departments/year

To → From ↓	A	B	C	D	E	F	Total
A		Q = 30			P = 40 R = 20		90
B	R = 20		Q = 30	P = 40			90
C		R = 20		Q = 30		P = 40	90
D			P = 40		Q = 30	R = 20	90
E		P = 40		R = 20		Q = 30	90
F							0
Total	20	90	70	90	90	90	450

Total number of trips = 450

Problem 9.2: A facility that will be used to produce a single product has three departments A, B and C that must be used in the configuration shown in Fig. 9.10. The workload flows and travel distances between work centers are given in Table 9.4. In addition, two trial and error optional layouts are shown in Figs. 9.10b and 9.10c. Evaluate the two layouts on a load distance basis and identify the preferred layout. Assume that the cost to transport this product is ₹ 100 per 1 unit of load distance.

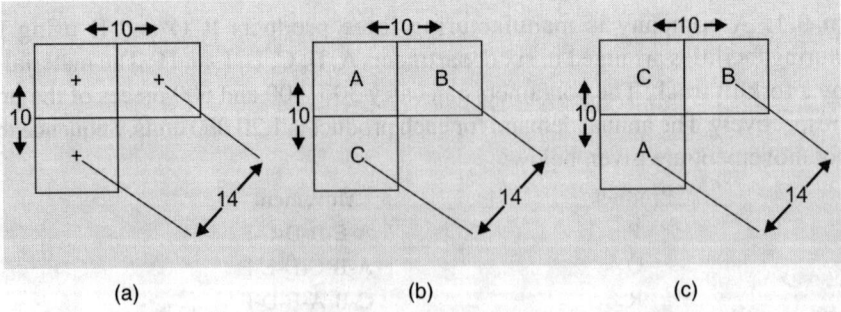

Fig. 9.10 (a) Configuration, (b) Option 1 and (c) Option 2

Table 9.4 Interdepartmental workload per week

To → From ↓	A	B	C
A	–	30	25
B	20	–	40
C	15	50	–

Solution:

d_{ij} = Distance from departments i to j

l_{ij} = Load moves from departments i to j

Interdepartmental workload per week is given in Table 9.5.

Table 9.5 Interdepartmental workload per week (l_{ij})

To → From ↓	A	B	C
A	–	30	25
B	20	–	40
C	15	50	–

Distance matrix for option 1 is given in Table 9.6.

Table 9.6 Distance matrix for option 1 (d_{ij})

To → From ↓	A	B	C
A	–	10	10
B	10	–	14
C	10	14	–

Distance matrix for option 2 is given in Table 9.7.

Table 9.7 Distance matrix for option 2 (d_{ij})

To → From ↓	A	B	C
A	–	14	10
B	14	–	10
C	10	10	–

Load-Distance matrix for option 1 is given in Table 9.8.

Table 9.8 Load-Distance matrix for option 1 ($l_{ij}xd_{ij}$)

To → From ↓	A	B	C	Total
A	–	300	250	550
B	200	–	560	760
C	150	700	–	850
Total	350	1,000	810	2,160

Load-Distance matrix for option 2 is given in Table 9.9.

Table 9.9 Load-Distance matrix for option 2 ($l_{ij} \times d_{ij}$)

To → From ↓	A	B	C	Total
A	–	420	250	670
B	280	–	400	680
C	150	500	–	650
Total	430	920	650	2,000

Cost of transport per 1 unit of load-distance = ₹ 100
Total material handling cost for option 1 = 2,160 × 100 = ₹ 2,16,000
Total material handling cost for option 2 = 2,000 × 100 = ₹ 2,00,000
Total material handling cost is less for option 2. So layout 2 is preferred.

Problem 9.3: The present layout is shown in Fig. 9.11. The manager of the department is intending to interchange the departments C and F in the present layout. The handling frequencies between the departments are given in Table 9.10. All the departments are of the same size and configuration. The material handling cost per unit length travel between departments is same. What will be the effect of interchange of departments C and F in the layout?

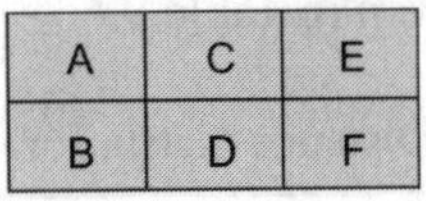

Fig. 9.11 Present layout

Table 9.10 Handling frequencies (l_{ij})

To → From ↓	A	B	C	D	E	F
A	–	0	90	160	50	0
B	–	–	70	0	100	150
C	–	–	–	20	0	0
D	–	–	–	–	180	10
E	–	–	–	–	–	40
F	–	–	–	–	–	–

Solution: Assume the distance between the adjacent departments is equal to 1. Handling frequencies are given in Table 9.11.

Table 9.11 Handling frequencies (l_{ij})

To → From ↓	A	B	C	D	E	F
A	–	0	90	160	50	0
B	–	–	70	0	100	150
C	–	–	–	20	0	0
D	–	–	–	–	180	10
E	–	–	–	–	–	40
F	–	–	–	–	–	–

Distance matrix for the existing layout is given in Table 9.12.

Table 9.12 Distance matrix for existing layout (d_{ij})

To → From ↓	A	B	C	D	E	F
A	-	1	1	2	2	3
B	-	-	2	1	3	2
C	-	-	-	1	1	2
D	-	-	-	-	2	1
E	-	-	-	-	-	1
F	-	-	-	-	-	-

Load-Distance matrix for initial layout is given in Table 9.13.

Table 9.13 Load-Distance matrix for initial layout ($l_{ij} \times d_{ij}$)

To → From ↓	A	B	C	D	E	F	Total
A	-	0	90	320	100	0	510
B	-	-	140	0	300	300	740
C	-	-	-	20	0	0	20
D	-	-	-	-	360	10	370
E	-	-	-	-	-	40	40
F	-	-	-	-	-	-	0
Total	0	0	230	340	760	350	1680

Interchange departments F and C. The resultant layout is shown in Fig. 9.12.

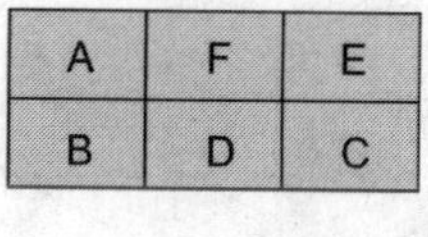

Fig. 9.12 The resultant layout after interchanging departments F and C as shown in Fig. 9.11.

The distance matrix for modified layout is given in Table 9.14.

Table 9.14 The distance matrix for modified layout (d_{ij})

To → From ↓	A	B	C	D	E	F
A	-	1	3	2	2	1
B	-	-	2	1	3	2
C	-	-	-	1	1	2
D	-	-	-	-	2	1
E	-	-	-	-	-	1
F	-	-	-	-	-	-

Load-Distance matrix for modified layout is given in Table 9.15.

Table 9.15 Load-distance matrix for modified layout ($l_{ij} \times d_{ij}$)

To → From ↓	A	B	C	D	E	F	Total
A	–	0	270	320	100	0	690
B	–	–	140	0	300	300	740
C	–	–	–	20	0	0	20
D	–	–	–	–	360	10	370
E	–	–	–	–	–	40	40
F	–	–	–	–	–	–	0
Total	0	0	410	340	760	350	1,860

Interchange of departments increase the total material handling. Thus, it is not a desirable modification.

9.6.4 Relationship Chart

The relative importance of having one department near another department is displayed in relationship chart. Entry in each diamond shaped cell (Fig. 9.13) of the chart shows the relationship between two departments. The entry in the diamond cell indicates the degree of closeness of the relationship between departments. The relationship diagram recognizes the need for exploring relationship rather than calculating exact flow and costs. The closeness (Table 9.16) is expressed on scale. These pairwise interdepartmental closeness rating is then used to develop a suitable layout which satisfies as many pairwise relationship as possible. Example of relationship chart is shown in Fig. 9.13.

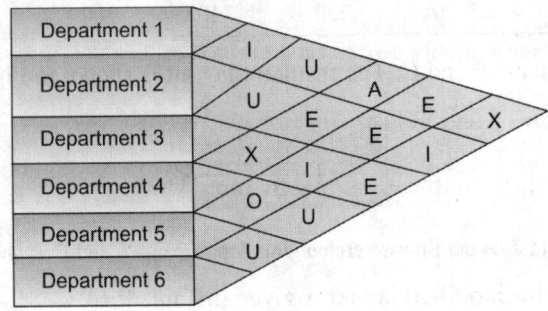

Fig. 9.13 Relationship chart

Table 9.16 Closeness ratings

Letters	Degrees of closeness	Values
A	Absolutely essential	64
E	Essential	16
I	Important	4
O	Ordinary closeness	1
U	Unimportant	0
X	Not desirable	–1,024

9.6.5 Solved Problems

Problem 9.4: In given relationship chart (Fig. 9.14), arrange the departments into a grid given in Fig. 9.15.

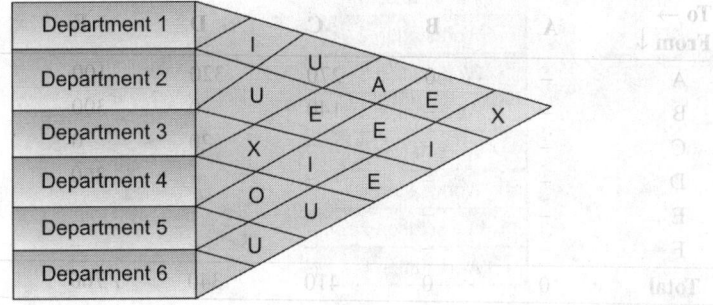

Fig. 9.14 Relationship chart

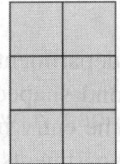

Fig. 9.15 Grid

Solution:

Place the departments 1 and 4 close to each other which are having highest closeness rating 'A'.

Place the departments 4 and 2 close to each other which are having second highest closeness rating 'E'.

Place the departments 2 and 5 close to each other which are having second highest closeness rating 'E'.

Place the departments 5 and 3 close to each other which are having third highest closeness rating 'I'.

Place the departments 2 and 6 close to each other which are having third highest closeness rating 'I'.

The resultant layout is shown in Fig. 9.16.

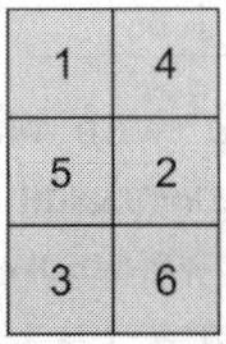

1	4
5	2
3	6

Fig. 9.16 Resultant layout

This is only heuristic procedure. Here Layout is developed by satisfying as many closeness ratings as possible.

Problem 9.5: A chemical fertilizer facility has eight work centers that must be processed into a 2 row by 4 column facility. The closeness rating for absolutely necessary 'a', very important' and undesirable 'x' indexes are given in Fig. 9.17. Develop a suitable layout.

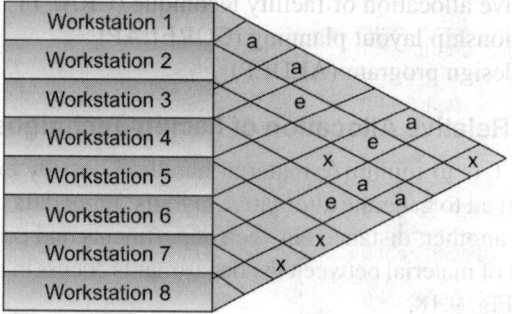

Fig. 9.17 Relationship chart

Solution: Order of closeness ratings:

'a' absolutely necessary
'e' very important
'x' undesirable

Workstations 2 and 3 have Closeness rating 'a' with Workstation 3. The resultant Layout is shown below.

1	2		
3			

Workstation 8 has Closeness rating 'a' with Workstation 3. The resultant Layout is shown below.

1	2		
3	8		

Workstation 6 has Closeness rating 'a' with Workstation 1. The resultant Layout is shown below.

6	1	2	
	3	8	

Workstation 4 has Closeness rating 'e' with Workstation 2. The resultant Layout is shown below.

6	1	2	4
	3	8	

Workstation 7 has Closeness rating 'e' with Workstation 4. Workstation 5 doesn't have closeness ratings with any of the remaining workstations. Workstation 5 can be placed anywhere in the Layout. The resultant Layout is shown below.

6	1	2	4
5	3	8	7

This is only heuristic procedure. Here Layout is developed by satisfying as many closeness ratings as possible.

9.7 COMPUTER PACKAGES FOR LAYOUT ANALYSIS

Several computerized approaches are available for developing and analyzing process layouts. The important computerized packages are:

- Computerized relative allocation of facility technique (CRAFT)
- Computerized relationship layout planning (CORELAP)
- Automated layout design program (ALDEP)

9.7.1 Computerized Relative Allocation of Facility Technique (CRAFT)

The objective of CRAFT is to minimize material handling cost by exchanging departments. Departments are exchanged to generate alternative layouts. Input data includes material moved from one department to another, distance between departments and cost of moving 1 kg-metre material. The movement of material between the departments occurs in a straight line. Flowchart for CRAFT is given in Fig. 9.18.

Limitations:

- It is a heuristic method.
- It is step by step method (Iterative method).
- Optimum solution is not guaranteed.
- Distance between the departments is linear.

Advantages:

- It is a Quantitative technique.

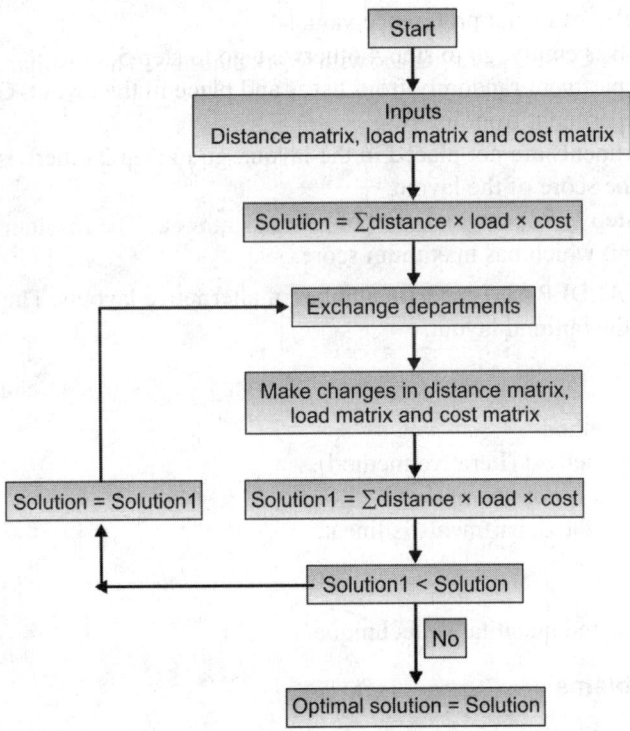

Fig. 9.18 Flowchart for CRAFT

9.7.2 Automated Layout Design Program (ALDEP)

ALDEP is a construction type algorithm. Layout is built by placing the departments one by one in the layout from top to bottom and left to right. After placing all the departments in the layout, a score is computed. This is nothing but some of the closeness rating values of different neighboring departments in the layout. The algorithm is repeated for a prescribed number of times and the best layout is selected on the maximum layout score.

The basic data required for the Automated Layout Design Program:
- Total number of departments
- Area of each department
- Length and width of layout
- Relationship chart
- Closeness rating of various departments
- Minimum departmental preference value
- Number of iterations to be performed

Steps:

Step 1: Select a department randomly and place in the layout.

Step 2: From the relationship chart, determine relationship ratings of selected department with unselected departments. Classify the unselected departments into two lists namely List A and List B.

List A contains the unselected departments whose relationship values are less than minimum departmental preference value.

List B contains the unselected departments whose relationship values are greater than minimum departmental preference value.

Step 3: If the List B is empty, go to step 4 otherwise go to step 5.

Step 4: Select a department randomly from list A and place in the layout. Go to step 6.

Step 5: Select a department from list B.

Step 6: If all departments are not placed in the layout, go to step 2 otherwise go to step 7.

Step 7: Compute the score of the layout.

Step 8: Go to the step 1. Repeat n number of times.

Step 9: Select layout which has maximum score.

The objective of ALDEP is to generate number of alternative layouts. The layout which has maximum score is the optimal layout.

Limitations:
- It is a heuristic method.
- It is step by step method (Iterative method).
- Optimum solution is not guaranteed.
- Distance between the departments is linear.

Advantages:
- It is a qualitative and quantitative technique.

9.7.3 Solved Problems

Problem 9.6: Apply ALDEP algorithm for the following data to design a suitable Layout.
Number of departments in the layout = 7

Table 9.17 Areas of departments

Departments	Functions	Areas (Sq m)
1	Receiving	12,000
2	Milling	8,000
3	Press	6,000
4	Screw machine	12,000
5	Assembly	8,000
6	Painting	12,000
7	Shipping	12,000
Total		70,000

Table 9.18 Relationship chart

To → From ↓	1	2	3	4	5	6	7
1	–	E	O	I	O	U	U
2	E	–	U	E	I	I	U
3	O	U	–	U	U	O	U
4	I	E	U	–	I	U	U
5	O	I	U	I	–	A	I
6	U	I	O	U	A	–	E
7	U	U	U	U	I	E	–

Table 9.19 Quantitative values of degree of closeness

Degrees of closeness	Notations	Values
Absolutely essential	A	64
Essential	E	16
Important	I	4
Ordinary closeness	O	1
Unimportant	U	0
Not desirable	X	–1,024

Solution: Input data:

Number of departments = 7

Minimum departmental preference value = 4

Areas of the departments are given in Table 9.20.

Table 9.20 Areas of departments

Departments	Functions	Areas (Sq.m)
1	Receiving	12,000
2	Milling	8,000
3	Press	6,000
4	Screw machine	12,000
5	Assembly	8,000
6	Painting	12,000
7	Shipping	12,000
Total		70,000

Relationship chart is given in Table 9.21.

Table 9.21 Relationship chart

To → From ↓	1	2	3	4	5	6	7
1	–	E	O	I	O	U	U
2	E	–	U	E	I	I	U
3	O	U	–	U	U	O	U
4	I	E	U	–	I	U	U
5	O	I	U	I	–	A	I
6	U	I	O	U	A	–	E
7	U	U	U	U	I	E	–

Assumed layout of building is given in Fig. 9.19.

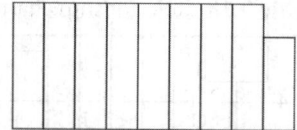

Each Rectangle area is approximately 8000 Sq.m

Fig. 9.19 Layout of building

Iteration 1:

Step 1: Randomly department 4 is selected. Place the department 4 in the layout as shown in Fig. 9.20a.

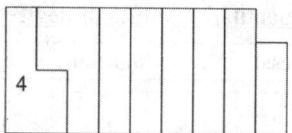

Fig. 9.20 (a) Layout after placing the department 4

Step 2:

Selected department ↓		Unselected departments				
4	1	2	3	5	6	7
Closeness ratings	I	E	U	I	U	U

List A: [3 6 7]

List B: [1 2 5]

Step 3: List B is not empty. So go to step 5.

Step 5: Department 2 is selected randomly from the list B and place department 2 in the layout. The resultant Layout is shown in Fig. 9.20b.

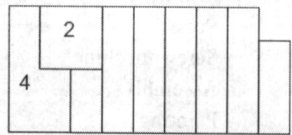

9.20 (b) Layout after placing the departments 4 and 2

Step 6: All the departments are not placed in the Layout. Go to step 2.

Iteration 2:

Step 2:

Selected department ↓	Unselected departments				
2	1	3	5	6	7
Closeness ratings	E	U	I	I	U

List A: [3 7]
List B: [1 5 6]
Step 3: List B is not empty. So go to step 5.
Step 5: Department 1 is selected randomly from the list B and place department 1 in the layout. The resultant Layout is shown in Fig. 9.20c.

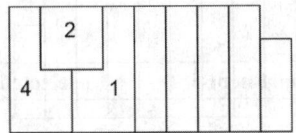

Fig. 9.20 (c) Layout after placing the departments 4, 2 and 1

Step 6: All the departments are not placed in the Layout. Go to step 2.

Iteration 3:

Step 2:

Selected department ↓	Unselected departments			
1	3	5	6	7
Closeness ratings	O	O	U	U

List A: [3 5 6 7]
List B: [Empty]
Step 3: List B is empty. So go to step 4.
Step 4: Department 6 is selected randomly from the list A and place department 6 in the layout. The resultant Layout is shown in Fig. 9.20d. Go to step 6.

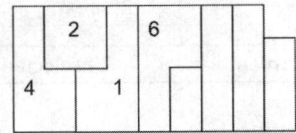

Fig. 9.20 (d) Layout after placing the departments 4, 2, 1 and 6

Step 6: All the departments are not placed in the Layout. Go to step 2.

Iteration 4:

Step 2:

Selected department ↓	Unselected departments		
6	3	5	7
Closeness ratings	O	O	U

List A: [3 5 7]
List B: [Empty]

Step 3: List B is empty. So go to step 4.

Step 5: Department 5 is selected randomly from the list A and place department 5 in the layout. The resultant Layout is shown in Fig. 9.20e. Go to step 6.

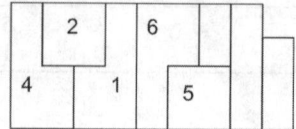

Fig. 9.20 (e) Layout after placing the departments 4, 2, 1, 6 and 5

Step 6: All the departments are not placed in the Layout. Go to step 2.

Iteration 5:

Step 2:

Selected department ↓	Unselected departments	
6	3	7
Closeness ratings	U	E

List A: [3]

List B: [7]

Step 3: List B is not empty. So go to step 5.

Step 5: Department 7 is selected randomly from the list B and place department 7 in the layout. The resultant Layout is shown in the Fig. 9.20f. Go to step 6.

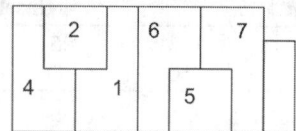

Fig. 9.20 (f) Layout after placing the departments 4, 2, 1, 6, 5 and 7

Step 6: All the departments are not placed in the Layout. Go to step 2.

Iteration 6:

Step 2:

Selected department ↓	Unselected department
7	3
Closeness rating	U

List A: [3]

List B: [Empty]

Step 3: List B is empty. So go to step 4.

Step 4: Select the department 3 from the list A and place department 3 in the layout. The resultant Layout is shown in Fig. 9.20g. Go to step 6.

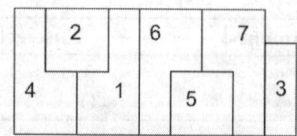

Fig. 9.20 (g) Layout after placing the departments 4, 2, 1, 6, 5, 7 and 3

Step 6: All the departments are placed in the Layout. Go to step 7.

Step 7: Compute the score from the relationship chart. Computation of score for the generated Layout is shown in Table 9.22.

Table 9.22 Computation of score for the generated layout

Neighboring pair of departments	REL Grades	REL values
4-2	E	16
4-1	I	4
2-1	E	16
1-6	U	0
6-7	E	16
6-5	A	64
5-7	I	4
7-3	U	0
		Total = 120

Step 8: Go to step 1. Repeat the process n number of times.

Step 9: Select the layout which has maximum score.

The objective of ALDEP is to generate number of alternative layouts. The layout which has maximum score is optimal layout.

9.7.4 Computerized Relationship Layout Planning (CORELAP)

Input requirements for computerized relationship layout planning (CORELAP) are given below:

- Total number of departments
- Area of each department
- Length and width of layout
- Relationship chart
- Closeness ratings of various departments

General approach is to select the most critical department first and place it at the centre of the layout. After the first department is placed, the department having closeness relationship with the department already placed is selected and place best location adjacent to the previously placed department. CORELAP builds the layout from the centre. Final score of the layout is computed by using the closeness values from the relationship chart.

Limitations:

- It is a heuristic method.
- It is step by step method (Iterative method).
- Optimum solution is not guaranteed.
- Distance between the departments is linear

Advantages:

- It is a qualitative and quantitative technique.

9.7.5 Solved Problems

Problem 9.7: Apply ALDEP algorithm for the data given in the Problem 9.6 to design a suitable Layout.

Solution: Each Rectangle area is approximately 8,000 sq m.

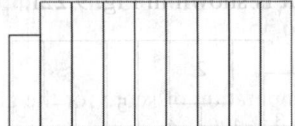

Fig. 9.21 Layout of Building

Department 2 has high closeness ratings with many departments. So Place department 2 at the centre. The resultant Layout is shown in Fig. 9.22a.

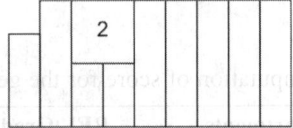

Fig. 9.22 (a) Layout after placing department 2

Departments 1 and 4 have 'E' closeness rating with department 2. So place departments 1 and 4 adjacent to department 2. The resultant Layout is shown in Fig. 9.22b.

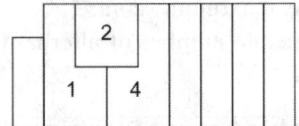

Fig. 9.22 (b) Layout after placing departments 1 and 4

Department 5 has 'I' closeness rating with department 4. So place department 5 adjacent to department 4. The resultant Layout is shown in Fig. 9.22c.

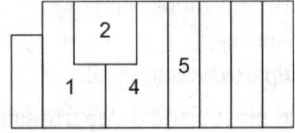

Fig. 9.22 (c) Layout after placing department 5

Department 6 has 'A' closeness rating with department 5. So place department 6 adjacent to department 5. The resultant Layout is shown in Fig. 9.22d.

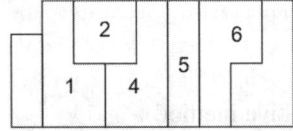

Fig. 9.22 (d) Layout after placing department 6

Department 7 has 'E' closeness rating with department 6. So place department 7 adjacent to department 6. The resultant Layout is shown in Fig. 9.22e.

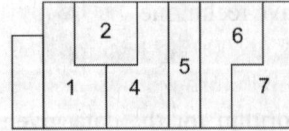

Fig. 9.22 (e) Layout after placing department 7

Department 3 has 'O' closeness rating with department 1. So place department 3 adjacent to department 1. The resultant Layout is shown in Fig. 9.22f.

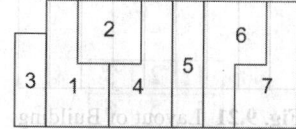

Fig. 9.22 (f) Layout after placing department 3

Compute the score from the relationship chart. Computation of score for the generated Layout is shown in Table 9.23.

Table 9.23 Computation of score for the generated layout

Neighboring pair of departments	REL Grades	REL values
3-1	O	1
1-2	E	16
1-4	I	4
2-4	E	16
4-5	I	4
5-6	A	64
6-7	E	16
	Total = 121	

9.8 QUESTIONS AND PROBLEMS

1. Explain the various tools and techniques of plant layout.
2. Identify and describe the different models/methods used to assist layout designers.
3. Explain various types of layouts.
4. What are the advantages and disadvantages of product and process layouts?
5. Explain the essential features of fixed position layout.
6. Explain the essential features of Group technology/Cellular layout.
7. What is REL chart? How it is used for layout design?
8. What is Travel chart? How it is used for layout design?
9. Explain computer packages for layout analysis.
10. Explain different types of plant layouts used in manufacturing.
11. Explain the essential features of CRAFT layout model.
12. Explain the essential features of CORELAP layout model.
13. Explain ALDEP. What are the advantages and limitations?
14. Discuss applications of REL charts and Travel chats.
15. Which type of layout is preferred for ship manufacturing?
16. Which type of layout is used in workshops of any technical institution?
17. A company is considering three different locations (A,B and C) for a new facility that will have annual costs ₹ 27,500, ₹ 25,400 and ₹ 29,000 respectively. The annual revenues are ₹ 40,000 for location A, ₹ 36,000 for location B and ₹ 42,000 for location C. Which location will maximize the net return per year?

Answer: Net return per year = Annual revenue – Annual cost.

C is the optimal location.

18. XYZ management has purchased a plant has shown in the figure below. Firm proposes to locate six departments A, B, C, D, E and F, which have the number of moves per day between departments as shown in Travel chart. Develop a layout that minimizes the material handling.

1	2	3
4	5	6

Fig. Q.18 Layout for six production centers

Travel chart

To → From ↓	A	B	C	D	E	F
A	–	7	–	–	–	5
B	–	–	–	4	10	–
C	–	7	–	–	2	–
D	–	–	8	–	–	3
E	4	–	–	–	–	3
F	–	6	–	–	10	–

Answer:

A	B	D
F	E	C

10

Material Handling

10.1 DEFINITION

There are many ways by which material handling has been defined but one simple definition is, "Material handling is the movement and storage of material at the lowest possible cost through the use of proper method and equipment". Material handling ranges from movement of raw material, work-in-progress, finished goods, rejected, scraps, packing material, etc. These materials are of different shape and sizes as well as weight. The movement of materials takes place from one processing area to another or from one department to another department of the plant. The cost of material handling contributes significantly to the total cost of manufacturing. The importance of material handling function is greater in those industries where the ratio of handling cost to the processing cost is large. Logistics describe the general movement of materials. Material handling is also known as internal logistic; it involves the movement and storage of materials inside a given production environment.

10.2 OBJECTIVES OF MATERIAL HANDLING

The objectives of material handling are as follows:
- Reduce manufacturing cycle time
- Improve material flow control
- Creation and encouragement of safe and hazard-free work condition
- Reduce delays, and damage
- Maintain or improve product quality
- Increase productivity of labor and facilities
- Maximize the equipment utilization and facilities
- Reduce tare weight
- Control inventory

10.3 DETERMINATION OF COST OF HANDLING

The total cost of handling per unit is the sum of the following:
1. Cost of material handling equipment/unit = C_{MH}

$$C_{MH} = \frac{\text{Cost of equipment}}{\text{Number of units produced over the working life of the equipment}}$$

115

2. Cost of labor/unit: C_L

$$C_L = \frac{\text{Total labor cost to operate material handling equipment in period } t}{\text{Number of units produced in period } t}$$

3. Cost of maintenance of material handling equipment/unit: C_M

$$C_M = \frac{\text{Total maintenance cost in period } t}{\text{Number of units produced in period } t}$$

$\therefore$ The total cost of material handling per unit $= C_{MH} + C_L + C_M$

10.4 MANUFACTURING CYCLE TIME

The total time required to make a product from the receipt of its raw material to the finished state is called manufacturing cycle time. Manufacturing cycle time can be reduced using an efficient and effective material handling system. The movement of the material can be faster and handling distance could be reduced with the adoption of an appropriate material handling system. In any organization, a considerable amount of material handling is done in one form or the other. This movement is either done manually or through an automated process.

10.5 MATERIAL HANDLING PRINCIPLES

These principles serve as rough guides or rules of thumb for material handling system design. The designers of material handling systems are usually advised to follow principles. However, in some cases they might not be able to apply them to the fullest extent because of factors such as the limitation on capital, physical characteristics of the building, and capability of the equipment.

Material handling principles are as follows:

- **Orientation principle:** It encourages study of all available system relationships before preliminary planning. The study includes looking at existing methods, problems, etc.
- **Planning principle:** It establishes a plan which includes basic requirements, desirable alternates and planning for contingency.
- **Systems principle:** It integrates handling and storage activities, which are cost effective into integrated system design.
- **Unit load principle:** Handles product in a unit load as large as possible
- **Space utilization principle:** Encourages effective utilization of all the space available
- **Standardization principle:** It encourages standardization of handling methods and equipment.
- **Ergonomic principle:** It recognizes human capabilities and limitations by designing effective handling equipment.
- **Energy principle:** It considers consumption of energy during material handling.
- **Ecology principle:** It encourages minimum impact upon the environment during material handling.
- **Mechanization principle:** It encourages mechanization of handling process wherever possible as to encourage efficiency.
- **Flexibility principle:** Encourages methods and equipment which are possible to utilize in all types of conditions
- **Simplification principle:** Encourages simplification of methods and processes by removing unnecessary movements

- **Gravity principle:** Encourages usage of gravity principle in movement of goods
- **Safety principle:** Encourages provision for safe handling equipment according to safety rules and regulation
- **Computerization principle:** Encourages computerization of material handling and storage systems
- **System flow principle:** Encourages integration of data flow with physical material flow
- **Layout principle:** Encourages preparation of operational sequence of all systems available
- **Cost principle:** Encourages cost benefit analysis of all solutions available
- **Maintenance principle:** Encourages preparation of plan for preventive maintenance and scheduled repairs
- **Obsolescence principle:** Encourages preparation of equipment policy as to enjoy appropriate economic advantage.

10.6 NEED TO DESIGN A MATERIAL HANDLING SYSTEM

The need to design a material handling system arises when:
- a new product is being planned for manufacture.
- change in the existing product design requires a corresponding change in the layout.
- obsolescence of facilities taken place.
- frequent accidents occurred.
- adoption of new safety standards considered.

10.7 FACTORS FOR CONSIDERATION OF MATERIAL HANDLING SYSTEM DESIGN

Material handling equipment consists of cranes, conveyors and industrial trucks. The major factors for consideration of material handling system design are:
- Material form: Gas, liquid, semi-liquid, solid
- Characteristics: Chemical, electrical, mechanical
- Quantity
- Production process and equipments
- Building construction
- Layout
- The existing material handling equipment
- Packaging
- Production planning and control

10.8 MATERIAL HANDLING EQUIPMENT

Material handling equipment is an industrial device that is used to move, transport, store, protect, handle and/or dispose goods, products and materials. This equipment is generally used in large industries such as shipping and logistics, warehousing, pharmaceuticals, food industry, construction and manufacturing. Small and medium industries also use material handling equipment that is intended for smaller workload and applications. Material handling equipment is mainly used to improve workers' performance and efficiency. These devices help greatly in decreasing unwanted accidents and muscle strains to laborers doing manual work such as lifting and moving bulk and heavy materials.

10.9 TYPES OF MATERIAL HANDLING EQUIPMENT

Material handling equipment is divided into two major types:
- Conventional material handling equipment
- Computer-aided/Automated material handling equipment

10.9.1 Conventional Material Handling Equipment

Conventional material handling equipment is a non-automated device that is run by traditional motorized system. Conventional material handling equipment is divided into following categories.

- Conveyors
- Cranes and hoists
- Industrial Trucks

Trucks are used in process layout for material handling. Product layout is used in mass production system. Conveyors are used in product layout for material handling.

Conveyor: Conveyors are designed to move large quantities of materials over fixed paths using chains, belts, rollers or other mechanical devices.

- Roller conveyors: Pathway consists of a series of rollers that are perpendicular to direction of travel. Powered rollers rotate to drive the loads forward. A Roller conveyor is shown in Fig. 10.1.

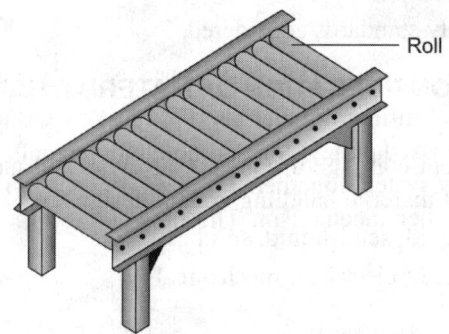

Fig. 10.1 Roller conveyor

- Belt conveyor: Belt is used to move loads. Belt is made of reinforced elastomeric. Supports slider or rollers used to support forward loop. Flat belt (shown) and V-shaped belt are two common forms of belts. A Belt conveyor is shown in the Fig. 10.2.

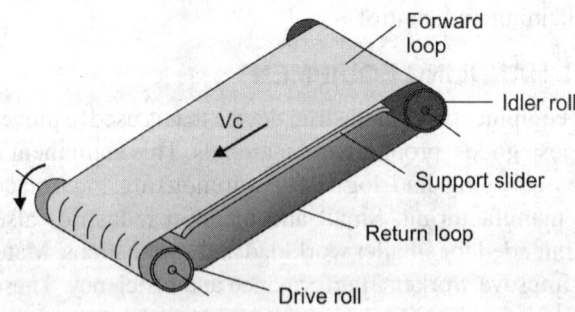

Fig. 10.2 Belt conveyor

Cranes and hoists: Handling devices for lifting, lowering and transporting materials.
Cranes: Cranes (Fig. 10.3) are used for horizontal movement of materials. Cranes consist
of a hoist supported on a horizontal beam that is cantilevered from a vertical column or
wall support. The horizontal beam can pivot about the vertical axis formed by the column
or wall support; this provides a horizontal sweep for the crane. This forms a semi-circular
or circular area in which the horizontal beam can move. The horizontal beam also serves
as the track along which the hoist trolley moves. The hoist provides vertical lift and lower
motions.

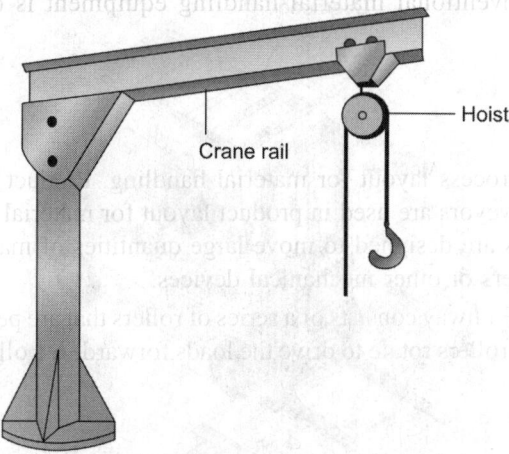

Hoist

Crane rail

Fig. 10.3 Crane

- Hoists: Used for vertical lifting of materials (Fig. 10.4). Used to raise and lower loads, it
  consists of one or more fixed pulleys, one or more moving pulleys, and a rope/chain/cable
  that connects the pulley system together. The load is attached to the moving pulley(s) by
  means of a hook, or other mechanism. The more pulleys a hoist has, the greater the

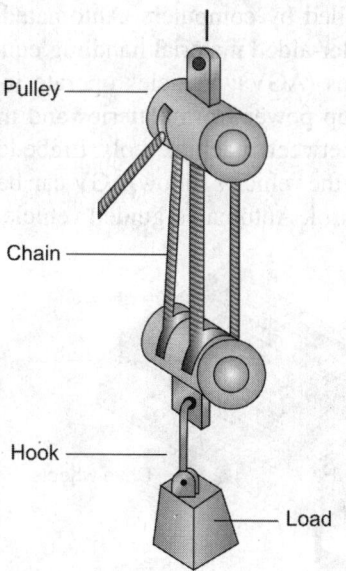

Pulley

Chain

Hook

Load

Fig. 10.4 Hoist

mechanical advantage it can display; whereby mechanical advantage is formulated as the ratio of the load weight to the driving force required to lift the weight. The driving force is applied either manually, or by electric or pneumatic motor.

Industrial trucks: Industrial trucks are motorized vehicles that are used to move and transport products, goods and materials. These trucks can run through electrically powered mechanism and can be used for different applications. Examples of industrial trucks are pallet trucks and forklift trucks. Forklift truck is shown in Fig. 10.5.

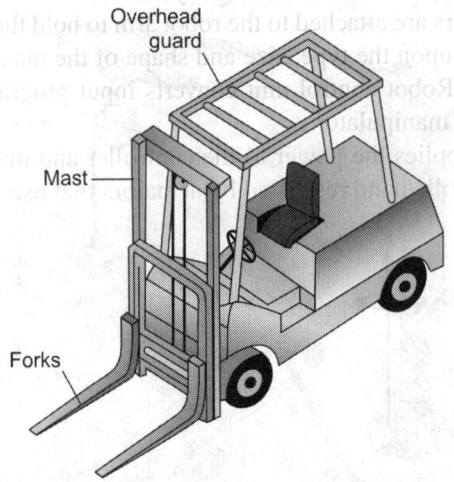

Fig. 10.5 Forklift truck

10.9.2 Computer-aided/Automated Material Handling Equipment

Computer-aided material handling equipment features more technologically advanced mechanism. Devices that are included in this type of material handling equipment are those that are connected and controlled by computers. Automated guided vehicles and industrial robots are examples of computer-aided material handling equipment.

- Automated guided vehicles (AGV): Vehicles operate independently and are driven by electric motors that pick up power from batteries and moves generally in a fixed path. Used to move work parts between machine tools. Embedded guide wires in the floor emit electromagnetic signal that the vehicles follow. AGV can be controlled by central computers or on-board computer control. Automated guided vehicle is shown in Fig. 10.6.

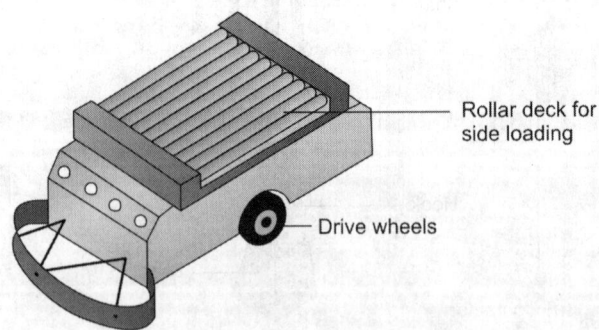

Fig. 10.6 Automated guided vehicle

- Industrial robots: Robot is a programmable machine which possesses human like characteristics. Robot is programmed to form the task repeatedly until it is reprogrammed to perform some other task. Industrial robots are very useful material handling devices in an automated environment. An industrial robot is programmed to move materials, parts, tools, etc. An industrial robot is shown in Fig. 10.7.

Robot components: The following are basic components of an industrial robot.

- Manipulator: It is a mechanical unit that provides motions similar to those of human arm and hand. Manipulator consists of links and joints.
- End effecter: Grippers are attached to the robot arm to hold the workpiece. The design of the gripper depends upon the type, size and shape of the material to be handled.
- Robot control unit: Robot control unit converts input program into signals to perform various functions of manipulator.
- Power supply: It supplies the power to the controller and manipulator. Each motion of manipulator is controlled and regulated by actuators that use an electrical, pneumatic or hydraulic power.

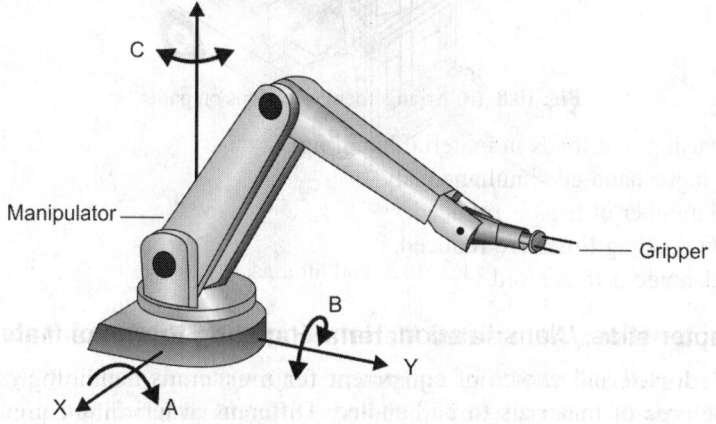

Fig. 10.7 Industrial robot

10.9.3 Unit Load Principle

A unit load is the mass that is to be moved or handled at one time. Multiple items are placed in the container or pallets and or moved at one time. Container, Pallet box and Tote box are shown in Fig. 10.8.

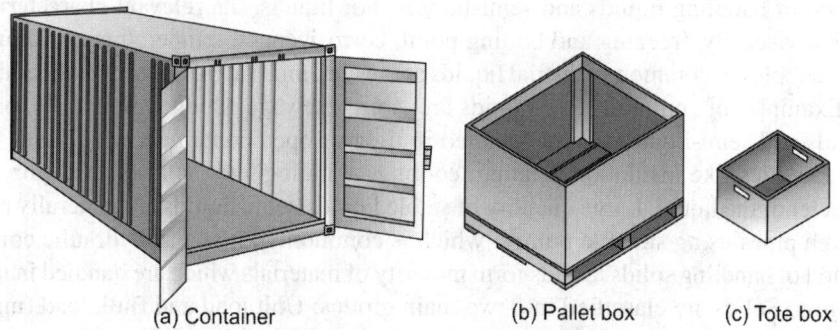

(a) Container   (b) Pallet box   (c) Tote box

Fig. 10.8 (a) Container (b) Pallet box (c) Tote box (*Contd...*)

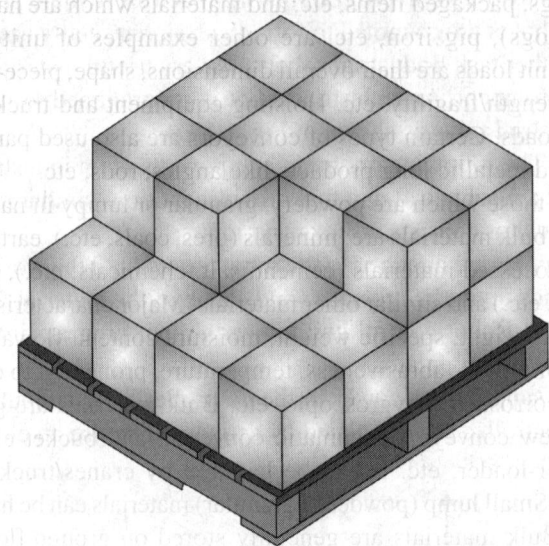

Fig. 10.8 (d) Arrangement of cartons on pallet

Reasons for using unit loads in material handling:
- Multiple items handled simultaneously
- Required number of trips is reduced.
- Loading/unloading times are reduced.
- Product damage is decreased.

10.9.4 Characteristics, Classification, Handling and Storage of Materials

Method to be adopted and choice of equipment for a materials handling system primarily depends on the type of materials to be handled. Different materials are used in industries. Basic classification of material is made on the basis of forms, which are: i. Gases, ii. Liquids, iii. Semi-liquids and iv. Solids.

  i. Method of handling gases: For gases it is primarily pressure, high or low. Chemical properties are also important. Gases are generally handled in tight and where required, pressure resisting containers. However, most common method of handling of large volume of gas is through pipes by the help of compressor, blower, etc. This process is known as pneumatic conveying.

 ii. Method of handling liquids and semi-liquids: For liquids, the relevant characteristics are density, viscosity, freezing and boiling point, corrosiveness, temperature, inflammability, etc. Examples of common industrial liquids are: water, mineral oils, acids, alkalis, chemicals, etc. Examples of common semi-liquids are: slurry, sewage, sludge, mud, pulp, paste, etc. Liquids and semi-liquids can be handled in tight or open containers which may be fitted with facilities like insulation, heating, cooling, agitating, etc. as may be required by the character of the liquid. Large quantity of stable liquids/semi-liquids are generally conveyed through pipes using suitable pumps, which is commonly known as hydraulic conveying.

iii. Method of handling solids: Solids form majority of materials which are handled in industrial situation. Solids are classified into two main groups: Unit load and Bulk load (materials). Unit loads are formed solids of various sizes, shapes and weights. Some of these are counted by number of pieces like machine parts, molding boxes, and fabricated items. Tared goods like containers, bags, packaged items, etc. and materials which are handled en-masses like forest products (logs), pig iron, etc. are other examples of unit loads. The specific characteristics of unit loads are their overall dimensions, shape, piece-weight, temperature, inflammability, strength/fragility, etc. Hoisting equipment and trucks are generally used for handling unit loads. Certain types of conveyors are also used particularly for cartons/packaged items and metallic long products like angles, rods, etc.

- Bulk materials are those which are powdery, granular or lumpy in nature and are stored in heaps. Example of bulk materials are: minerals (ores, coals, etc.), earthly materials (gravel, sand, clay, etc.) processed materials (cement, salt, chemicals, etc.), agricultural products (grain, sugar, flour, etc.) and similar other materials. Major characteristics of bulk materials are lump size, bulk weight, specific weight, moisture content, flowability (mobility of its particles), angles of repose, abrasiveness, temperature, proneness to explosion, stickiness, fuming or dusty, corrosively, hygroscopic, etc. Bulk materials are generally handled by belt conveyor, screw conveyor, pneumatic conveyor, and bucket elevator, grab bucket, skip hoist, dumper-loader, etc. It can be handled by cranes/trucks when collected in containers or bags. Small lump (powdered/granular) materials can be handled pneumatically or hydraulically. Bulk materials are generally stored on ground/floor in open or under shed, and also in bunkers/silos.

10.9.5 Factors Affecting the Selection of Material Handling Equipment

The factors affecting the selection of material handling equipment are:
- Adaptability: The load carrying and movement characteristics of the equipment should fit the material handling problem.
- Flexibility: Where possible the equipment should have flexibility to handle more than one material, referring either to class or size.
- Load capacity: Equipment selected should have great enough load-carrying characteristics to do the job effectively, yet should not be too large and result in excessive operating costs.
- Power: Enough power should be available to do the job.
- Speed: Rapidity of movement of material, within the limits of the production process or plant safety, should be considered
- Space requirements: The space required to install or operate material handling equipment is an important factor in its selection.
- Supervision required: As applied to equipment selection, this refers to the degree of automaticity designed into the equipment.
- Ease of maintenance: Equipment selected should be easily maintained at reasonable cost.
- Environment: Equipment selected must conform to any environment regulations.
- Cost: The consideration of the cost of the equipment is an obvious factor in its selection.

10.9.6 Advantages of Material Handling System

- Improve efficiency of a production system by ensuring the right quantity of materials delivered at the right place at the right time most economically.
- Cut down indirect labor cost.
- Reduce damage of materials during storage and movement.
- Maximize space utilization by proper storage of materials and thereby reduce storage and handling costs.
- Minimize accident during material handling.

- Reduce overall cost by improving material handling.
- Improve customer services by supplying materials in a manner convenient for handlings.

10.9.7 Disadvantages of Material Handling System

- Additional capital cost involved in any material handling system.
- Once a material handling system get implemented, flexibility for further changes gets greatly reduced.
- With an integrated material handling system installed, failure/stoppage in any portion of it leads to increased downtime of the production system.
- Material handling system needs maintenance, hence any addition to material handling means additional maintenance facilities and costs.

10.9.8 Benefits of Automated Handling System (AHS)

- Total elimination of manual handling, leading to dramatic and immediate improvement of product quality
- Safe and efficient product flow in a compact and fluid system configuration
- Sharp reduction of labor costs guaranteeing prompt return on investment (3–4 years payback)
- Considerable factory floor space gain
- Clean, orderly, efficient production environment
- Fully computerized operation affording real-time system control, comprehensive data processing facilities, and just-in-time flexibility
- Product traceability resulting in better customer service
- Full compatibility with trend towards larger and heavier product units
- High efficiency packing process
- Last but not the least safe and physically effortless operation

10.10 QUESTIONS

1. Explain principles of material handling.
2. Define material handling.
3. How the cost of material handling per unit is determined?
4. Describe the factors affecting the selection of materials handling equipment.
5. Explain the principle of unit load.
6. Describe the following material handling equipment:
   a. Cranes
   b. Trucks
   c. Automatic guided vehicle
   d. Robotics
7. How will you classify material handling devices? Give examples of each type.
8. What are the advantages of material handling system?
9. What are the disadvantages of material handling system?
10. What are the benefits of an automated handling system (AHS)?

11

Assembly Line Balancing

11.1 INTRODUCTION

The assembly line is a production line where the material moves continuously at a uniform average rate through a sequence of workstations where assembly work is performed. Typical assembly lines are car assembly, washer and dryer assembly, computer assembly, toy assembly.

A diagrammatic sketch of assembly line of car is shown in Fig. 11.1.

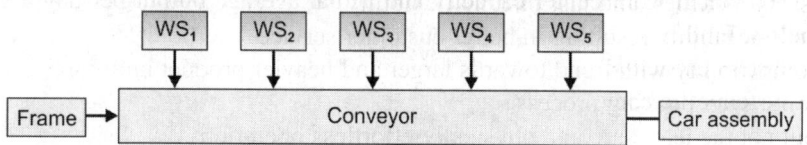

Fig. 11.1 Assembly line of car

Objective of the line balancing problem is to group the tasks at workstations so that the total time required at each workstation is approximately same. In a perfectly balanced assembly line, total time taken by all operations assigned to each workstations is equal. Then efficiency of the assembly line is 100%. However, in most practical situations it is very difficult to achieve perfect balance. When the workstation times are unequal, the slowest station determines the overall production rate of the line. The slow operations are often called the bottleneck operations, since they restrict the flow of parts in a line.

Production line balancing: Production line balancing is the main problem in designing a production line in product type layout. Let us assume that there is a production line consists of machine x, machine y, and machine z to produce a product (Fig. 11.2). First operation is performed on machine x, second operation is performed on machine y and third or final operation is performed on machine z. Machine x, machine y and machine z can produce 200 units, 100 units and 50 units per hour respectively. Number of units per hour at the end of the line is equal to 50. For every cycle machine z idle time is zero, idle time for the machine y is 50% of time and idle time for the machine x is 75%.

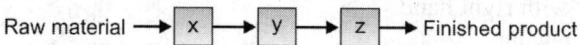

Fig. 11.2 Unbalanced production line

The unbalanced line (Fig. 11.2) is balanced by adding 3 more machines of type *z* and 1 more machine of type *y*. The balanced line is shown in Fig. 11.3. This is an example of balancing production line by adding facilities.

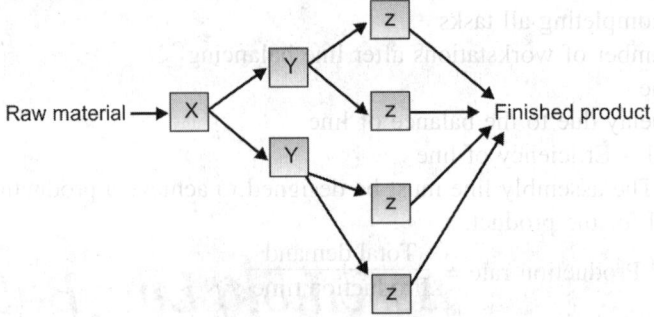

Fig. 11.3 Balanced production line

If any production line remains unbalanced, machinery utilization may be poor. In most practical situations, it is very difficult to achieve perfect balance. Generally production line is balanced by adding or deleting production facilities.

11.1.1 Solved Problem

Problem 11.1: A company produces a product sequentially through four centers A,B ,C and D. The individual work center capacity and actual average output per day are shown in Fig. 11.4 below. Find:

    a. The system capacity

    b. The system efficiency

Solution:

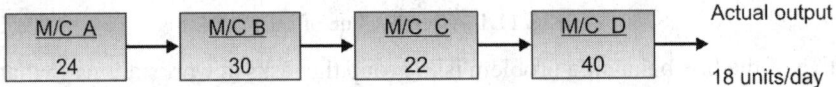

Fig. 11.4 The individual work center capacity and actual average output per day

a. System capacity= Capacity of the most limited machine in the line = 22 units/day

b. System efficiency = $\dfrac{\text{Actul output}}{\text{System capacity}} = \dfrac{18}{22} = 0.82 = 82\%$

11.2 TERMS USED IN ASSEMBLY LINE

Line: An assembly line composed of several workstations, at which specific operations are performed.

Work element or Task: It is the smallest task into which the job can be divided.

    Example: In bolt and nut assembly

    Tasks are:

    • Picking the bolt with left hand

    • Picking the nut with right hand

    • Assembly or screwing

Workstation: A workstation is a position along the assembly line where one or more assembly tasks are performed by one or more workers.

Line balance efficiency:
$$\text{Efficiency of line} = \frac{\Sigma t_i}{nc}$$
Where
 Σt_i = Time of completing all tasks
 n = Actual number of workstations after line balancing
 c = Cycle time
Balance delay: Delay due to the balance of line
Balance delay = 1 – Efficiency of line
Production rate: The assembly line must be designed to achieve a production rate sufficient to satisfy demand for the product.
$$\text{Production rate} = \frac{\text{Total demand}}{\text{Production time}}$$
Cycle time: The time between two consecutive units coming out of the production line.
$$\text{Cycle time} = \frac{\text{Production time}}{\text{Total demand}}$$
Minimum cycle time: Longest task time

11.2.1 Solved Problems

Problem 11.2: If the demand is 480 units for 8 hours. Determine production rate.
Solution:
$$\text{Production rate} = \frac{\text{Total demand}}{\text{Production time}}$$
$$\text{Production rate} = \frac{480}{8} = 60 \text{ units/hour}$$

Problem 11.3: If the demand is 480 units for 8 hours. Determine cycle time.
Solution:
$$\text{Cycle time} = \frac{\text{Production time}}{\text{Total demand}} = \frac{8 \times 60}{480} = 1 \text{ minute}$$

11.3 LINE BALANCING ALGORITHMS

The objective of line balancing algorithms is to distribute the total workload on the assembly line as evenly as possible among the workstations.

There are several methods to solve the line balancing problem. The important methods are:
 • Ranked positional weight technique
 • COMSOAL (Computer Method for Sequencing Operations for Assembly Lines)
 • Largest candidate rule
 • Kilbridge-Wester method

All these methods are heuristic. Heuristic methods are based on common sense and experimentation rather than mathematical optimization.

11.3.1 Ranked Positional Weight Technique (RPW)

Ranked positional weight technique generates feasible solutions using the following steps:

Step 1: Draw the precedence diagram and determine positional weight for each task.
Positional weight for task i = Time to perform task i + Total time of all the tasks which follow the task i

Step 2: Arrange the tasks in descending order as per weight.

Step 3: Determine the cycle time.

$$\text{Cycle time} = \frac{\text{Production time}}{\text{Total demand}}$$

Step 4: Determine minimum number of workstations.

$$\text{Minimum number of workstations} = \frac{\text{Sum of the times of all tasks}}{\text{Cycle time}}$$

Step 5: $j = 1$

Select the task with the highest positional weight and assign to workstation j.

Step 6: Remove the assigned task from the list. Calculate unassigned cycle time.

Unassigned cycle time for the workstation j = Cycle time – (Sum of the times of all the tasks which are assigned to workstation j)

Step 7: Select the task with next highest positional weight and assign to the workstation j, if the conditions 'a' and 'b' are satisfied. Go to step 6. Otherwise repeat step 7. If no task in the list satisfies 'a' and 'b' then $j = j + 1$ and go to step 8.

Conditions:

    a. Precedence relationship should be satisfied.

    b. Task time $\leq$ Unassigned cycle time

Step 8: Unassigned cycle time for the workstation j = Cycle time

Go to step 7.

Step 9: Continue steps 6 to 8 until all the tasks have been assigned

Step 10: Determine efficiency of line.

$$\text{Efficiency of line} = \frac{\Sigma t_i}{nc}$$

Σt_i = Time of completing all tasks

n = Actual number of workstations after line balancing

c = Cycle time

Step 11: Determine Balance delay.

Balance delay = 1 – Efficiency of line

11.3.2 Solved Problems

Problem 11.4: Assembly of a job should be done as per the information given in Fig. 11.5 and the cycle time is 45 seconds. Calculate number of workstations and balance delay using RPW method.

Solution:

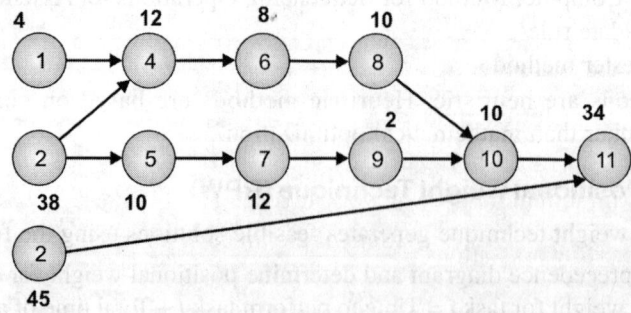

Fig. 11.5 Precedence diagram

Step 1: Determine positional weight for each task.

Positional weight for task i = Time to perform task i + Total time of all the tasks which follow the task i

Weight for task 1 = Time to perform task 1 + Total time of all the tasks which follow the task 1

$$= 4 + (12 + 8 + 10 + 10 + 34) = 78$$

Weight for task 2 = $38 + (12 + 8 + 10 + 10 + 12 + 2 + 10 + 34) = 136$

Weight for task 10 = $10 + (34) = 44$

Similarly, weight of the remaining tasks are calculated and is given in Table 11.1.

Table 11.1 Positional weight of each task

Task number	Time (seconds)	Positional weight
1	4	78
2	38	136
3	45	79
4	12	74
5	10	68
6	8	62
7	12	58
8	10	54
9	2	46
10	10	44
11	34	34
	$\Sigma t_i = 185$	

Step 2: Arrange the tasks in descending order as per positional weight.

Table 11.2 Tasks in descending order as per positional weight

Task number	Time (seconds)
2	38
3	45
1	4
4	12
5	10
6	8
7	12
8	10
9	2
10	10
11	34

Step 3: Determine the cycle time.

$$\text{Cycle time} = \frac{\text{Production time}}{\text{Total demand}}$$

Given cycle time = 45 seconds

Step 4: Determine minimum number of workstations.

$$\text{Minimum number of workstations} = \frac{\text{Sum of the times of all tasks}}{\text{Cycle time}} = \frac{\Sigma t_i}{45} = \frac{185}{45} = 4.1 = 5$$

Step 5: $j = 1$. Select the task with the highest positional weight and assign to workstation j (Table 11.3).

Table 11.3 Assigning the tasks to the workstations

Workstation 1			Workstation 2			Workstation 3			Workstation 4			Workstation 5		
Task	Time (sec)	UAC (sec)	Task	Time (sec)	UAC (sec)	Task	Time (sec)	UAC (sec)	Task	Time (sec)	UAC (sec)	Task	Time (sec)	UAC (sec)
2	38	7	3	45	0	4	12	33	8	10	35	11	34	0
1	4	3				5	10	23	10	10	25			
						6	8	15						
						7	12	3						
						9	2	1						

*UAC = Unassigned cycle time

Step 6: Remove the assigned task (i.e. task 2) from the Table 11.2. Calculate unassigned cycle time.

Unassigned cycle time for the workstation j = Cycle time – (Sum of the times of all the tasks which are assigned to workstation j)

Unassigned cycle time for the workstation 1 = 45 – 38 = 7

Step 7: Select the task with next highest positional weight (i.e. task 1) from the Table 11.2. Conditions 'a' and 'b' are satisfied. Assign the task 1 to the workstation. Go to step 6.

Unassigned cycle time for the workstation 1 = 45 – 42 = 3

Select the task with next highest positional weight from the Table 11.2. No task in Table 11.2 satisfies conditions 'a' and 'b'.

So $j = j + 1 = 1 + 1 = 2$.

Step 8: Now Unassigned cycle time for the workstation 2 = 45

Step 9: Continue steps 6 to 8 until all the tasks have been assigned (Table 11.3).

From Table 11.3,

actual number of workstations = n = 5

Step 10: Determine efficiency of line.

$$\text{Efficiency of line} = \frac{\Sigma t_i}{nc} = \frac{185}{45 \times 5} = 0.822$$

Σt_i = Time of completing all tasks

n = Actual number of workstations after line balancing

c = Cycle time

Step 11: Determine Balance delay

Balance delay = 1 – Efficiency of line = 1 – 0.822 = 0.177

Problem 11.5: Toy assembly should be done as per the information given in Table 11.4 and the cycle time is 1 minute.

Table 11.4 Given information

Element	1	2	3	4	5	6	7	8	9	10
Time (minutes)	0.5	0.3	0.8	0.2	0.1	0.6	0.4	0.5	0.3	0.6
Immediate predecessors	–	1	1	2	2	3	4,5	3,5	7,8	6,9

i. Construct precedence diagram.

ii. Calculate number of workstations and balance delay using RPW method.

Solution:

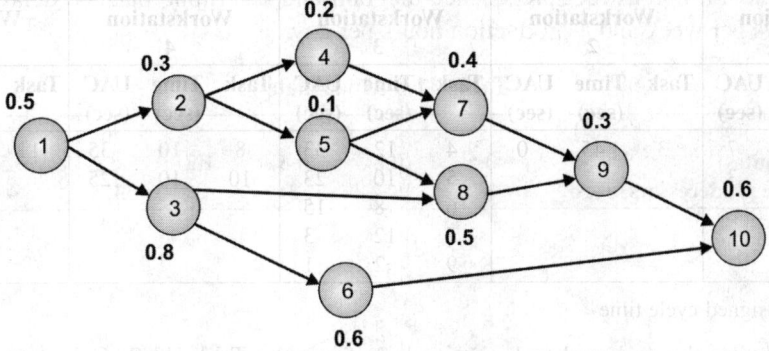

Fig. 11.6 Precedence diagram

Step 1: Determine positional weight for each task.

Table 11.5 Positional weight for each task

Element	1	2	3	4	5	6	7	8	9	10
Time (minutes)	0.5	0.3	0.8	0.2	0.1	0.6	0.4	0.5	0.3	0.6
Positional weight	4.3	2.4	2.8	1.5	1.9	1.2	1.3	1.4	0.9	0.6

Step 2: Arrange the tasks in descending order as per positional weight.

Table 11.6 Tasks in descending order as per positional weight

Element	1	3	2	5	4	8	1	6	9	10
Time (minutes)	0.5	0.8	0.3	0.1	0.2	0.5	0.4	0.6	0.3	0.6

Step 3: Given cycle time = 1 minute

Step 4:

$$\text{Minimum number of workstations} = \frac{\text{Sum of the times of all tasks}}{\text{Cycle time}} = \frac{\Sigma t_i}{1} = \frac{4.3}{1} = 4.3 \approx 5$$

Steps 5 to 11:

Table 11.7 Assigning the tasks to the workstations

Workstation 1			Workstation 2			Workstation 3			Workstation 4			Workstation 5		
Task	Time (min)	UAC (min)	Task	Time (min)	UAC (min)	Task	Time (min)	UAC (min)	Task	Time (min)	UAC (min)	Task	Time (min)	UAC (min)
1	0.5	0.5	3	0.8	0.2	8	0.5	0.5	6	0.6	0.4	10	0.6	0.4
2	0.3	0.2	4	0.2	0	7	0.4	0.1	9	0.3	0.1			
5	0.1	0.1												

*UAC = Unassigned cycle time

From Table 11.7,

Actual number of workstations = $n = 5$

Efficiency of line $= \dfrac{\Sigma t_i}{nc} = \dfrac{4.3}{5 \times 1} = 0.86$

Balance delay = 1 – Efficiency of line = 1 – 0.86 = 0.14

Problem 11.6: The demand of the assembly line is 1,600 units/week. As per information given in Table 11.8, construct precedence diagram and determine balance delay. Assume 5 working days per week and 8 production hours per day.

Table 11.8 Given data

Element	1	2	3	4	5	6	7	8
Time (minutes)	1	0.5	0.8	0.3	1.2	0.2	0.5	1.5
Immediate predecessors	–	–	1, 2	2	3	3, 4	4	5, 6, 7

Solution:

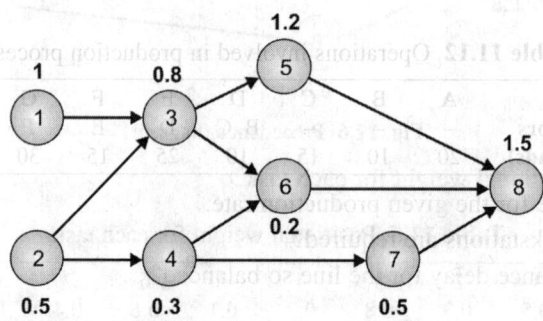

Fig. 11.7 Precedence diagram

Step 1: Determine positional weight for each task.

Table 11.9 Positional weight for each task

Element	1	2	3	4	5	6	7	8
Time (minutes)	1	0.5	0.8	0.3	1.2	0.2	0.5	1.5
Positional weight	4.7	5	3.7	2.5	2.7	1.7	2	1.5

Step 2: Arrange the tasks in descending order as per positional weight.

Table 11.10 Tasks in descending order as per weight

Element	2	1	3	5	4	7	6	8
Time (minutes)	0.5	1	0.8	1.2	0.3	0.5	0.2	1.5

Step 3:
$$\text{Cycle time} = \frac{\text{Production time}}{\text{Total demand}} = \frac{8 \times 5 \times 60}{1600} = 1.5 \text{ minutes}$$

Step 4:
$$\text{Minimum number of workstations} = \frac{\text{Sum of the times of all tasks}}{\text{Cycle time}} = \frac{\Sigma t_i}{1.5} = \frac{6}{1.5} = 4$$

Step 5 to 11

Table 11.11 Assigning the asks to the workstations

Workstation 1			Workstation 2			Workstation 3			Workstation 4			Workstation 5		
Task	Time (min)	UAC (min)	Task	Time (min)	UAC (min)	Task	Time (min)	UAC (min)	Task	Time (min)	UAC (min)	Task	Time (min)	UAC (min)
2	0.5	1	3	0.8	0.7	5	1.2	0.3	7	0.5	1	8	1.5	0
1	1	0	4	0.3	0.4									
			6	0.2	0.2									

*UAC = Unassigned cycle time

From Table 11.11,

Actual number of workstations = $n = 5$

Efficiency of line $= \dfrac{\Sigma t_i}{nc} = \dfrac{6}{5 \times 1.5} = 0.8$

Balance delay = 1 – Efficiency of line = 1 – 0.8 = 0.2

Problem 11.7: A shoe exporter produces 70 pairs of shoes per hour on its production line. The operations involved in the production process are listed in Table 11.12 along with the sequencing requirements.

Table 11.12 Operations involved in production process

Task	A	B	C	D	E	F	G	H	I	J
Immediate predecessors	–	–	–	B, C	D	E	F	A, G	H	I
Estimated time (seconds)	20	10	15	10	25	15	30	30	20	25

i. Balance the line for the given production rate.
ii. How many workstations are required?
iii. What is the balance delay for the line so balanced?

Solution:

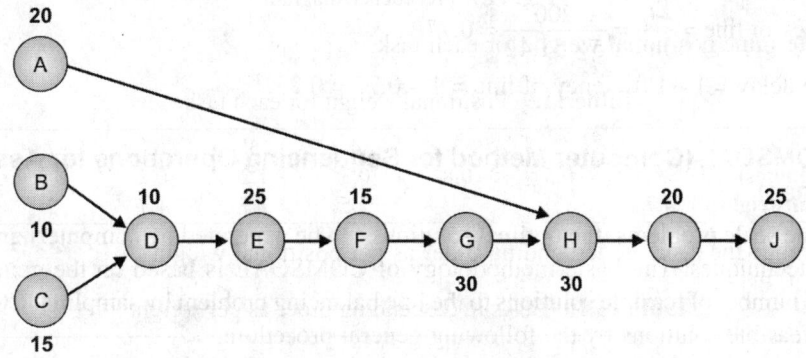

Fig. 11.8 Precedence diagram

Step 1: Determine positional weight for each task.

Table 11.13 Positional weight for each task

Task	A	B	C	D	E	F	G	H	I	J
Estimated time (seconds)	20	10	15	10	25	15	30	30	20	25
Positional weight	95	165	170	155	145	120	105	75	45	25

Step 2: Arrange the tasks in descending order as per positional weight.

Table 11.14 Tasks in descending order as per positional weight

Task	C	B	D	E	F	G	A	H	I	J
Estimated time (seconds)	15	10	10	25	15	30	20	30	20	25

Step 3:

$$\text{Cycle time} = \frac{\text{Production time}}{\text{Total demand}} = \frac{60 \times 60}{70} = 51.42 \text{ seconds}$$

Step 4:

$$\text{Minimum number of workstations} = \frac{\text{Sum of the times of all tasks}}{\text{Cycle time}} = \frac{\Sigma t_i}{51.42} = \frac{200}{51.42}$$

$$= 3.88 \approx 4$$

Steps 5 to 11:

Table 11.15 Assigning the tasks to the workstations

Workstation 1			Workstation 2			Workstation 3			Workstation 4			Workstation 5		
Task	Time (sec)	UAC (sec)	Task	Time (sec)	UAC (sec)	Task	Time (sec)	UAC (sec)	Task	Time (sec)	UAC (sec)	Task	Time (sec)	UAC (sec)
C	15	36.42	E	25	26.42	G	30	21.42	H	30	21.42	J	25	26.42
B	10	26.42	F	15	11.42	A	20	1.42	I	20	1.42			
D	10	16.42												

*UAC = Unassigned cycle time

From Table 11.15,

Actual number of workstations = $n = 5$

$$\text{Efficiency of line} = \frac{\Sigma t_i}{nc} = \frac{200}{5 \times 51.42} = 0.77$$

Balance delay = 1 – Efficiency of line = 1 – 0.77 = 0.23

11.3.3 COMSOAL (Computer Method for Sequencing Operations for Assembly Lines)

For the large scale problems, the optimal solutions can be generated by computerizing the line balancing techniques. The basic methodology of COMSOAL is based on the generation of fairly large number of feasible solutions to the line balancing problem by sampling. COMSOAL generates feasible solutions by the following general procedure.

COMSOAL generates feasible solutions by using the following procedure:

Step 1: Form Table A. Table A tabulates the total number of tasks that immediately precede each task.

Step 2: Scan Table A and identify all the tasks that have no preceding tasks and place them in Table B.

Step 3: Transfer the tasks in Table B to Table C whose task time is less than or equal to unassigned cycle time.

Step 4: Assignment is now made to the workstation 1 by selecting the task at random from Table C.

Step 5: Eliminate the assigned task from the list.

Step 6: Determine unassigned cycle time.

Step 7: Repeat the steps 1 to 6 until all the elements are assigned.

11.3.4 Solved Problem

Problem 11.8: Assembly line balance should be done as per the information given in Fig. 11.9 and the cycle time is 10 minutes. Solve the problem using COMSOAL.

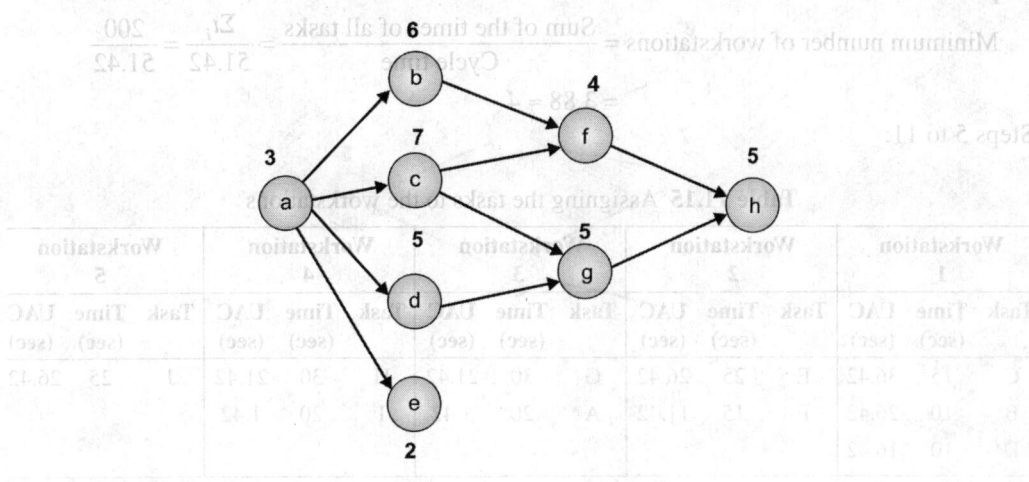

Fig. 11.9 Precedence diagram

Solution:

Iteration 1:

Step 1: Form Table A. Table A tabulates the total number of tasks that immediately precede each task.

Table A

Task	a	b	c	d	e	f	g	h
Time	3	6	7	5	2	4	5	5
Immediately preceding tasks	0	1	1	1	1	2	2	2

Step 2: Scan Table A and identify all the tasks that have no preceding tasks and place them in Table B.

Table B

Task	Time
a	3

Step 3: Transfer the tasks in Table B to Table C, whose task time is less than or equal to unassigned cycle time.

Table C

Task	Time
a	3

Step 4: Assignment is now made from Table 11.16 to the workstation 1 by selecting the task at random from Table C.

Step 5: Eliminate the assigned task 'a' from Fig. 11.9.

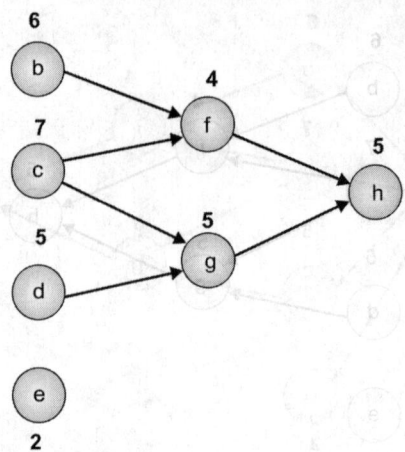

Fig. 11.9 (a) Figure after removing task 'a'

Step 6: Determine unassigned cycle time.

Unassigned cycle time = 10 – 3 = 7 minutes

Step 7: Repeat the steps 1 to 6 until all the elements are assigned.

Iteration 2:

Table A

Task	b	c	d	e	f	g	h
Time	6	7	5	2	4	5	5
Immediately preceding tasks	0	0	0	0	2	2	2

Table B

Task	b	c	d	e
Time	6	7	5	2
Immediately preceding tasks	0	0	0	0

Table C

Task	b	c	d	e
Time	6	7	5	2

Assign the task 'c' to the workstation 1.

Unassigned cycle time = 7 – 7 = 0

Open workstation 2.

Remove task 'c' form the Fig. 11.9a.

Iteration 3:

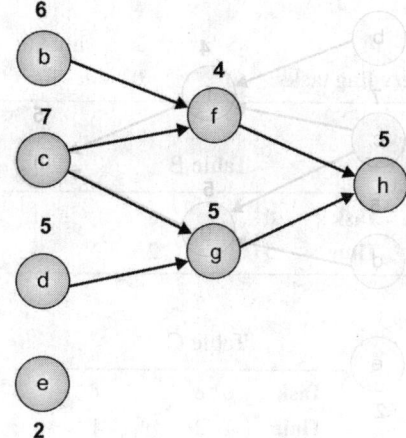

Fig. 11.9 (b) Figure after removing task 'c'

Table A

Task	b	d	e	f	g	h
Time	6	5	2	4	5	5
Immediately preceding tasks	0	0	0	1	1	2

Table B

Task	b	d	e
Time	6	5	2

Table C

Task	b	d	e
Time	6	5	2

Assign task 'b' to workstation 2.

Iteration 4:

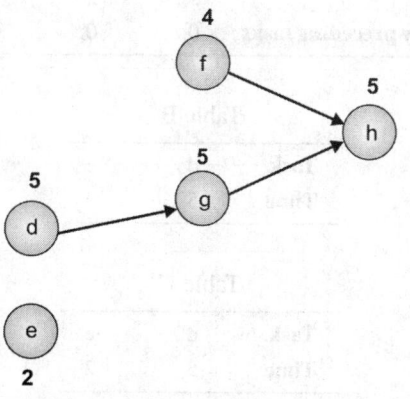

Fig. 11.9 (c) Figure after removing task 'b'

Table A

Task	d	e	f	g	h
Time	5	2	4	5	5
Immediately preceding tasks	0	0	0	1	1

Table B

Task	d	e	f
Time	5	2	4

Table C

Task	e	f
Time	2	4

Assign task 'f' to the workstation 2.

Iteration 5:

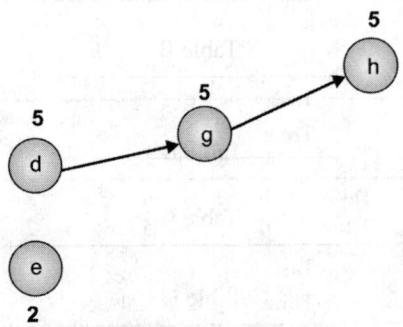

Fig. 11.9 (d) Figure after removing task 'f'

Table A

Task	d	e	g	h
Time	5	2	5	5
Immediately preceding tasks	0	0	1	1

Table B

Task	d	e
Time	5	2

Table C

Task	d	e
Time	5	2

Assign the task 'd' to the workstation 3.

Iteration 6:

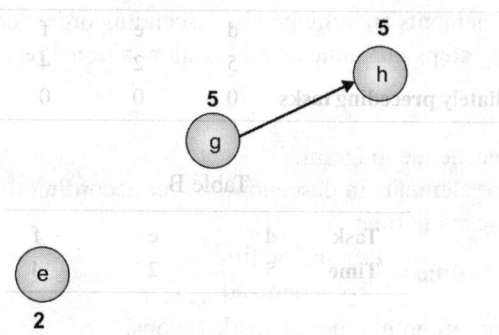

Fig. 11.9 (e) Figure after removing task 'd'

Table A

Task	e	g	h
Time	2	5	5
Immediately preceding tasks	0	0	1

Table B

Task	e	g
Time	2	5

Table C

Task	e	g
Time	2	5

Assign the task 'g' to workstation 3.

Remove the task g from Fig. 11.9e.

Assign tasks 'e' and 'h' to the workstation 4.

Table 11.16 Assigning the tasks to the workstations

Workstation 1			Workstation 2			Workstation 3			Workstation 4		
Task	Time (min)	UAC (min)	Task	Time (min)	UAC (min)	Task	Time (min)	UAC (min)	Task	Time (min)	UAC (min)
a	3	7	b	6	4	d	5	5	e	2	8
c	7	0	f	4	0	g	5	0	h	5	3

From Table 11.16,

Actual number of workstations = $n = 4$

Efficiency of line $= \dfrac{\Sigma t_i}{nc} = \dfrac{37}{4 \times 10} = 0.92$

Balance delay = 1 – Efficiency of line = 1 – 0.92 = 0.08

11.3.5 Largest Candidate Rule

In this method, work elements are arranged in descending order according to the activity times. Except this, remaining steps are same as ranked in positional technique.

Steps:

Step 1: Draw the precedence diagram.

Step 2: Arrange work elements in descending order according to the activity times.

Step 3: Determine the cycle time.

$$\text{Cycle time} = \frac{\text{Production time}}{\text{Total demand}}$$

Step 4: Determine minimum number of workstations.

$$\text{Minimum number of workstations} = \frac{\text{Sum of the times of all tasks}}{\text{Cycle time}}$$

Step 5: $j = 1$

Select the task at the top of the list and assign to the workstation j, if the conditions 'a' and 'b' are satisfied. Go to step 6. Otherwise repeat the step 5 for the next task.

a. Precedence relationship should be satisfied.

b. Task time $\leq$ Unassigned cycle time

Step 6: Remove the assigned task from the list. If all the tasks are assigned then go to step 9. Otherwise calculate unassigned cycle time.

Unassigned cycle time for the workstation j = Cycle time – (Sum of the times of all tasks which are assigned to workstation j)

Step 7: Select the task at the top of the list and assign to the workstation j, if the conditions 'a' and 'b' are satisfied. Go to step 6. Otherwise repeat the step 7 for the next task. If no task in the list satisfies conditions 'a' and 'b' then go to step 8.

Step 8: $j = j + 1$ and go to step 7.

Step 9: Determine efficiency of line.

$$\text{Efficiency of line} = \frac{\Sigma t_i}{nc}$$

Σt_i = Time of completing all tasks

n = Actual number of workstations after line balancing

c = Cycle time

Step 10: Determine Balance delay.

Balance delay = 1 – Efficiency of line

11.3.6 Solved Problem

Problem 11.9: Toy assembly should be done as per the information given in Table 11.17 and the cycle time is 1 minute.

Table 11.17 Given information

Element	1	2	3	4	5	6	7	8	9	10
Time (minutes)	0.5	0.3	0.8	0.2	0.1	0.6	0.4	0.5	0.3	0.6
Immediate predecessors	–	1	1	2	2	3	4,5	3,5	7,8	6,9

  i. Construct precedence diagram.

 ii. Calculate number of workstations and balance delay using largest candidate rule.

Solution:

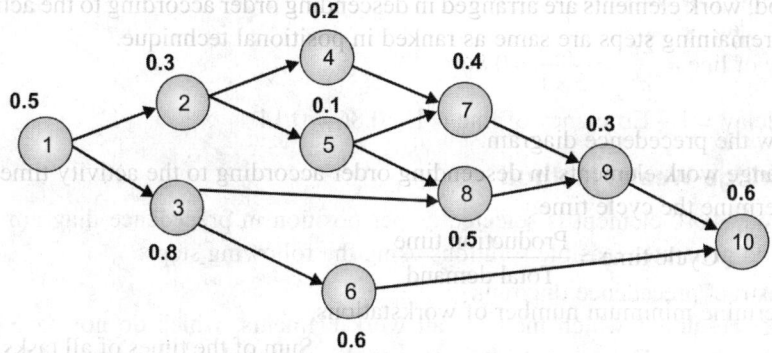

Fig. 11.10 Precedence diagram

Step 1: Draw precedence diagram (Fig. 11.10) using information from Table 11.17.

Step 2: Arrange the tasks in descending order as per task time.

Table 11.18 Tasks in descending order as per task time

Element	3	6	10	1	8	7	2	9	4	5
Time (minutes)	0.8	0.6	0.6	0.5	0.5	0.4	0.3	0.3	0.2	0.1

Step 3:

Given cycle time = 1 minute

Step 4:

$$\text{Minimum number of workstations} = \frac{\text{Sum of the times of all tasks}}{\text{Cycle time}} = \frac{\Sigma t_i}{1} = \frac{4.3}{1} = 4.3 \approx 5$$

Step 5: $j = 1$

Select the task 3 at the top of the list. Task 3 does not satisfy condition 'a'. Repeat the step 5 for the next tasks. Task 1 satisfies conditions 'a' and 'b'. Go to step 6.

Step 6: Remove the assigned task 1 from the list.

Unassigned cycle time for the workstation 1= 1 – 0.5 = 0.5.

Select the task 3 at the top of the list. Task 3 does not satisfy condition 'a'. Repeat the step 5 for the next tasks. Task 2 satisfies conditions 'a' and 'b'. Assign the task 2 to the workstation 1.

Unassigned cycle time for the workstation 1 = 0.5 – 0.3 = 0.2. Repeat the procedure until all the tasks are assigned.

Table 11.19 Assigning the tasks to the workstations

Workstation 1			Workstation 2			Workstation 3			Workstation 4			Workstation 5		
Task	Time (min)	UAC (min)	Task	Time (min)	UAC (min)	Task	Time (min)	UAC (min)	Task	Time (min)	UAC (min)	Task	Time (min)	UAC (min)
1	0.5	0.5	3	0.8	0.2	6	0.6	0.4	8	0.5	0.5	10	0.6	0.4
2	0.3	0.2	5	0.1	0.1	7	0.4	0	9	0.3	0.2			
4	0.2	0												

*UAC = Unassigned cycle time

From Table 11.19,

Actual number of workstations = n = 5

Efficiency of line = $\dfrac{\Sigma t_i}{nc} = \dfrac{4.3}{5 \times 1} = 0.86$

Balance delay = 1 − Efficiency of line = 1 − 0.86 = 0.14

11.3.7 Kilbridge-Wester Method

In this method, work element is selected as per position in precedence diagram. Kilbridge-Wester method generates feasible solutions using the following steps:

Step 1:   Construct precedence diagram.

Step 2:   Make column 1 which includes all work elements, which do not have precedence work element. Delete work elements which are included in column 1. Make column 2 in which list all elements, which follow elements in column 1. Delete work elements which are included in column 2. Continue till all work elements are exhausted.

Step 3:   Determine the cycle time.

$$\text{Cycle time} = \frac{\text{Production time}}{\text{Total demand}}$$

Step 4:   Determine minimum number of workstations.

$$\text{Minimum number of workstations} = \frac{\text{Sum of the times of all tasks}}{\text{Cycle time}}$$

Step 5:   Assign the work element to the workstations as per position in precedence diagram so that the total station time is ≤ Cycle time.

Step 6:   Repeat step 5 for unassigned work elements.

11.3.8 Solved Problem

Problem 11.10: The demand of the assembly line is 1,600 units/week. As per information given in Table 11.20, construct precedence diagram and balance delay.

Table 11.20 Given information

Element	1	2	3	4	5	6	7	8
Time (minutes)	1	0.5	0.8	0.3	1.2	0.2	0.5	1.5
Immediate predecessors	–	–	1, 2	2	3	3, 4	4	5, 6, 7

Solution:

Step 1:   Construct precedence diagram.

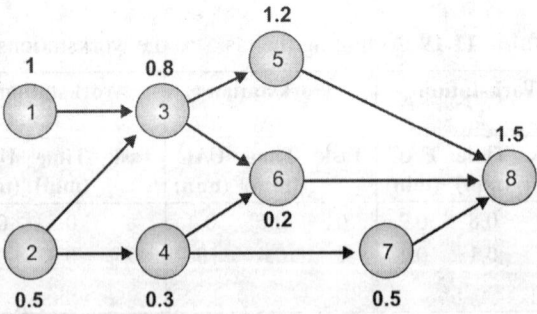

Fig. 11.11 Precedence diagram

Step 2: Make column 1 which includes work elements 1 and 2, which do not have precedence work element. Delete work elements 1 and 2.

Make column 2 which includes work elements 3 and 4, which do not have precedence work element. Delete work elements 1 and 2.

Make column 3 which includes work elements 5, 6 and 7, which do not have precedence work element. Delete work elements 5, 6 and 7.

Make column 4 which includes work element 8, which do not have precedence work element.

Resultant is shown in Fig. 11.12.

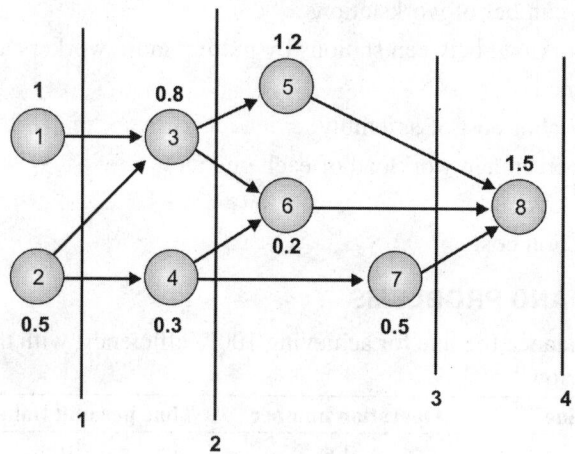

Fig. 11.12 Making columns

Step 3:

$$\text{Cycle time} = \frac{\text{Production time}}{\text{Total demand}} = \frac{8 \times 5 \times 60}{1600} = 1.5 \text{ minutes}$$

Step 4: Determine minimum number of workstations.

$$\text{Minimum number of workstations} = \frac{\text{Sum of the times of all tasks}}{\text{Cycle time}} = \frac{\Sigma t_i}{1.5} = \frac{6}{1.5} = 4$$

Step 5: Assign the work elements to the workstations as per column position so that the total station time is ≤ 1.5 minutes.

Step 6: Repeat step 5 for unassigned work element.

Table 11.21 Assigning the tasks to the workstations

Workstation 1			Workstation 2			Workstation 3			Workstation 4			Workstation 5		
Task	Time (min)	UAC (min)	Task	Time (min)	UAC (min)	Task	Time (min)	UAC (min)	Task	Time (min)	UAC (min)	Task	Time (min)	UAC (min)
1	1	0.5	3	0.8	0.7	5	1.2	0.3	7	0.5	1	8	1.5	0
2	0.5	0	4	0.3	0.4	6	0.2	0.1						

*UAC = Unassigned cycle time

From Table 11.21,

Actual number of workstations = $n = 5$

$$\text{Efficiency of line} = \frac{\Sigma t_i}{nc} = \frac{6}{5 \times 1.5} = 0.8$$

$$\text{Balance delay} = 1 - \text{Efficiency of line} = 1 - 0.8 = 0.2$$

11.4 OBJECTIVES OF LINE BALANCING

- To equalize the workload among the assemblers
- To identify the bottleneck operation
- To establish the speed of the assembly line
- To determine the number of workstations
- To balance the workload between stations by putting more workers at the slower stations
- To reduce idle time
- To determine the labor cost of assembly
- To establish the percentage workload of each operator
- To assist in plant layout
- To reduce production cost

11.5 QUESTIONS AND PROBLEMS

1. How do you balance the line for achieving 100% efficiency, with the information given in the Table below?

Machine	Operation number	Time per unit (minutes)
A	1	10
B	2	15
C	3	20

2. What is line balancing?
3. Explain the methods of line balancing.
4. Explain line balancing with an example.
5. Explain the following with respect to line balancing:
   i.   Station time
   ii. Cycle time
   iii. Line efficiency
   iv. Balance delay
6. Explain various terms associated with respect to line balancing.
7. What is the purpose of assembly line balancing?
8. Production rate is 3 units per hour. What is cycle time?   **Answer:** 20 minutes
9. A plant is capable of producing 1,60,000 good parts per year. 20% of production will have to be scrapped because of defects. What is the required system capacity in parts per year?   **Answer:** 2,00,000 parts/year
10. A line balancing problem involves 10 workstations having a Σtimes = 24 (where the shortest is 2.1 minutes and the longest is 3minutes). Assuming one worker is located at each station and using the longest time as the cycle time, what would be the balance delay.   **Answer:** Efficiency of line = 80%, Balance delay = 20%
11. A line balancing analysis resulted in a precedence grouping as shown in the Table. Find the balance efficiency, assuming the longest actual time is the cycle time.

Workstations	Work element assigned	Actual time (minutes)
A	1, 2	1.2
B	3, 5, 6	1.4
C	4, 7	0.9
D	8, 10, 11	1.3
E	9	1.5

Answer: Cycle time = 1.5 and efficiency = 84%

12. Given following data describing a line balancing problem, develop a solution allowing a cycle time of 3 minutes. What is the efficiency of a line? Use COMSOAL.

Task element	A	B	C	D	E	F	G	H	I
Time (minutes)	1	1	2	1	3	1	1	2	1
Element predecessor	–	A	B	B	C, D	A	F	G	E, H

Answer: 86.6%

13. Given the following tasks, time and sequence, develop a balanced line capable of operating with a 10 minutes cycle time, using Rank positional weight method and calculate efficiency of line.

Task element	A	B	C	D	E	F	G	H
Time (minutes)	3	5	7	5	3	3	5	6
Element predecessor	–	A	B	–	C	B, D	D	G

Answer:

One of the solutions

Workstation	1	2	3	4
Task element	A, D	B, G	C, E	F, H

Efficiency = 92.5

14. Find the number of workstations required to achieve a line balance for the following 10 workelements using positional weight method. Assume cycle time of 10 minutes.

Task element	1	2	3	4	5	6	7	8	9	10
Time (minutes)	5	10	5	2	7	5	10	2	5	7

The precedence diagram is

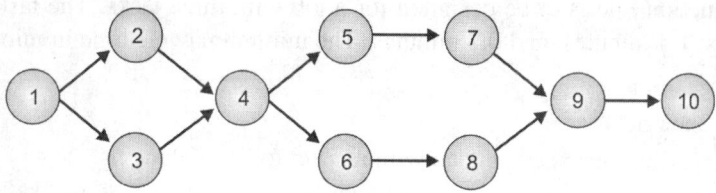

Determine the line efficiency.

Answer:

One of the solutions

Workstation	1	2	3	4	5	6	7
Task element	1, 3	2	4, 6, 8	5	7	9	10

Efficiency = 82.8%

15. The assembly line for XYZ product has the following work elements:

Task element	A	B	C	D	E	F	G	H	I	J
Time (minutes)	4	4	3	2	2	3	5	8	1	1
Element predecessor	–	–	B	A, C	D	D	E, F	E, F	G, H	I

    a. Draw the diagram for that assembly.

    b. Using Kilbridge and Wester method for a cycle time of 8 minutes, find the following:

       i. Line efficiency

       ii. Balance delay

    c. What is the monthly output at the line?

Answer:

One of the solutions

Workstation	1	2	3	4	5
Task element	A, B	C, D, F	E, G	H	I, J

Efficiency = 82.5%; Balance delay = 17.5%; Output 7.5 units/hour

16. The individual workstations in a toy production line layout have design capacities (units/day) as shown in the figure below. If the actual output of the system is 80 Toys/day, what is the system efficiency?

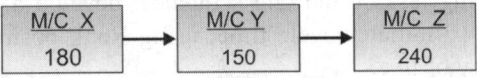

                                                                   Answer: 53%

17. A furniture manufacturing requires the time shown in figure to perform five tasks in an assembly line. Operations are to be scheduled for producing 6 units per hour and each employee can contribute 48 minutes per hour of productive work.

    a. What is the cycle time in minute per unit?

    b. What is the theoretical minimum number of personnel?

    c. Combine the task into most efficient grouping of workstations using the longest time rule. What is the resulting efficiency of balance?

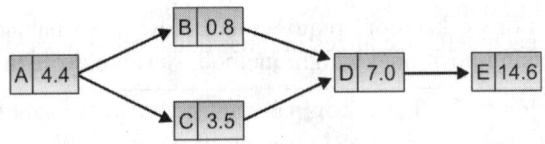

                 Answer: a. 8 minutes per unit b. 3.78 employees c. 94.7%

18. A production line is to be designed for a job with three tasks. The task times are 0.3 minutes, 1.4 minutes, and 0.7 minutes. The minimum cycle time in minutes, is:

    A. 0.3

    B. 0.7

    C. 1.4

    D. 2.4

    E. 0.8

                                          Answer: C

19. Students can answer a multiple choice question in 2 minutes but given a test with 30 questions and is given only 30 minutes. What is the minimum number of students to collaborate to answer all the questions in the exam?

    Hint: Total operation (task) time = 60 minutes = 30 × 2 minutes

    Operating time = 30 minutes

    60/3 = 2 students must collaborate.

Part IV | Manufacturing

12

Group Technology and Cellular Manufacturing

12.1 INTRODUCTION

Group technology is a manufacturing philosophy in which similar parts are identified and grouped together to take advantage of their similarities in design and production. Similar parts are arranged into part families, where each part family possesses similar design and/or manufacturing characteristics. For example, a plant producing 10,000 different part numbers may be able to group the vast majority of these parts into 30–40 distinct families. It is reasonable to believe that the processing of each member of a given family is similar, and this should result in manufacturing efficiencies. The efficiencies are generally achieved by arranging the production equipment into machine groups, or cells. Grouping the production equipment into machine cells, where each cell specializes in the production of a part family, is called cellular manufacturing.

There are two major tasks that a company must undertake when it implements group technology. These two tasks represent significant obstacles to the application of GT (Group Technology).

- Identifying the part families: Grouping the parts into families consumes a significant amount of time.
- Rearranging production machines into machine cells: It is time consuming and costly to plan and accomplish this rearrangement and the machines are not producing during the changeover.

12.2 BENEFITS OF GROUP TECHNOLOGY

The benefits include:

- GT promotes standardization of tooling, fixtures, and setups.
- Material handling is reduced because parts are moved within a machine cell rather than within the entire factory.
- Process planning and production scheduling are simplified. Setup times are reduced, resulting in lower manufacturing lead times.
- Work-in-process is reduced.
- Worker satisfaction usually improves.
- Higher quality work is accomplished using group technology.

12.3 PART FAMILY

A part family is a collection of parts that are similar either because of geometric shape and size or because similar processing steps are required in their manufacture. Figure 12.1 consists of eight parts produced in a machine shop. Grouping of these parts into three part families is shown in the Fig. 12.2.

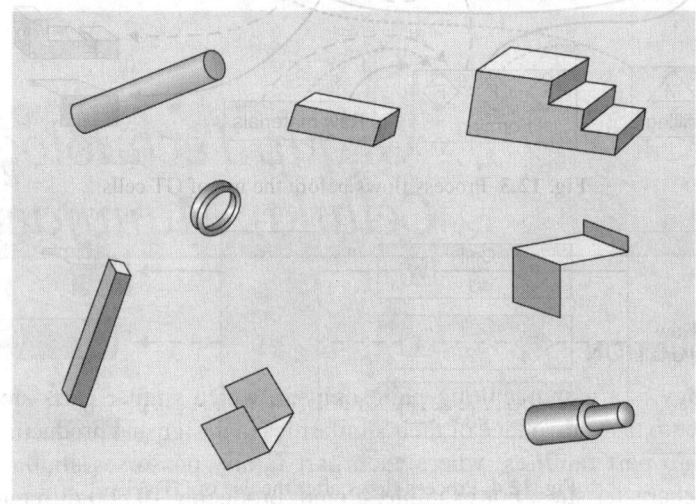

Fig. 12.1 Unorganized parts

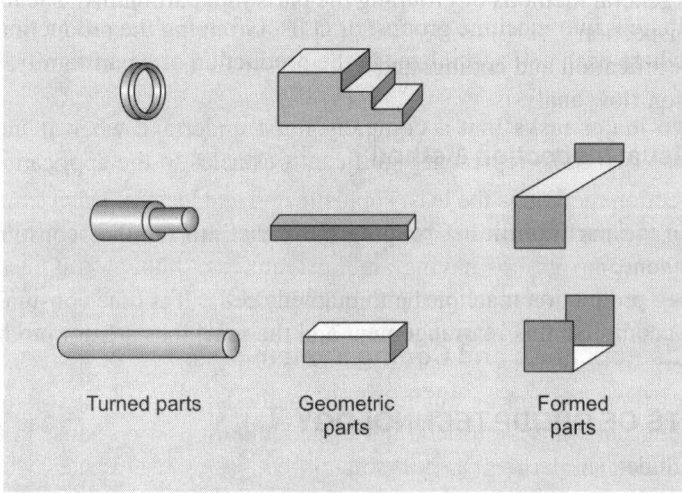

| Turned parts | Geometric parts | Formed parts |

Fig. 12.2 Parts organized into part families

12.3.1 Group Technology Layout or Cellular Layout

The efficiencies are generally achieved by arranging the production equipment into machine groups, or cells. Grouping the production equipment into machine cells, where each cell specializes in the production of a part family, is called cellular manufacturing. Figure 12.3 shows process layout before the use of GT cells. Figure 12.4 shows cellular layout where the machines are arranged into cells. Cell is capable of producing Welding parts. Cell 2 is capable of producing Turning parts. Cell 3 is capable of producing Milling parts.

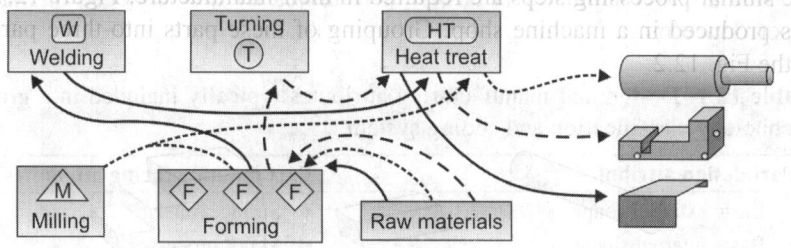

Fig. 12.3 Process flows before the use of GT cells

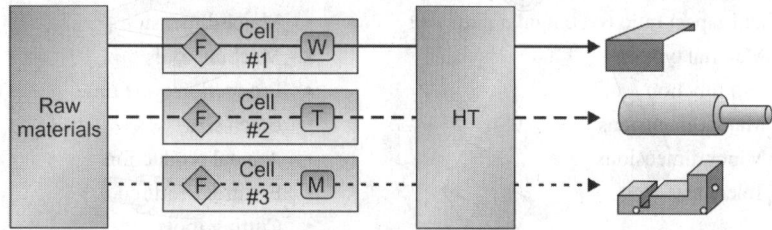

Fig. 12.4 Process flows after the use of GT cells

12.4 METHODS OF GROUPING PARTS INTO PART FAMILY

There are three general methods of grouping the parts into part family. The three methods are:
1. Visual inspection
2. Parts classification and coding, and
3. Production flow analysis

12.4.1 The Visual Inspection Method

The visual inspection method is the least sophisticated and least expensive method. It involves the classification of parts into families by looking at either the physical parts or their photographs and arranging them into groups having similar features. Although this method is generally considered to be the least accurate of the three methods.

12.4.2 Part Classification and Coding System

This is the most time consuming among three methods. In parts classification and coding, similarities among parts are identified, and these similarities are related in a coding system. Two categories of part similarities can be distinguished: 1. Design attributes, which are concerned with part characteristics such as geometry, size and material; and 2. Manufacturing attributes, which consider the sequence of processing steps required to make a part. While the design and manufacturing attributes of a part are usually correlated, the correlation is less than perfect. Accordingly, classification and coding systems are devised to include both a part's design attributes and its manufacturing attributes.

Parts classification systems fall into one of three categories:
1. Systems based on part design attributes
2. Systems based on part manufacturing attributes
3. Systems based on both design and manufacturing attributes

Table 12.1 presents a list of common design and manufacturing attributes typically included in classification schemes. A certain amount of overlap exists between design and manufacturing attributes, since a part's geometry is largely determined by the sequence of manufacturing processes performed on it.

Table 12.1 Design and manufacturing attributes typically included in a group technology classification and coding system

Part design attributes	Part manufacturing attributes
• Basic external shape	• Surface finish
• Basic internal shape	• Major processes
• Rotational or Rectangular	• Minor operations
• Length to diameter ratio (rotational parts) and aspect ratio (rectangular parts)	• Operation sequence
	• Major dimension
• Material type	• Machine tools
• Part function	• Production cycle time
• Major dimensions	• Batch size
• Minor dimensions	• Annual production
• Tolerances	• Fixtures required
	• Cutting tools

12.4.2.1 MICLASS coding and classification system

MICLASS is a 12-digit coding system based on the design attributes of parts. The general layout of MICLASS 12-digit coding system is shown in the Table 12.2. In this approach, each attribute of a part has a specific position in the code. This structure is simple to implement, but a large number of digits may be required to represent characteristics of a part.

Table 12.2 The general layout of MICLASS 12-digit coding system

Digit	1	2	3	4	5	6	7	8	9	10	11	12
Attribute of a part	Main shape	Shape elements		Position of shape elements	Main dimension		Dimen-sion ratio	Auxiliary dimension	Tolerance code		Material code	

12.4.2.2 Opitz classification system

H. Opitz of the University of Aachen in Germany developed this system. The Opitz coding scheme uses the following digit sequence:

$$12345\ 6789\ ABCD$$

The basic code consists of nine digits, which can be extended by adding four more digits. The first nine digits are intended to convey both design and manufacturing data. The first five digits, 12345, are called the form code. It describes the primary design attributes of the part, such as external shape (e.g. rotational *vs* rectangular) and machined features (e.g. holes, threads, gear teeth, etc.). The next four digits, 6789, constitute the supplementary code, which indicates some of the attributes that would be of use in manufacturing (e.g. dimensions, work material, starting shape and accuracy). The extra four digits, ABCD, are referred to as the secondary code and are intended to identify the production operation type and sequence.

Part coding using Opitz part classification and coding system is shown in Fig. 12.5.

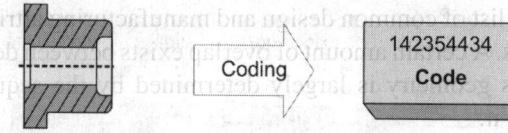

Part to be coded

Fig. 12.5 Part coding using Opitz part classification and coding system

Basic structure of Opitz System is shown in Fig. 12.6. Digits 1 through 5 of Opitz form code is given in Table 12.3.

Table 12.3 Opitz form code (Digits 1 through 5)

Digit 1			Digit 2			Digit 3			Digit 4		Digit 5	
Part class			External shape, external shape elements			Internal shape, internal shape elements			Plane surface machining		Auxiliary holes and gear teeth	
0		L/D ≤ 0.5	0		Smooth, no shape elements	0		No hole, no breakthrough	0	No surface machining	0	No auxiliary hole
1		0.5 < L/D < 3	1	Stepped to one end or smooth	No shape elements	1	Smooth or stepped to one end	No shape elements	1	Surface plane and/or curved in one direction, external	1	Axial, not on pitch circle diameter
2	Rotational parts	L/D ≥ 3	2		Thread	2		Thread	2	External plane surface related by graduation around the circle	2	Axial on pitch circle diameter
3			3		Functional groove	3		Functional groove	3	External groove and/or slot	3	Radial, not on pitch circle diameter
4			4	Stepped to both ends	No shape elements	4	Stepped to both ends	No shape elements	4	External sp line (polygon)	4	No gear teeth	Axial and/or radial and/or other direction
5			5		Thread	5		Thread	5	External plane surface and/or slot, external sp line	5	Axial and/or radial on PCD and/or other directions
6			6		Functional groove	6		Functional groove	6	Internal plane surface and/or slot	6	Spur gear teeth
7	Nonrotational parts		7	Functional cone	7	Functional cone	7	Internal sp line (polygon)	7	Bevel gear teeth
8			8	Operating thread	8	Operating thread	8	Internal and external polygons, grooves and/or slots	8	With gear teeth	Other gear teeth
9			9	All others	9	All others	9	All others	9	All others

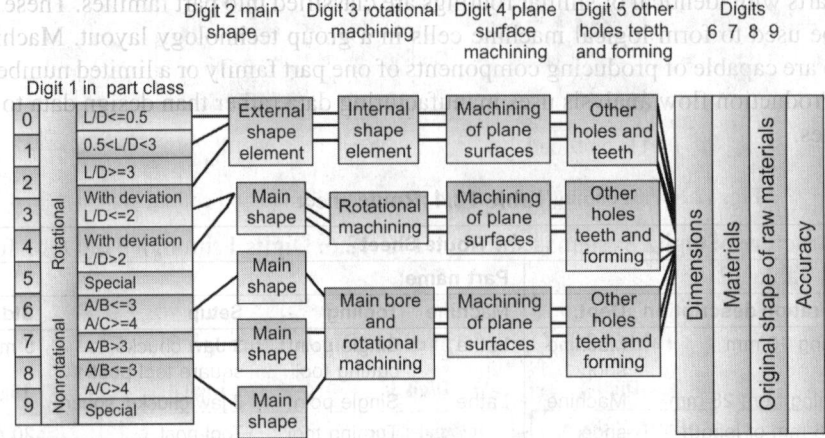

Fig. 12.6 Basic structure of Opitz system

12.4.3 Production Flow Analysis

Production flow analysis is a method for identifying part families by using the information contained in production route sheets rather than on part drawings. A typical route sheet, includes the following information: 1. Sequence of operations to be performed on the work part 2. A brief description of each operation indicating the processing to be accomplished, with references to dimensions and tolerances on the part drawing; 3. The specific machines on which the work is to be done; and 4. Any special tooling, such as dies, molds, cutting tools, jigs or fixtures, and gages. Some companies also include setup times, cycle time standards, and other data. It is called a route sheet because the processing sequence defines the route. Example of route sheet to obtain finished product (Fig. 12.8) from the raw material (Fig. 12.7) is given in Table 12.4.

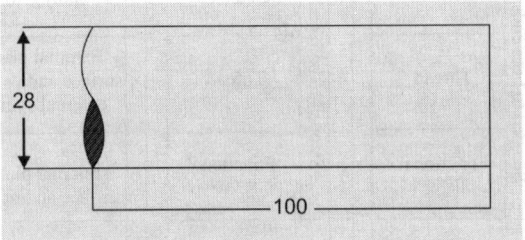

Fig. 12.7 Raw material

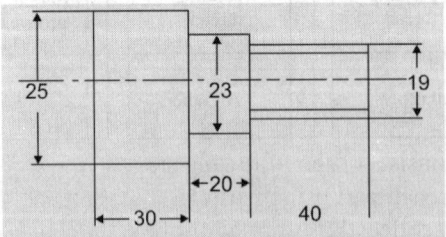

Fig. 12.8 Finished product

Work parts with identical or similar routings are classified into part families. These families can then be used to form logical machine cells in a group technology layout. Machine cells (Fig. 12.9) are capable of producing components of one part family or a limited number of part families. Production flow analysis uses manufacturing data rather than design data to identify part families.

Table 12.4 Route sheet

Route Sheet						
Part no.:			**Part name:**			
No.	**Operation description**	**Dept.**	**Machine**	**Tooling**	**Setup**	**Std. time**
1.	Facing 10 mm	Machine shop	Lathe	Single point cutting tool	3 Jaw chucks, square tool post	5 minutes
2.	Turning from 28 mm to 25 mm of length 90 mm	Machine shop	Lathe	Single point Turning tool	3 jaw chucks, square Tool post	20 minutes
3.	Turning from 25 mm to 23 mm of length 60 mm	Machine shop	Lathe	Single point Turning Tool	3 Jaw chucks, square tool post	10 minutes
4.	Turning from 23 mm to 19 mm of length 60 mm	Machine shop	Lathe	Single point Turning Tool	3 Jaw chucks, square tool post	5 minutes
5.	Threading of length 40 mm	Machine shop	Lathe	Thread cutting Tool	3 Jaw chucks, square tool post	10 minutes

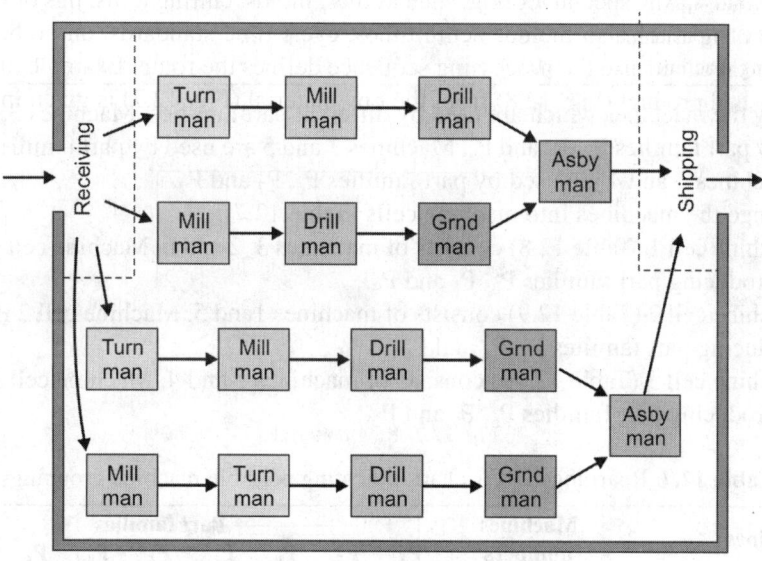

Fig. 12.9 Cellular layout based on GT

Production flow analysis consists of the following steps:
a. Develop the code for each part using part classification and coding system.
b. From the route sheets, collect the information about the sequence of operations performed on each part and machines used to perform each operation. Available machines in a machine shop are given in Table 12.5.

Table 12.5 Available machines in machine shop of a XYZ company

Name of the machines	Machine numbers
Cutting machine	01
CNC lathe	02
Turret lathe	03
CNC milling machine	04
Drilling machine	05
CNC drilling machine	06
Grinding machine	07

c. Group the parts into part families. Each part family consists of one or more parts which are similar in sequence of operations. In Table 12.6, P_1, P_2, P_3, P_4, P_5, P_6, P_7, P_8 and P_9 are part families.

d. Draw production flow analysis chart. The sequence of machines used by each part family is then displayed on PFA chart (Table 12.6). This chart is also known as part machine incidence chart.

Table 12.6 Production flow analysis chart

Machines	Machine numbers	Part families								
		P_1	P_2	P_3	P_4	P_5	P_6	P_7	P_8	P_9
Cutting machine	01	1			1				1	
CNC lathe	02						1			1
Turret lathe	03			1		1				1
CNC milling machine	04		1					1		
Drilling machine	05	1							1	
CNC drilling machine	06			1						1
Grinding machine	07		1				1	1		

e. Identify the machines which are used by different part families. Machines 3, 2 and 6 are used by part families P_3, P_5 and P_9. Machines 1 and 5 are used by part families P_1, P_4 and P_8. Machines 7 and 4 are used by part families P_6, P_7 and P_2.

f. Rearrange the machines into machine cells (Table 12.7).
  • Machine cell 1 (Table 12.8) consists of machines 3, 2 and 6. Machine cell 1 is capable of producing part families P_3, P_5 and P_9.
  • Machine cell 2 (Table 12.9) consists of machines 1 and 5. Machine cell 2 is capable of producing part families P_1, P_4 and P_8.
  • Machine cell 3 (Table 12.10) consists of machines 7 and 4. Machine cell 3 is capable of producing part families P_6, P_7 and P_2.

Table 12.7 Rearranged PFA chart indicating possible machine groupings

Machines	Machines numbers	Part families								
		P_3	P_5	P_9	P_1	P_4	P_8	P_6	P_7	P_2
Turret lathe	03	1	1	1						
CNC lathe	02		1	1						
CNC drilling machine	06	1		1						
Cutting machine	01				1	1	1			
Drilling machine	05				1		1			
Grinding machine	07							1	1	1
CNC milling machine	04							1		1

Table 12.8 Machine cell 1

Machines	Machine numbers	Part families		
		P_3	P_5	P_9
Turret lathe	03	1	1	1
CNC lathe	02		1	1
CNC drilling machine	06	1		1

Table 12.9 Machine cell 2

Machines	Machine numbers	Part families		
		P_1	P_4	P_8
Cutting machine	01	1	1	1
Drilling machine	05	1		1

Table 12.10 Machine cell 3

Machines	Machine numbers	Part families		
		P_6	P_7	P_2
Grinding machine	07	1	1	1
CNC milling machine	04	1		1

12.4.4 Solved Problem

Problem 12.1: Given the rotational part design in Fig. 12.10, determine the form code in the Opitz parts classification and coding system.

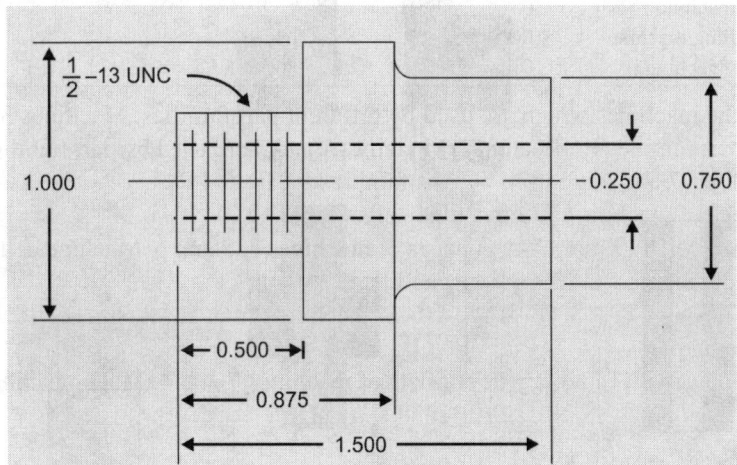

Fig. 12.10 Part design

Solution: With reference to Fig. 12.10 and using Table 12.3, the five-digit code is developed as follows:

Length-to-diameter ratio, L/D = 1.5, so Digit 1 = 1

External shape: Stepped on both ends with screw thread on one end, so Digit 2 = 5

Internal shape: Part contains a through-hole, so Digit 3 = 1

Plane surface machining: None. So Digit 4 = 0

Auxiliary holes, gear teeth, etc.: None. So Digit 5 = 0

Design code: 15100.

12.5 COMPOSITE PART

Composite part or hypothetical part for a given family includes all the design and manufacturing attributes of the family. In general, individual part in the family will have some features of composite part, but the composite parts possess all of the features. A composite part is shown in Fig. 12.11. It represents a family of rotation parts with features of individual parts of part family shown in Fig. 12.12. A machine cell to produce the composite part is also capable of producing individual parts of a part family.

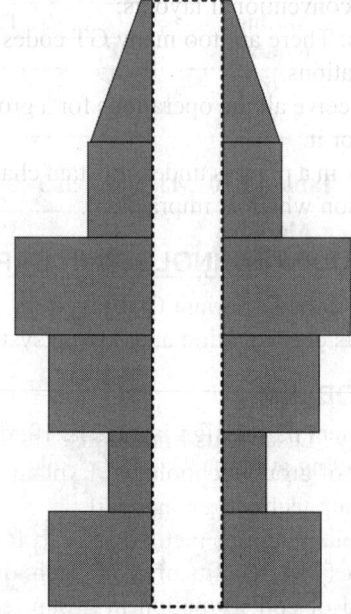

Fig. 12.11 Composite part

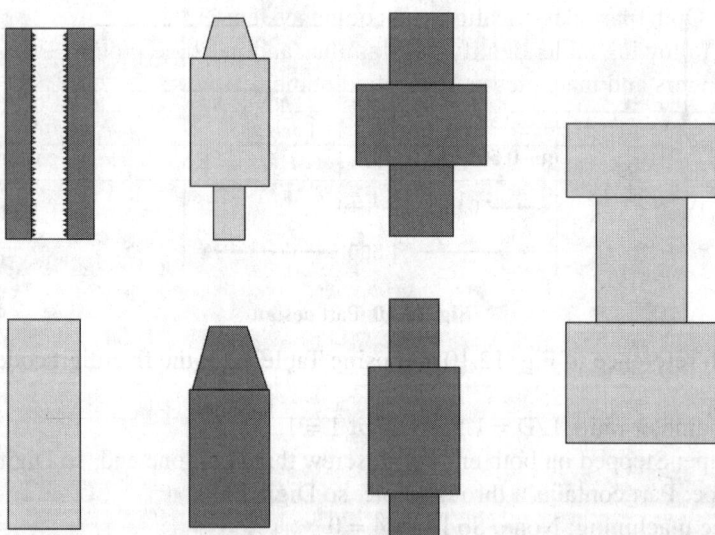

Fig. 12.12 Individual parts of a part family

12.6 LIMITATIONS OF GROUP TECHNOLOGY OR CELLULAR MANUFACTURING

Group Technology is a great concept. But all good concepts have their own limitations and need proper care in their applications for results to be realized in practice. The limitations of cellular manufacturing may be as follows:

1. High cost: The cost of implementation is generally high.
2. Not suitable for large variety of products: May not be suitable for a factory with a very large variety of products.
3. The entire production of the company cannot be put under the GT and hence GT will have to coexist with the conventional layouts.
4. Not suits all applications: There are too many GT codes in use and there is no one GT code that suits all applications.
5. It is often difficult to conceive all the operations for a group of components being taken care of the cell created for it.
6. The range of product mix in a plant is under constant change in which case the GT cells may need constant revision which is impractical.

12.7 APPLICATION OF GROUP TECHNOLOGY IN CAPP

A retrieval CAPP system is also called a variant CAPP system. It is based on the principles of group technology (GT) and parts classification and coding system.

12.8 QUESTIONS AND PROBLEM

1. What is group technology? Discuss its limitations.
2. What are the advantages of group technology? Explain any one method of coding and classification used in group technology in detail.
3. What is part family? Explain various methods of part family formation.
4. List out the characteristics and benefits of group technology.
5. Explain any one method of cell formation in group technology with the help of an example.
6. Explain MICLASS part classification and coding system.
7. Explain Opitz part classification and coding system.
8. For the following table identify part families and machine groups. Parts are identified by the letters and machines are identified numerically.

Machines	Parts								
	A	B	C	D	E	F	G	H	I
01			1	1	1				
02	1	1					1	1	1
03						1	1	1	
04	1	1		1					
05			1		1				
06		1						1	1
07	1		1	1					
08		1				1		1	1

Computer-aided Manufacturing (CAM)

13.1 DEFINITION

Computer-aided manufacturing (CAM) refers to the use of computers in converting engineering designs into finished products. Production requires the creation of process plans and production schedules, which explain how the product will be made, what resources will be required, and when and where these resources will be deployed. Production also requires the control and coordination of the necessary physical processes, equipment, materials and labor. In CAM, computers assist managers, manufacturing engineers, and production workers by automating many production tasks. CAM is often used with computer-aided design (CAD). A direct link between product design and manufacturing can be established using CAD/CAM. Product engineers use a CAD system to establish the part geometry, dimensions, and tolerances. This design data can be transferred to the CAM system where the part programers develop the CNC program to machine the part.

13.2 COMPUTERS IN MANUFACTURING

Computer-aided manufacturing can be used as a general term to describe a variety of industrial automation technologies. CAM software allows the entire production process to become automated. Nearly every aspect of the manufacturing process, with the exception of skilled computer programers and operators, could be controlled by CAM technologies.

Some common types of computer-aided manufacturing include:
- Numerical control (NC) machines
- Industrial robots
- Flexible manufacturing systems (FMS)
- Computerized facility layout
- Material requirements planning
- Manufacturing resource planning
- Computer-aided process planning (CAPP)
- Tool data management

Numeric control (NC) machines: Numeric control (NC) machines are the most common type of computer-aided manufacturing. It applies specific formulas to processing raw materials. For example, if a circle must be cut from a sheet of metal, a NC machine can determine, using mathematical calculations and numeric input, exactly where and how to cut to get a perfect circle. Additionally, using the same algorithms, the computer can determine the exact placement of cuts to produce the largest number of circles per sheet, as well as exactly how to position the metal for optimal cutting. Numerical control of machine tools is defined as a method of automation in which various functions of machine tools are controlled by letters, numbers and symbols. Basically, a NC machine runs on a part program. The part program consists of instructions written in numerical codes. These instructions are entered into the input medium. The controller translates these numerical codes into the machine actual details, which are used to control machine functions such as feed, speed, tool change, movement of axes, etc. Each of the machine axes is connected to the servo motor or a stepper motor. The movement of the cutting tool with respect to the workpiece is given in terms of coordinates. A typical numerical control system for a milling machine is given in Fig. 13.1.

In NC machine tools, one or more of the following functions may be automatic:
- Changing of tools in the spindle
- Starting and stopping of machine tool spindle
- Controlling the spindle speed
- Positioning the tool tip at desired locations and guiding it along desired paths by automatic control of the motion of slides
- Controlling the rate of movement

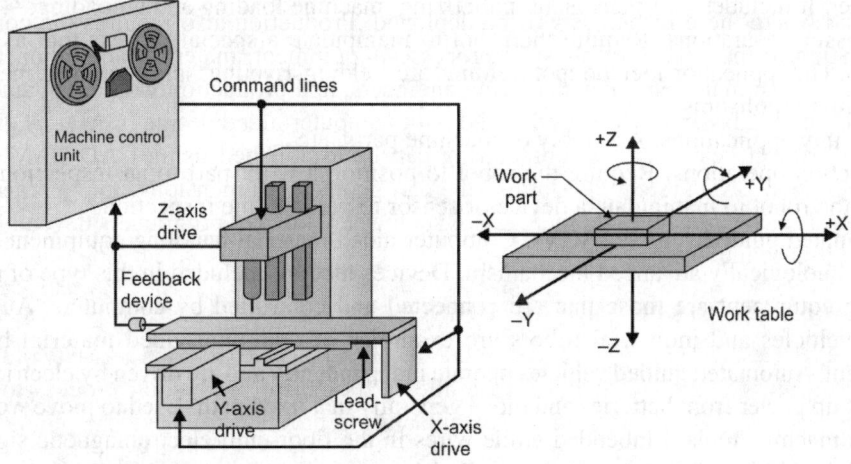

Fig. 13.1 A typical numerical control system for a milling machine

The applications of CNC include both for machine tool as well as non-machine tool areas. In the machine tool category, CNC is widely used for lathe, drill press, milling machine, grinding unit, laser, sheet-metal press working machine, tube bending machine, etc. Highly automated machine tools such as turning center and machining center which change the cutting tools automatically under CNC control have been developed. In the non-machine tool category, CNC applications include welding machines (arc and resistance), coordinate measuring machine, electronic assembly, tape laying and filament winding machines for composites, etc. The benefits of CNC are: 1. High accuracy in manufacturing, 2. Short production time, 3. Greater manu-facturing flexibility, 4. Simple fixtures, 5. Contour machining (2 to 5 axis machining), 6. Reduces human error. The drawbacks include high cost, maintenance, and the requirement of skilled part programer.

Industrial robots: Industrial robots are another example of computer-aided manufacturing. Robot is a computerized mechanical device that performs a series of complicated and complex tasks automatically. The robot can be programmed to perform sequence of mechanical motions, and it can repeat that motion sequence over and over until it is reprogrammed to perform some other job. Industrial robots are robots that are used in an industrial manufacturing environment. These are articulated arms specifically developed for such applications as welding, material handling, painting and others. Industrial robots are found in a variety of locations including the automobile and manufacturing industries. Programers and other computer experts thus replace human workers, who now run the computer system rather than performing the manufacturing tasks. An industrial robot is shown in Fig. 13.2.

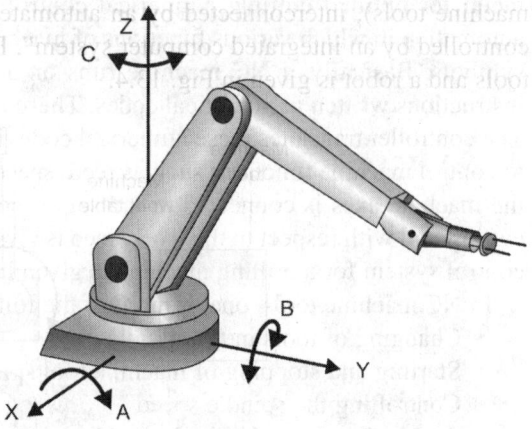

Fig. 13.2 Industrial robot

Industrial robot applications can be divided into:

Material handling applications: Involve the movement of material or parts from one location to another. It includes part placement, palletizing, machine loading and unloading.

Processing operations: Require the robot to manipulate a special process tool as the end effectors. The application include spot welding, arc welding, riveting, spray painting, machining, metal cutting, polishing.

Assembly applications: Assembly of machine parts, etc.

Inspection operations: Require the robot to position a work part to an inspection device. Involve the robot to manipulate a device or sensor to perform the inspection.

Automated guided vehicle (AGV): Computer-aided material handling equipment features more technologically advanced mechanism. Devices that are included in this type of material handling equipment are those that are connected and controlled by computers. Automated guided vehicles and industrial robots are examples of computer-aided material handling equipment. Automated guided vehicles operate independently and are driven by electric motors that pick up power from batteries and move generally in a fixed path. Used to move work parts between machine tools. Embedded guide wires in the floor emit electromagnetic signal that the vehicles follow. AGV can be controlled by central computers or on-board computer control. Automated guided vehicle is shown in Fig. 13.3.

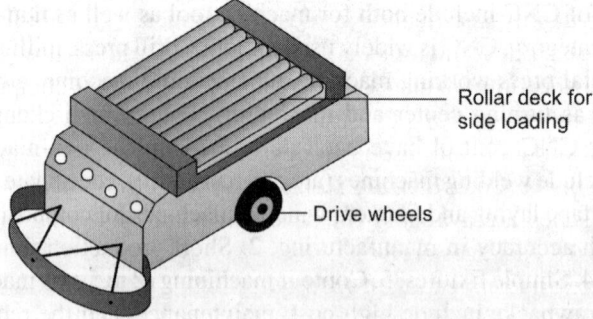

Rollar deck for side loading

Drive wheels

Fig. 13.3 Automated guided vehicle

Flexible manufacturing systems: Flexible systems can produce several similar products with the same equipment, assisted by CAM software. Flexible manufacturing system is "A highly automated GT machine cell, consisting of a group of processing stations (usually CNC machine tools), interconnected by an automated material handling and storage system, and controlled by an integrated computer system". Flexible manufacturing cell with two machine tools and a robot is given in Fig. 13.4.

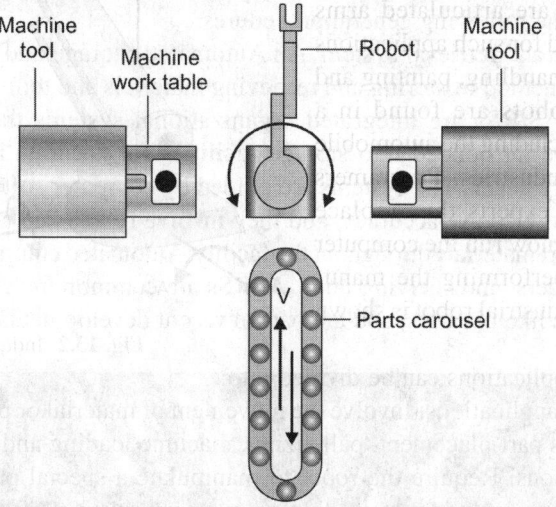

Fig. 13.4 Flexible manufacturing cell with two machine tools and robot

Applications of flexible manufacturing systems:
- Sheet metal press working
- Forging and assembly
- Welding
- Assembly of parts
- Manufacturing of parts
- Fabrication of sheet metal
- Textile machinery manufacture

Computerized facility layout: Computerized facility layout is a plan of an optimum arrangement of personnel, operating equipment, storage space, material handling equipment and all other supporting services along with the design of the best structure to contain all these facilities. The overall objective of plant layout is to design a physical arrangement that meets the required output quantity and quality in most economical way. In all manufacturing and service industries the layout of facilities is an important determinant of operating efficiency and costs. Whenever the flow of materials or people is complex, computerized procedures offer only the feasible means of developing and evaluating alternative arrangements. Computerized layout programs are useful in devising good processing layouts. The original program is CRAFT, or the Computerized Relative Allocation of Facility Technique. Several computerized approaches are available for developing and analyzing process layouts. The important computerized packages are:
- Computerized relative allocation of facility technique (CRAFT)
- Computerized relationship layout planning (CORELAP)
- Automated layout design program (ALDEP)

COMSOAL (Computer method for sequencing operations for assembly lines): The assembly line is a production line where the material moves continuously at a uniform average rate through a sequence of workstations where assembly work is performed. Typical assembly lines are car assembly, washer and dryer assembly, computer assembly, toy assembly. For the large scale problems, the optimal solutions can be generated by computerizing the line balancing techniques. The basic methodology of COMSOAL is based on the generation of fairly large number of feasible solutions to the line balancing problem by sampling. COMSOAL generates feasible solution by the following general procedures.

Automated storage and retrieval system: An Automated Storage and Retrieval System is a computer controlled method of storing and retrieving materials and tools using racks, bins and stackers. ASRSs or AS/RSs are integrated organizational systems that are partly or fully automated and can be introduced into storage facilities for accurate and efficient material handling and inventory needs. The systems are guided by computer software that monitors all on-hand inventories to increase accuracy, and they involve robots and automated pickers that can locate and transport materials throughout the facility. Automated equipment includes pickers, robotic transport, loaders, and shelving units. ASRSs are common in various kinds of storage facilities and archives like libraries, and are part of recent developments in material handling.

Fig. 13.5 Automated storage and retrieval system

Computer-integrated warehousing (CIW) is an organizational principle that incorporates computer management of inventory in storage facilities. Using radio frequency identifying (RFID) barcodes and radio frequency data terminals (RFDT), the larger ASRS is managed by a computer. The RFID barcodes allow close to perfect inventory statistics and the RFDT allow workers to access the system at multiple points in a facility. Each time an item's RFID barcode is scanned, computer records are updated as to the item's location, stock number and other data. This system is called Real Time Warehouse Control (RTWC). The RTWC system integrates with the ASRS to optimize and streamline the entire material handling process in a warehouse.

Material requirements planning: Material requirements planning is a computational technique that converts master production schedule for the end product into detail purchasing and manufacturing schedules of raw materials and components used in end product. Each end product may contain hundreds of individual components. These components are produced from raw material or purchased from the supplier. Some of these components are common for several end products. These components are assembled into simple assemblies and these subassemblies are put together into more complex subassemblies and so on until final products are assembled. Each step in manufacturing, purchasing and assembly takes time. All these factors must be incorporated into MRP calculation. Each calculation is simple. But the magnitude of data is so large that the manual application of MRP is practically impossible. That is why MRP system is computerized. Some of the vendors of MRP systems were computer manufacturers such as IBM, Honeywell, Digital Equipment, and Siemens. If a manufacturing company is looking for better plan stock resources, machine scheduling, and labor loads, material requirements planning (MRP) software can help you improve these tasks and lower operational costs. One of the most commonly required modules for manufacturers is an MRP software application. Manufacturers use material requirements planning to calculate material requirements, coordinate personnel and machine workloads, and optimally planned for purchasing to meet customer demands. For manufacturers looking to shorten delivery order times and eliminate inefficiency in the production process, MRP software is a critical technology asset.

Manufacturing resource planning (MRP II): Manufacturing resource planning II is a computer modeling technique for planning, analyzing and controlling manufacturing operations. MRP II includes software for many areas of manufacturing such as purchasing, inventory, material requirements planning, shop floor scheduling, capacity planning, customer order entry and accounting. Advantages are standardization and automation of business processes leading to improvements in cost control and revenue. The goal of MRP II is to provide consistent data to all players (Fig. 13.6) in the manufacturing process as the product moves through the production line. MRP II includes feedback from the shop floor on how the work has progressed, to all levels of the schedule so that the next run can be updated on a regular basis. For this reason, it is sometimes called 'Closed Loop MRP'. It is a family of high performance business modules integrated into a single solution sharing critical data across the network of a manufacturing company.

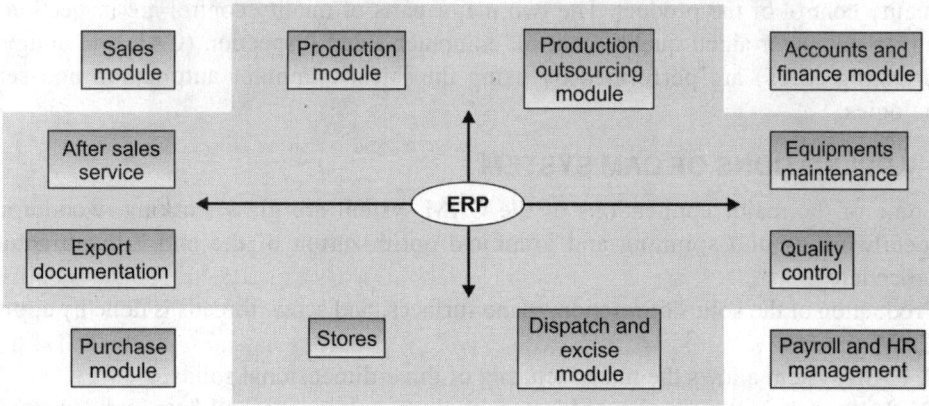

Fig. 13.6 Enterprise resource planning

E-manufacturing: E-manufacturing is sharing real-time data with trading partners and customers and making decisions about production, through internet.

Computer-aided process planning (CAPP) or Automated process planning: To convert the product design into physical entity, manufacturing plan is needed. The activity of developing such plan is called process planning.

| Design | Process planning | Manufacturing |

Fig. 13.7 Computer-aided process planning

Process planning involves in determining the most appropriate manufacturing and assembly processes and the sequence in which they should be accomplished to produce a given part or product according to the specifications set forth in the product design documentation. The scope and variety of process plans are generally limited by the available processing equipment of the company or plant. The initial form of the workpiece may be casting, forging, etc. The process planner has to prepare a list of processes after selecting raw workpiece to convert it into a predetermined final shape. If different planners were asked to develop a process plan for the same part, they would probably come up with different plans. Use of computers in process planning is usually called computer-aided process planning (CAPP) or automated process planning. Computer-aided process planning is a means of automatically develop the process plan from geometric image of the component. CAPP is usually considered to be the part of computer-aided manufacturing (CAM).

Tool data management: Tool Data Management involves in keeping track of your cutting tool's offsets—length and diameter, as well as managing all sorts of other aspects. It can include software for offline tool setting, crib inventory, wear tracking, and so on. Used properly, TDM can ensure your tools are set up right for the jobs, help you evaluate the performance of the tools, reduce setup times, and provide you with a lot of insights into your tooling inventory you never had before.

Computer-aided quality control: Computer-aided quality control makes use of computers for quality control of the product. The two major parts of quality control are inspection and testing. In computer-aided quality control, computer-aided inspection (CAI) and computer-aided testing (CAT) are performed by using the latest computer automation and sensor technologies.

13.3 APPLICATIONS OF CAM SYSTEM

- Some of the major applications of the CAM system are glass working, woodturning, metalworking and spinning and graphical optimization of the entire manufacturing procedure.
- Production of the solids of rotation, plane surfaces, and screw threads is done by applying CAM systems.
- A CAM system allows the manufacturing of three-dimensional solids.
- Products such as candlestick holders, table legs, bowls, baseball bats, crankshafts, and cam shafts can be manufactured using the CAM system.
- CAM system can also be applied to the process of diamond turning to manufacture diamond tipped cutting materials.
- Aspheric optical elements made from glass, crystals, and other metals can also be produced using CAM systems.
- Computer-aided manufacturing can be applied to the fields of mechanical, electrical, industrial and aerospace engineerings.

- Applications such as thermodynamics, fluid dynamics, solid mechanics, and kinematics can be controlled using CAM systems.
- Other applications such as electromagnetism, ergonomics, aerodynamics, and propulsion and material science may also use computer-aided manufacturing.

13.4 CAM SOFTWARE

CAD/CAM software is expensive, and each package comes with hundreds of features and add-ons. CAM Software is used to generate machining programs corresponding to part designs. If the workpiece geometry is complex and the machined features are numerous, then generally CAM software is essential for generating the program both accurately and quickly. In general, CAM development has proceeded along two different tracks. Geometry-based CAM systems used in die/mold machining and other applications focus on complex milled geometries. Algorithms provide specific and efficient ways for machining complex geometries with both high precision and long tool life. By contrast, feature-based CAM systems are generally used in the production machining of geometrically simpler components. The CAM software includes face milling, planar milling, Z-level milling, pocketing, generalized roughing (including high speed machining capabilities), hole making and feature-based machining.

13.5 CAM SOLUTIONS FOR MACHINING PROCESSES

CAM solutions are for NC programers involved in milling, drilling, turning, punching, wire EDM, laser and plasma cutting. In combination with CAD, CAM provides fully associative, integrated tools for product design and NC. CAM uses the same CAD geometry to generate tool paths to ensure the part you machined is the same part you have modeled. Tool path and machine simulation and verification in CAM help manufacturing engineers quickly improve NC program quality and machine efficiency. CAM also includes post processing, tool editors and industry-specific machining solutions that automate interrelated tasks in the manufacturing process.

Lot of companies are providing software for CAD/CAM/CAE solutions worldwide. Each software has its own importance and specialization. Some details of the widely used software are as below.

Mastercam: Mastercam consists of set of predefined tool paths. They include contour, drill, pocketing, face, peel mill, engraving, surface high speed, and many more enable machinists to cut parts efficiently and accurately. Mastercam users can create and cut parts using one of many supplied machine and control definitions or they can use Mastercam's advanced tools to create their own customized definitions.

Gibbs cam: Gibbs cam offers extremely powerful solutions for programing CNC machine tools.

SIEMENS NX (CAD/CAM/CAE): NX is the commercial CAD/CAM/CAE PLM software suite developed by Siemens PLM Software. NX is widely used in the engineering industry, especially in the automotive and aerospace sectors. NX has some presence in the consumer goods design sector. NX is a parametric solid/surface feature-based modeler. It uses the parasolid geometric modeling kernel.

Edgecam: Edgecam is a market leading computer-aided manufacturing (CAM) system. Edgecam is capable of programing, milling, turning and mill-turn machines. Edgecam generates sophisticated tool path generation

WorkNC: WorkNC is a Computer-aided manufacturing (CAM) software developed by Sescoi for 2, 2.5, 3, 3+2 and 5-axis machining.

Pro/ENGINEER: Pro/ENGINEER is a parametric, integrated 3D CAD/CAM/CAE solution created by Parametric Technology Corporation (PTC). It was the first to market with parametric, feature-based, associative solid modeling software on the market. The application runs on Microsoft Windows and Unix platforms, and provides solid modeling, assembly modeling and drafting, finite element analysis, and NC and tooling functionality for mechanical engineers.

CATIA (Computer-aided three-dimensional interactive application): CATIA is a multi-platform CAD/CAM/CAE commercial software suite developed by the French company Dassault Systemes. CATIA is written in C++ programing language.

BobCAD-CAM: BobCAD-CAM is a developer of CAD/CAM software for the manufacturing industry. They have grown to be one of the most popular and most affordable CAD/CAM solutions because of their web-based direct sales approach.

SolidWorks: SolidWorks is one of the leading software packages in the CAD/CAM industry. It's powerful, fully developed and has a number of features.

Types of CAM software are virtually unlimited. Major CAM Softwares are given below:

Softwares	Companies
NX (Unigraphics)	Siemens PLM Software
Mastercam	CNC Software
CATIA	Dassault Systèmes
Cimatron	Cimatron group
Edgecam	Planit (Pathtrace)
Powermill	Delcam
T-FLEX	Top Systems Ltd.
Solid CAM	Solid CAM
Pro/Tool Maker	PTC
Space-E/CAM	NDES (Hitachi Zosen)
Tebis	Tebis AG
WorkNC	Sescoi
CAMWorks	Geometric Tech.
ESPRIT	DP Technology
GibbsCAM	Cimatron Group
MazaCAM	Yamazaki Mazak Corp.
OneCNC	QARM Pty. Ltd.
SUM3D	CIM Solutions
SurfCAM	Surfware Inc.
TopSolid	Missler Software

13.6 BENEFITS OF CAM

- Production efficiency via increased production speeds, raw material consistency and more precise tool accuracy.
- CAD/CAM can improve productivity, product quality, and profitability.
- Computers can eliminate redundant design and production tasks, improve the efficiency of workers, increase the utilization of equipment.
- Reduces inventories, waste, and scrap.
- Decreases the time required to design and make a product.
- Improves ability of the factory to produce different products.
- Reduces engineering persons requirements.
- Improves accuracy of product.
- Reduces training time for routine drafting tasks and NC part programing.
- Fewer errors in NC part programing.
- Saves material and machining time by optimizing algorithm.

- Provides operational results on the status of work in progress.
- Assistance in inspection of complicated parts.
- Better communication interfaces and greater understanding among engineers, designers, drafters, management and different project groups.

13.7 COMPUTER-AIDED ENGINEERING (CAE)

Engineers also use computer programs to estimate the performance and cost of design prototypes and to calculate the optimal values for design parameters. These programs supplement and extend traditional hand calculations and physical tests. When combined with CAD, these automated analysis and optimization capabilities are called computer-aided engineering (CAE).

13.8 QUESTIONS

1. Define CAM.
2. Explain role of computers in manufacturing.
3. Explain material requirements planning.
4. Explain MRP II.
5. Define flexible manufacturing system.
6. Explain role of indusial robot in manufacturing.
7. What is automated guided vehicle?
8. What do you meant by e-manufacturing?
9. What is an automated storage and retrieval system?
10. What is computer-aided process planning?
11. What is CAM software used in CAM?
12. What is computer-aided engineering?
13. What are the benefits of CAM?
14. What are the applications of CAM?

14

Industrial Robots

14.1 INTRODUCTION

Robot is a system that contains sensors, control systems, manipulators, power supplies and softwares, all working together to perform a task. Designing, building, programing and testing a robot is a combination of physics, mechanical engineering, electrical engineering, structural engineering, mathematics and computing. In some cases, biology, medicine, chemistry might also be involved. A study of robotics means study of all of these disciplines. Robot is basically a computerized mechanical device that performs a series of complicated and complex tasks automatically. The robot can be programed to perform sequence of mechanical motions, and it can repeat that motion sequence over and over until it is reprogramed to perform some other job.

14.2 CHARACTERISTICS OF ROBOT

- **Sensing:** A robot is required to sense its surroundings using light sensors (eyes), touch and pressure sensors (hands), chemical sensors (nose), hearing and sonar sensors (ears), and taste sensors (tongue).
- **Movement:** A robot is required to move around its environment.
- **Energy:** A robot is required to power itself. A robot might be solar powered, electrically powered, or battery powered. The way your robot gets its energy will depend on what your robot needs to do.
- **Intelligence:** A robot needs some kind of intelligence. This is where programing comes into the pictures. The robot will have to have some way to receive the program so that it knows what it is to do.

14.3 INDUSTRIAL ROBOTS

Industrial robots are robots used in an industrial manufacturing environment. These are articulated arms specifically developed for such applications as welding, material handling, painting and others. Industrial robots are found in a variety of locations including the automobile and manufacturing industries.

14.4 ANATOMY OF ROBOT OR ELEMENTS OF ROBOT SYSTEM

A Robot is a system, consists of the following elements, which are integrated to form a whole:

1. **Axis/Axes:** An axis is a line across which a rotating body turns. Two axes are required to reach any point in a straight plane, while three axes (X, Y, and Z) are needed to reach any point in space. Three further axes (roll, pitch and yaw) are needed to control the orientation of the end of the robot arm or wrist. A robot with three axes is shown in the Fig. 14.1.

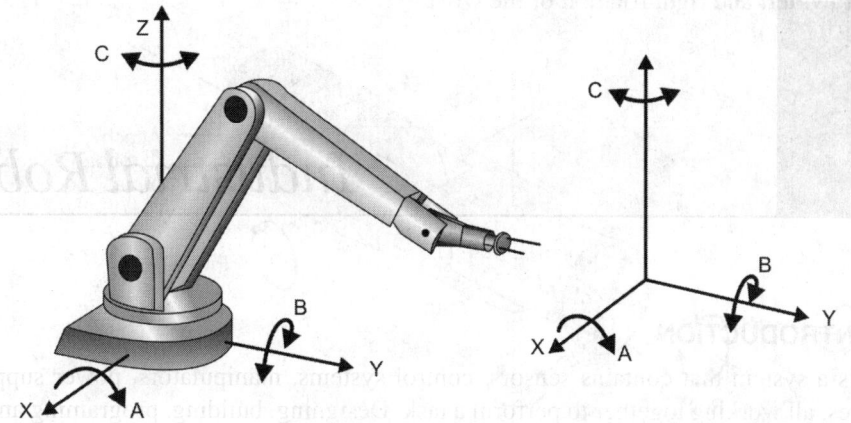

Fig. 14.1 Space coordinates

2. **Manipulators:** This is the main body of the robot. Robot manipulators are created from a sequence of link and joint combinations. The links are the rigid members connecting the joints, or axes. The axes are the movable components of the robotic manipulator that cause relative motion between adjoining links. Robotic manipulators can be divided into two sections, each with a different function:

   - **Arm and body**: The arm and body of a robot are used to move and position parts or tools within a work envelope. They are formed from the joints connected by large links.
   - **Wrist:** The wrist is used to orient the parts or tools at the work location. It consists of two or three compact joints.

   Robot manipulator is shown in Fig. 14.2.

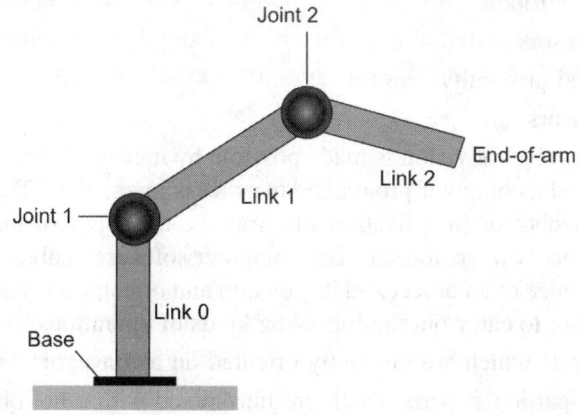

Fig. 14.2 Robot manipulator

3. **End effector:** This is the part that generally handles objects, makes connection to other machines, or performs the required tasks. End effectors include grippers, vacuum cups, spray guns, welding tools and electro magnetic pick-ups, their performance being vital to precision and repeatability. End effectors are mounted on the wrist. Wrist has 3 degrees of freedom (Fig. 14.3).

   Roll: Rotating the wrist about the arm axis

   Pitch: Up and down rotation of the wrist

   Yaw: left and right rotation of the wrist

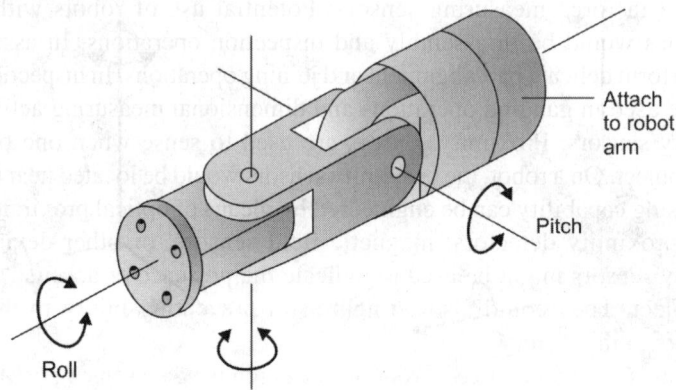

Fig. 14.3 Wrist motions

4. **Joints:** Robot joints are described as either rotational or translational. Rotational joints have a rotary action along the joint axis and are also referred to as revolute. Translational joints have a linear or sliding motion along the joint axis and are also known as prismatic.

5. **Actuators:** Actuators are also known as drives. Drives are devices that convert electrical, hydraulic and pneumatic energy into robot motion. Common types of actuators are servomotors, stepper motors, pneumatic cylinders, etc.

6. **Sensors:** Sensors are used to collect information about the internal state of the robot or to communicate with the outside environment. Robots are often equipped with external sensory devices such as a vision system, touch and tactile sensors, etc. which help to communicate with the environment.

   Sensors used in robots:

   i. Vision sensors

   ii. Tactile and proximity sensors

   iii. Voice sensors

   Vision sensors: Robot vision is made possible by means of video camera, a sufficient light source and a computer programed to process image data. The camera is mounted either on the robot or in a fixed position above the robot so that its field of vision includes the robots work volume. The computer software enables the vision system to sense the presence of an object and its position and orientation. Vision capability would enable the robot to carry out the following kinds of operations.

   • Retrieve parts which are randomly oriented on a conveyor

   • Recognize particular parts which are intermixed with other objects

   • Perform assembly operations which require alignment

Tactile and proximity sensor: Tactile sensors provide the robot with the capability to respond to contact forces between itself and other objects within its work volume. Tactile sensors can be divided into two types:

• Touch sensors

• Stress sensors

Touch sensors are used simply to indicate whether contact has been made with an object. A simple micro switch can serve the purpose of a touch sensor. Stress sensors are used to measure the magnitude of the contact force. Strain gauge devices are typically employed in force measuring sensors. Potential use of robots with tactile sensing capabilities would be in assembly and inspection operations. In assembly, the robot could perform delicate part alignment and joining operations. In inspection, touch sensing would be used in gauging operations and dimensional measuring activities.

Proximity sensors: Proximity sensors are used to sense when one object is close to another object. On a robot, the proximity sensors would be located near the end effectors. This sensing capability can be engineered by means of optical proximity devices, eddy-current proximity detectors, magnetic field sensors, or other devices. In robotics, proximity sensors might be used to indicate the presence or absence of a work part or other object. They could also be helpful in preventing injury to the robots human coworkers in the factory.

Voice sensors: Another area of robotics research is voice sensing or voice programing. Voice programing can be defined as the oral communication of commands to the robot or other machine. The robot controller is equipped with a speech recognition system which analyzes the voice input and compares it with a set of stored word patterns. When a match is found between the input and the stored vocabulary word the robot performs some actions which corresponds to the word. Voice sensors could be useful in robot programing to speed up the programing procedure just as it does in NC programing. It would also be beneficial in especially in hazardous working environments for performing unique operations such as maintenance and repair work. The robot could be placed in hazardous environment and remotely commanded to perform the repair chores by means of step by step instructions.

7. **Controller:** Controller controls the motions. The controller receives its data from the computer, controls the motions of actuator and coordinates motions with the sensory feedback information. Suppose that in order for the robot to pick up a part from the bin it is necessary that its first joint be at 35°. If the joint is not already at this magnitude, the controller will send signal to the actuator causing to move. It will measure the change in the joint angle through the feedback sensor attached to the joint (a potentiometer). When the joint reaches to the desired value the signal is stopped. In more sophisticated robots, the velocity and the force exerted by the robot are also controlled by the controller. Control is of two types. They are:

• Point-to-point control

• Continuous path control

Point-to-point control

• Only the end points are programed, and the path used to connect the end points are computed by the controller.

• User can control velocity, and may permit linear or piecewise linear motion.

- Feedback control is used during motion to ascertain that individual joints have achieved desired location.

Continuous path control

- In addition to the control over the end points, the path taken by the end effectors can be controlled.
- Path is controlled by manipulating the joints throughout the entire motion, via closed loop control.
- Applications: Spray painting, polishing, grinding, and arc welding.

14.5 DEGREES OF FREEDOM (DOF)

Degree of freedom is the number of independent movements the end effecter can make along the axes of its coordinate system. For example, movement along the X, Y and Z coordinates only constitutes three degrees of freedom, while adding rotation around the Z-axis is four degrees of freedom. Thus, an industrial robot has a maximum of six degrees of freedom.

14.6 KINEMATICS

Kinematics is the study of motion in robotics.

14.7 WORK ENVELOPE OR WORK VOLUME

This is the total volume of space that the end effectors of the manipulator can reach and is also known as workspace and work volume. The size and shape is determined by the robot kinematics and the number of degrees of freedom; it should be large enough to accommodate all the points the end effector needs to reach. Depending on the configuration and size of the links and wrist joints, robots can reach a collection of points called a workspace.

14.8 OPERATING AND PERFORMANCE PARAMETERS

- Payload (kg): The maximum load, including weight of the end effector
- Reach (mm): The maximum a robot can extend its arm
- Speed (mm/second): How fast it can position its end effector or rotate an axis (deg/second)
- Acceleration (mm/second): Defines how quickly an axis can accelerate to top speed
- Accuracy (± mm): How closely a robot can move to a specified place in the work envelope
- Repeatability (± mm): How precisely a robot can return repeatedly to a given position
- Cycle Time (seconds): Cumulative time for completing one full set of process operations

14.9 TYPES OF ROBOTS

Types of robots are:

- Cartesian
- Cylindrical
- Spherical
- SCARA
- Articulated arm

Cartesian robot/Gantry robot: A robot which is constructed around this configuration consists of three orthogonal slides, as shown in Fig. 14.4. The three slides are parallel to the *x*, *y*, and *z* axes of the cartesian coordinate system. By appropriate movement of these slides, the robot is capable of moving its arm at any point within its three dimensional rectangular spaced workspace. For a cartesian robot (like an overhead crane), the workspace is a prism as shown in Fig. 14.4.

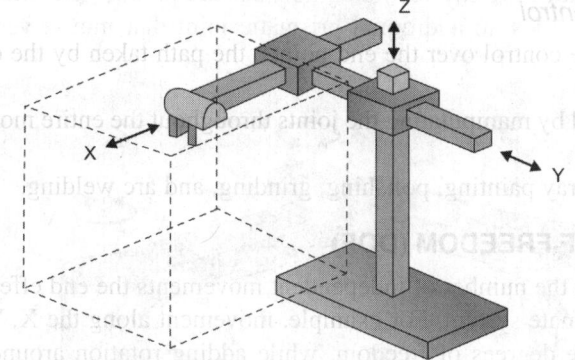

Fig. 14.4 Cartesian robot

Cylindrical coordinate robots: In this configuration, the robot body is a vertical column that swivels about a vertical axis. The arm consists of several orthogonal slides which allow the arm to be moved up or down and in and out with respect to the body. This is illustrated schematically in Fig. 14.5. For a cylindrical robot (like an overhead crane), the workspace is a cylinder as shown in Fig. 14.5.

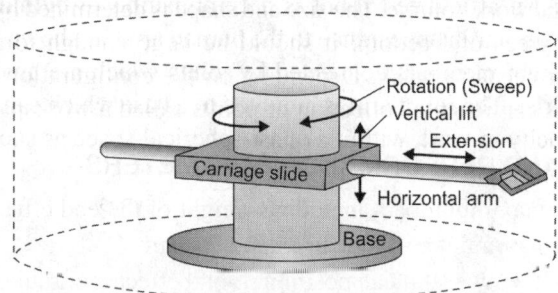

Fig. 14.5 Cylindrical coordinate robot

Spherical/Polar robot: This configuration also goes by the name "spherical coordinate" because the workspace within which it can move its arm is a partial sphere as shown in Fig. 14.6. The

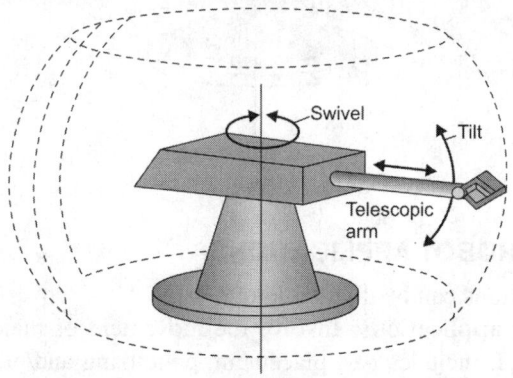

Fig. 14.6 Spherical/Polar robot

robot has a rotary base and a pivot that can be used to raise and lower a telescoping arm. For a spherical robot, the workspace is a sphere as shown in Fig. 14.6.

SCARA robot: They have two revolute joints that are parallel and allow the robot to move in a horizontal plane, plus an additional prismatic joint that moves vertically as shown in Fig. 14.7.

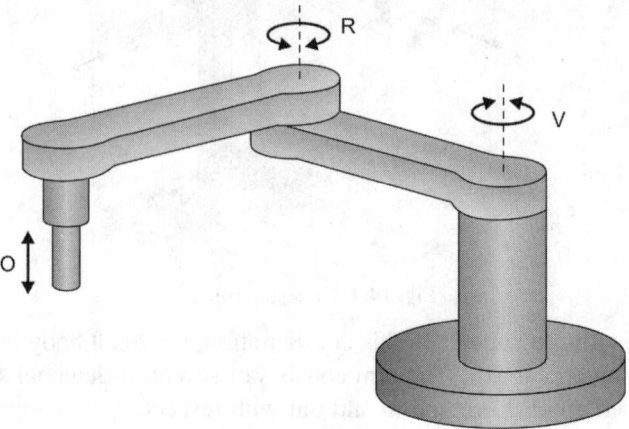

Fig. 14.7 SCARA robot

Articulated robot: An articulated robot is a combination of cylindrical and articulated configurations. This is similar in appearance to the human arm, as shown in Fig. 14.8. The arm consists of several straight members connected by joints which are analogous to the human shoulder, elbow, and wrist. The robot arm is mounted to a base which can be rotated to provide the robot with the capacity to work within a quasi-spherical space as shown in Fig. 14.8.

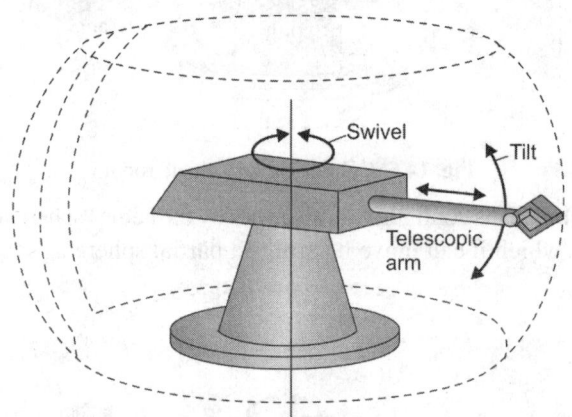

Fig. 14.8 Articulated robot

14.10 INDUSTRIAL ROBOT APPLICATIONS

Industrial robot applications can be divided into:

- **Material-handling applications:** Involve the movement of material or parts from one location to another. It includes part placement, palletizing and/or depalletizing, machine loading and unloading.

- **Processing operations:** Require the robot to manipulate a special process tool as the end effectors. The application includes spot welding, arc welding, riveting, spray painting, machining, metal cutting, deburring, polishing, etc.
- **Assembly applications:** Assembly of machine parts, etc.
- Inspection operations

14.11 ROBOT PROGRAMING

Programing is accomplished by:
- setting desired sequence of moves.
- adjusting end stops for each axis accordingly.
- the sequence of moves is controlled by a "sequencer", which uses feedback received from the end steps to index to next step in the program.

Typically performed using one of the following:
- Lead through programing
- Robot programing languages
- Simulation and offline programing

Lead through programing: Work cycle is taught to robot by moving the manipulator through the required motion cycle and simultaneously entering the program into controller memory for later playback.

Robot programing languages: Use textual programing language to enter commands into robot controller

Simulation and offline programing: Program is prepared at a remote computer terminal and downloaded to robot controller for execution without need for lead through methods. In conventional usage, robot programing languages still require some production time to be lost in order to define points in the workspace that are referenced in the program. They therefore involve offline programing. Advantage of true offline programing is that the program can be prepared beforehand and downloaded to the controller with no lost production time. Graphical simulation is used to construct a 3-D model of the robot cell in which locations of the equipment in the cell have been defined previously.

14.12 ADVANTAGES OF ROBOTS

- Robotics increase productivity, safety, efficiency, quality, and consistency of products.
- Robots can work in hazardous environments.
- Robots need no environmental comfort.
- Robots work continuously without any humanity needs and illnesses.
- Robots have repeatable precision at all times.
- Robots can be much more accurate than humans; they may have micro inch accuracy.
- Robots and their sensors can have capabilities beyond that of humans.
- Robots can process multiple stimuli or tasks simultaneously, humans can only one.
- Robots replace human workers who can create economic problems.

14.13 DISADVANTAGES OF ROBOTS

- Robots lack capability to respond in emergencies.
- Robots may have limited capabilities in:
  i. Degrees of freedom
  ii. Sensors

   iii. Vision systems
   iv. Real-time response
- Robots are costly, due to:
   i. Initial cost of equipment
   ii. Installation costs
   iii. Need for peripherals
   iv. Need for training
   v. Need for programing

14.14 ROBOT ACCURACY AND REPEATABILITY

Three terms used to define precision in robotics, similar to numerical control precision:
1. Control resolution: Capability of robot's positioning system to divide the motion range of each joint into closely spaced points.
2. Accuracy: Capability to position the robot's wrist at a desired location in the work space, given the limits of the robot's control resolution.
3. Repeatability: Capability to position the wrist at a previously taught point in the work-space.

14.15 QUESTIONS

1. What are the disciplines required for designing, building, programing and testing a robot?
2. Describe different types of robots with neat sketches.
3. What are the various applications of robot?
4. Describe characteristics of a robot.
5. Explain various types of sensors used in the robot.
6. Explain applications of sensors in robots.
7. Explain operating and performance parameters of robot.
8. Explain the role of robot in production processes.
9. What are the advantages and disadvantages of robot?
10. Define the tasks that the robot will perform.
11. Describe commonly used configurations in robotics.

12. Draw the approximate workspace for robot as shown in the figure. Assume that the dimensions of the base and other parts of the structure of the robot are as shown in the figure.

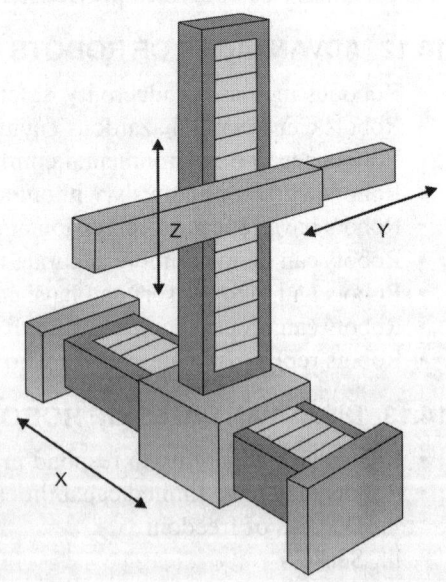

13. Draw the approximate workspace for robot shown in the figure. Assume that the dimensions of the base and other parts of the structure of the robot are as shown in the figure.

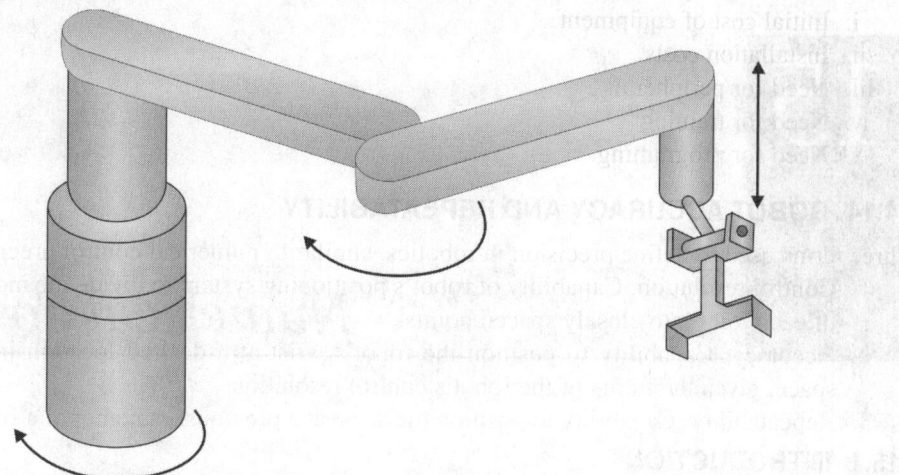

14. Draw the approximate workspace for robot shown in the figure. Assume that the dimensions of the base and other parts of the structure of the robot are as shown in the figure.

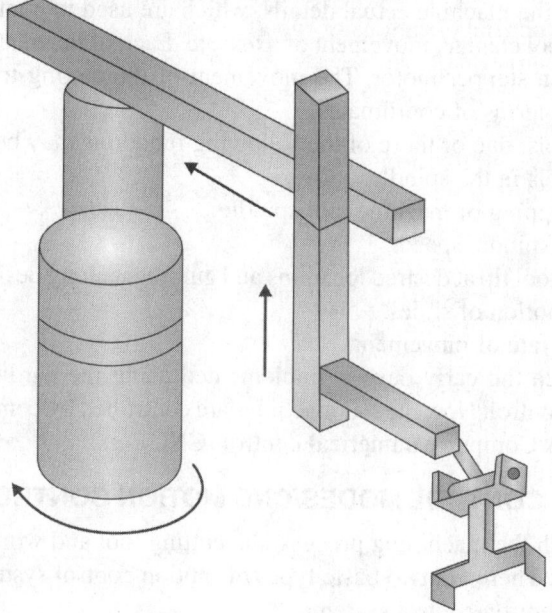

15. Define work volume and degrees of freedom.

15

Numerical Control

15.1 INTRODUCTION

Numerical control of machine tools is defined as a method of automation in which various functions of machine tools are controlled by letters, numbers and symbols. Basically, a NC machine runs on a part program. The part program consists of instructions written in numerical codes. These instructions are entered into the input medium. The controller translates these numerical codes into the machine actual details, which are used to control machine functions such as feed, speed, tool change, movement of axes, etc. Each of the machine axes is connected to the servomotor or a stepper motor. The movement of the cutting tool with respect to the workpiece is given in terms of coordinates.

In NC machine tools, one or more of the following functions may be automatic:

- Changing of tools in the spindle.
- Starting and stopping of machine tool spindle.
- Controlling the spindle speed.
- Positioning the tool tip at desired locations and guiding it along desired paths by automatic control of the motion of slides.
- Controlling the rate of movement.

The NC machine in the early days is implemented using the hardware logic. It is called hardware numerical control. Now the NC machines are controlled by computer. The new control systems are termed as Computer numerical control (CNC).

15.2 NUMERICAL CONTROL MODES/CNC MOTION CONTROL SYSTEMS

In order to accomplish the machining process, the cutting tool and workpiece must be moved relative to each other. There are two basic types of motion control systems for machining.

- Point-to-point motion control system
- Continuous path motion control system

Most control units are able to handle both point-to-point and continuous path machining. Knowledge of both motion control systems is necessary to understand what applications each has in CNC.

Point-to-point numerical control mode/Point-to-point motion control system: In this mode (Fig. 15.1), the control has the capability to operate in all the three axes, but not necessarily simultaneously. As a result, it is possible to move the tool to any point (in X- and Y-axes) in the fastest possible speed and carryout the machine operation in one axis (Z-axis) at that point. The tool is not always in contact with the workpiece. The cutting tool performs operations on the workpiece at specific points. This is useful for drilling, boring, tapping, reaming and punching operations.

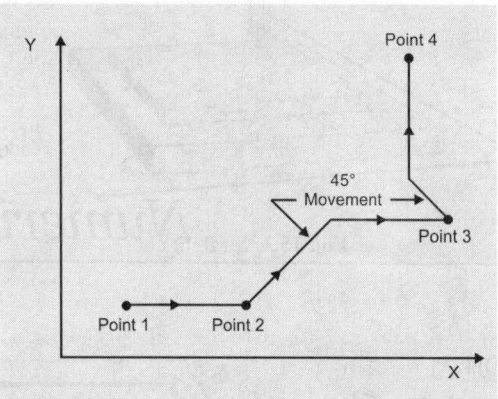

Fig. 15.1 Point-to-point mode

Cutting tool (Fig. 15.1) moves from point 1 to point 2 in a straight line and the cutting tool moves only along the X-axis. To move cutting tool from point 2 to point 3 requires that motion along both the X- and Y-axes. A similar motion takes place between points 3 and 4.

Continuous path motion control system/Contouring numerical control: Contouring, or continuous path machining, involves work such as that produced on a lathe or milling machine, where the cutting tool is in contact with the workpiece continuously. Continuous path positioning is the ability to control motions on two or more machine axes simultaneously to keep a constant cutter-workpiece relationship. The programed information in the CNC program must accurately position the cutting tool from one point to the next and follow a predefined accurate path at a programed feed rate in order to produce the form or contour required (Fig. 15.2).

Types of contour machining:
- Straight cut
- Complex contour

Straight cut: Cutting tool (Figs. 15.2 and 15.3) moves parallel to one of the major axes. This would help in obtaining the milling in a straight line along any of the axes. The cutting tool is

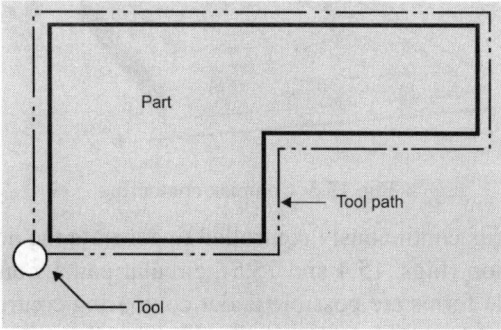

Fig. 15.2 Straight cut

always in contact with the workpiece. Straight milling is an example of continuous path machining.

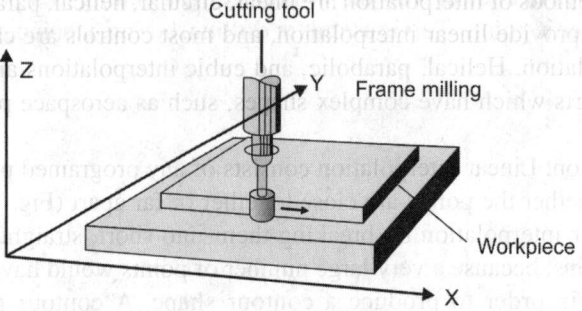

Fig. 15.3 Straight cut

Complex contour:

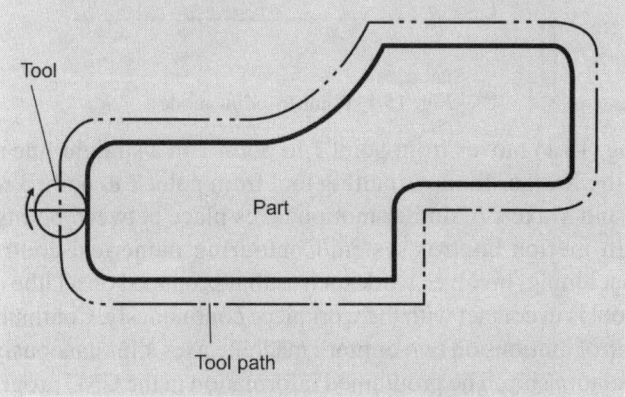

Fig. 15.4 Complex contouring

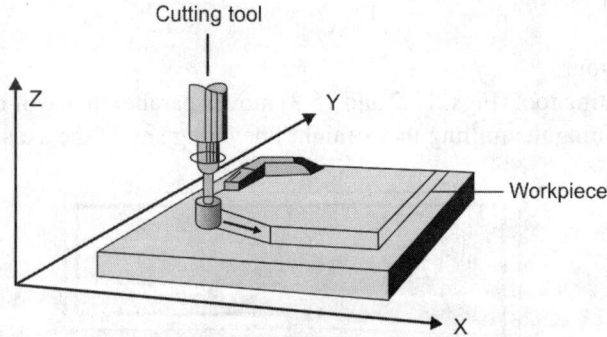

Fig. 15.5 Complex contouring

The path of the cutter is continuously controlled to generate the desired geometry. Straight surfaces at any orientation (Figs. 15.4 and 15.5), circular paths, conical shapes or any other mathematically definable forms are possible under contouring control. In order to machine a curved path in a numerical control contouring system the direction of the feed rate must be continuously changed so as to follow the path.

15.3 INTERPOLATION

Interpolation is the method by which contouring machine tools move from one programed point to the next. Methods of interpolation are linear, circular, helical, parabolic, and cubic. All contouring controls provide linear interpolation, and most controls are capable of both linear and circular interpolation. Helical, parabolic, and cubic interpolations are used by industries that manufacture parts which have complex shapes, such as aerospace parts and dies for car bodies.

Linear interpolation: Linear interpolation consists of any programed points linked together by straight lines, whether the points are close together or far apart (Fig. 15.6). Curves can be produced with linear interpolation by breaking them into short, straight line segments. This method has limitations; because a very large number of points would have to be programed to describe the curve in order to produce a contour shape. A contour programed in linear interpolation requires the coordinate positions (X- and Y-coordinates in two-axis work) for the start and finish of each line segment. Therefore, the end point of one line or segment becomes the start point for the next segment, and so on, throughout the entire program.

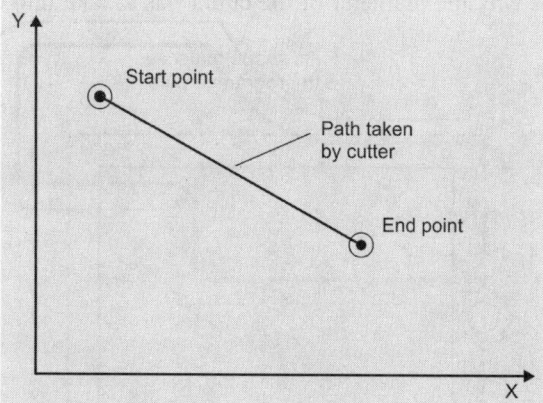

Fig. 15.6 An example of two-axis linear interpolation

Circular interpolation: For two-dimensional circular interpolation (Figs. 15.7, 15.8 and 15.9), the MCU must be supplied with the X- and Y-coordinates, radius, start point, end point, and direction of cut.

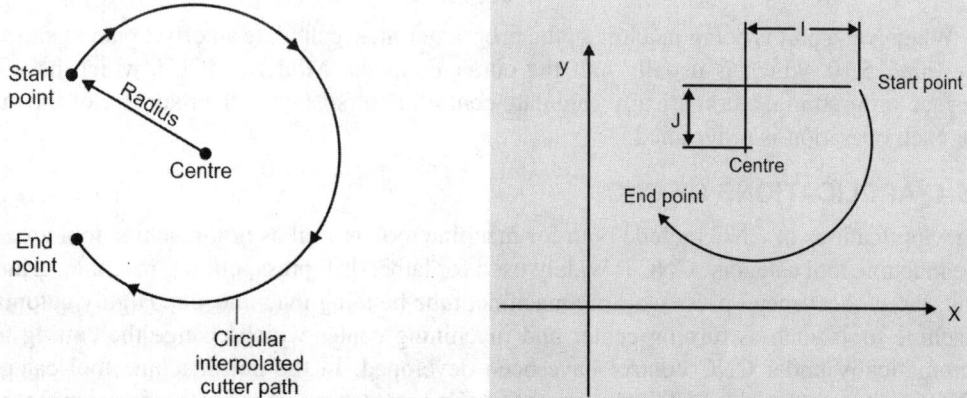

Fig. 15.7 For two-dimensional circular interpolation   **Fig. 15.8** Clockwise circular interpolation

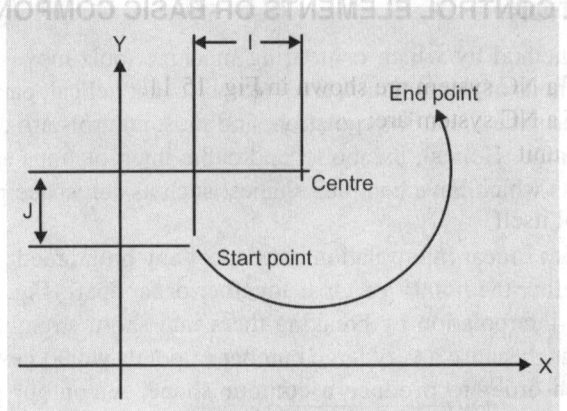

Fig. 15.9 Counter clockwise circular interpolation

Tool path with cutter compensation: In NC machining, cutter axis is moving along the programed path. That is why the diameter of the cutter has to take into consideration.

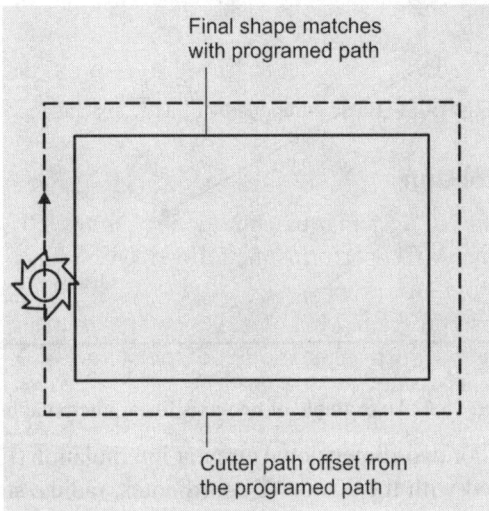

Fig. 15.10 Tool path with cutter compensation

Whenever a part is being machined, the programer must calculate an offset path as shown in the Fig. 15.10, which is usually half the cutter diameter. Modern MCUs, which have part surface programing, automatically calculate centreline offsets once the diameter of the cutter for each operation is programed.

15.4 APPLICATIONS OF CNC

The applications of CNC include both for machine tool as well as non-machine tool areas. In the machine tool category, CNC is widely used for lathe, drill press, milling machine, grinding unit, laser, sheet-metal press working machine, tube bending machine, etc. Highly automated machine tools such as turning center and machining center which change the cutting tools automatically under CNC control have been developed. In the non-machine tool category, CNC applications include welding machines (arc and resistance), coordinate measuring machine, electronic assembly, tape laying and filament winding machines for composites, etc.

15.5 NUMERICAL CONTROL ELEMENTS OR BASIC COMPONENTS OF NC SYSTEM

Basic components of a NC system are shown in Fig. 15.11.

Basic components of a NC system are:

- Machine control unit
- Part program
- The machine tool itself
- Cutting tool

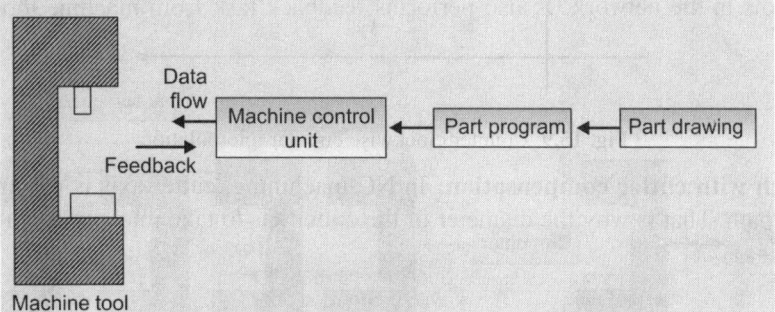

Fig. 15.11 Basic components of a NC machine tool

15.5.1 Machine Control Unit

The basic component of the NC system is machine control unit (MCU). The MCU is considered as the brain of the machine. MCU performs the following two functions:

   a. It reads the part program.

   b. After reading the part program, MCU decodes part program and converts into mechanical actions of machine tool such as the motion of the cutting tool, spindle speeds, feed rate, tool changes and several other functions.

The elements of machine control unit includes:

1. The tape reader
2. A data buffer
3. Signal output channels to the machine tool
4. Feedback channels from the machine tool

The tape reader: The tape reader is an electromechanical device for reading the program instructions.

A data buffer: The data contained on the tape are read into the data buffer. The purpose of this device is to store the input instructions in logical blocks of information.

Signal output channels to the machine tool: The signal output channels are connected to servomotors and other controls in the machine tool. Through these channels the instructions are sent to machine tool from controlling unit.

Feedback channels from the machine tool: To know that the instructions have been properly executed by the machine, feedback data are sent back to the controller via the feedback channel.

All modern NC systems today are sold with computer as the controller unit. This type of control is called computer numerical control.

Considering the structure of the controller unit, machine tools are categorized as NC, CNC (computer numerical control) and DNC (distributed numerical control) machines.

In a NC machine tool, the data processing unit is a tape reader. Punched tape is input for the tape reader (Fig. 15.12a).

In a CNC machine tool (Fig. 15.12b), the data processing unit is a computer. If the same CNC program is used on various machine tools, it has to be loaded separately into each machine tool. The computer reads and stores the NC program.

A DNC system (Fig. 15.12c) consists of central computer to which a group of CNC machine tools are connected via a common network. The communication in a DNC system is usually achieved using a standard protocol such as TCP/IP or MAP. The central computer system has various functions. It stores NC programs. It downloads these programs to any number of CNC machine tools in the network. It also performs feedback task from machine tools to central computer.

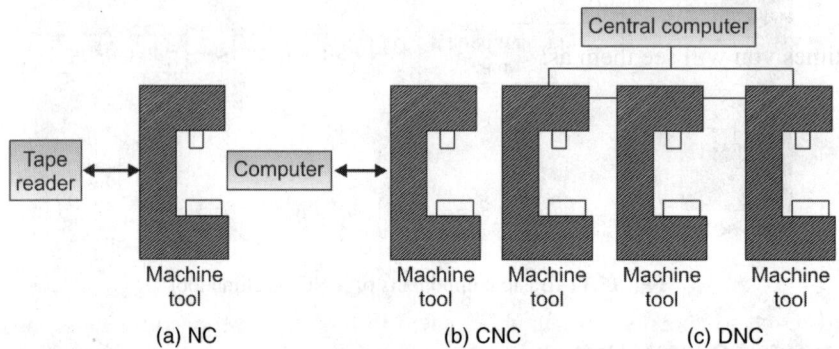

Fig. 15.12 NC, CNC and DNC

15.5.2 Part Program

Part program is a very important software element in the NC manufacturing system. It is the detailed plan of manufacturing instructions required for machining the part as per drawing. Part programing can be done manually or by using computers. The following are the methods that are used to create the NC program:

 a. **Manual part programing:** The NC programer writes the code manually and stores it in a file similar to C, C++ and java programings.
 b. **Computer assisted programing:** The NC programer uses NC programing languages such as APT (Automatically programed tools), COMPACT, SPLIT, PROMPT, etc. to generate NC program.
 c. **Part programing using CAD/CAM systems:** This approach is similar to approach 2. However, the programer uses CAD database of part directly. The most popular Cam programs in use today are Mastercam, Gibbscam and BobCad. These Cam programs can create the NC program (G codes and M codes) from the geometry data. Often, this program can be several thousand lines long.
 d. **Manual data input:** The NC programer uses the controller of the machine tool to input the NC data directly. The controller also allows the programer to save the data and its program.

15.5.2.1 Manual part programing

The NC programer writes the code manually and stores it in a file similar to C, C++ and java programings.

Codes: Irrespective of the method that is used to create the NC part program, the instruction statement must be in a low level language understood by the machine controller. The syntax of instructional statement of a NC program is in the form: *Codeword data.*

The most common codes used, when programing CNC machine tools are G-codes, and M-codes. Other codes such as F, S, D, and T are used for machine functions such as feed, speed, cutter diameter offset, tool number, etc.

Existing code words are:

N-word: N-word specifies sequence number of statements. N-word is at the beginning of every line. Sequence numbers are in increasing order. Instructions are executed as per the order of sequence number.

N1

N2

N3

Sometimes you will see them as:

N2

N4

N6

Or

N10

N20

N30

G-word: G-word specifies instruction related to the cutting tool movement on geometry. G-words are from G00 to G99. The functions of a few common G-codes are shown in Fig. 15.13. Examples of G-codes are given below:

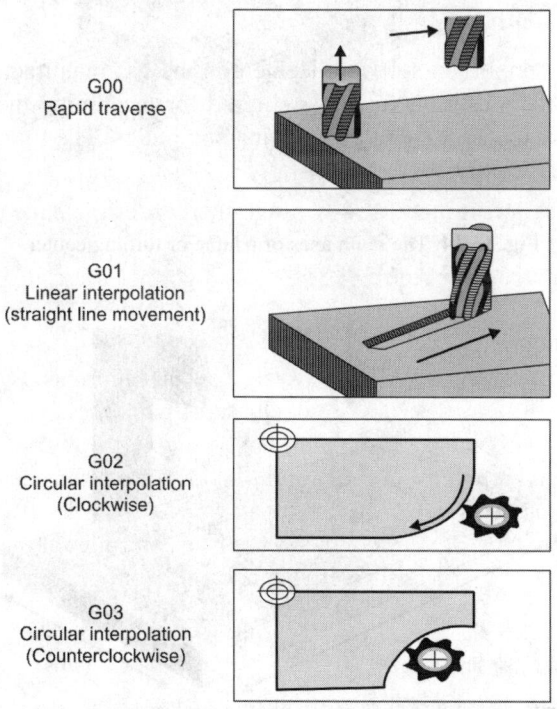

G00
Rapid traverse

G01
Linear interpolation
(straight line movement)

G02
Circular interpolation
(Clockwise)

G03
Circular interpolation
(Counterclockwise)

Fig. 15.13 The functions of a few common G-codes

G00: G00 specifies rapid point to point movement. G00 code rapidly positions the cutting tool while it is above the workpiece from one point to another point on a job. During the rapid traverse movement, either the X- or Y-axis can be moved individually or both axes can be moved at the same time. Although the rate of rapid travel varies from machine to machine, it ranges between 5 and 20 m/min.

G01: Specifies linear motion between points

G02: Specifies clockwise circular motion (circular interpolation)

G03: Specifies counterclockwise circular motion (circular interpolation)

G17: Specifies XY-plane selection

G18: Specifies XZ-plane selection

G19: Specifies YZ-plane selection

Work coordinates: X, Y, and Z define coordinates. A coordinate word specifies the target point of the tool movement or the distance to be moved. X = −150 Y = −250 represents the movement to (150, −250). Whether the dimensions are absolute or incremental will have to be defined previously using G-codes. The main axes of a lathe or turning center is shown in Fig. 15.14. The main axes of a vertical machining center are shown in Fig. 15.15.

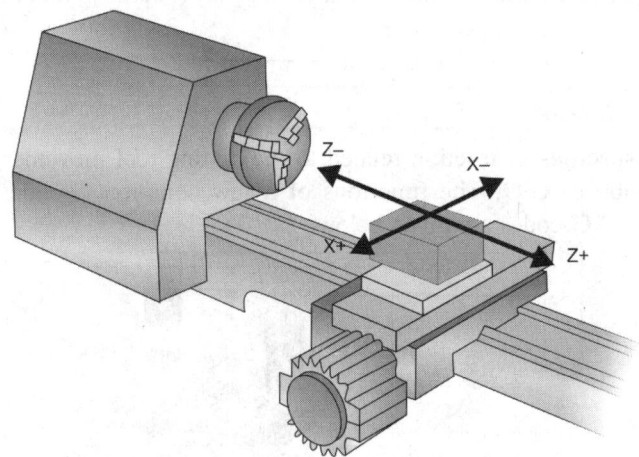

Fig. 15.14 The main axes of a lathe or turning center

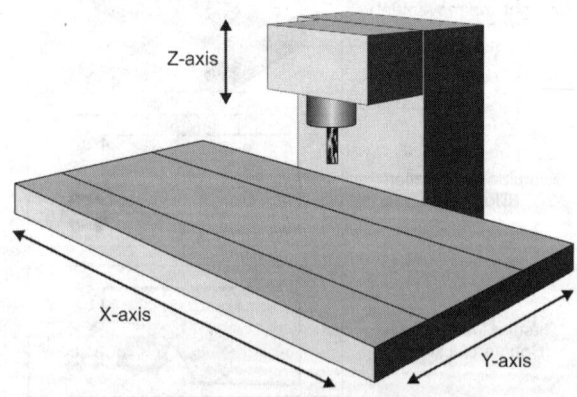

Fig. 15.15 The main axes of a vertical machining center

Z-dimension raises the cutter above the work surface.

Z-dimension feeds the cutter into the work surface.

Zero point setting: The machine zero point can be set by the operator to suit the holding fixture or the part to be machined. The operator can use the MCU controls to locate the spindle over the desired part zero and then set the X- and Y-coordinates registered on the console to zero. N1 G28 X0 Y0 Z0 sends spindle to home zero position.

F-word: F-word specifies feed rate for the cutting tool. F100 represents a feed rate of 100 mm/min.

Two types of programing modes, the incremental system and the absolute system, are used for CNC. Both systems have applications in CNC programing, and no system is either right or wrong all the time. Most controls on machine tools today are capable of handling either incremental or absolute programing. Incremental program locations are always given as the distance and direction from the immediately preceding point. In incremental programing (Fig. 15.16), the G91 command indicates to the computer and MCU (Machine control unit) that programing is in the incremental mode. Absolute program locations are always given from a single fixed zero or origin point (Fig. 15.17). The zero or origin point may be a position on the machine table, such as the corner of the work table or at any specific point on the workpiece.

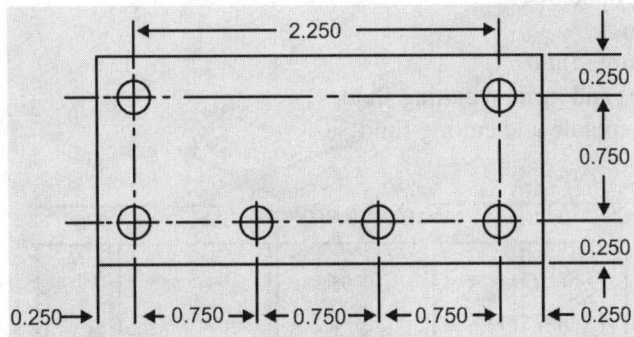

Fig. 15.16 A workpiece dimensioned in the incremental system mode

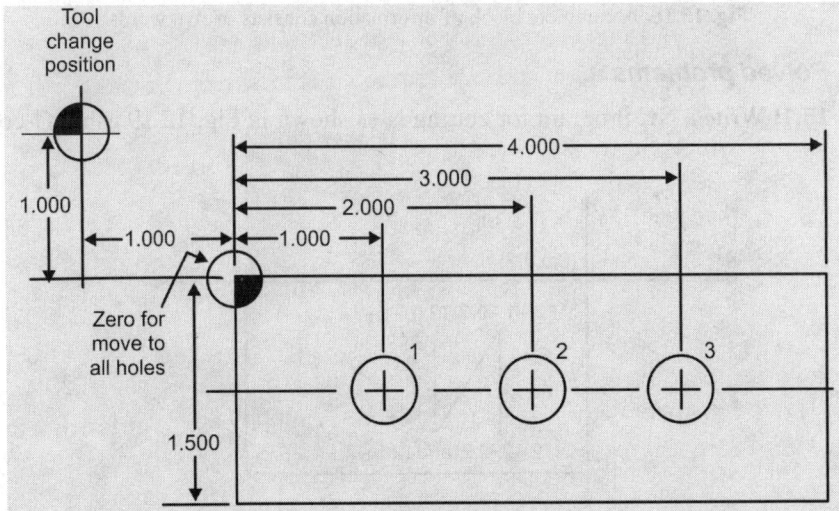

Fig. 15.17 Determination of coordinates of part for machining holes

In absolute dimensioning and programing, each point or location on the workpiece is given as a certain distance from the zero or reference point. In absolute programing, the G90 command indicates to the computer and MCU that the programing is in the absolute mode.

In absolute programing, all dimensions must be taken from the top left-hand corner of the part. Top left-hand corner of the part is taken as origin (X = 0 and Y = 0).

S-word: S-word specifies the speed of the spindle of the machine tool. S1000 represents a spindle speed of 1,000 rpm.

T-word: T-word specifies a tool from the tool library of the machine tool. The selection of tool is commanded under a T-address. T04 represents tool number 4.

M-word: M-word specifies instruction related to the machine tool.

EOB (end of the block) specifies the end of a NC program.

M or miscellaneous codes are used to either turn ON or OFF different functions which control certain machine tool operations. M-codes are not grouped into categories, although several codes may control the same type of operations such as M03, M04, and M05 which control the machine tool spindle. Examples of M-codes are given below:

M03 turns the spindle on in a clockwise direction

M04 turns the spindle on in a counterclockwise direction

M05 turns the spindle off

M06 tool change

M09 turn off cutting fluid

M13 spindle start and turn on cutting fluid

M17 turn of the spindle and cutting fluid

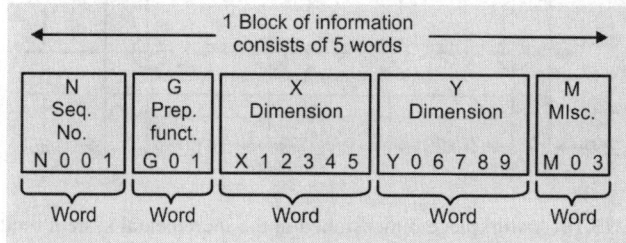

Fig. 15.18 A complete block of information consists of five words

15.5.2.2 Solved problems

Problem 15.1: Write a NC program for cutting N as shown in Fig. 15.19 using G-codes and M-codes.

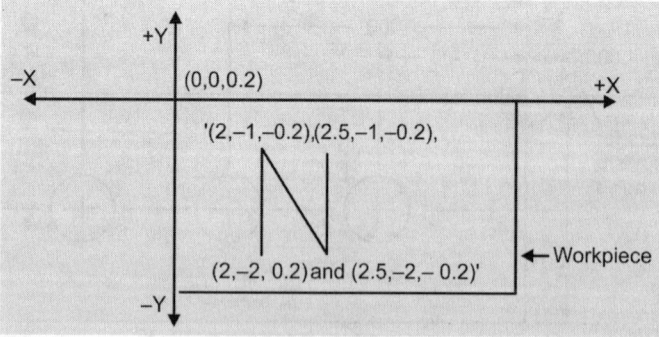

Fig. 15.19 Geometry of N

Solution:

NC program for cutting N using G-code and M-code is listed below:

N01 G00 X2 Y–2 Z0.2 M13	/Go to the bottom left corner of N
N02 G01 Z–0.2 F10.0	/Cut the letter depth
N03 G01 Y–1	/Cut left leg of the N
N04 G01 X2.5 Y–2	/Cut the diagonal side of N
N05 G01 Y–1	/Cut the right leg of N
N06 G00 Z0.200	/withdrawing the cutting tool from material
N07 M17	/Turn off spindle and cutting fluid

Problem 15.2: A square of 2.0 in. × 2.0 in. is to be milled using a 1/2 in. end milling cutter. Write a NC part program to make the square as shown in Fig. 15.20.

Solution: Let us set up the lower left corner of the square at (6.0, 6.0). Using tool radius compensation, the square can be produced.

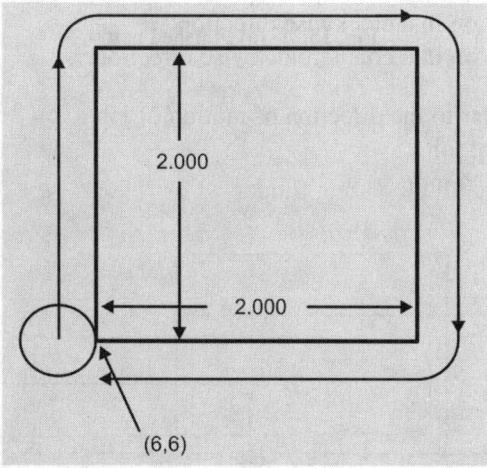

2.000

2.000

(6,6)

Fig. 15.20 Geometry

Cutter compensation: Shifting tool path so that the actual finished cut is either moved to the left or right of the programed path. Normally, shifted exactly by tool radius. The compensation is effective from the start. G41: Cutter compensation.

Table 15.1 Part program

Part program	Explanation
N10 G41 S1000 F5	Begin compensation, set feed and speed
N20 G00 X6.0000 Y6.000 Z1.000	Move to lower left corner
N30 G01 Z–1.000	Plunge down the tool
N40 Y8.000	Cut to upper left hand corner
N50 X8.000	Cut to upper right corner
N60 Y6.000	Cut to lower right corner
N70 X6.000	Cut to lower left corner
N80 Z1.000	Lift the tool
N90 G40 M30	End compensation and stop the machine

Problem 15.3: Prepare part program to produce the component as shown in Fig. 15.21a from raw material of size 60 mm diameter and 70 mm length using turning operation.

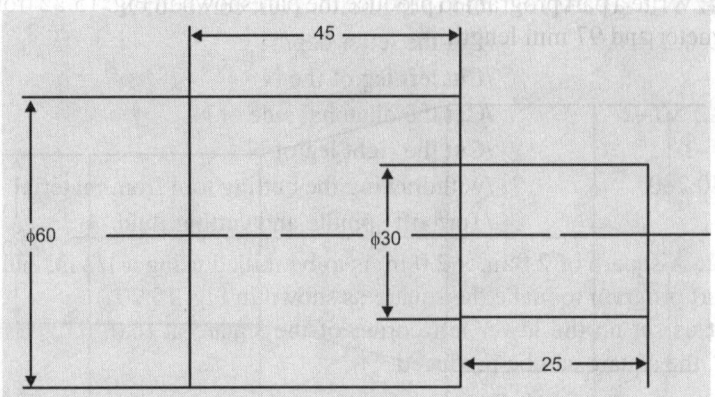

Fig. 15.21 (a) Finished part

Solution:

X-axis: Axis perpendicular to the direction of motion of job

Z-axis: Axis of rotation of job

Maximum depth of cut is 4 mm

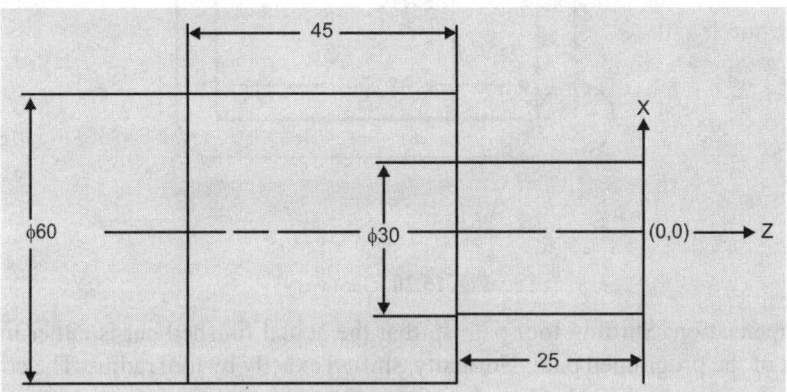

Fig. 15.21 (b) Creation of origin

Program:

N10 X32 Z1.0	Position at the starting point
N20 G97 F0.3	
N30 G90 X26 Z–25	Reduce the diameter from 60 to 52 of length 25
N40 X22 Z–25	Reduce the diameter from 52 to 44 of length 25
N50 X 18 Z–25	Reduce the diameter from 44 to 36 of length 25
N60 X 15 Z–25	Reduce the diameter from 36 to 30 of length 25

Taper turning

The (X, Z) coordinate of the small diameter, large diameter, and a feed rate must be programed.

–Z: Moves the cutting tool along the length of the workpiece.

X: Moves the cutting tool away from the work diameter.

–X: Moves the cutting tool into the work diameter.

Problem 15.4: Write a part program to produce the part shown in Fig. 15.22 from raw material of 30 mm diameter and 97 mm length.

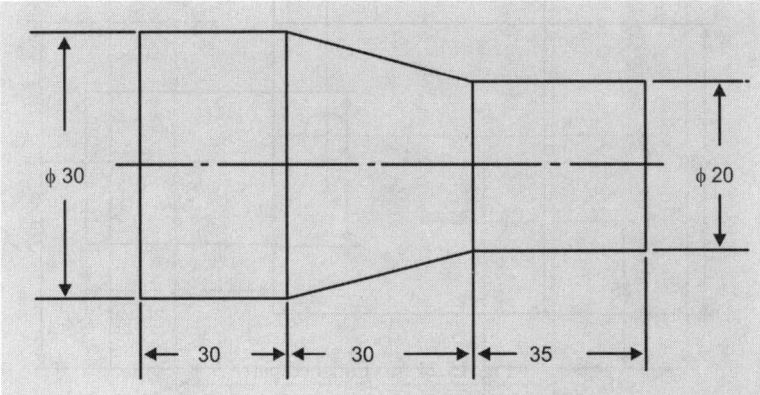

Fig. 15.22 Finished part

Solution: The layout of the figure to produce the part from the given raw material is shown in Fig. 15.23.

Sequence of operations:

Facing 2 mm length

Turning of 35 mm length

Taper turning of 30 mm length

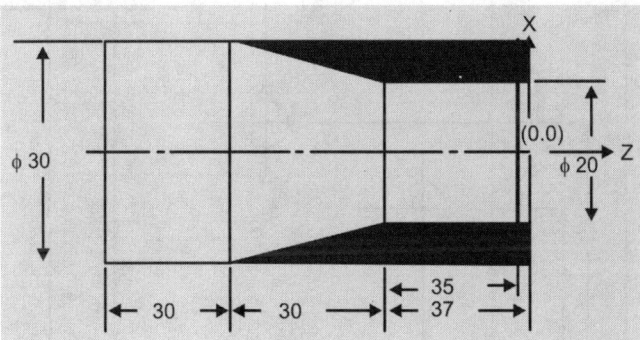

Fig. 15.23 The layout of the figure to produce the part from the given raw material

N01 G54 G91 G71 G94 M03 S800 (Parameters setting)

N05 G00 X15 Z1 (Initial tool position)

N10 G01 X15 Z–3 F2 (Facing of 2 mm length)

N15 G01 X10 Z–35 (Turning operation of 35 mm length)

N20 G01 X5 Z–30 (Taper turning operation of 30 mm length)

N25 G00 X1 Z66 (Final position of tool)

N30 M02 (End of program)

Problem 15.5: Write a part program to drill 3 holes. The finished part is shown in Fig. 15.24a. Diameter of the hole is 10 mm and depth of the hole is 15 mm.

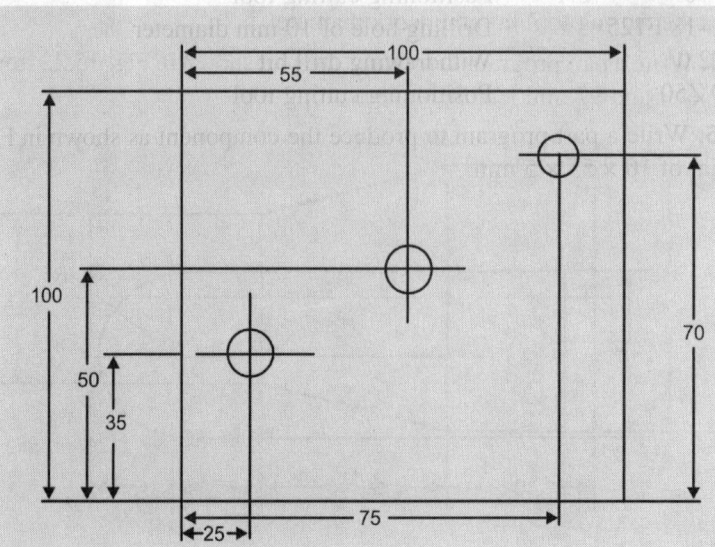

Fig. 15.24 (a) Finished part

Solution:

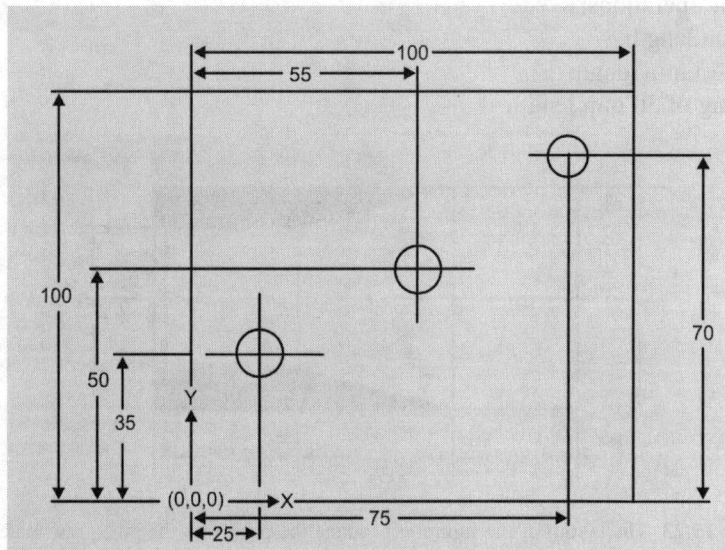

Fig. 15.24 (b) Finished part with origin

Program:

N10 G00 X25 Y35 Z2	Positioning cutting tool
N20 G01 Z–18 F125	Drilling hole of 10 mm diameter
N30 G00 Z2.0	Withdrawing drill bit
N40 X55 Y50	Positioning cutting tool
N50 G01 Z–18 F125	Drilling hole of 10 mm diameter
N60 G00 Z2.0	Withdrawing drill bit

N70 X75 Y70	Positioning cutting tool
N80 G01 Z–18 F125	Drilling hole of 10 mm diameter
N90 G00 Z2.0	Withdrawing drill bit
N65 X0 Y0 Z50	Positioning cutting tool

Problem 15.6: Write a part program to produce the component as shown in Fig. 15.25 from the raw material of $10 \times 55 \times 5$ mm^3.

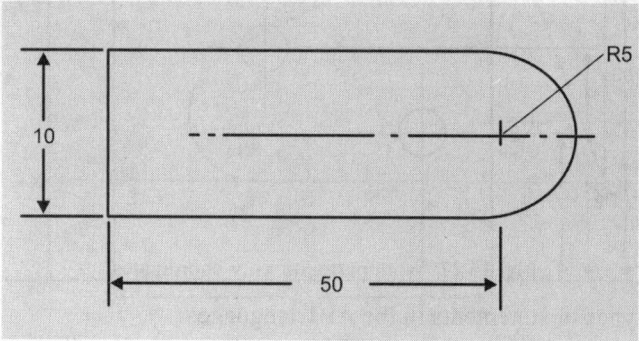

Fig. 15.25 Finished part

Solution:

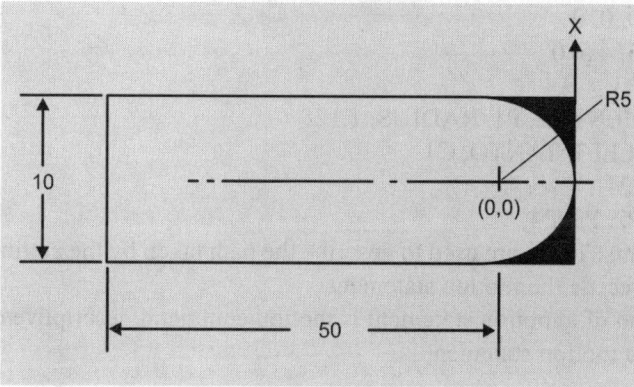

Fig. 15.26 Finished part with origin

Program:

N01 G91 G71 G94 M03 S800 (Parameters setting)

N05 G01 X–5 Z0 F1 (Initial position of cutting tool)

N10 G02 X5 Z–5 I0 R5 (Circular Interpolation: Final point is (X = 5, Z = –5), centre is origin, radius = 5)

N15 G00 X6 Z6 (Final position of tool)

15.5.2.3 Computer-aided part programing

It is difficult to write the programs for complex shapes. Then computer-aided part programing methods are used. APT (Automatically programed tools) is the first computer-aided part programing.

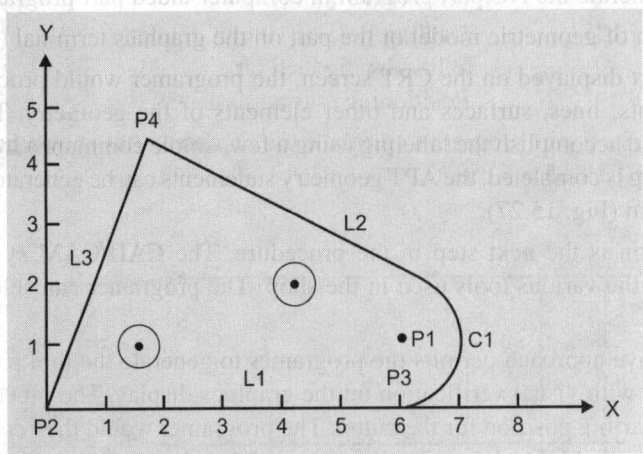

Fig. 15.27 Work part with x-, y-coordinates

There are four types of statements in the APT language:

Geometric statements: The first step in CNC program is to describe work part geometry. Geometry statements describe work part geometry. Geometry statements for the work parts in Fig. 15.27 are given below.

P1 = POINT/6.0, 1.125, 0

P2 = POINT/0, 0, 0

P3 = POINT/6.0, 0, 0

P4 = POINT/1.5, 4.5, 0

Ll = LINE/P2, P3

C1 = CIRCLE/CENTRE, P1, RADIUS, 1.125

L2 = LINE/P4, LEFT, TANTO, C1

L3 = LINE/P2, P4

PLl = PLANE/P2, P3, P4

Motion statements: These are used to describe the path taken by the cutting tool. Geometric statements must precede the motion statements.

The general form of a motion statement is motion command/descriptive data.

An example of a motion statement is,

GOTO/P1

For the selected drilling operation, to be specified are: Initial height, Retract height, feed plane, Top of stock, and Depth of hole. Once all the relevant parameters are specified, Mastercam generates the program for the machining of the holes.

Contour milling: Contour tool paths are used to remove material along a curve. Similar to the drilling operation, the parameters to be specified are: Initial height, Retract height, feed plane, Top of stock, and maximum depth of the contour.

Post processor statements: These are applied to the specific machine tool and control system. They are used to specify feeds, speeds and other features of the machine.

FEDRAT/3.55

SPEED/500

Auxiliary statements: These are the statements to control the operations of machine tool.

COOLANT/ON

COOLANT/OFF

Steps used to generate the NC part program in computer-aided part programing:

i. Construction of geometric model of the part on the graphics terminal (Fig. 15.27).

ii. With the part displayed on the CRT screen, the programer would proceed to label the various points, lines, surfaces and other elements of the geometry. The CAD/CAM system would accomplish the labeling using a few simple commands by the programer. After labeling is completed, the APT geometry statements can be generated automatically by the system (Fig. 15.27).

iii. Tool selection is the next step in the procedure. The CAD/CAM system has a tool library with the various tools used in the shop. The programer can select one of these tools.

iv. The interactive approach permits the programer to generate the tool path in a step-by-step manner with visual verification on the graphics display. The procedure begins by defining a starting position for the cutter. The programer would then command the tool to move along the defined geometric surfaces of the part. As the tool is being moved on the CRT screen, the corresponding motion commands are automatically prepared by the CAD/CAM system. The interactive mode provides the user with the opportunity to insert postprocessor statements at appropriate points during program creation. These postprocessor statements would consist of machine tool instructions such as feed rates, speeds, and control of the cutting fluid.

15.5.2.4 *Part programing using CAD/CAM systems/Automatic part programing*

NC programs can be generated by Cam software automatically using the information from the geometric model of the part in the CAD database. Often this program can be several thousand lines long. The most popular Cam software in use today is Mastercam, Gibbscam and BobCad. Cam software can create the NC program (G-codes and M-codes) from the geometry data.

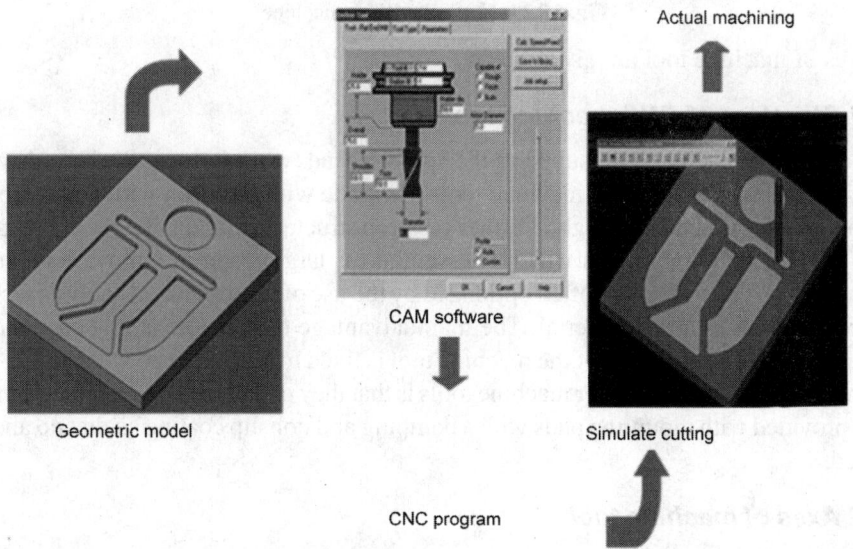

Actual machining

CAM software

Geometric model

Simulate cutting

CNC program

Fig. 15.28 NC program generation using CAD

15.5.2.5 Manual data input

The NC programer uses the controller of the machine tool to input the NC data directly. The controller also allows the programer to save the data and its program.

15.5.3 Machine Tool/Machining Center

The third basic component of NC system is the machine tool. It is the part of the NC system which performs useful work. In the beginning, milling machines were converted to CNC by simply replacing the motion elements by automated devices. Most of the modern CNC milling machines have expanded machining capabilities by the addition of accessory devices, making them more versatile. Now these milling machines are called machining center rather than milling machines.

The machine tool consists of work table and spindle as well as the motors and controls to drive them. It also includes cutting tools, work fixtures and other auxiliary equipment needed in the manufacturing operation. NC machines range from drilling machine to highly sophisticated and versatile machining centers. A vertical milling machine is shown in Fig. 15.29.

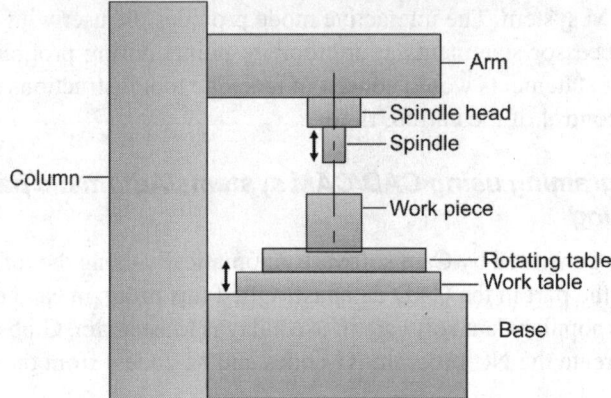

Fig. 15.29 Vertical milling machine

Features of machine tool are given below.

15.5.3.1 Structure of CNC machine tools

The Structure of a machine tool supports the spindle head, work holding devices, drives, etc. Machine structures used in CNC machine tools are made with cast iron with heavy ribbing to provide high stiffness and low weight. Further cast iron structure provides the necessary material damping to reduce the vibrations which is essential for large material removal rate and high speed machining. The frames are often optimized by the use of finite element analysis technique.

Concrete is used as a bed material. The main advantage of concrete is low cost and better damping capacity. The structure of the machine tool is fixed to concrete bed with studs. Another innovation present in many heavier machine tools is that they do not require separate foundation. They are provided with mounting pads with a damping and nonslip coating so that no anchoring is required.

15.5.3.2 Axes of machine tool

Most of the machines come with three axes. However, there are some versions, which come with more than three axes. For example, the spindle head can be swiveled in one or two axes (about X- or Y-axis). These are required for machining sculptured surfaces. In a machine tool spindle provides the necessary motion and power for machining. The machining force is directly transmitted to the spindle as axial and radial forces. In CNC machine tools, because of the larger material removal rates the magnitude of the cutting forces will also be larger. Hence, it is necessary that the spindle deflection is minimized.

Work holding: Conventional devices such as vices and chucks are used for holding simple components. Grid plates are used as one of the fixing devices for holding complex shape components. The grid plates are provided with drilled and tapped hoes for holding the workpiece. Angular plates are also used for holding workpieces.

Rotary table: Rotary table is one of the most common accessory used in horizontal milling machine. The rotary table can have more than one axis of rotation capability. As shown in Fig. 15.30, all four faces of the workpiece can be machined in horizontal machining center using rotary table.

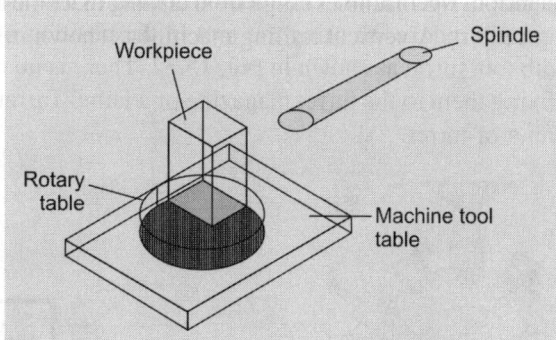

Fig. 15.30 Rotary table used in horizontal machining center for machining all four faces

15.5.3.3 Drives

Following are two drives used in CNC machines:

Spindle drives: Generally DC motors or AC motors are used to drive the spindle.

Feed drives: Feed drives are used to drive the axis as per the NC program. The feed drives that are used in CNC machine are:

- DC servomotors
- AC servomotors

Stepper motors: A stepper motor rotates in fixed angular increments. Typical step resolution is 1.8°. However, micro step motors are capable of 0.014° steps.

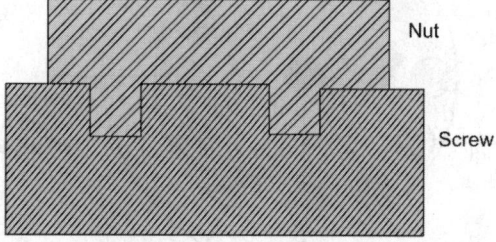

Fig. 15.31 Lead screw with nut

Lead screw: Lead screw converts the rotary motion from the drive motor to the linear motion to move various axes of the machine tool. In conventional machine tools, square thread is used. In view of the metal to metal contact between nut and the screw and in order to increase the speed of movement, most of the machine tools use lead screw with recirculating ball nut. In the case of recirculating ball nut, a series of balls circulate in the channel in form of thread. This results a highly efficient rolling motion of the balls in space between screw shaft and nut.

15.5.3.4 Automatic tool changer

The machining center is capable of milling, drilling, and reaming, tapping, boring, facing and other similar operations. A variety of machining operations means that variety of tools is required. The main feature of machining center is automatic tool changing capability. The tools are stored in a tool magazine or drum. Types of tool magazine are given below.

Turrets: The simplest type of tool magazine is a turret. Turrets are used on CNC lathes and turning centers. Most of the turning centers are provided with a tool turret, which may have a capacity of 8 to 12 tools of various types (Fig. 15.32). Generally cutting tools are positioned on the turret based on sequence of operations. This method combines tool storage with tool change procedure. The turret simply indexes to bring the tool in the position of machining since the spindle is combined with tool turret as shown in Fig. 15.32. There is no need to load or unload the cutting tools after fixing them to the turret magazine of a lathe. Turret indexing can only be referred to the positioning of turret.

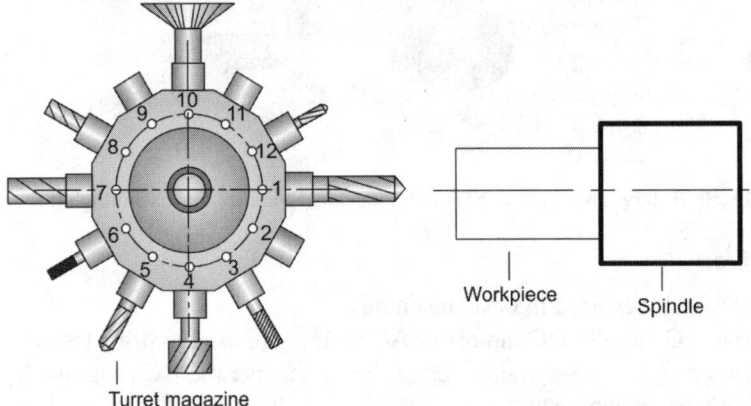

Fig. 15.32 Turret

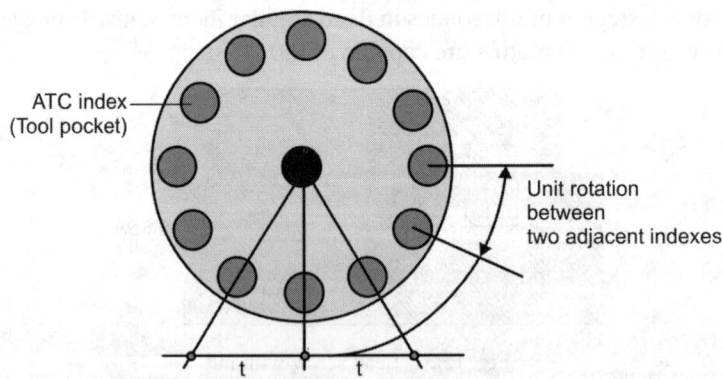

t (indexing time) is the time elapsed for one unit rotation

Fig. 15.33 Drum or Disc type tool magazine

Drum or Disc type tool magazine: The next type of tool magazine in most of machine tools is the drum or disc type magazine. Drum or Disc type tool magazine is shown in Fig. 15.33. Drum contains number of pockets in which cutting tools are stored. An automatic tool changing mechanism operating under program control exchanges the tool in the spindle and the tool in the drum. ATCs are used on CNC milling machines and milling centers. Considerable amount of time is spent in the idle movement of tool during the operation of machine tool change and tool setup. Procedure of tool changing is given below:

1. The tool drum rotates to the proper position such that empty pocket is nearer to the spindle.
2. Cutting tool is unloaded from the spindle to the empty pocket.
3. The tool drum rotates to the proper position such that the pocket which contains tool be loaded is nearer to the spindle.
4. Loading of the cutting tool into the spindle.

15.5.3.5 Machine tool

CNC turning centre: Majority of the components machined in the industry are the cylindrical shape. Cylindrical components are produced by CNC lathes called turning center. Most of the tuning centers are also provided with tool turrets which have a capacity of 8 to 12 tools of various types. The turret head is directly mounted on the saddle and the saddle slides over the bed ways. The major categories of CNC turning center are:

- Turn mill center
- Multiple axes turning center
- Vertical turning center
- Twin turret turning center
- Multiple spindles turning center

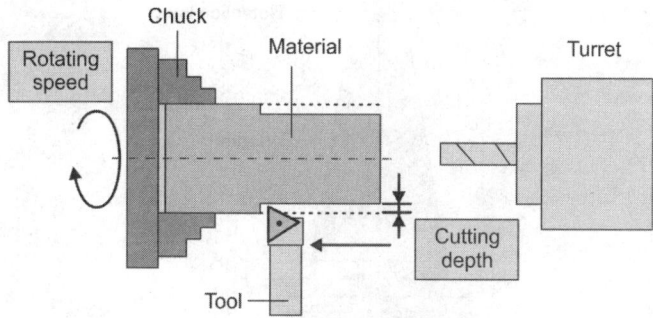

Fig. 15.34 CNC turning center

Turn mill center: Figure 15.34 is a CNC turning center. It is a machine tool which performs the turning as well as milling operation in the same machine tool. In order to do the milling operation in a lathe, the main spindle will not rotate. Instead tool will rotate in the tool turret. The tools such as twist drills, end mill, etc. will be rotating in the tool turret. The driven tool can therefore move in the X- and Y-directions like the normal turning center and can also move in Z-axis.

Twin turret turning center: Two turrets are used in a turning center in the place of one. Both turrets are capable of moving independently. The main advantage is that the machining can be done by two tools. Since two tools will be machining simultaneously, a large amount of chips would be generated.

Multiple spindles turning center: For large volume production of small and medium sized parts, it is necessary to use multiple spindles turning center. In this, the head stock consists of spindles which are arranged parallel to each other at fixed distance. These machines have the capability of simultaneous machining of more than one part.

Vertical turning center: For very large diameter workpieces, the turning center with horizontal spindle becomes difficult. The vertical turning center of the spindle is in the vertical direction. The heavy and larger workpieces are clamped on to the chuck. To facilitate quick changes automatic pallet changer is used in the place of chuck. The vertical turning center is useful for large volume of production of small automobile components. Vertical production center can perform the following operations.

- Turning
- Drilling
- Milling
- Grinding
- Gear cutting

15.5.3.6 Cutting tool

Actual machining operation is performed by the cutting tool. A milling tool is an assembly of number of parts. The assembly is fixed to the machine spindle. The assembly consists of adopter to suit the spindle taper. A power operating draw bar is employed to pull the tooling at the retention knob (hydraulic) for clamping or releasing the tool from the spindle. The preset tool usually needs to be removed from the machine for adjustment during batch production. The tool is stored on a drum. Using automatic tool changer the tools are automatically change in the spindle.

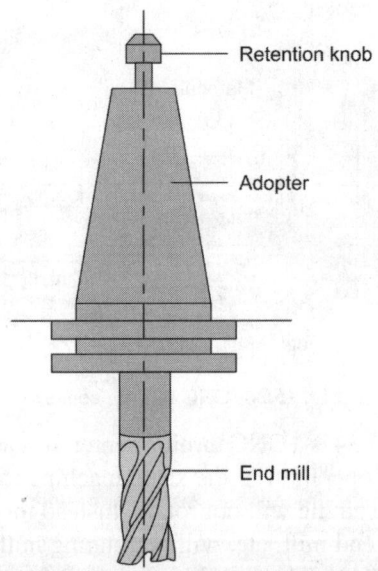

Retention knob

Adopter

End mill

Fig. 15.35 Typical spindle tooling holding an end mill

15.6 ADVANTAGES OF CNC

- Easier to program
- Easy storage of existing programs

- Easy to change a program
- Avoids human errors
- NC machines are safer to operate
- Complex geometry is produced as cheaply as simple ones
- Usually generates closer tolerances than manual machines

15.7 QUESTIONS

1. Explain NC modes.
2. What is NC machining center? Explain its features. Give some examples.
3. What do you mean by NC?
4. Explain linear and circular interpolations.
5. Explain the structure and features of CNC machine tool.
6. What is part program? What are the methods that are used to create the NC program?
7. Explain various codes used in NC program.
8. Explain the function of automatic tool changer.
9. Explain computer-aided part programing.
10. Explain part programing using CAD/CAM system.

15.8 G- AND M-CODES

The total number of these codes are 100, out of which some of the important codes are given below with their functions.

G-Codes (Preparatory Functions):

Codes	Functions
G00	Rapid positioning
G01	Linear interpolation
G02	Circular interpolation clockwise (CW)
G03	Circular interpolation counterclockwise (CCW)
G20	Inch input (in.)
G21	Metric input (mm)
G24	Radius programing
G28	Return to reference point
G29	Return from reference point
G32	Thread cutting
G40	Cutter compensation cancel
G41	Cutter compensation left
G42	Cutter compensation right
G43	Tool length compensation positive (+) direction
G44	Tool length compensation minus (−) direction
G49	Tool length compensation cancels
G53	Zero offset or M/C reference
G54	Settable zero offset
G84	Canned turn cycle
G90	Absolute programing
G91	Incremental programing

Note: On some machines and controls, some may be differ.

M-Codes (Miscellaneous Functions):

Codes	Functions
M00	Program stop
M02	End of program
M03	Spindle start (forward CW)
M04	Spindle start (reverse CCW)
M05	Spindle stop
M06	Tool change
M08	Coolant on
M09	Coolant off
M10	Chuck-clamping
M11	Chuck-unclamping
M12	Tailstock spindle out
M13	Tailstock spindle in
M17	Tool post rotation normal
M18	Tool post rotation reverse
M30	End of tape and rewind or main program end
M98	Transfer to subprogram
M99	End of subprogram

Note: On some machines and controls, some may be differ.

16

Flexible Manufacturing System

16.1 INTRODUCTION

"A highly automated GT machine cell, consisting of a group of processing stations (usually CNC machine tools), interconnected by an automated material handling and storage system, and controlled by an integrated computer system". The FMS relies on the principles of GT. An FMS is capable of producing a single part family or a limited range of part families.

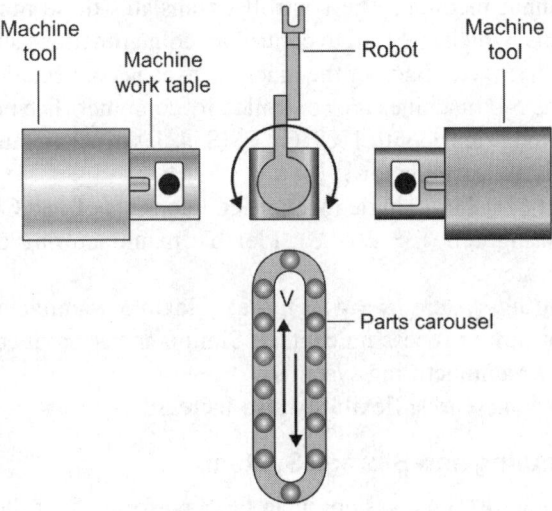

Fig. 16.1 Flexible manufacturing cell with two machine tools and robot

Figure 16.1 is a flexible manufacturing system consisting of two CNC machine tools that are loaded and unloaded by an industrial robot from a parts carousel. Periodically a worker must unload completed parts from carousel and replace them with new parts.

16.2 TYPES OF FLEXIBILITY

- Machine flexibility: Capability of machine to perform variety of operations.
- Product flexibility: It processes different part styles.

- Schedule flexibility: It accepts changes in production schedule.
- Routing flexibility: It accepts the parts through alternative workstations' sequences in response to equipment malfunctions and breakdowns.
- Expansion flexibility: It accommodates introduction of new part designs, adding of workstations, etc.
- Volume flexibility: It accepts change in volume of each part produced.

16.3 FMS COMPONENTS

There are several basic components of FMS. They are:
- Workstations
- Material handling and storage system
- Computer control system
- Human labor

16.3.1 Workstations

Load and unload station(s): Raw work parts enter the system at this point and the finished parts exit the system from here. Loading and unloading can be accomplished either manually or by material handling system.

Machining centers: The workstations used in FMS are CNC machine tools. Numerical control of machine tools is defined as a method of automation in which various functions of machine tools are controlled by letters, numbers and symbols. Basically a NC machine runs on a part program. The part program consists of instructions written in numerical codes. These instructions are entered into the input medium. The controller translates these numerical codes into the machine's actual details, which are used to control machine functions such as feed, speed, tool change, movement of axes, etc. Each of the machine axes is connected to the servomotor or a stepper motor. Now the NC machines are controlled by computer. The new control systems are termed as Computer Numerical Control (CNC). FMS is classified by number of machines. The different machine cells are given below:

Single machine cell ($n = 1$): A single machine cell consists of one CNC machine center.

Flexible manufacturing cell ($n = 2$ or 3): Flexible manufacturing cell consists of two or three CNC machine centers.

Flexible manufacturing system ($n = 4$ or more): Flexible manufacturing system ($n = 4$ or more) consists of 4 or more processing centers. Simultaneous productions of different parts are possible in flexible manufacturing systems.

Number of machines increases, flexibility also increases.

16.3.2 Material Handling and Storage System

The Material handling and storage system in an FMS performs the following functions:
- It moves the parts between stations.
- It handles a variety of part styles.
- Temporary storage of parts at the workstation for processing.
- Loading the work parts into machine and unloading the work parts from the machine.
- The handling system is controlled by the computer system.

Type of the layout decides the type of the material handling equipment. Process layout is used for batch and job production systems. Forklift trucks and AGVs (Automated guided vehicles) are used in process layout for material handling. Product layout is used in mass production system. Conveyors are used in product layout for material handling.

Types of material handling equipment:
- Conveyors
- Cranes and hoists
- Automated guided vehicles
- Industrial robots

Conveyor systems: Conveyors are designed to move large quantities of materials over fixed paths using chains, belts, rollers or other mechanical devices.

- Roller conveyors: Pathway consists of a series of rollers that are perpendicular to direction of travel. Powered rollers rotate to drive the loads forward. Figure 16.2 is a Roller conveyor.

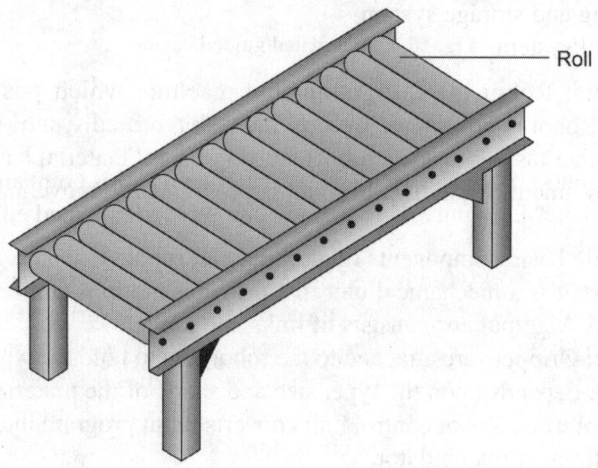

Fig. 16.2 Roller conveyor

- Belt conveyor: Belt is used to move loads. Belt is made of reinforced elastomeric. Support slider or rollers used to support forward loop. Flat belt (shown) and V-shaped belt are two common forms of belts. Figure 16.3 is a Belt conveyor.

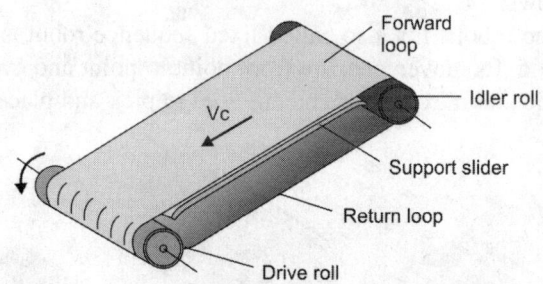

Fig. 16.3 Belt conveyor

Cranes and hoists: Handling devices for lifting, lowering and transporting materials.

Cranes: Used for horizontal movement of materials

Hoists: Used for vertical lifting of materials

Automated guided vehicles: Vehicles operate independently and are driven by electric motors that pick up power from batteries and moves generally in a fixed path. Used to move work parts between machine tools. Embedded guide wires in the floor emit electro-magnetic signal that the vehicles follow. Figure 16.4 is an Automated guided vehicle.

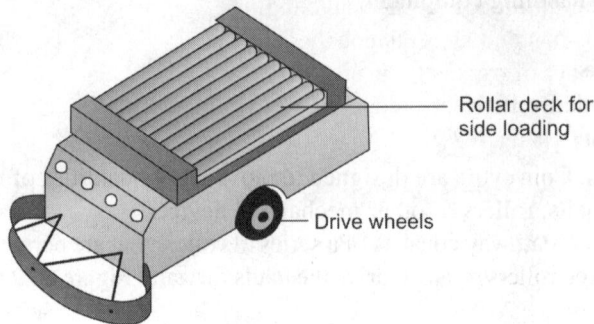

Fig. 16.4 Automated guided vehicle

Industrial robots: Robot is a programmable machine which possesses human like characteristics. Robot is programed to form the task repeatedly until it is reprogramed to perform some other task. Industrial robots are very useful material handling devices in an automated environment. An industrial robot is programed to move materials, parts, tools, etc.

The following are basic components of an industrial robot:

Manipulator: It is a mechanical unit that provides motions similar to those of human arm and hand. Manipulator consists of links and joints.

End effecter: Grippers are attached to the robot arm to hold the workpiece. The design of the gripper depends upon the type, size and shape of the material to be handled.

Robot control unit: Robot control unit converts input program into signals to perform various functions of manipulator.

Power supply: It supplies the power to the controller and manipulator. Each motion of manipulator is controlled and regulated by actuators that use an electrical, pneumatic or hydraulic power.

Robot types: Robots are generally classified as cartesian or rectilinear, cylindrical, polar or spherical jointed arms. They are also classified, from material handling point of view, as follows:

Pick and place robot: It is also called fixed sequence robot and is programed for a specific operation. Its movements are from point to point and cycle is repeated. These robots are simple and inexpensive and are used to pick and place materials.

Unit load:

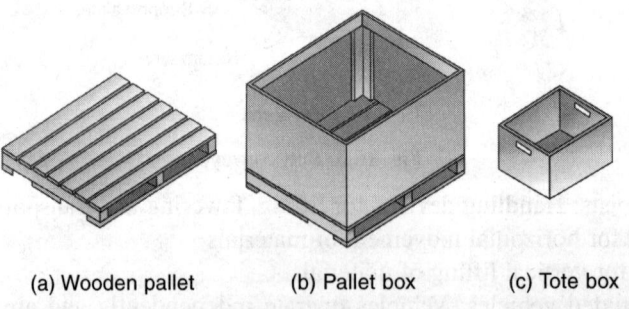

(a) Wooden pallet   (b) Pallet box   (c) Tote box

Fig. 16.5 Pallets

A unit load is the mass that is to be moved or handled at one time. Multiple items are placed in the container or pallets (Fig. 16.5) and/or moved at one time.

Reasons for using unit loads in material handling:
- Multiple items handled simultaneously
- Required number of trips is reduced
- Loading/unloading times are reduced
- Product damage is decreased

16.3.3 FMS Computer Control System

FMS computer system consists of central computer and CNC machines and other components. The central computer coordinates overall operation of the system. Functions performed by the FMS computer control system are given below:

- Workstation control: CNC is used to control the individual machine tools. Numerical control of machine tools is defined as a method of automation in which various functions of machine tools are controlled by letters, numbers and symbols. Basically a NC machine runs on a part program. Part program is stored in the computer.
- Distribution of control instructions to workstations: Central computer is required to coordinate processing at individual stations. A DNC system (Fig. 16.6) consists of central computer to which a group of CNC machine tools are connected via a common network. The communication in a DNC system is usually achieved using a standard protocol such as TCP/IP or MAP. The central computer system has various functions. It stores NC programs. It downloads these programs to any number of CNC machine tools in the network. It also performs feedback task from machine tools to central computer.
- Production control: Computer controls various production planning functions such as product mix, machine scheduling and other planning functions. Product mix means number of units of each product is to be produced. This is controlled by central computer. Machine scheduling means when to start a particular operation on particular product on a particular machine.
- Traffic control: This refers to the Management of the primary handling system to move parts between workstations, stopping parts at workstations, controlling the speed of the movement of the material handling equipment and moving the parts to load and unload stations.
- Shuttle control: Coordination of secondary handling system with primary handling system, i.e. loading of the material from material handling equipment to machine tool and unloading of material from machine tool to material handling equipment.
- Workpiece monitoring: Monitoring the status of each part in the system, i.e. number of operations performed, etc.
- Tool control: Tool control is concerned with management of cutting tools. Tool controls consist of tool location and tool life monitoring.
- Tool location: This involves in keeping the track of each tool in the system, i.e. whether the cutting tool is in the machine tool or in the tool storage.
- Tool life monitoring: Maintaining the record of the machining time usage. If the usage life is more than cutting tool life, it is replaced with new tool.
- Performance monitoring and reporting: The computer control system is programed to collect the data on utilization of machine tools and material handling equipment, volume of each component produced per week, quality of production, cutting tool life, etc.
- Diagnostics: Identifying the probable source of problem when malfunction occurs. It is also used to plan preventive maintenance and replacement of worn parts. The purpose of diagnostics' is to reduce break downs and increase availability of system.

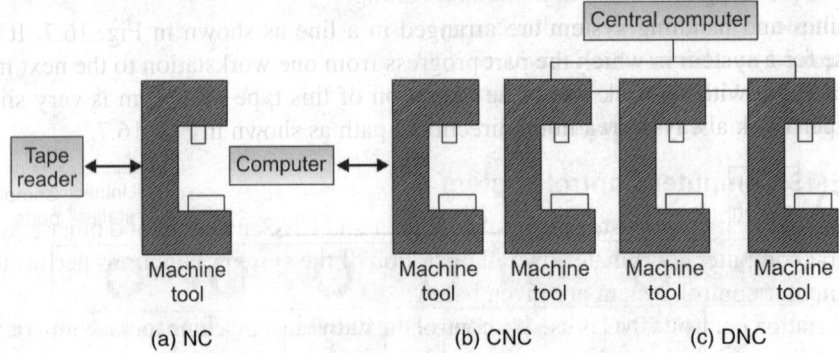

Fig. 16.6 NC, CNC and DNC

16.3.4 Human Resources or Human Labor in Flexible System

One additional component in the FMS is human labor. The flexible manufacturing system is a highly automated production facility. However, human resources are required to manage, maintain, and service the FMS.

Duties performed by human labor:

- Load man: Responsible for loading raw material into the system. This is typically done according to instructions and schedules generated by the computer. The load/unload area is at a convenient central location in the manufacturing system.
- Unload man: Responsible for unloading finished parts from the system. This is typically done according to instructions and schedules generated by the computer. The load/unload area is at a convenient central location in the manufacturing system.
- Tool setter: Tool setter is responsible for changing and setting cutting tools, inventory of cutting tools, design of cutting tools and jigs and fixtures.
- Mechanical/Hydraulic technician: He is responsible for maintenance and repair of equipment.
- Programer: Responsible for NC part programing in a machine system and operating the computer system.
- System manager: This person has overall responsibility for the operation and overall management of the system. The functions include production planning, responding to deviations and exceptions to normal operations and supervision of the other human resources which support the system.
- Rover operator: The duties of the rover operator include reacting to unscheduled machine stops, identifying broken tools or tools in need of immediate replacement, tool adjustments, and so forth. This person may also be responsible for certain manual production tasks or inspection operations.

16.4 TYPES OF FMS LAYOUT

The different types of FMS layout are:

- Progressive or Line type layout
- Loop type layout
- Ladder type layout
- Open field type layout
- Robot-centered type layout

16.4.1 Progressive or Line Type Layout

The machines and handling system are arranged in a line as shown in Fig. 16.7. It is most appropriate for a system in which the part progress from one workstation to the next in a well defined sequence with no back flow. The operation of this type of system is very similar to transfer type. Work always flows in unidirectional path as shown in Fig. 16.7.

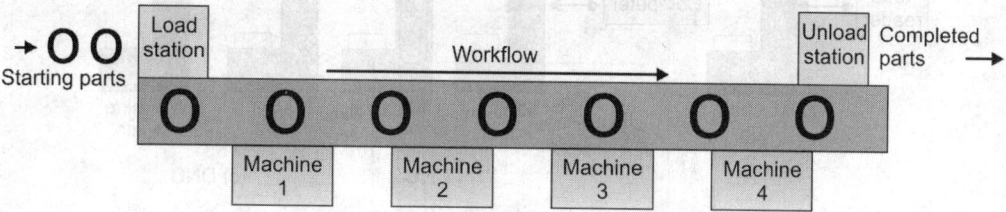

Fig. 16.7 Progressive or line type layout

16.4.2 Loop Type Layout

The basic loop configuration is as shown in Fig. 16.8. The parts usually move in one direction around the loop, with the capability to stop and be transferred to any station. FMS loop layout with secondary part handling system at each workstation to allow unobstructed flow on loop. The loading and unloading stations are typically located at one end of the loop as shown in Fig. 16.8.

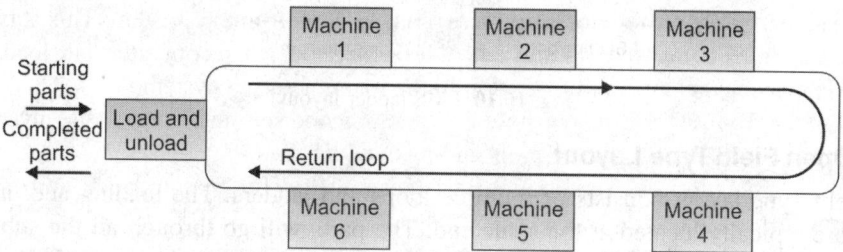

Fig. 16.8 FMS loop layout

An alternative form of loop layout is shown in Fig. 16.9. In this type, load and unload stations are different. This arrangement is used to return empty pallets to the starting position.

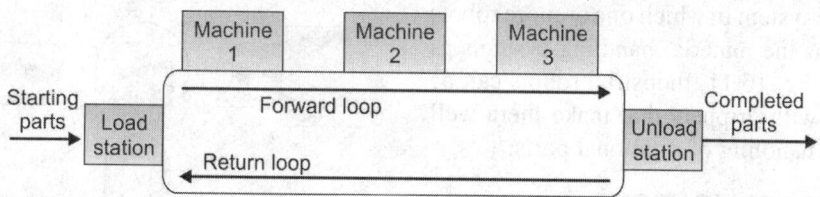

Fig. 16.9 Alternative form of FMS loop layout

16.4.3 Ladder Type Layout

The configuration is shown in Fig. 16.10. The loading and unloading stations are typically located at the same end. The sequence to the operation/transfer of parts from one machine tool to another is in the form of ladder steps as shown in Fig. 16.10.

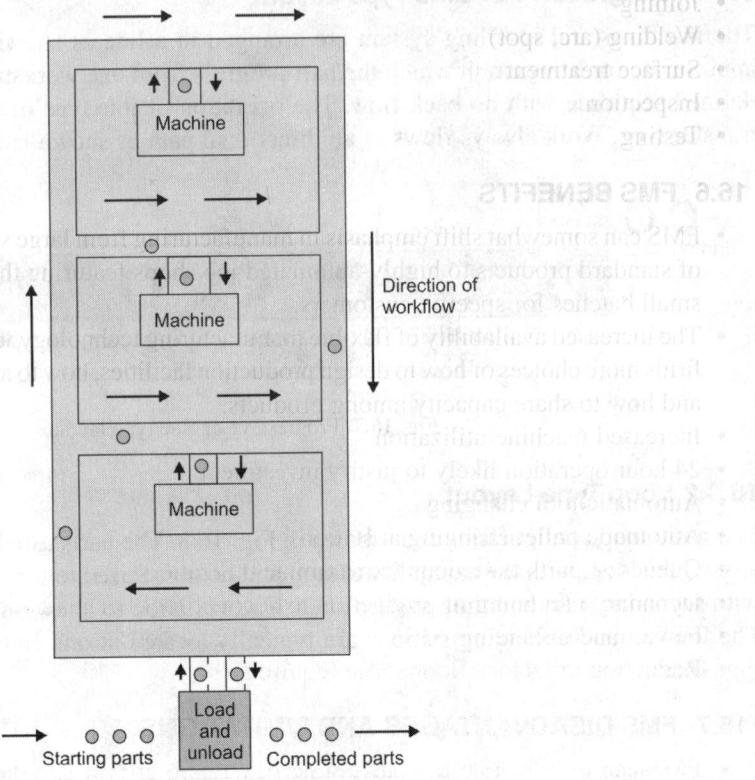

Fig. 16.10 FMS ladder layout

16.4.4 Open Field Type Layout

Open Field Type layout consists of multiple loops and ladders. The loading and unloading stations are typically located at the same end. The parts will go through all the substations, such as CNC machines, coordinate measuring machines and workstation by the help of AGVs from one substation to another.

16.4.5 Robot-centered Type Layout

Robot-centered cell is a relatively new form of flexible system in which one or more robots are used as the material handling systems as shown in Fig. 16.11. Industrial robots can be equipped with grippers that make them well suited for handling of rotational parts.

16.5 FMS APPLICATIONS

- Sheet metal processing (punching, shearing, bending, and forming)
- Forging
- Metal cutting
- Metal forming
- Assembly

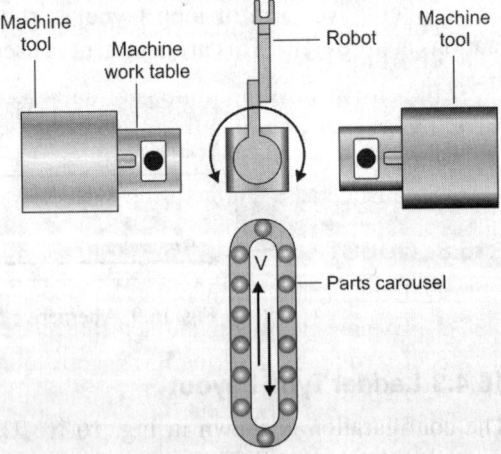

Fig. 16.11 Robot-centered type layout

- Joining
- Welding (arc, spot)
- Surface treatment
- Inspection
- Testing

16.6 FMS BENEFITS

- FMS can somewhat shift emphasis in manufacturing from large scale, repetitive production of standard products to highly-automated job shops featuring the manufacture of items in small batches for specific customers.
- The increased availability of flexible manufacturing technology will also give multi-product firms more choices of how to design production facilities, how to assign products to facilities, and how to share capacity among products.
- Increased machine utilization
- 24 hour operation likely to justify investment
- Automatic tool changing
- Automatic pallet changing at stations
- Queues of parts at stations to maximize utilization
- Dynamic scheduling of production to account for changes in demand
- Fewer machines required
- Reduction in factory floor space required.

16.7 FMS DISADVANTAGES AND LIMITATIONS

- FMS can only handle a relatively narrow range of part varieties, so it must be used for similar parts (family of parts) that require similar processing.
- Due to increased complexity and cost, a FMS also requires a longer planning and development period than traditional manufacturing equipment.
- FMS may not be appropriate for some firms. Since new technology is costly and requires several years to install and become productive, it requires a supportive infrastructure and the allocation of scarce resources for implementation. Frankly, many firms do not possess the necessary resources. Economically justifying a FMS can be a difficult task.
- Personnel training costs may be relatively high.
- For some firms, worker resistance is a problem. Workers tend to perceive automation as an effort to replace them.
- Expensive, costing millions of dollars.
- Technological problems of exact component positioning and precise timing necessary to process a component.
- Sophisticated manufacturing systems.

16.8 QUESTIONS

1. Explain flexible manufacturing systems.
2. Explain flexible manufacturing system layout configurations.
3. List out and describe FMS workstations.
4. Explain types of flexibility in manufacturing.
5. Discuss various components of FMS in detail.
6. Explain flexible manufacturing system applications.
7. What are the disadvantages of flexible manufacturing systems?

Process Selection

17.1 INTRODUCTION

Process selection refers to the strategic decisions of selecting the kind of production process to have in a manufacturing plant. Manufacturing process is the process of converting raw materials into products.

17.2 FACTORS THAT AFFECT SELECTION OF MANUFACTURING PROCESS

- Shape of the final product and initial raw material
- Product design specification: A product design specification is a document that is converted into a salable product.
- Quality: A process that consistently fails to produce items that are in specification is clearly inappropriate for the task in hand. But this factor also highlights the need to get the specification right and not to over specify unnecessarily. Producing products to unnecessarily tight specifications can lead to the wrong manufacturing process being selected and will invariably lead to increased production costs.
- Number of parts or products required and desired production rate
- Flexibility: How easy to adapt the manufacturing process for required changes to the product in terms of shape, materials or finishing. A flexible process means that changes can be made relatively easily. An example of a manufacturing process that is highly flexible and can be easily adapted is CNC machining.
- Type of material and its basic properties: Brittle and hard materials cannot be formed easily, but they can be cast or machined.
- Size, thickness, and shape complexity: Parts with thin cross-sections cannot be cast properly; complex parts cannot be formed easily
- Tolerances and surface finish
- Additional operations: Grinding, polishing (better finishing but more expensive)
- Operational and cost considerations
- Design and cost of tooling
- Lead time required to begin production
- Effect of workpiece material on tool and die life

- Minimize scrap (e.g. machining produces too much scrap)
- Availability of machines and equipment
- Environmental concerns

17.3 MANUFACTURING PROCESSES FOR METALS

- Casting
- Forming and shaping: Rolling, forging, extrusion, drawing, sheet forming, powder metallurgy, molding
- Machining: Turning, boring, drilling, milling, planning, shaping, broaching, grinding, ultrasonic machining, chemical machining, electrical discharge machining (EDM), electro-chemical machining, high-energy beam machining
- Joining: Welding, brazing, soldering, diffusion bonding, adhesive bonding, mechanical joining
- Finishing operation: Honing, lapping, polishing, burnishing, surface treating, coating, plating

17.4 MATERIAL REMOVAL PROCESS OR MACHINING PROCESS

The starting material is solid (commonly a metal, ductile or brittle) from which excess material is removed from the starting workpiece using machining operations. The resulting part has the described geometry. Most important machining operations are turning, drilling and milling accomplished by using cutting tools that are harder and stronger than the work metal. Grinding is a common process in this category, which is used to remove material.

17.5 CUTTING PARAMETERS

Process planning includes selection of the cutting parameters for various machining operations. The cutting parameters consist of the speed, feed, and depth of cut. Depth of cut is usually predetermined by the workpiece geometry and operation sequence. Therefore, the problem is determining the proper speed and feed combination. Surface finish, force and power are directly affected by the cutting parameters' feed, speed and depth of cut. Selection of the cutting parameters, also affects the time and cost required to produce a component. Several methods are currently marketed that recommend sets of parameters that either optimize machining cost, time or production rate or simply retrieve data table or calculated values.

17.6 DIFFERENT APPROACHES FOR SOLVING SPEED/FEED SELECTION PROBLEM

The approaches or the methods of solving the speed/feed selection problem are:
- Experience and judgment of process planner, foreman, or machine operator
- Handbook recommendations
- Computerized machining data systems

Experience and judgment of process planner, foreman, or machine operator: In this approach, process planner determines cutting parameters based on judgment and experience. This method is a least scientific approach and carries the greatest risk. The risk lies in the potential loss of the individual who has acquired the experience and judgment over many years in the shop. Cutting conditions derived from personal experience are not based on economic criteria.

Handbook recommendations: Handbooks of machining data are generally developed from a systematic analysis of large quantities of machining data. In this method, the cutting parameters, speed and feed are based on laboratory experiments. The best known handbook is the "Machining

Data Handbook". The handbook approach represents a definite improvement over personal judgment.

Drawbacks:

- Suggested feeds and speeds are based on worst-case conditions.
- Handbook's recommendations are considered as general guidelines. These recommendations may not be suitable for the particular product line and machine tools of a given shop.
- Handbooks are not suitable for process planning in CNC machines.

Computerized machining data system: To overcome these difficulties, computerized machining data systems are developed. In this method, large quantity of cutting data is generated over a range of feasible conditions from laboratory experiments. For each set of conditions, computations are made to determine the cost of the operation. The data generated is stored in a computerized storage file that can be accessed by process planner. The importance of these systems has grown with increase in the use of NC machines and the economic need to operate these machines as efficiently as possible. To access these files, the user would have to enter certain descriptive data that would identify the types of machining operation, work material, tooling, and so on. The printout would consist of a listing of the machining recommendations corresponding to the input data.

17.7 ELEMENTS OF COST IN MACHINING OPERATION

C_{pc} = Cost per workpiece in ₹/piece
C_o = Cost to operate the machine tool (labor, machine and applicable overhead), ₹/minute
T_m = Machining time, minutes
T_h = Workpiece handling time in minutes
T = Tool life in minutes
C_t = Cost of tooling in ₹/cutting edge
T_{tc} = Tool change time in minutes

The total cost per piece is given in the following equation:

$$C_{pc} = C_o T_m + C_o T_h + \frac{T_m}{T}(C_t + C_o T_{tc}) \longrightarrow 1$$

17.8 OPTIMIZATION MODEL TO PREDICT OPTIMUM CUTTING SPEED

Optimization models predict the optimum cutting speed for a given feed rate for an operation that minimizes cost or maximizes the production rate.

A common mathematical model to predict optimum cutting speed relies on the familiar Taylor equation for tool life:

$$VT^n = C \longrightarrow 2$$

Where

V = Surface speed in m/minute
T = Tool life in minutes
C and n are constants

By combining Equations (1) and (2), the equation for minimum cost cutting speed can be derived as

$$\text{Cutting speed for minimum cost} = \frac{C}{\left(\dfrac{(1-n)}{n}\dfrac{(C_o T_{tc} + C_t)}{C_o}\right)^n}$$

In a similar way, the cutting speed that yields maximum production rate can also be derived:

$$\text{Cutting speed for maximum production rate} = \frac{C}{\left(\dfrac{(1-n)T_{tc}}{n} \right)^{n}}$$

17.9 BREAKEVEN ANALYSIS IN SELECTION OF PROCESS

There may be number of alternative machines which may satisfy the requirement. In such cases breakeven analysis is helpful for selection among the alternatives. For example, center lathe, capstan lathe and numerical controlled machine may be capable of doing a particular work. Breakeven analysis can be an effective tool in determining selection of a machine.

17.10 SOLVED PROBLEMS

Problem 17.1: Company XYZ has to choose between two machines to purchase. The selling price is $10 per unit. Machine A: Annual cost of $3,000 with variable cost per unit of $5. Machine B: Annual cost of $8,000 with variable cost per unit of $2. Find the range of quantities at which Machine A and Machine B are preferred.

Solution: Total cost of production on Machine A = Total cost of production on Machine B

$$
\begin{aligned}
\text{FC} + \text{VC(Q)} &= \text{FC} + \text{VC(Q)} \\
\$3{,}000 + \$5Q &= \$8{,}000 + \$2Q \\
\$3Q &= \$5{,}000 \\
Q &= 1{,}667 \text{ units}
\end{aligned}
$$

Machine A is preferred when quantity demanded is < 1,667.
Machine B when quantity demanded exceeds ≥ 1,667

Problem 17.2: A company has a choice of two machines to purchase. They both make the same product which sells for $10. Machine A has fixed cost of $5,000 and a variable cost per unit of $5. Machine B has fixed cost of $15,000 and variable cost per unit of $1. Under what conditions would you select Machine A?

Solution: Total cost of production on Machine A = Total cost of production on Machine B

$$
\begin{aligned}
\text{FC} + \text{VC(Q)} &= \text{FC} + \text{VC(Q)} \\
\$5{,}000 + \$5Q &= \$15{,}000 + \$1Q \\
\$4Q &= \$10{,}000 \\
Q &= 2{,}500 \text{ units}
\end{aligned}
$$

Machine A should be purchased, if expected demand is < 2,500 units/year.
Machine A should be purchased, if expected demand is ≥ 2,500 units/year.

Problem 17.3: Determine the machine to be purchased for a particular component by using the following data:

	Capstan lathe	Auto lathe
Setup time	1 hour/batch	6 hours/batch
Operating time	10 minutes/piece	2 minutes/piece
Setup labor cost	₹ 70,000/hour	₹ 70,000/hour
Operating labor cost	₹ 100/hour	60/hour
Tooling cost	Nil	₹ 20,000/batch
Machine overheads	₹ 1,500/hour	₹ 2,800/hour
Lot size	1,500	1,500

Solution:

Capstan lathe:

Total cost = Setup cost + Operating cost + Overhead cost + Tooling cost

$$\text{Total cost per batch} = 1 \times 70,000 + \frac{10}{60} \times 1,500 \times 100$$

$$+ \left(1 + \frac{10}{60} \times 1,500\right)1,500 + 0 = ₹4,71,500$$

$$\text{Total cost per piece} = \frac{4,71,500}{1,500} = ₹314.3$$

Auto lathe:

Total cost = Setup cost + Operating cost + Overhead cost + Tooling cost

$$\text{Total cost per batch} = 6 \times 70,000 + \frac{2}{60} \times 1,500 \times 60$$

$$+ \left(6 + \frac{2}{60} \times 1,500\right)2,800 + 20,000 = ₹5,99,800$$

$$\text{Total cost per piece} = \frac{5,99,800}{1,500} = ₹399.86$$

Hence, Capstan lathe is economical.

17.11 QUESTIONS AND PROBLEMS

1. What is meant by process selection?
2. What are the factors influencing selection of manufacturing processes?
3. List out different types of manufacturing processes.
4. What are cutting parameters?
5. Explain various approaches for solving the speed and feed selection problem.
6. Explain breakeven analysis in selection of a process.
7. Determine the machine to be purchased for a particular component by using the following data:

	Capstan lathe	**Auto Lathe**
Setup time	0.5 hour	6 hours
Operating time	35 minutes	2 minutes
Setup labor cost	₹ 70/hour	₹ 70/hour
Operating labor cost	₹ 60/hour	₹ 8,000/hour
Tooling cost	₹ 1,500/batch	₹ 2,000/batch
Machine overheads	₹ 150 /hour	₹ 280/hour
Lot size	1,500	1,500

Answer: Total cost for Capstan lathe = 1,85,360,
Total cost for Auto lathe = 66,680

18

Process Planning

18.1 INTRODUCTION

To convert the product design into physical entity, manufacturing plan is needed. The activity of developing such plan is called process planning (Fig. 18.1).

Design   Process planning   Manufacturing

Fig. 18.1 Process planning

Process planning involves in determining the most appropriate manufacturing and assembly processes and the sequence in which they should be accomplished to produce a given part or product according to specifications set forth in the product design documentation. The scope and variety of process plans are generally limited by the available processing equipment of the company or plant. The initial form of the work piece may be casting, forging, etc. The process planner has to prepare a list of processes after selecting raw work piece to convert it into a predetermined final shape. If different planners were asked to develop a process plan for the same part, they would probably come up with different plans.

18.2 INFORMATION REQUIRED FOR PROCESS PLANNING SYSTEM

To convert the product design into physical entity, manufacturing plan is needed. The activity of developing such plan is called process planning. Process planning is concerned with the engineering and technological issues of how to make the product and its parts and type of equipment and tooling are required to fabricate the parts and assemble the product. Figure 18.2 depicts the information flow in the development of process plan. Portions of information in process plan are sent to other departments within the company.

18.3 STEPS IN PROCESS PLANNING

 a. *Interpretation of design drawings:* The part or product design must be analyzed (materials, dimensions, tolerances, surface finishes, etc.) at the start of the process planning procedure.

b. Selection of processes: The process planner must select which processes are required.
c. Determination of sequence of operations
d. Equipment selection for each operation
e. Selection of tools, dies, molds, fixtures, and gages for each operation.
f. Selection of material handling equipment to move material from one machine center to other
g. Determination of standard time for each operation
h. Generation of route sheet

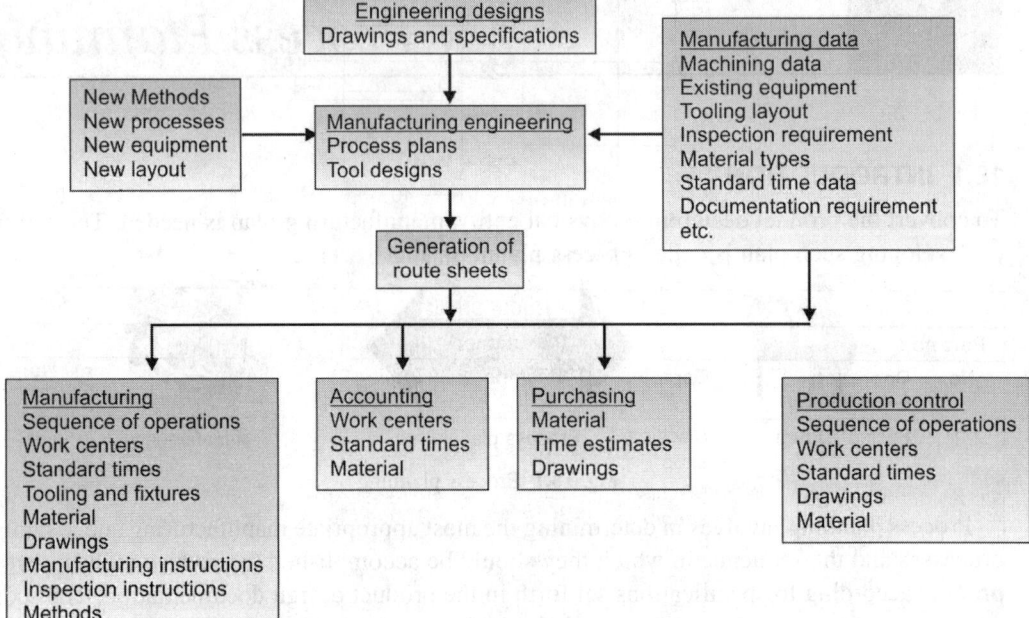

Fig. 18.2 Process plan information flow

18.4 ROUTE SHEET

Process plans or route sheets should not necessarily remain static. If there is change in volume of production, change in equipment, and change in processes then the most effective way to manufacture a particular part also changes. For individual parts, the processing sequence is documented on a form called, a route sheet. Engineering drawings are used to specify the product design. Route sheets are used to specify the process plan. They are counterparts, one for product design, and the other for manufacturing.

A typical route sheet, includes the following information: 1. Sequence of operations to be performed on the work part 2. A brief description of each operation 3. The specific machines on which the work is to be done; and 4. Any special tooling, such as dies, molds, cutting tools, jigs or fixtures, and gages.

Some companies also include setup times, cycle time standards, and other data. It is called a route sheet because the processing sequence defines the route. Example of route sheet to manufacture the product (Fig. 18.4) from raw material (Fig. 18.3) is given in Fig. 18.5.

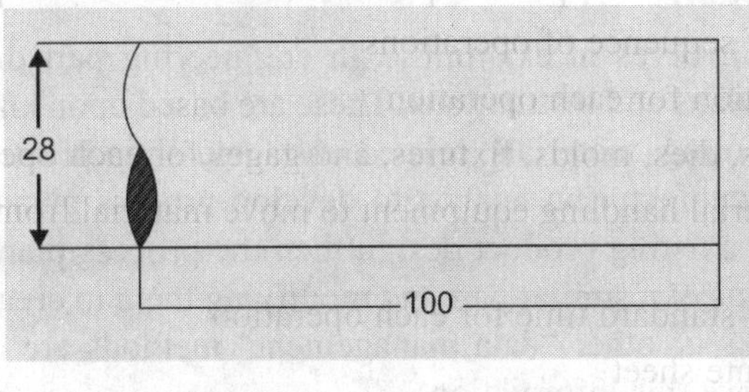

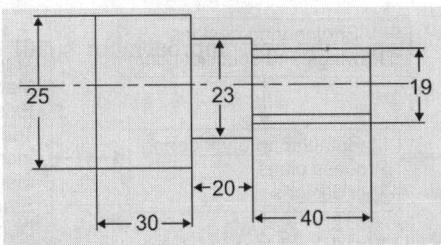

Fig. 18.3 Raw material

Fig. 18.4 Finished product

Route Sheet						
Part no.:			**Part name:**			
No.	**Operation description**	**Dept.**	**Machine**	**Tooling**	**Setup**	**Std. time**
1	Facing 10 mm	Machine shop	Lathe	Single point cutting Tool	3 Jaw chuck, Square Tool post	5 minutes
2	Turning from 28 mm to 25 mm of length 90 mm	Machine shop	Lathe	Single point Turning Tool	3 Jaw chuck, Square Tool post	20 minutes
3	Turning from 25 mm to 23 mm of length 60 mm	Machine shop	Lathe	Single point Turning Tool	3 Jaw chuck, Square Tool post	10 minutes
4	Turning from 23 mm to 19 mm of length 60 mm	Machine shop	Lathe	Single point Turning Tool	3 Jaw chuck, Square Tool post	5 minutes
5	Threading of length 40 mm	Machine shop	Lathe	Thread cutting Tool	3 Jaw chuck, Square Tool post	10 minutes

Fig. 18.5 Typical route sheet

18.5 APPROACHES TO PROCESS PLANNING

1. Manual approach
2. Computer-aided process planning (CAPP) or automated process planning
   Computer-aided process planning is a means of automatically develop the process plan
   from geometric image of the component. Computer-aided process planning systems are
   designed around three approaches. These approaches are called:
   a. Retrieval CAPP systems
   b. Generative CAPP systems
   c. Hybrid CAPP systems

18.5.1 Manual Approach

The manual approach involves in examining an engineering part drawing and developing manufacturing process plans and instructions. These are based upon knowledge of process and machine capabilities, tooling, materials, related costs and shop practices, etc. This approach requires very skilled manufacturing analyst to develop process plans. If the new part to be produced belongs to an existing product design then the process planning involves recalling that existing process plans of a similar part and modifying them to create a new routing for the new product. Workbooks or other "data management" methods are often used manually to classify, store, and retrieve this information.

Advantages:

- The manual approach often is the best approach for small companies and processing alternatives are small.
- Good flexibility
- Low investment costs

Disadvantages:

- Manual process planning is highly subjective, labor intensive, time consuming and tedious.
- Manual process planning requires well-trained and experienced personnel in manufacturing shop floor practices.
- The limitations of manual process planning really surface when the number of process plans and revisions to those plans increase. Consequently, inconsistent plans and large time requirements for planning often result.
- Manually generated plans always reflect the personal experiences, preferences, and prejudices of the analyst.

18.5.2 Computer-aided Process Planning (CAPP) or Automated Process Planning

Use of computers in process planning is usually called computer-aided process planning (CAPP) or automated process planning. Computer-aided process planning is a means of automatically develop the process plan from geometric image of the component. CAPP is usually considered to be the part of computer-aided manufacturing (CAM).

The benefits derived from computer-automated process planning include the following:

- Process rationalization and standardization: Automated process planning leads to more logical and consistent process plans than when process planning is done completely manually. Standard plans tend to result in lower manufacturing costs and higher product quality.
- Increased productivity of process planners: The systematic approach and the availability of standard process plans in the data files permit more work to be accomplished by the process planners.
- Reduced lead-time for process planning: Process planners working with a CAPP system can provide route sheets in a shorter lead-time compared to manual preparation.
- Improved legibility: Computer-prepared route sheets are neater and easier to read than manually prepared route sheets.
- Incorporation of other application programs: The CAPP program can be interfaced with other application programs, such as cost estimating and work standards.

Computer-aided process planning systems are designed around three approaches. These approaches are called:
- Retrieval CAPP systems and
- Generative CAPP systems
- Hybrid CAPP systems

18.5.2.1 Retrieval CAPP system

A retrieval CAPP system is also called a variant CAPP system. It is based on the principles of group technology (GT) and parts classification and coding system. In this type of CAPP, a standard process plan (route sheet) is stored in computer files for each part code number. The standard route sheets are based on current part routings in use in the factory or on an ideal process plan that has been prepared for each family. It should be noted that the development of the database of these process plans require substantial effort.

A retrieval CAPP system operates is illustrated in Fig. 18.6. Before the system can be used for process planning, a significant amount of information must be compiled and entered into the CAPP data files. This is referring to as the "preparatory phase".

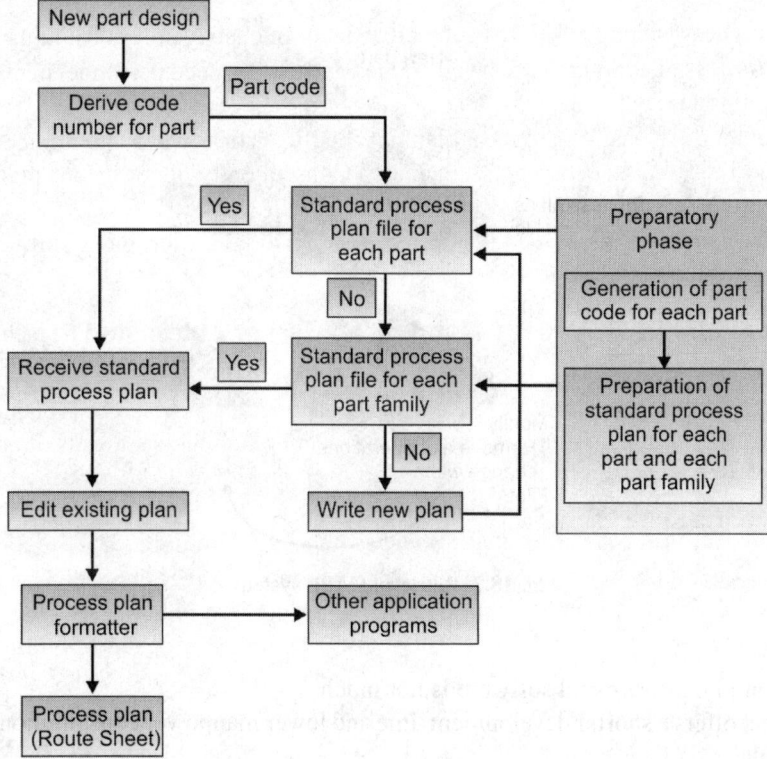

Fig. 18.6 Structure of retrieval CAPP system

Preparatory phase consists of the following steps:
1. Selecting an appropriate classification and coding scheme for the company.
2. Forming part families for the parts produced by the company.
3. Preparing standard process plan for each part family or for each part code number.

After the preparatory phase has been completed, the system is ready for use.

For new part for which the process plan is to be determined, the first step is to derive the GT code number for the part using classification and coding scheme of the company. With this code number, a search is made in computer database to determine, if a standard route sheet exists for the given part code. If the database contains a process plan for the part, it is retrieved (hence, the word "retrieval" for this CAPP system) and displayed for the user. This retrieved standard process plan is examined to determine whether any modifications are necessary. Then user edits the standard plan accordingly.

If the file does not contain a standard process plan for the given code number, the user may search the database for a similar or related code number for which a standard route sheet does exist. By editing an existing process plan the user prepares the route sheet for the part.

If the file does not contain a similar process plan then the user prepares the route sheet for the new part by starting from scratch. This route sheet becomes the standard process plan for the new part code number.

The process planning session concludes by taking printout of the route sheet in the proper format.

Example of the retrieval CAPP system: Example of the retrieval CAPP system is shown in Fig. 18.7.

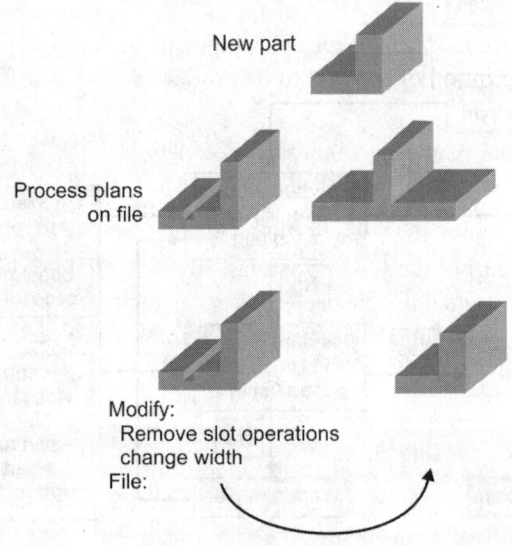

Fig. 18.7 Retrieval CAPP system

Advantages:
- Investment in hardware and software is not much.
- The system offers a shorter development time and lower manpower consumption to develop process plan.
- The system is very reliable and reasonable in real production environments for small and medium size companies.
- Quality of process plan depends on knowledge and background of process planner.

Disadvantages:
- Inconsistency in editing.
- Quality is dependent on knowledge and skill of planner.

- Optimization of variables such as material, geometry, size, precision, quality, alternative processing sequences and machine loading is difficult.

18.5.2.2 Generative CAPP system

Process plans are automatically generated by means of logical procedures, formulae, technology algorithms and geometry based data. In generative approach, a process plan is created from scratch for each component without human intervention and without a set of predefined standard plans.

Generative CAPP system is usually considered part of the field of expert systems, a branch of artificial intelligence. An expert system may be defined as a computer program that relies on knowledge and reasoning to perform a difficult task like CAPP. Truly universal system is not developed.

18.5.2.2.1 Elements in a generative CAPP system

Elements of generative CAPP system are:

- Computer comparable part description
- Knowledge base
- Inference engine

Computer comparable part description: The input to generative process planning is a computer comparable description of the part to be produced. Two possible ways of providing this description are:

i. The geometric model of the part that is developed on a CAD system during product design.

ii. A GT code number of the part that defines the part features in significant detail.

Knowledge base: The heart of the CAPP system is the knowledge base. First, the technical knowledge of manufacturing and the logic used by successful process planners must be captured and coded into a computer program. The knowledge and logic of the human process planners is incorporated into a "knowledge base". The generative CAPP system then uses knowledge base to create route sheets.

Inference engine: The part of the CAPP system that carries out reasoning function is called the inference engine. Inference is a kind of search technique, where a given pattern is matched against a set of stored patterns. Here the rules are scanned until one is found whose conditions match with the data in the working memory. Once such a rule is found the actions in the consequent part of the rule is performed.

Knowledge representation: Knowledge is represented by using IF and THEN statements.

IF < CONDITIONS > THEN < ACTION >

Example 1:
From Table 18.1,
If $C_1 = Y$ and $C_2 = Y$ then choose Action = A

Example 2:
If
Feature is a hole
Diameter <= 50 mm
Surface finish 0.8 μm to 1.6 μm
Then machining operations: Drilling, rough reaming, finish reaming

A typical generative CAPP with expert system has been shown in Fig. 18.8. Generative approach eliminates disadvantages of variant approach and bridges the gap between the CAD and CAM.

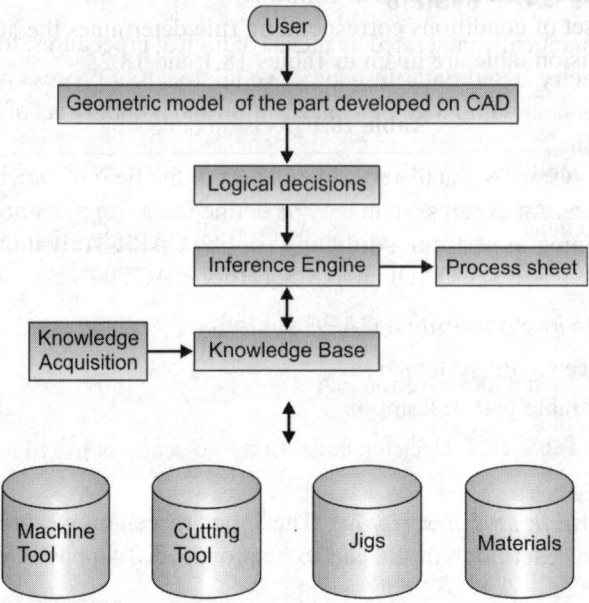

Fig. 18.8 Generative approach to CAPP

Operation or implementation of generative CAPP systems: In generative approach, process plans are generated by means of decision logics, formulae, technology algorithms and geometry based data to perform many processing decisions for converting a part from raw material to a finished state. The rules of manufacturing and the equipment capabilities are stored in a computer system.

Decision logic in CAPP or Knowledge representation techniques/methods: The generative system is implemented through the development of decision rules appropriate for the parts to be processed. The Knowledge is represented by:
- Decision tables
- Decision trees (if-then, if-then else)
- Artificial intelligence approaches.

Decision tables: Decision tables are simple form of knowledge representation technique.

Figure 18.9 shows the structure of a decision table. A decision table is partitioned into conditions and actions and is represented in a tabular form. A decision table is a program-structuring tool, which provides readable documentation. Also decision table can be used with

Conditions and actions	Rules
Conditions	Condition alternative
Actions	Actions entries

Fig. 18.9 The standard format used for presenting a decision table

preprocessor to eliminate some program coding and to provide automatic checks for completeness, contradiction and redundancy.

The decision making process works as follows:

For a particular set of conditions corresponding rule determines the action.

Examples of decision table are given in Tables 18.1 and 18.2.

Table 18.1 Decision table

Conditions	C_1	Y	Y	N	N
	C_2	Y	N	Y	N
Actions		A	B	C	D

$$\text{IF} < \text{CONDITIONS} > \text{THEN} < \text{ACTION} >$$

Example 3:

From Table 18.1,

If $C_1 = Y$ and $C_2 = Y$ then choose Action = A

Table 18.2 Decision table for the selection of machine

Condi-tions	Cylindrical shape	Y	Y	Y	Y	Y
	External threads	Y	Y	N	Y	Y
	Internal threads	N	Y	N	N	Y
	Slot on external surface	N	N	Y	Y	Y
Actions		Select Lathe machine	Select Lathe machine and Drilling machine	Select Milling machine	Select Lathe and Milling machine	Select Lathe machine, Drilling machine and Milling machine

If Cylindrical shape = Y, External threads = Y, Internal threads = Y
and Slot on External surface = N then choose Action = Select Lathe machine and Drilling machine

Decision tree: A decision tree is a graph with a single root and branches emanating from the root. Decision trees can be converted into computer program. The starting node is the root and every branch represents a decision statement, which is either false or true. Decision trees are easier to update, maintain, visualize and develop. Figure 18.10 illustrates decision tree.

From the Fig. 18.10,

If $C_1 = Y$ and $C_2 = N$ then choose Action = B

Artificial intelligence techniques and expert systems:

Artificial intelligence is to construct computer systems that perform tasks that are considered to require intelligence. It is a branch of computer science dealing with symbols. An expert system is a computer program that depends on

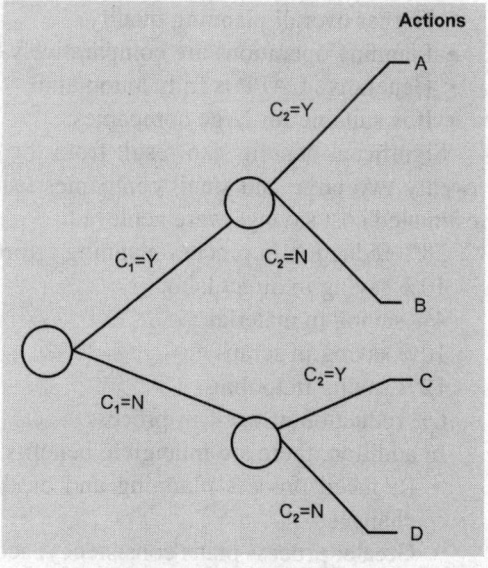

Fig. 18.10 Decision tree

knowledge and reasoning. Expert systems are based on decision logics, formulae, technology algorithms and geometry based data to perform many processing decisions for converting a part from raw material to a finished state. Human expert decides conclusions based on personal knowledge. An expert system decides conclusions based on the knowledge it possesses. The steps to be taken to build expert systems involve in building up the knowledge base from the simplest elements to the most complex, i.e. building up the concepts first, then rules, then models and strategies. The CAPP systems in which Artificial intelligence techniques have been applied are called Expert systems. Such an expert system can be defined as a tool, which has the capability to understand problem specific knowledge and use domain knowledge intelligently to suggest alternative paths of action. Artificial intelligence techniques and expert systems are considered to have the largest potential for the development of CAPP system.

Development of generative CAPP system:

Two types of planning are available: forward and backward planning.

Using forward planning, we begin with the stock as the initial state and the part features are removed until the final part is obtained.

Using backward planning, we begin with the final part as the initial state and part features are building up until the stock is obtained.

Generative CAPP implementation:

- The logic of process planning must be identified and captured.
- The part to be produced must be clearly and precisely defined in a computer compatible format.
- The captured logic of process planning and part description data must be incorporated into a unified manufacturing database.

Advantages:

- The generative CAPP has all the advantages of Variant CAPP, however, it has an additional advantage that it is fully automatic and up-to-date process plan is generated at each time.
- Flexibility and consistency for process planning for new parts
- Higher overall planning quality.
- Planning operations are comparatively fast.
- Generative CAPP is fully automatic.
- It is suitable for large companies.

Significant benefits can result from the implementation of CAPP. In a detailed survey of twenty two large and small companies using generative type CAPP systems, the following estimated cost savings were achieved:

58% reduction in process planning effort

10% saving in direct labor

4% saving in material

10% saving in scrap

12% saving in tooling

6% reduction in work-in-process

In addition, there are intangible benefits as follows:

- Reduced process planning and production lead-time; faster response to engineering changes
- Greater process plans consistency; access to up-to-date information in a central database
- Improved cost estimating procedures and fewer calculation errors

- More complete and detailed process plans
- Improved production scheduling and capacity utilization
- Improved ability to introduce new manufacturing technology and rapidly update process plans to utilize the improved technology

Disadvantages:

- It requires more extensive setup.
- It requires major revisions, if a new equipment or processing capabilities became available.

Applications:

The majority of generative CAPP systems implemented to date have focused on process planning for fabrication of sheet metal parts and less complex machined parts. In addition, there has been significant recent effort with generative process planning for assembly operations, including PCB assembly.

18.5.2.3 Hybrid approach

Hybrid approach is an approach to perform the task of process planning which combines both variant and generative types. Each approach is appropriate under certain conditions.

18.6 CAPP APPLICATIONS

CAPP is used in almost all industries like:

- Sheet metal forming
- Aircraft manufacturing
- Paper industries
- Automobile industries, etc...

18.7 FACTS ABOUT CAPP TECHNOLOGY

Many CAPP systems use many advanced techniques and approaches such as feature-based modeling, object oriented programing methods, expert system and artificial intelligence.

Though tremendous effort has been made in developing CAPP systems, the effectiveness of these systems is not fully satisfactory.

In spite of the benefits promised by the various developed CAPP systems, their adaptation by industry is painfully slow.

18.8 LIMITATIONS OF CAPP

- Collection of huge database is difficult.
- Automatic reorganization of database is difficult.
- Design of CAPP for complex applications is difficult and costly.

18.9 QUESTIONS

1. What is process planning?
2. Explain route sheet with an example.
3. What are the steps in process planning?
4. What are the advantages of computer-aided process planning over manual process planning?
5. Explain retrieval CAPP system with an example.

6. Explain generative CAPP system.
7. Explain elements in generative CAPP system.
8. What are the various steps in developing a process plan?
9. Why the need for CAPP arises?
10. What are different approaches to CAPP? Describe briefly.
11. Briefly describe the "Knowledge based Process Planning".
12. Write short notes on the following:
    i. Manual experience-based process plan
    ii. Computer-aided process plan

19

Forecasting

19.1 INTRODUCTION

Success in business depends greatly on the ability to estimate future values of sales, output, price, etc. Forecasting is a tool used for predicting future sales, output, and price, etc. using qualitative and quantitative techniques.

Applications of forecasting:
- Strategic planning (long range planning)
- Finance and accounting (budgets and cost controls)
- Marketing (future sales, new products)
- Production and operations

Organizations employ forecasting techniques to determine future demand, inventory, costs, capacities, and interest rate changes.

19.2 TYPES OF FORECASTING

There are two types of forecasting according to forecasting range. They are:
- Short-term forecasting
- Long-term forecasting

Short-term forecasting: This type of forecasting can be defined when it covers a period of less than one year. The period is dependent upon the nature of business. Power production companies use hourly forecast of power demand. Production planning of manufacturing companies is usually based on monthly forecast of sales.

Purpose of short-term forecasting:
- The problem of overproduction and shortage of production can be avoided.
- To reduce the total cost of production
- To reduce inventory costs
- Planning for working on capital requirement
- To prepare master production schedule
- Capacity planning
- Material requirements planning

Long-term forecasting: This type of forecast can be defined when it covers a period of 1 year to 20 years. The period is dependent upon the type of business. In many industries like ship building, petroleum refineries and paper making industries use long-term forecasting. In these type of companies long-term forecasting is needed because total capital investment is high.

Purpose of long-term forecasting:
- To plan for the new unit
- To expand existing unit to meet future demand
- To plan for long-term financial requirements
- To train the personnel so that manpower requirement can be met in future to meet the demand.

19.3 FORECASTING TECHNIQUES

There are two types of forecasting techniques:
- Qualitative techniques
- Quantitative techniques

19.3.1 Qualitative Techniques

These techniques are primarily based upon judgment and intuition and especially when sufficient information and data is not available so that complex quantitative techniques cannot be used. The widely used qualitative methods are:

a. **Jury of executive opinion:** This is a method by which the relevant opinions of experts are taken, combined and averaged. These opinions could be taken on an individual basis or there could be a brain storming group session in which all members participate in generating new ideas that can later be evaluated for their feasibility and profitability.

b. **Opinions of the sales person:** The sales people being closer to consumers can estimate future sales in their own territories, more accurately. Based on these and the opinions of sales managers, reasonable trends of the future sales can be calculated. These forecasts are good for short range planning since sales people are not sufficiently sophisticated to predict long-term trends.

c. **Consumers' expectations:** This method involves a survey of the customers as to their future needs. This method is especially useful where the industry serves a limited market. Based on the future needs of the customers, a general overall forecast for the demand can be made.

d. **The Delphi method:** The Delphi method originally developed by Rank Corporation in 1969 for forecasting military events, has become a useful tool in other areas also. It is basically a more formal version of the jury of opinion method. A panel of experts is given a situation and asked to make initial predictions, on the basis of a prescribed questionnaire, these experts develop written opinions. These responses are analyzed and summarized and submitted back to the panel for further considerations. All these responses are anonymous so that no member is influenced by others opinions. This process is repeated until a consensus is obtained.

19.3.2 Quantitative Techniques

Quantitative techniques rely on time series data and analytical techniques.

Time series: A time series is just a collection of past values of the variable being predicted.

Example: Sales by month for the last ten years. Time series data (Demand over a period) is usually plotted on a graph to determine the various characteristics of the time series data. The time series data has following patterns:

Nonrecognizable pattern: In this pattern, demand does not follow any recognized pattern. Figure 19.1 shows purely nonrecognizable pattern.

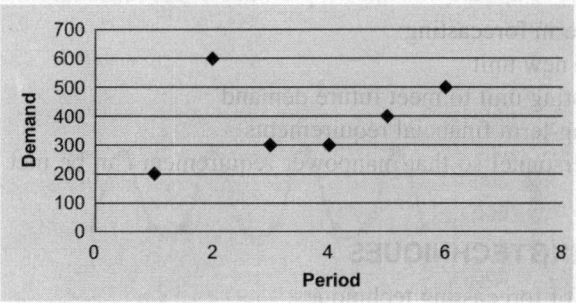

Fig. 19.1 Purely nonrecognizable pattern

Constant pattern: In this pattern, demand remains constant throughout the period (Fig. 19.2).

Trend pattern: Increases or decreases of demand linearly or nonlinearly over time (Figs. 19.3 and 19.4).

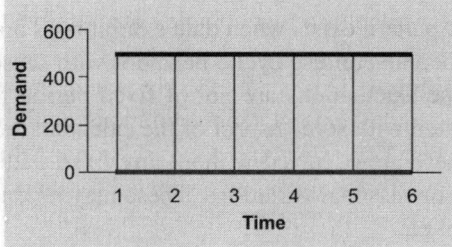

Fig. 19.2 Constant pattern

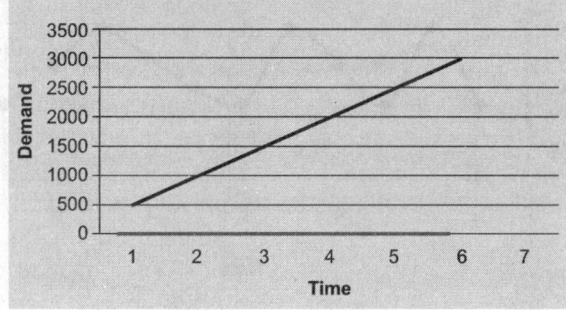

Fig. 19.3 Increasing linear trend

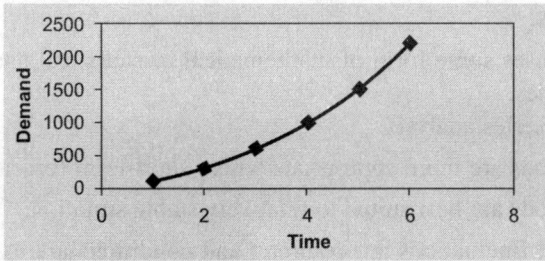

Fig. 19.4 Increasing nonlinear trend

Seasonal pattern: Seasonality corresponds to a pattern in the data that repeats at regular intervals. Demand fluctuations may be caused by weather, tradition, festival, etc. (Fig. 19.5). Seasonality is always of a fixed and known period. Hence, seasonal time series are sometimes called periodic time series.

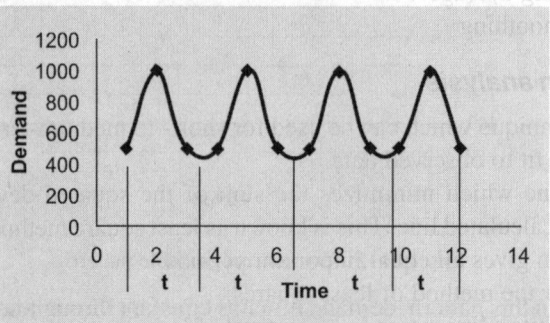

Fig. 19.5 Seasonal variation

Cyclic pattern: A cyclic pattern exists when data exhibit rises and falls that are not of fixed period (Fig. 19.6). Many people confuse cyclic behavior with seasonal behavior, but they are really quite different. If the fluctuations are not of fixed period then they are cyclic; if the period is fixed and associated with some aspect of the calendar, then the pattern is seasonal.

Irregular variations: These are occurred without any fixed pattern. The variations are not explained by trend, cyclic or seasonal variations. These may be due to natural calamities like flood, draught, earthquake, etc.

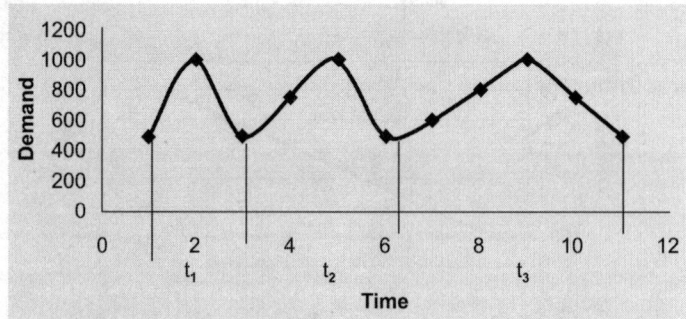

Fig. 19.6 Cyclic pattern

Time series analysis:

Time series analysis uses some form of mathematical or statistical analysis on the past data arranged in a time series.

Suitability of time series analysis:

- Time series methods are more appropriate where short-term forecasts are required.
- Time series methods are best suited to relatively stable situation.
- Where substantial fluctuations are common and conditions are expected to change then time series methods may give poor results.

Time series analysis techniques:

i. Regression analysis
ii. Smoothing methods
 - Moving average
 - Weighted moving average
 - Exponential smoothing

19.3.2.1 Regression analysis

This is a statistical technique which can be used for short- to medium-terms forecasting which seeks to establish best fit to observed data.

Best fit line is a line which minimizes the sum of the squared deviations of the actual observations from the calculated line. This is known as least square method of linear regression. Least square regression gives all equal important periods.

Linear regression or the method of least squares:

If a time series exhibits a linear trend, the method of least squares may be used to determine a trend line (projection) for future forecasts. Figure 19.7 shows line of best fit using the method of least squares.

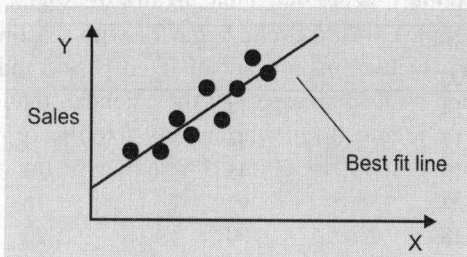

Fig. 19.7 Line of best using the method of least squares

The general form of the equation of any straight line on a graph is

$$y = a + bx$$

The independent variable (x) is the time period and the dependent variable (y) is the actual observed value (sales) in the time series.

Regression analysis is similar, except the independent variable is not restricted to time. We could forecast sales based upon the price of the product, income, etc.

$$\Sigma y = na + b\Sigma x \qquad \qquad ...(1)$$
$$\Sigma xy = a\Sigma x + b\Sigma x^2 \qquad \qquad ...(2)$$

To find values of constants a and b, it is necessary to solve two simultaneous equations 1 and 2 known as normal equations.

Accuracy of the regression line: Coefficient of determination (r^2) is calculated to determine how good the line of best fit is.

$$r^2 = \frac{\Sigma(y_e - y_m)^2}{\Sigma(y - y_m)^2}$$

Where

y_e = Estimation of y given by the regression equation for each value of x
y_m = Mean of the actual values of y
y = Individual actual values of y

Standard deviation of error: There are several formulae for finding out error. One formula is given below:

$$S_e = \sqrt{\frac{\Sigma y^2 - a\Sigma y - b\Sigma xy}{n-2}}$$

Where n = Number of observations

19.3.2.2 Solved problems

Problem 19.1: For the data given in Table 19.1, determine the best line of fit and determine the forecasts for 8th and 9th years and also determine coefficient of determination and standard deviation.

Table 19.1 Sales data

Year	1	2	3	4	5	6	7
Sales of books in 000's	14	17	15	23	18	22	27

Solution: Computations to determine the line of best fit is given in Table 19.2.

Table 19.2 To determine the line of best fit

Years (x)	Sales (y)	xy	x^2	y^2
1	14	14	1	196
2	17	34	4	289
3	15	45	9	225
4	23	92	16	529
5	18	90	25	324
6	22	132	36	484
7	27	189	49	729
$\Sigma x = 28$	$\Sigma y = 136$	$\Sigma xy = 596$	$\Sigma x^2 = 140$	$\Sigma y^2 = 2776$

$$y = a + bx$$
$$\Sigma y = na + b\Sigma x$$
$$136 = 7a + 28b \qquad \qquad ...(1)$$
$$\Sigma xy = a\Sigma x + b\Sigma x^2$$
$$596 = 28a + 140b \qquad \qquad ...(2)$$

Equation 2 $596 = 28a + 140b$

Equation 1 × 4 $544 = 28a + 112b$

Equation 2 – Equation 1 × 4 $52 = 0 + 28b$

$$b = 1.86$$
$$a = 12$$

Best fit equation

$$y_e = 12 + 1.86x$$

Forecast for 8th year $= y_8 = 12 + 1.86 \times 8 = 26.88$

Forecast for 9th year $= y_9 = 12 + 1.86 \times 9 = 28.74$

$$y_m = \frac{136}{7} = 19.42$$

A computation to find out r^2 is given in Table 19.3.

Table 19.3 To find out r^2

x	y	y_e	$y_e - y_m$	$(y_e - y_m)^2$	$y - y_m$	$(y - y_m)^2$
1	14	13.86	−5.57	31.02	−5.43	29.48
2	17	15.72	−3.71	13.76	−2.53	5.90
3	15	17.58	−1.81	3.42	−4.43	19.62
4	23	19.44	−0.01	0.00	3.57	12.74
5	18	21.30	1.87	3.49	−1.43	2.04
6	22	23.16	3.73	13.91	2.37	6.60
7	27	25.02	5.59	31.24	7.57	57.30
$\Sigma x = 28$	$\Sigma y = 136$			$\Sigma(y_e - y_m)^2 = 96.84$		$\Sigma(y - y_m)^2 = 133.68$

$$r^2 = \frac{\Sigma(y_e - y_m)^2}{\Sigma(y - y_m)^2} = \frac{96.84}{133.68} = 0.72$$

To find standard error:

$$S_e = \sqrt{\frac{\Sigma y^2 - a\Sigma y - b\Sigma xy}{n-2}} = \sqrt{\frac{2776 - 12(136) - 1.86(596)}{7-2}} = 2.66$$

Problem 19.2: A company believes in its sales according to a linear trend. Demand during the last nine months is given in Table 19.4.

Table 19.4 Sales data

Month	Jan	Feb	Mar	Apr	May	June	July	Aug	Sep
Sales	5	19	18	15	31	22	27	30	28

Using the least square method, determine a linear trend line and also forecast the sales for each of the remaining months in the year.

Solution: Computations to determine the line of best fit is given in Table 19.5.

Table 19.5 To determine the line of best fit

x	y	xy	x^2
1	5	5	1
2	19	38	4
3	18	54	9
4	15	60	16
5	31	155	25
6	22	132	36
7	27	189	49
8	30	240	64
9	28	252	81
$\Sigma x = 45$	$\Sigma y = 195$	$\Sigma xy = 1125$	$\Sigma x^2 = 285$

$$y = a + bx$$
$$\Sigma y = na + b\Sigma x$$
$$195 = 9a + 45b \qquad \text{...(1)}$$
$$\Sigma xy = a\Sigma x + b\Sigma x^2$$
$$1125 = 45a + 285b \qquad \text{...(2)}$$

Equation 1 × 5 $975 = 45a + 225b$
Equation 2 $1125 = 45a + 285b$
Equation 1 × 5 – Equation 2 $150 = 0 + 60b$
$$b = 2.5$$
$$a = 9.16$$

Line of best fit

$$y = 9.16 + 2.5x$$

Forecast for the month of October $= y = 9.16 + 2.5 \times 10 = 34.16$
Forecast for the month of November $= y = 9.16 + 2.5 \times 11 = 36.66$
Forecast for the month of December $= y = 9.16 + 2.5 \times 12 = 39.16$

Problem 19.3: A company manufactures tractors finds that there exists a relationship between sales of tractors and index of agriculture income. The following data (Table 19.6) has been collected by the company for the last five years.

Table 19.6 Sales data

Years	1988	1989	1990	1991	1992
Demands in 000's	100	112	130	150	280
Indexes of agriculture income	125	140	180	190	220

i. Fit a regression line.
ii. Estimate the sales of tractors for the year 1993, for the given index 250.

Solution: Computations to determine the line of best fit is given in Table 19.7.

Table 19.7 To determine the line of best fit

x	y	xy	x^2
125	100	12500	15625
140	112	15680	19600
180	130	23400	32400
190	150	28500	36100
220	280	61600	48400
$\Sigma x = 855$	$\Sigma y = 772$	$\Sigma xy = 141680$	$\Sigma x^2 = 152125$

$$y = a + bx$$
$$\Sigma y = na + b\Sigma x$$
$$772 = 5a + 855b \qquad \text{...(1)}$$
$$\Sigma xy = a\Sigma x + b\Sigma x^2$$
$$141680 = 855a + 152125b \qquad \text{...(2)}$$

Equation 1 × 171 $132012 = 855a + 146205b$
Equation 2 $146180 = 855a + 152125b$
Equation 1 × 171 – Equation 2 $-9668 = -5920b$
$$b = 1.63$$
$$a = -124.86$$

Best fit equation

$$y = -124.86 + 1.63x$$

At $x = 250$
$$y = 282.6$$

Exponential growth

A linear trend line is suited to data which is expected to increase by the same amount in each period. Where this relationship does not apply some form of nonlinear representation must be used. An important nonlinear relation is exponential growth. In exponential growth, the data is expected to grow by some proportion. A typical example of exponential growth is shown in Fig. 19.8.

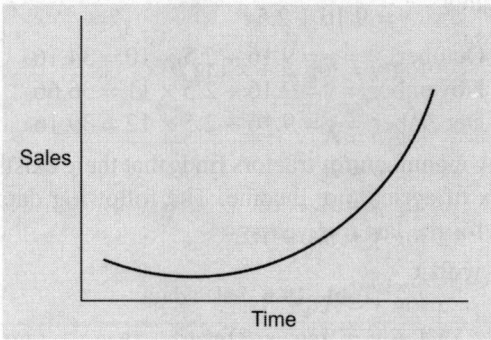

Fig. 19.8 Exponential growth curve

The exponential function takes in the form

$$y = ab^x$$

Where

y = Variable to be predicted

x = Number of periods

a and b are constants

Taking log on both sides

$$\log y = \log a + x \log b$$

Let

$$\log a = A$$

$$\log b = B$$

$$\log y = A + xB \qquad \qquad ...(1)$$

$$\Sigma \log y = nA + B\Sigma x \qquad \qquad ...(2)$$

Multiplying Equation 1 with x

$$x \log y = Ax + Bx^2$$

$$\Sigma x \log y = A\Sigma x + B\Sigma x^2 \qquad \qquad ...(3)$$

Constants A and B are obtained by solving Equations 2 and 3.

Problem 19.4: For the data given in Table 19.8, determine sales forecast for the year 2010 and 2011 using exponential growth.

Table 19.8 Sales data

Year	Sales (₹ in millions)
2005	100
2006	150
2007	225
2008	337.5
2009	506.25

Solution:

The exponential function takes in the form

$$y = ab^x$$

Where

    y = Variable to be predicted

    x = Number of periods

    a and b are constants

Taking log on both sides

$$\log y = \log a + x \log b$$

Let

$$\log a = A$$
$$\log b = B$$
$$\log y = A + xB \qquad \text{...(1)}$$
$$\Sigma \log y = nA + B\Sigma x \qquad \text{...(2)}$$

Multiplying Equation 1 with x

$$x \log y = Ax + Bx^2$$
$$\Sigma x \log y = A\Sigma x + B\Sigma x^2 \qquad \text{...(3)}$$

Constants A and B are obtained by solving Equations 2 and 3.

Computation to determine constants a and b is given in Table 19.9.

Table 19.9 Determining constants a and b

Year	Sales ₹ in millions y	x	$\log y$	$x\log y$	x^2
2005	100.00	1	2	2	1
2006	150.00	2	2.1761	4.3522	4
2007	225.00	3	2.3522	7.0566	9
2008	337.50	4	2.5282	10.1128	16
2009	506.25	5	2.7045	13.5225	25
		$\Sigma x = 15$	$\Sigma \log y = 11.7610$	$\Sigma x \log y = 37.0441$	$\Sigma x^2 = 55$

$$11.76 = 5A + 15B \qquad \text{...(4)}$$
$$37.004 = 15A + 55B \qquad \text{...(5)}$$

Equation 4 × 3

$$35.28 = 15A + 45B \qquad \text{...(6)}$$
$$37.004 = 15A + 55B \qquad \text{...(5)}$$

Equation 6 – Equation 5

$$-1.76 = 0 - 10B$$
$$B = 0.176$$
$$\log b = 0.176$$
$$b = 10^{0.176}$$
$$b = 1.5$$

Substituting $B = 0.176$ in Equation 4

$$11.76 = 5A + 15 \times 0.176$$
$$5A = 9.12$$
$$A = 1.824$$
$$\log a = 1.824$$
$$a = 10^{1.824} = 66.66$$

Forecasted equation
$$y = 66.66 \ (1.5^x)$$
Forecast for the year 2010 = 66.66 (1.5^x) = 66.66 (1.5^6) = 759.2
Forecast for the year 2011 = 66.66 (1.5^x) = 66.66 (1.5^7) = 1138.9

Logarithmic function: A linear trend line is suited to data which is expected to increase by the same amount in each period. Where this relationship does not apply some form of nonlinear representation must be used. An important nonlinear relation is logarithmic function.

The logarithmic function takes in the form:
$$y = ax^b$$
Where

y = Variable to be predicted

x = Number of periods

a and b are constants

Taking log on both sides
$$\log y = \log a + b \log x$$
Let

$$\log a = A$$
$$\log y = A + b \log x \qquad \qquad ...(7)$$
$$\Sigma \log y = nA + b \ \Sigma \log x \qquad \qquad ...(8)$$
Multiplying Equation 7 with x
$$x \log y = xA + xb \log x$$
$$\Sigma x \log y = A\Sigma x + b\Sigma x \log x \qquad \qquad ...(9)$$
Constants A and b are obtained by solving Equations 8 and 9.

Problem 19.5: For the data given in Table 19.10, determine sales forecast for the period 6 and period 7 using Logarithmic function.

Table 19.10 Sales data

Period	Sales ₹ in millions
1	50
2	200
3	450
4	800
5	1250

Solution:

The Logarithmic function takes in the form
$$y = ax^b$$
Where

y = Variable to be predicted

x = Number of periods

a and b are constants

Taking log on both sides
$$\log y = \log a + b \log x$$
Let

$$\log a = A$$
$$\log y = A + b \log x \qquad \qquad ...(1)$$
$$\Sigma \log y = nA + b \ \Sigma \log x \qquad \qquad ...(2)$$

Multiplying Equation 1 with x

$$x \log y = xA + x\, b \log x$$
$$\Sigma x \log y = A\Sigma x + b\Sigma x \log x \qquad \qquad ...(3)$$

Constants A and b are obtained by solving Equations 2 and 3.

Computation to determine constants a and b is given in Table 19.11.

Table 19.11 Determining constants a and b

Year (x)	Sales (y)	*log y*	*log x*	*xlogx*	*xlogy*
1	50	1.6990	0	0	1.6990
2	200	2.3010	0.3010	0.602	4.602
3	450	2.6531	0.4771	1.4313	7.9593
4	800	2.9031	0.6021	2.4084	11.6124
5	1250	3.0969	0.6990	3.495	15.4845
$\Sigma x = 15$		$\Sigma \log y = 12.6531$	$\Sigma \log x = 2.0791$	$\Sigma x \log x = 7.9367$	$\Sigma x \log y = \Sigma 41.3572$

$$12.6531 = 5A + 2.0791b \qquad \qquad ...(4)$$
$$41.3512 = 15A + 7.9367b \qquad \qquad ...(5)$$

Equation 4 × 3

$$37.9593 = 15A + 6.2376b \qquad \qquad ...(6)$$

Equation 5 – Equation 6

$$3.3979 = 0 + 1.6991b$$
$$b = 2$$

Substituting $b = 2$ in Equation 4

$$12.6531 = 5A + 2.0791 \times 2$$
$$5A = 8.4949$$
$$A = 1.698$$
$$\log a = 1.698$$
$$a = 10^{\,1.698} = 50$$

Forecast equation $= y = ax^b = 50x^2$

Forecast for the period 6 $= y = 50x^2 = 50\,(6^2) = 1800$

Forecast for the period 6 $= y = 50x^2 = 50\,(7^2) = 2450$

19.3.2.3 Smoothing methods

In cases in which the time series data of sales of a company can have fluctuations high or low because of the seasonal variations and random variations, one can use smoothing methods to average out the irregular components of the time series.

Three common smoothing methods are:

- Moving average
- Weighted moving average
- Exponential smoothing

19.3.2.3.1 Moving average forecasting

Moving average method is one of the smoothing techniques. A simple moving average is calculated as follows:

$$\text{Moving average} = \frac{\text{Sum of the demand for } n \text{ periods}}{n}$$

$$\text{Moving average} = \frac{D_1 + D_2 + D_3 + ... + D_n}{n}$$

Where D_i = Demand for the period i

Characteristics of moving average:
- Different moving averages produce different forecasts.
- The greater the number of periods in the moving average the greater the smoothing effect.
- Equal weight is given to each of the values used in the moving average.

19.3.2.3.2 Solved problem

Problem 19.6: A company sales is given in Table 19.12.

Table 19.12 Time series sales data

Month	Actual sales
Jan	450
Feb	440
Mar	460
Apr	410
May	380
June	400
July	370
Aug	360
Sep	410
Oct	450
Nov	470
Dec	490
Jan	460

Forecast using 3 months moving average and 6 months moving average. Which of the forecasting method is better?

Solution: Forecast using 3 months moving average and 6 months moving average is given in Table 19.13.

Table 19.13 Forecast using 3 months moving average and 6 months moving average

Month	Actual sales	Forecast using 3 months moving average	Forecast using 6 months moving average
Jan	450		
Feb	440		
Mar	460		
Apr	410	450	
May	380	437	
June	400	417	
July	370	397	423
Aug	360	383	410
Sep	410	377	397
Oct	450	380	388
Nov	470	407	395
Dec	490	443	410
Jan	460	470	425

From Table 19.13, forecasted demand using 3 months moving average:

$$\text{July} = 397$$
$$\text{Aug} = 383$$

Forecasted demand using 6 months moving average:

$$July = 423$$
$$Aug = 410$$

A computation of mean square error is given in Table 19.14.

Table 19.14 To determine mean square error

Month	Actual sales	Forecast using 3 months moving average	Forecast using 6 months moving average	Square of error (3 months moving average)	Square of error (6 months moving average)
July	370	397	423	729	2,809
Aug	360	383	410	529	2,500
Sep	410	377	397	1,089	169
Oct	450	380	388	4,900	3,844
Nov	470	407	395	3,969	5,625
Dec	490	443	410	2,209	6,400
Jan	460	470	425	100	1,225
Sum				13,525	22,632

$$\text{Mean square error (3 months moving average)} = \frac{13,525}{7} = 1,932$$

$$\text{Mean square error (6 months moving average)} = \frac{22,632}{7} = 3,233.14$$

Mean square error for 3 months moving average is small. Hence, 3 months moving average is preferred.

19.3.2.3.3 Weighted moving average

The more recent observations are typically given more weight than older observations. For convenience, the weights usually sum to 1. The regular moving average gives equal weight to past data values when computing a forecast for the next period. The weighted moving average allows different weights to be allocated to past data values.

19.3.2.3.4 Solved problem

Problem 19.7: The past data on the load on the weaving machine is given in Table 19.15.

Table 19.15 Load on the weaving machine in hours

Month	June	July	Aug	Sep	Oct	Nov
Load	585	610	675	750	860	970

i. Compute the load on the weaving machine using 5 months moving average for the month of December

ii. Compute a weighted 3 months moving average for December, where the weights are 0.5 for the latest month 0.3 and 0.2 for others months respectively.

Solution: Forecast for the month of December using 5 months moving average

$$= \frac{610 + 675 + 750 + 860 + 970}{5} = 773$$

Forecast for the month of December using weighted 3 months moving average

$$= 0.5 \times 970 + 0.3 \times 860 + 750 \times 0.2 = 893$$

19.3.2.3.5 Exponential smoothing method

Simple moving average method gives equal weight to all the periods. Exponential smoothing method is distinguished by the fact that it assigns weight to all previous data and pattern of weights assigned are of exponential form, i.e. more weight is given to recent data and the weights assigned to older period decrease exponentially.

$$F_t = \alpha D_{(t-1)} + (1 - \alpha)F_{(t-1)}$$

is called smoothing constant $0 < \alpha < 1$

F_t = Forecasted demand for the period t

$D_{(t-1)}$ = Actual demand for the period $(t-1)$

$F_{(t-1)}$ = Forecasted demand for the period $(t-1)$

$$F_{(t-1)} = \alpha D_{(t-2)} + (1- \alpha)F_{(t-2)}$$

$$F_t = \alpha D_{(t-1)} + (1- \alpha)(\alpha D_{(t-2)} + (1- \alpha)F_{(t-2)})$$

$$F_t = \alpha D_{(t-1)} + \alpha (1- \alpha) D_{(t-2)} + (1- \alpha)^2 F_{(t-2)}$$

$$F_t = \alpha D_{(t-1)} + \alpha (1- \alpha) D_{(t-2)} + (1- \alpha)^2 (\alpha D_{(t-3)} + (1- \alpha)F_{(t-3)})$$

$$F_t = \alpha D_{(t-1)} + \alpha (1- \alpha) D_{(t-2)} + \alpha (1- \alpha)^2 D_{(t-3)} + (1- \alpha)^3 F_{(t-3)}$$

For n periods

$$F_t = \alpha D_{(t-1)} + \alpha (1- \alpha) D_{(t-2)} + \alpha (1 - \alpha)^2 D_{(t-3)} + ... + \alpha(1- \alpha)^{(n-1)} D_{(t-n)} + (1- \alpha)^n F_{(t-n)}$$

Pattern of weights in exponential method is given in Fig. 19.9

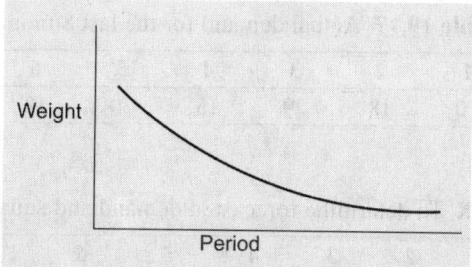

Fig. 19.9 Weights in exponential smoothing follows exponential curve

The weight for each of the demands in the past is discounted by a factor $(1 - \alpha)$. The value of the α selected is small (0.05 to 0.1), if the demand pattern is smooth or stable and large value of α is used for the fluctuating demand.

Degree of smoothing as moving average for n periods is $\alpha = \dfrac{2}{(n+1)}$

Comparison of moving average and exponential smoothing:

Exponential smoothing carries all past history (forever)

Moving average eliminates "bad" data after n periods

Moving average requires all n past data points to compute new forecast estimate while exponential smoothing only requires last forecast and last observation of 'demand' to continue.

19.3.2.3.6 Solved problem

Problem19.8: The demand for the disposal plastic tubing for general hospital is 300 units and 350 units for September and October respectively. Using 200 units as forecasted demand for September compute the forecast for the month of November. Assume the value of alpha = 0.7.

Solution:

Table 19.16 Given data

Month	Sep	Oct
Actual demand D_t	300	350
Forecasted demand F_t	200	–

$$\alpha = 0.7$$

$$F_t = \alpha D_{(t-1)} + (1-\alpha) F_{(t-1)}$$

$$F_{oct} = \alpha D_{sep} + (1-\alpha) F_{sep}$$

$$= 0.7 \times 300 + 0.3 \times 200$$

$$= 270 \text{ units}$$

$$F_{nov} = \alpha D_{oct} + (1-\alpha) F_{oct}$$

$$= 0.7 \times 350 + 0.3 \times 270$$

$$= 326 \text{ units.}$$

Problem 19.9: The demand for the particular period is given in Table 19.17. Compute forecasted demand for 8 periods using exponential smoothing method. Take $\alpha = 0.1$ and 0.3. Which of the forecast is better? Assume forecasted demand for period 1 is 15.

Table 19.17 Actual demand for the last 8 months

Period	1	2	3	4	5	6	7	8
Demand	10	18	29	15	30	12	16	8

Solution:

Table 19.18 To determine forecasted demand and square of error

Period	1	2	3	4	5	6	7	8	Sum	Mean
Actual Demand	10	18	29	15	30	12	16	8		
Forecasted demand $\alpha = 0.1$	15	14.5	14.85	16.26	16.14	17.52	16.97	16.87		
Forecasted demand $\alpha = 0.3$	15	13.5	14.85	19.09	17.86	21.50	18.65	17.85		
Square of error $\alpha = 0.1$	25	12.25	200.2	1.58	192.09	30.4		0.94	78.6	541.06	67.6
Square of error $\alpha = 0.3$	25	20.25	200.2	16.72	147.3	90		7.0	97.6	604.07	75.5

Square of error is less for $\alpha = 0.3$. So $\alpha = 0.3$ is a better method.

Problem 19.10: The actual demand for an item is as shown in Table 19.19. Using the 9 months moving average determine forecasted demand for the month 32. By using exponential forecast determine forecasted demand for the month 33.

Table 19.19 The actual demand for an item

Month	24	25	26	27	28	29	30	31	32
Demand	78	65	90	71	80	101	84	60	73

Solution:

Forecast for the month of 32 using 9 months moving average $= \dfrac{\Sigma \text{Demand}}{9} = F_{32}$

$$F_{32} = \frac{78 + 65 + 90 + 71 + 80 + 101 + 84 + 60 + 73}{9} = 78$$

$$\alpha = \frac{2}{(n+1)} = \frac{2}{9+1} = 0.2$$

$$F_{33} = D_{32} + (1 - \alpha)F_{32} = 0.2\,(73) + (1 - 0.2)\,78 = 77 \text{ units}$$

19.3.2.3.7 Forecast for seasonal variations

Seasonality corresponds to a pattern in the data that repeats at regular intervals. Demand fluctuations may be caused by weather, tradition, festival, etc. (Fig. 19.10). Seasonality is always of a fixed and known period. Hence, seasonal time series is sometimes called periodic time series. The corrections for such seasonal peaks and valleys in demand can be made by comparing these peaks/valleys with the general average demand during unseasonal periods. The multiplication of corresponding seasonal indices with forecast average should give forecast for the different seasonal periods.

The following are the methods used to forecast for seasonal variations:

Moving average method: In a moving average method of calculation monthly (or quarterly) data is used to compute 12 months (or 4 quarters) moving average. This dampens out all seasonal fluctuations.

Simple average method: Finding simple average for each month for a series of years. Then determine average of averages. From average of averages determine seasonal index.

$$\text{Month or period average} = \frac{\text{Sum of the sales of a particular month or period for a series of years}}{\text{Number of years}}$$

$$\text{Average of averages} = \frac{\text{Sum of the averages}}{\text{Number of averages}}$$

$$\text{Seasonal index for a particular month or period} = \frac{\text{Month or period average}}{\text{Average of averages}}$$

Ratio to trend method:

Determine the trend using forecasting equation.

For linear trend forecasting equation will be of the form

$$Y = a + bX$$

Where

a = Intercept

b = Slope

X = Time value (years, quarters, months, etc.)

Y = Trend value

A seasonal index (SI) is a ratio that relates a recurring seasonal variation to the corresponding trend value at the given time.

In a ratio to trend method, the actual value is divided by the trend value to obtain the ratio for the each period. Period may be month or quarter.

$$\text{Ratio for period '}i\text{'} = \frac{\text{Actual value for the period '}i\text{'}}{\text{Trend value for the period '}i\text{'}}$$

Average the ratio to obtain seasonal index value

$$\text{Seasonal index value} = \frac{\text{Sum of the ratios}}{\text{Number of years}}$$

The indexes can be used to obtain forecast of the actual (seasonalized) sales.

The actual (seasonalized) sales = (Seasonal index) Trend value

19.3.2.3.8 Solved problems

Problem 19.11: The Forecasting equation in July of year 2004 with X units in months was $Y = 1800 + 20X$. Determine forecasted sales for year 2005 to year 2012. Actual sales in the month of July is given in Table 19.20. Determine seasonal index values for July and explain its meaning. Forecast of actual sales (seasonalized sales) in July of year 2012. Use ratio to trend method.

Table 19.20 Actual sales in the month of July

Year	2005	2006	2007	2008	2009	2010	2011
July month actual sales	22	30	18	26	45	36	40

Solution:

Given Forecasting equation in July of year 2004 with X units in months was $Y = 1800 + 20X$. Computation of ratio for the month of July is given in Table 19.21.

Table 19.21 Computation of ratio for the month of July

Year	2005	2006	2007	2008	2009	2010	2011	2012
July month actual sales	22	30	18	26	45	36	40	
X (months)	12	24	36	48	60	72	84	96
Yearly forecasted sales (Trend value) ($Y = 1800 + 20X$)	2040	2280	2520	2760	3000	3240	3480	3720
Monthly forecasted sales = $\dfrac{\text{Yearly sales}}{12}$	170	190	210	230	250	270	290	310
Ratio for July = $\dfrac{\text{July month actual sales}}{\text{Monthly forecasted sales}}$	0.13	0.16	0.09	0.11	0.18	0.13	0.14	

$$\text{Seasonal index value} = \frac{\text{Sum of the ratios}}{\text{Number of periods}}$$

$$= \frac{0.13 + 0.16 + 0.09 + 0.11 + 0.18 + 0.13 + 0.14}{7}$$

$$= \frac{0.94}{7} = 0.13.$$

This means that July is seasonal 13 percent of the trend value.

The actual (seasonalized) forecast for the month of July of year 2012

$$= (\text{Seasonal index value}) \times \text{Trend value} = 0.13 \times 310 = 40 \text{ units}$$

Problem 19.12: A quarterly sales figure of company for 4 years is given in Table 19.22. Assuming that the trend line is absent, determine if there is any seasonality in the sales data.

Table 19.22 A quarterly sales figure of company for 4 years

Year	I quarter	II quarter	III quarter	IV quarter
2006	7.4	8.2	6.6	7.0
2007	7.4	7.8	7.2	7.2
2008	8.00	8.2	6.6	6.2
2009	6.6	8.8	8.0	8.0

What are the seasonal indices for various quarters? Use Simple average method.

Solution: Calculation of seasonal index is given in Table 19.23.

Table 19.23 Calculation of seasonal index

Year	I quarter	II quarter	III quarter	IV quarter
2006	7.4	8.2	6.6	7.0
2007	7.4	7.8	7.2	7.2
2008	8.0	8.2	6.6	6.2
2009	6.6	8.8	8.0	8.0
Sum of the sales of a particular period for 4 years	29.4	33.0	28.4	28.4
Period average				
$= \dfrac{\text{Sum of the sales of a particular period for 4 years}}{\text{Number of years}}$	7.35	8.25	7.10	7.10
Average of averages $= \dfrac{\text{Sum of the averages}}{\text{Number of averages}}$				

Average of averages $= \dfrac{\text{Sum of the averages}}{\text{Number of averages}}$ $\dfrac{7.35+8.25+7.10+7.10}{4} = 7.45$

| Seasonal index for a period $= \dfrac{\text{Period average}}{\text{Average of averages}}$ | $\dfrac{7.35}{7.45}$ $=0.987$ | $\dfrac{8.25}{7.45}$ $=1.107$ | $\dfrac{7.10}{7.45}$ $=0.953$ | $\dfrac{7.10}{7.45}$ $=0.953$ |

Problem 19.13: Quarterly trend values for units demanded have been computed as Q1 = 620, Q2 = 655, Q3 = 690 and Q4 = 725. The corresponding seasonal indexes for the quarters are 0.72, 1.33, 1.05 and 0.90 respectively. Forecast the actual (seasonalized) sales for Q3 and Q4.

Solution: The actual (seasonalized) sales for Q3 = (Seasonal index for Q3) Trend value for Q3

The actual (seasonalized) sales for Q3 = (1.05)690 = 725 units

The actual (seasonalized) sales for Q4 = (Seasonal index for Q4) Trend value for Q4

The actual (seasonalized) sales for Q4 = (0.90)725 = 653 units

19.3.2.4 Measures of forecast accuracy or forecast error

Demand for the product is forecasted using many forecasting methods. The effectiveness of forecasting method is determined by using forecast error. Forecast error is the numerical difference between forecasted demand and the actual demand. The error should be as small as possible.

Methods of finding forecast error:

- Mean absolute deviation
- BIAS
- Mean squared error (MSE)
- Standard deviation of error

19.3.2.4.1 Mean absolute deviation

$$\text{Mean Absolute deviation} = \frac{\text{Sum of absolute values of forecast errors for all periods}}{\text{Number of periods}}$$

$$= \frac{\Sigma|F_i - D_i|}{n}$$

It is a measure or forecast error for each period and the average forecast error. It expresses the magnitude but not the direction of error. The absolute value is referred as mean absolute deviation.

19.3.2.4.2 BIAS

In this method direction of forecast error is determined. This is also calculated as sum of the forecast errors for all periods divided by the total number of periods.

$$\text{Bias} = \frac{\Sigma(F_i - D_i)}{n}$$

Bias indicates the directional tendency of forecast errors.

19.3.2.4.3 Mean squared error (MSE)

The average of the squared forecast errors for the historical data is calculated. The forecasting method or parameter(s) which minimize this mean squared error is then selected.

19.3.2.4.4 Standard deviation of error

There are several formulae for finding out error. One formula is given below:

$$S_e = \sqrt{\frac{\Sigma y^2 - a\Sigma y - b\Sigma xy}{n-2}} .$$

19.3.2.4.5 Tracking signal

$$\text{Mean absolute deviation} = \text{MAD} = \frac{\Sigma|\text{Actual demand} - \text{Forecasted demand}|}{n}$$

$$\text{Tracking signal} = \frac{\Sigma(\text{Actual demand} - \text{Forecasted demand})}{\text{MAD}} .$$

Action limits for tracking signals commonly range from three to eight. When the signal goes beyond the range, corrective action may be required.

19.3.2.4.6 Solved problems

Problem 19.14: A dealer for electrical appliances forecasts the demand for Geyser at the rate of 500 per month for the next 3 months. The actual demands turned out to be 400, 560 and 700. Calculate the forecast error, using MAD and BIAS.

Solution: $\text{Mean absolute deviation} = \dfrac{\Sigma|F_i - D_i|}{n}$

$$\text{Mean absolute deviation} = \frac{|(500-400)| + |(500-560)| + |(500-700)|}{3} = 120 \text{ units}$$

$$\text{Bias} = \frac{\Sigma(F_i - D_i)}{n} = \frac{(500-400) + (500-560) + (500-700)}{3} = -53.33$$

Problem 19.15: An item has forecast as shown in the Table19.24. Compute tracking signal.

Table 19.24 Forecasted sales for 6 periods

Period	Actual sales	Forecasted sales
1	80	78
2	92	79
3	71	83
4	83	79
5	90	80
6	102	83

Solution: Calculation of absolute deviation for each period is shown in Table 19.25.

Table 19.25 Calculation of absolute deviation for each period

Period	Actual sales	Forecasted sales	Actual – Forecast	|Actual – Forecast|
1	80	78	2	2
2	92	79	13	13
3	71	83	–12	12
4	83	79	4	4
5	90	80	10	10
6	102	83	19	19
			$\Sigma(\text{Actual} - \text{Forecast}) = 36$	$\Sigma\|\text{Actual} - \text{Forecast}\| = 60$

$$\text{Mean absolute deviation} = MAD = \frac{\Sigma|\text{Actual} - \text{Forecast}|}{n} = \frac{60}{6} = 10$$

$$\text{Tracking signal} = \frac{\Sigma(\text{Actual} - \text{Forecast})}{MAD} = \frac{36}{10} = 3.6$$

19.4 QUESTIONS AND PROBLEMS

1. Define forecasting. Explain its importance.
2. State the objectives of forecasting.
3. Sate the advantages and limitations of forecasting.
4. Name the various methods of forecasting.
5. Describe the least square method of forecasting.
6. Describe moving average method.
7. Describe various qualitative forecasting techniques.
8. Describe various quantitative forecasting techniques.
9. Describe various methods of determining forecasting errors.
10. Explain tracking signal.
11. Use least square method to develop a linear trend equation for the data given in the following Table. Determine the equation and also forecast a trend value for the year 16.

Year	1	2	3	4	5	6	7	8	9	10	11
Shipments (tons)	2	3	6	10	8	7	12	14	14	18	19

Answer: $Y = 10.3 + 1.6X$; 26.3 tons

12. A food processing company uses a moving average method to forecast next month demand. Past actual demand in units is shown in the Table.

Month	Actual demand
43	105
44	106
45	110
46	110
47	114
48	121
49	130
50	128
51	137

a. Compute 5 months moving average to forecast demand for the month 52.
b. Compute weighted 3 months moving average where the weights are highest for the latest three months and descend in order of 3, 2, and 1.

Answer: a. 126 units b. 133 units

13. The moving average forecast and actual demand for a hospital drug are shown in the Table. Compute the tracking signal.

Periods	Actual sales	Forecasted sales
27	71	78
28	80	75
29	101	83
30	84	84
31	60	88
32	73	85

Answer: 2.05

14. The demand over the past 9 months for a new breakfast cereal is shown in the Table. Develop forecast for November using a 5 period moving average where the weights are (from earliest to latest) 1,1 2,2 and 4.

Month	Feb	Mar	Apr	May	June	July	Aug	Sep	Oct
Units	70	76	75	80	92	87	93	114	105

Answer: 101 units

15. For $n = 7$ years of time series data $\Sigma Y = 56$, $\Sigma XY = 70$ and $\Sigma X^2 = 28$
a. Find intercept and slope of the linear trend line.
b. Forecast the Y value for 6 years distant from the origin.

Answer: a. Intercept = 8.0; slope = 2.5 b. 23

16. The annual sales of a company are as given below:

Year	2006	2007	2008	2009	2010
Sales in rupees	50,000	65,000	75,000	52,000	72,000

By the method of least square find the trend values for each of the five years. Also estimate the annual sales for the year 2011.

Answer: $a = 62.80$, $b = 3.1$ and Expected sales for 2006 = ₹ 66,100

17. Find the trend by least square method for data as follows:

Year	2006	2007	2008	2009	2010	2011	2012
Demand in 1,000 units	85	75	80	72	65	60	55

Also estimate demand for 2013.

Answer: $a = 70.28$, $b = -4.82$ and Expected sales for 2006 = 40,531 units

18. Demand for 10 months is given below. Determine forecast for the month of November using 3 months and 4 months moving averages.

Month	Jan	Feb	Mar	Apr	May	June	July	Aug	Sep	Oct
Sales in units	50	40	90	45	55	60	55	50	45	50

Answer: 3 months moving average = 48.3; 4 months moving average = 50

19. For the data given in the Table,
    a. Determine MAD
    b. Tracking signal

Period	Forecasted demand (units)	Actual demand (units)
1	900	1,000
2	1,000	1,100
3	1,050	1,000
4	1,010	960
5	980	970
6	985	970
7	980	995

Answer: MAD = 48.6; Tracking signal: +1.852

20. A firm has the following sales pattern during 2005 to 2009. Compute the sales for 2010 using $y = ab^x$

Year	2005	2006	2007	2008	2009
Sales (₹ lacs)	106	118	111	123	129

Answer: Sales for 2010 = 133.25 lacs

21. The production manager of a company has projected trend values for June, July and August are 586, 589 and 592 respectively. Seasonal indexes for June, July and August are 1.44, 1.22 and 1.06 respectively. Compute seasonalized production for June, July and August. **Answer:** 844, 719, 628

20

Production Planning and Control Systems

20.1 INTRODUCTION

Production planning is concerned with:
- Deciding which products to make
- Determining the quantity of each product to be produced and timing of output
- Determining resources required to accomplish the production plan

20.2 ACTIVITIES IN PRODUCTION PLANNING

Activities in production planning (Fig. 20.1) include the following:
- **Aggregate production planning:** This involves planning the production output levels for major product lines produced by the firm.

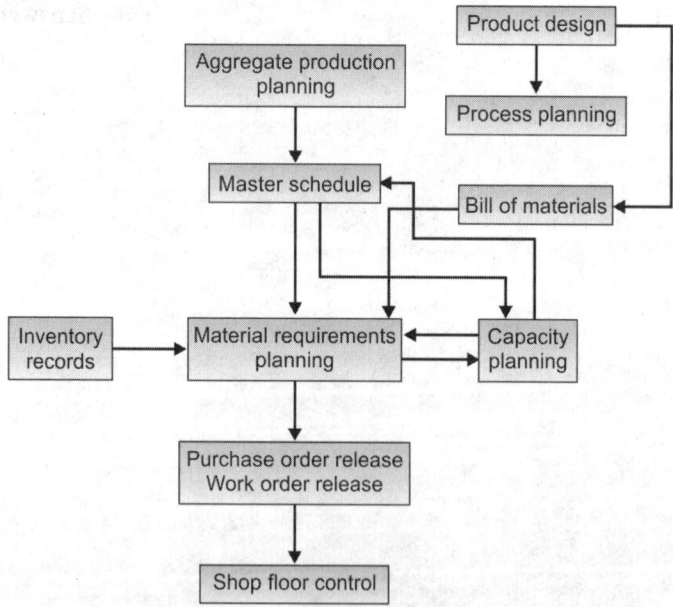

Fig. 20.1 Activities in production planning and control system

- **Master production schedule:** Aggregate production plan must be converted into a master production schedule which is a specific plan of the quantities produced of individual models within product line.
- **Material requirements planning:** Material requirements planning is a computational technique that converts master production schedule for the end product into detail purchasing and manufacturing schedule of raw materials and components used in end product.
- **Capacity planning:** Capacity planning is concerned with determining labor and equipment capacity required to meet the current master production schedule as well as the long-term future production needs of the firm.

20.2.1 Aggregate Production Planning

Aggregate production planning is an intermediate range planning. It is the process of planning the quantity and timing of output over the intermediate time horizon (3 months to 1 year). Table 20.1 is an example of aggregate production planning. Within this range the physical facilities and capacity of the plant are assumed to be fixed. Therefore, fluctuations in demand are met by varying labor and inventory schedules.

Table 20.1 Aggregate production plan

Month	Jan	Feb	Mar	Apr	May	Jun	Jul	Aug	Sep
Number of motors	30	45	50	30	60	30	30	40	40

Strategies by which fluctuations in demand are met:
- Vary the size of workforce: Demand is met by hiring and laying off workers in proportion to change in demand.
- Vary the number of hours worked: Maintain the stable workforce but permit idle time when demand is less and permit overtime when demand is peak.
- Vary inventory levels: Demand fluctuations can be met by producing more in earlier period to meet the demand of later period and allowing backlogs to meet this demand by producing more in later periods when capacity is available.
- Subcontract: Subcontracting excess demand to other companies.

Table 20.2 Costs involved in various strategies

Strategy	Cost
Vary the size of workforce:	
• Hire additional personnel as demand is more	• Employment cost for advertising and recruitment
• Layoff personnel as demand decreases	• Cost of compensation to workers for layoff
Vary the number of hours worked:	
• Work additional hours without changing the workforce	• Overtime premium wages
• Allowing slack period when demand is less	• Excess wages during slack period
Vary inventory levels:	
• Produce in earlier period and hold it until product is demanded	• Cost of holding inventory
• Offer to deliver the product when capacity is available	• Delay in receipt of revenues, loss of sales and customer dissatisfaction
Subcontract:	
• Subcontracting excess demand to other companies	• Reduce company overheads
	• Increase in subcontractor profits

Aggregate production planning seeks best combination of strategies to minimize the cost.

Costs involved in various strategies: Costs involved in various strategies is shown in Table 20.2.

20.2.1.1 Aggregate production planning methods

Aggregate production problems can be solved by using:
- Graphical method
- Heuristic approach
- Linear programing method

Graphical method: In this method, cumulative demand values and cumulative production capacities are plotted on the same graph (Fig. 20.3). This would help us to identify the gap between demand and production capacity in different periods. In this method, cost data is not taken into account.

20.2.1.2 Solved problem

Problem 20.1: A company has developed a forecast for an item that has the demand pattern as shown in the Table 20.3.

Table 20.3 Forecast for 8 months

Period	1	2	3	4	5	6	7	8
Forecast Demand	270	220	470	670	450	270	200	370

a. Plot the demand as a histogram. Determine average demand. Determine the production rate required to meet average demand.

b. Plot the actual cumulative forecast required and cumulative production over planning horizon. Indicate the excess inventory and backorder on the graph.

Solution: a. Draw histogram using the data from the Table 20.3. The resultant histogram is shown in the Fig. 20.2.

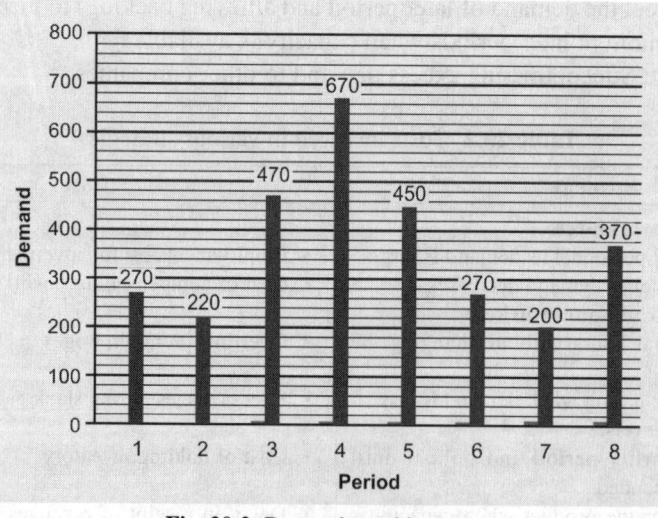

Fig. 20.2 Demand as a histogram

$$\text{Average demand} = \frac{270 + 220 + 470 + 670 + 450 + 270 + 200 + 370}{8} = 365$$

Assume production rate = Average demand = 365 units/period

b. Determine cumulative demand and cumulative production over planning horizon (Table 20.4). Draw the graph between cumulative demand and cumulative production over planning horizon using the data from the Table 20.4. The resultant graph is shown in Fig. 20.3.

Table 20.4 Cumulative demand and cumulative production over planning horizon

Period	1	2	3	4	5	6	7	8
Demand	270	220	470	670	450	270	200	370
Cumulative demand	270	490	960	1,630	2,080	2,350	2,550	2,920
Cumulative production	365	730	1,095	1,460	1,825	2,190	2,555	2,920

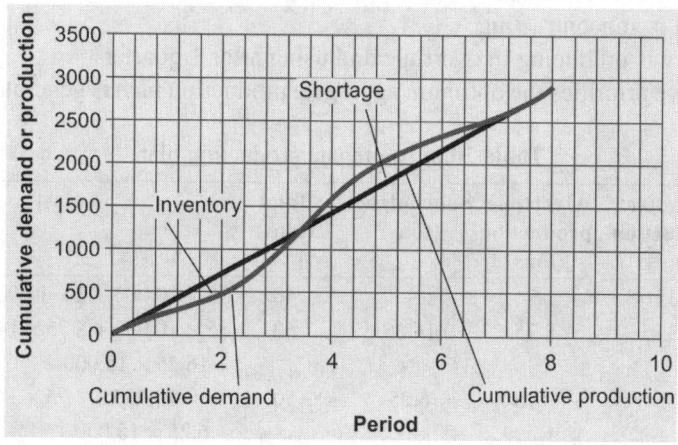

Fig. 20.3 Graph between cumulative demand and cumulative production over planning horizon

Heuristic approach

In this method, common sense and experience are used for determining production plan. This is the only heuristic method. Optimum solution is not guaranteed.

Solved problem

Problem 20.2: A company produces soaps. The aggregate production planning is done for a time horizon of one year. The planning horizon is divided into 4 quarters. The demand is given in Table 20.5.

Table 20.5 Demand in planning horizon

Quarter	I	II	III	IV
Demand (tons)	40	60	50	45

The company has a regular workforce which can produce 35 tons of output per quarter. If the workers are allowed to work overtime with a restriction that extra time cannot be more than 20% of regular time in any time. The output rate is 25% higher than regular time during overtime but the overtime expenses are 40% more than regular time. The company subcontract to other company at the cost of 50% premium than the cost of regular production. The regular time production cost is ₹ 10,000 per ton. Shortages are not allowed. Inventory carrying cost is ₹ 5,000 per ton per annum. Design the cost efficient production plan.

Solution:

Regular time production cost = ₹ 10,000/ton

Over time production cost = $10,000 + 0.4 \times 10,000 = ₹ 14,000$/ton

Subcontracting cost = $10,000 + 0.5 \times 10,000 = ₹ 15,000$/ton

Producing in overtime and using after 1 quarter = $14,000 + 5,000/4 = 15,250$/ton

Production capacity:

Regular time = 35 tons/quarter

Over time = $0.20 \times 35 \times 1.25 = 8.75$ tons/quarter

Priorities as per cost of production:

First priority is regular production

Second priority is overtime production

Third priority is subcontracting

Fourth priority is producing in overtime and using after 1 quarter

Based on above priorities the optimum aggregate production plan is generated in Table 20.6.

Table 20.6 Aggregate production plan

Period	Regular production (tons)	Overtime production (tons)	Subcontract (tons)	Total (tons)	Cost	
Quarter 1	35	5	–	40	$35 \times 10,000 + 5 \times 14,000$	= ₹ 4,20,000
Quarter 2	35	8.75	16.25	60	$35 \times 10,000 + 8.75 \times 14,000$ $+ 16.25 \times 15,000$	= ₹ 7,16,250
Quarter 3	35	8.75	6.25	50	$35 \times 10,000 + 8.75 \times 14,000$ $+ 6.25 \times 15,000$	= ₹ 5,66,250
Quarter 4	35	8.75	1.25	45	$35 \times 10,000 + 8.75 \times 14,000$ $+1.25 \times 15,000$	= ₹ 4,91,250
Total production cost to meet annual demand						**₹ 21,93,750**

Linear programing method

Aggregate planning is viewed as allocating capacity to meet demand or requirement. Transportation algorithm is used to solve the problem. In this case, supply consists of inventory on hand and units that can be produced through regular time (RT) and overtime (OT). Demand consists of individual period requirements plus any desired ending inventory. Costs associated are cost of producing units and carrying them inventory with later period. This approach can be used to include backorder costs. Production in a later period can be allocated to supply backordered demand from an earlier period at a shortage cost.

Solved problem

Problem 20.3: A production manager has a product which must be produced to meet fluctuating demand, He knows that the monthly requirements are 900 for the I month, 700 for the II month 1,100 for the III month and 1,000 for IV month. The product can be produced either on regular time or overtime. Regular production cannot exceed 900 items per month and over time production cannot exceed 500 items per month. Manufacturing cost per item in normal working time varies each month. The manufacturing costs for regular time are ₹ 3 in I month, ₹ 4 in II month, ₹ 2.50 in III month and ₹ 3 in IV month. The manufacturing costs for overtime are ₹ 4 in I month, ₹ 5 in II month, ₹ 3.5 in III month, and ₹ 4 in IV month. Monthly storage cost is

₹ 2 per unit. No inventory is to remain at the end of IV month. No shortages are allowed. Determine optimum production plan.

Solution: Demand of period I cannot be used from the production of period II, because shortages are not allowed. Hence cost of cell (II, I) is ∞. If the demand for the period II is met from the production of period I then add holding cost to the production cost. Similarly obtain costs in transportation table. These costs are shown in Table 20.7.

RT: Regular time

OT: Over time

Table 20.7 Converting the given data into transportation table

Period		I	II	III	IV	Dummy	Capacity
I	RT	3	5	7	9	0	900
	OT	4	6	8	10	0	500
II	RT	∞	4	6	8	0	900
	OT	∞	5	7	9	0	500
III	RT	∞	∞	2.5	4.5	0	900
	OT	∞	∞	3.5	5.5	0	500
IV	RT	∞	∞	∞	3	0	900
	OT	∞	∞	∞	4	0	500
Demand		900	700	1,100	1,000	1,100	5,600

The above transportation problem is solved by using Transportation algorithm. Optimal solution is given in Table 20.8.

Table 20.8 Optimal solution table

Period		I	II	III	IV	Dummy	Capacity
I	RT	3 (900)	5	7	9	0	900
	OT	4	6	8	10	0 (500)	500
II	RT	∞	4 (700)	6	8	0 (200)	900
	OT	∞	5	7	9	0 (500)	500
III	RT	∞	∞	2.5 (900)	4.5	0	900
	OT	∞	∞	3.5 (200)	5.5	0 (300)	500
IV	RT	∞	∞	∞	3 (900)	0	900
	OT	∞	∞	∞	4 (100)	0 (400)	500
Demand		900	700	1,100	1,000	1,900	5,600

Problem 20.4: Consider a three period model where regular and overtime productions are used. The production capacities for the three periods are given in Table 20.9.

Table 20.9 Production capacities

Period	Production capacity (units)	
	Regular	Overtime
1	15	10
2	15	0
3	20	15

The production cost per unit is ₹ 5 for regular production and ₹ 10 for overtime production. The holding cost per unit and shortage cost per unit are given by 1 and 2 respectively. The demand units for three periods are 20, 35 and 15 respectively. Determine optimum production schedule.

Solution: Demand of period I can be used from the production of period II by adding shortage cost. Hence cost of cell RT (II, I) is 7. If the demand for the period II is met from the production of period I then add holding cost to the production cost. Similarly obtain costs in transportation table. These costs are shown in Table 20.10.

RT: Regular time

OT: Over time

Table 20.10 Converting the given data into transportation table

Period		I	II	III	Dummy	Capacity
I	RT	5	6	7	0	15
	OT	10	11	12	0	10
II	RT	7	5	6	0	15
	OT	0	0	0	0	0
III	RT	9	7	5	0	20
	OT	14	12	10	0	15
Demand		20	35	15	5	75

Initial solution is given in Table 20.11.

Table 20.11 Initial solution

Period		I	II	III	Dummy	Capacity
I	RT	5 ⑮	6	7	0	15
	OT	10	11 ⑤	12	0 ⑤	10
II	RT	7 ⑤	5 ⑩	6	0	15
	OT	0	0	0	0	0
III	RT	9	7 ⑳	5	0	20
	OT	14	12	10 ⑮	0	15
Demand		20	35	15	5	75

Optimal solution is given in Table 20.12.

Table 20.12 Optimal solution

Period		I	II	III	Dummy	Capacity
I	RT	5 ⑮	6	7	0	15
	OT	10 ⑤	11 ⑤	12	0	10
II	RT	7	5 ⑮	6	0	15
	OT	0	0	0	0	0
III	RT	9	7 ⑤	5 ⑮	0	20
	OT	14	12 ⑩	10	0 ⑤	15
Demand		20	35	15	5	75

The associated minimum cost = ₹ 485

20.2.2 Master Production Schedule

Master production schedule follows aggregate production planning. It gives the detailed plan of each model. The example of master production plan is given in Table 20.14. If the company is producing only one model or product then aggregate production plan and master production schedule are same. Table 20.13 is an example of aggregate production plan.

Table 20.13 Aggregate production plan

Month	Jan	Feb	Mar	Apr	May	Jun	Jul	Aug	Sep
Number of motors	30	45	50	30	60	30	30	40	40

Master production schedule from Aggregate production plan is given in Table 20.14.

Table 20.14 Master production schedule from aggregate production plan

Month	Jan	Feb	Mar	Apr	May	Jun	Jul	Aug	Sep
5 h.p AC Motors	5	5	10	5	15	6	10	–	10
10 h.p AC Motors	10	7	10	5	10	4	5	–	20
5 h.p DC Motors	5	10	15	10	15	10	–	5	10
10 h.p DC Motors	10	23	15	10	20	10	–	35	–
Total number of motors	30	45	50	30	60	30	30	40	40

Time interval used in master production schedule depends upon the volume and manufacturing lead time of the products to be produced. Normally, weekly or monthly time intervals are used.

Main functions of master production schedule:

- It evaluates alternative schedules.
- It is the basic input for material requirements planning.
- Capacity requirements are directly derived from the master production schedule.

20.2.3 Material Requirements Planning (MRP I)

Material requirements planning is a computational technique that converts master production schedule for the end product into detail purchasing and manufacturing schedules of raw materials and components used in end product. Each end product may contain hundreds of individual components. These components are produced from raw material or purchased from the supplier. Some of these components are common for several end products. These components are assembled into simple assemblies and these subassemblies are put together into more complex subassemblies and so on until final products are assembled. Each step in manufacturing, purchasing and assembly takes time. All these factors must be incorporated into MRP calculation. Each calculation is simple. But the magnitude of data is so large that the manual application of MRP is practically impossible. That is why MRP system is computerized. Flowchart of material requirements planning is given in Fig. 20.4.

Objectives of MRP:

- To improve customer service by meeting delivery schedules
- To reduce inventory cost by reducing inventory levels
- To improve plant operating efficiency

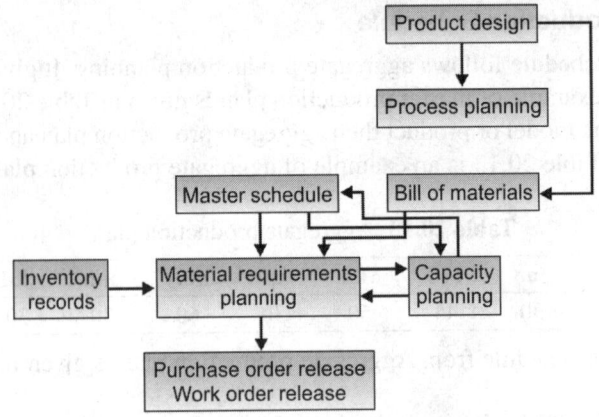

Fig. 20.4 Material requirements planning

MRP inputs:

MRP inputs are:

- Bill of materials
- Master production schedule
- Inventory records
- Lead times

Master production schedule: Master production schedule lists what end products and how many of each to be produced and when they are to be produced. The master production schedule provides overall production plan for the final product in terms of month by month deliveries.

Example of master production schedule for the product P_1 is shown in Table 20.15.

Table 20.15 Master production schedule for the product P_1

Month	Jan	Feb	Mar	Apr	May	Jun	Jul	Aug	Sep	Oct	Nov	Dec
Quantity	50	100	70	55	65	35	65	89	76	45	67	89

Bill of materials: The bill of materials is used to compute the raw material and component requirement for end product listed in the master production schedule. The bill of materials or product structure is listed in Fig. 20.5. Product P_1 is composed of two assemblies S_1 and S_2. S_1 is made up of components C_1, C_2 and C_3. S_2 is made up of components C_4, C_5 and C_6 respectively. The items at higher level are called parents of the items immediately below. For example, S_1 is

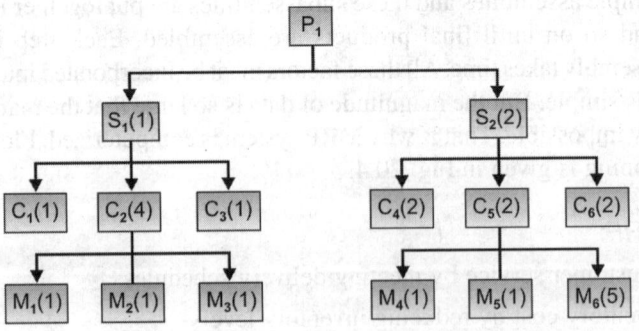

Fig. 20.5 Product structure or Bill of materials for product P_1

the parent of C_1, C_2 and C_3. The product structure must also specify the number of each subassembly and raw material that go into parent. These numbers are shown in the parenthesis in Fig. 20.5.

MRP computation for the product P_1 for the month of January using master production schedule and bill of materials is given in Table 20.16.

Table 20.16 MRP computation of product P_1 for the month of January

Part	P_1	S_1	S_2	C_1	C_2	C_3	C_4	C_5	C_6	M_1	M_2	M_3	M_4	M_5	M_6
Gross requirements	50	50	100	50	200	50	200	200	200	50	200	50	200	200	1000

Inventory record files: Other input to the MRP is inventory files. Computed quantities must be adjusted for inventories on hand or order.

Net requirement = Gross requirement – Inventory on hand + Ending inventory

Lead times are taken into consideration for calculation of material requirements planning. There are two types of lead times.

- Purchasing lead time is the time between placing an order and actual receipt of stock.
- Manufacturing lead time is the time taken in producing fabricated components. The manufacturing lead time is substantial.

MRP outputs: The MRP program generates variety of outputs.

The outputs include:

- Purchase order: Purchase order provides the authority to purchase raw materials or parts from outside vendors with quantity and delivery dates specified.
- Work order: Work order generates the authority to produce parts or subassemblies in the company's own factory.

Benefits of MRP:

- Reduction in inventory
- Better machine utilization
- Improved capacity utilization
- Aid in developing master production schedule
- Quicker response to change in demand

20.2.4 Capacity Planning

Plant capacity is used to define the maximum rate of output that the plant can produce under a given set of operating conditions. The operating conditions refer to number of shifts; number of days of plant operation per week and employment levels. Capacity for a production plant is measured in terms of output units/unit time.

Capacity planning (Fig. 20.6) is concerned with determining labor and equipment capacities required to meet the current master production schedule as well as the long-term future production needs of the firm. Master production schedule gives the information about output and timing of output. The master schedule is transformed into material and component requirements using MRP. Then these requirements are compared with available plant capacity over the planning horizon. If the schedule is incompatible with capacity, adjustments must make either in the master schedule or in plant capacity. Capacity adjustments can be accomplished in either short-term or long-term.

Short-term adjustments:

- Employment level: Hiring of the workers to meet the excess demand or laying off workers when demand is less

- Number of work shifts: Increasing or decreasing the number of shifts per week
- Labor overtime hours
- Inventory stockpiling
- Order backlogs
- Subcontracting

Long-term capacity requirements would include the following types of decisions:

- Investing in more productive machines or new types of machines
- New plant construction
- Purchasing of existing plants from other companies
- Closing down or selling off exiting facilities that will not be needed in future

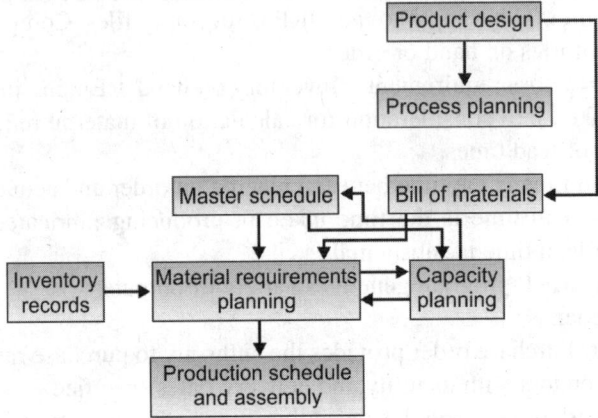

Fig. 20.6 Capacity planning

20.2.5 Production Control

Production control attempts to take corrective action when actual production is not progressing as per plan. Production control includes various techniques for controlling production and inventory. The important production control techniques are shop floor control, line of balance and inventory control.

The shop floor control :

The shop floor control (Fig. 20.7) is concerned with the following:

- Releasing production orders to the production shop
- Scheduling of production orders
- Monitoring and controlling progress of the production orders
- Acquiring current information on the status of the production

Three phases of shop floor control:

- Order release
- Order scheduling
- Order progress

Order release: Provides documentation to process a production order through the factory. Documentation for an order consists of:

1. Route sheet: Listing the sequence of operations, tools required, etc.
2. Material requisitions: To draw necessary raw materials from stores

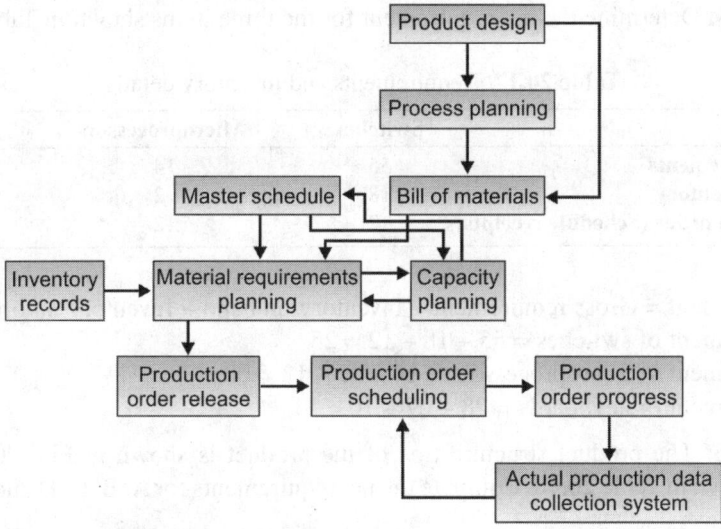

Fig. 20.7 Shop floor control

3. Job cards: Enough job cards to report the labor for each operation on the route sheet

4. Move tickets: To authorize transport of parts between work centers

5. Parts list: For assembly jobs

Order scheduling: The purpose of order scheduling is to make assignments of production orders to work centers in the plant. If the numbers of jobs are more than one for each production center then determining the sequence in which orders will be processed through each work center. Each job has certain priority determined by due date and other factors. Some of the methods of assigning priorities for production scheduling are:

• First come first serve

• Earliest due date

• Shortest processing time

• Least slack time (Time remaining until due date minus processing time remaining)

• Critical ratio (Ratio of time remaining until due date divided by processing time remaining)

Order progress: Order Progress monitors the status of the orders in the plant, work-in-process inventory, and other parameters that indicate production progress and performance. Factory reports generated by order progress module:

• Work order status reports: Whether orders are on schedule or behind

• Progress reports: Number of orders completed *vs* number that should have been completed

• Exception reports: These are designed to pinpoint deviations from the production schedule, overdue jobs and other exceptions.

These reports can be generated daily to achieve better control over the jobs in the plant.

20.2.6 Solved Problems

Problem 20.5: Determine the net requirement for the three items shown in Table 20.17.

Table 20.17 Requirements and inventory details

	Switches	Microprocessors	Keyboards
Gross requirements	55	14	28
On hand inventory	18	2	7
Inventory on order (Schedule receipts)	12	12	10

Solution:

Net requirement = Gross requirement – Inventory on hand – Inventory on order
Net requirement of switches = 55 – 18 – 12 = 25
Net requirement of microprocessors = 14 – 02 – 12 = 00
Net requirement of keyboards = 28 – 07 – 10 = 11

Problem 20.6: The product structure tree of the product is shown in Fig. 20.8. Inventory details are shown in Table 20.18. Compute the net requirements for A, B, C, D and E to produce 50 units of X.

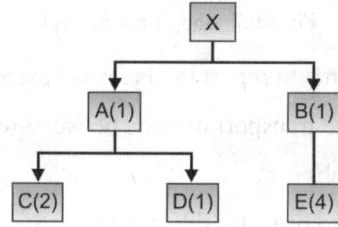

Fig. 20.8 Product structure

Table 20.18 Inventory details

Component	A	B	C	D	E
Inventory on hand	20	10	15	30	100

Solution: Gross requirements of components to produce 50 units of X are computed using product structure of product X. The resultant gross requirements of components are shown in Table 20.19.

Table 20.19 Gross requirements of components

Component	A	B	C	D	E
Gross requirements of components	50	50	60	30	160
Inventory on hand	20	10	15	30	100

Net requirement = Gross requirement – Inventory on hand
Net requirement of A = 50 – 20 = 30
Net requirement of B = 50 – 10 = 40
Net requirement of C = 60 – 15 = 45
Net requirement of D = 30 – 30 = 0
Net requirement of E = 160 – 100 = 60

Problem 20.7: Product structure tree to produce the product X is shown in Fig. 20.9. Compute the net requirements to produce 100 units of subassembly A. No stock is on hand or on order.

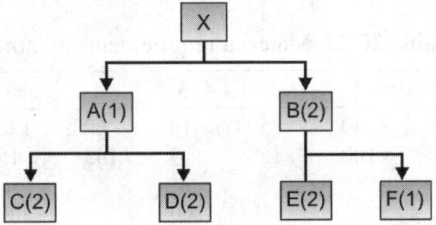

Fig. 20.9 Product structure

Solution: Gross requirements of components to produce 100 units of X are computed in Table 20.20.

Table 20.20 Gross requirements of components to produce 100 units of X

Component	A	B	C	D	E	F
Gross requirements of components	100	200	200	200	400	200
Inventory on hand	0	0	0	0	0	0

Net requirement = Gross requirement – Inventory on hand
Net requirement of A = 100 – 0 =100
Net requirement of B = 200 – 0 = 200
Net requirement of C = 200 – 0 = 200
Net requirement of D = 200 – 0 = 200
Net requirement of E = 400 – 0 = 200
Net requirement of F = 200 – 0 = 200

Problem 20.8: Complete the MRP format shown in Table 20.21. How many units are at end of the period 8?

Given the following Data:

Lot size = 200 units

Lead time = 3 weeks

Order the stock when the stock level is less than requirement taking lead time into consideration.

On hand available = 140 units at the beginning of week

Table 20.21 Gross requirement (units)

Week	1	2	3	4	5	6	7	8
Gross requirement (units)	40	85	10	60	130	110	50	170
On hand available (140 units at the beginning)								
Planned order receipts (at the end of week)								
Planned order release (at the beginning of week)								

Solution: Material requirements planning computation is shown in Table 20.22.

Table 20.22 Material requirements planning

Week	1	2	3	4	5	6	7	8
Gross requirement (units)	40	85	10	60	130	110	50	170
On hand available (at the end of period)	100	15	5	145	15	105	55	85
Planned order receipts (at the end of week)			200		200		200	
Planned order release (at the beginning of week)	200		200		200			

Problem 20.9: Forecast requirements of an item are shown in the Table 20.23. Complete the material requirement plan. Note that schedule receipt of 60 units is due in period 2 and safety stock of 25 units is to be maintained. Lead time is 2 weeks and order quantity is 60 units. 50 units are available at the beginning.

Table 20.23 Gross requirement (units)

Week	1	2	3	4	5	6	7	8	9	10
Gross requirement (units)	20	20	20	30	20	20	20	25	20	35
Scheduled receipts (at the end of period)										
On hand available (at the end of period)										
Planned order release (at the beginning of week)										

Solution: Order when stock is less than 25.

Material requirement planning computation is shown in Table 20.24.

Table 20.24 Material requirements planning

Week	1	2	3	4	5	6	7	8	9	10
Gross requirement (units)	20	20	20	30	20	20	20	25	20	35
Scheduled receipts (at the end of period)		60		60				60		60
On hand available (at the end of period)	30	70	50	80	60	40	80	55	35	60
Planned order release (at the beginning of week)		60			60			60		

Planned order release of 60 units in periods 2, 5 and 8.

20.3 LINE OF BALANCE

Line of balance can be used to schedule and control various processing steps. Line of balance is a graphical method of scheduling production. It is concerned with the number of parts passed through each processing step or operation.

Data required for drawing the line of balance:
- Production plan
- Planned delivery schedule
- Actual position at the review point

Methodology to draw line of balance at review point:

1. First construct cumulative delivery schedule chart with respect to time.
2. Draw the progress chart at the review point, i.e. actual number of parts passed through each processing step or operation.
3. Determination of planned number of finished units at the review point: Draw the vertical line parallel to y-axis at the review point until the cumulative delivery schedule curve is reached. The length of the vertical line from the review point to the point of intersection is the planned number of finished units at the review point.
4. Draw the horizontal line from the point of intersection to the corresponding step in the progress chart.
   a. If this line is above the actual number of units in the progress chart then actual schedule is behind the planned schedule.
   b. If this line is below the actual number of units in the progress chart then actual schedule is ahead of the planned schedule.
   c. If this line is same as actual number of units in the progress chart then actual schedule is matching the planned schedule.
5. Determination of position of the vertical line for the processing step 'i' on x-axis
   Position of the vertical line for the processing step 'i' on x-axis = Review point + Lead time for the processing step 'i'
6. To determine planned number of units for the processing step 'i', draw vertical line from the position determined above until cumulative schedule curve is reached. The length of the vertical line gives planned numbers of units for the processing step 'i'. If all the steps are completed go to the step 7 otherwise go to the step 5.
7. From the figure determine the difference between planned number of units and actual number of units.

20.3.1 Solved Problem

Problem 20.10: Production plan for the final assembly is given in Fig. 20.10. Lead times are given in days.

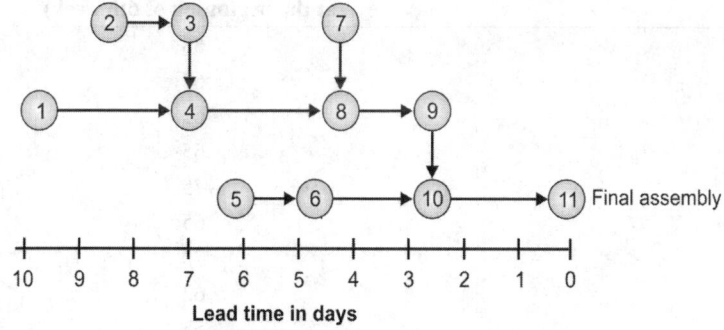

Fig. 20.10 Production plan

The production plan shows there are 11 operations or processing steps. Operation 11 is the delivery of finished units.

Operation 1 must be finished 10 days before final assembly
Operation 2 must be finished 8 days before final assembly
Operation 3 must be finished 7 days before final assembly

Operation 4 must be finished 7 days before final assembly

Operation 5 must be finished 6 days before final assembly

Operation 6 must be finished 5 days before final assembly

Operation 7 must be finished 4 days before final assembly

Operation 8 must be finished 4 days before final assembly

Operation 9 must be finished 2 days before final assembly

Operation 10 must be finished 2 days before final assembly

Planned delivery schedule of the finished units is given in Table 20.25.

Table 20.25 Planned delivery schedule

Week	Planned number of units (beginning of 6th week)
1	10
2	10
3	05
4	10
5	15
6	15
7	10
8	15
9	12
10	10
11	8

Actual number of units passed in each processing step at the beginning of 6th week is given in Table 20.26.

Table 20.26 Actual number of units processed in each step at the beginning of 6th week

Step	Number of units processed in each processing step (at the beginning of 6th week)
1	68
2	67
3	67
4	65
5	75
6	65
7	70
8	62
9	55
10	55
11	52

Construct Line of balance.

Solution: Cumulative planned delivery schedule is computed in Table 20.27.

Construct cumulative delivery schedule chart as shown in Fig. 20.11 with respect to time using the data in Table 20.27.

Table 20.27 Cumulative planned delivery schedule

Week	Cumulative number of units (At the beginning of week)
1	10
2	20
3	25
4	35
5	50
6	65
7	75
8	90
9	102
10	112
11	120

Draw the progress chart at the review point as shown in Fig. 20.11.

Determination of planned number of finished units at the review point for the step 11, i.e. final assembly:

Draw the vertical line parallel to y-axis at the beginning of 6th week until the cumulative delivery schedule curve is reached. The length of the vertical line from the review point to the point of intersection is the planned number of finished units (65 units) at the review point. Draw the horizontal line from the point of intersection to the 11th step in the progress chart. This line is above the bar chart of 11th step (52 units). So step 11 is 13 units behind the planned schedule (Fig. 20.11).

Determination of planned number of parts for the step 10:

Position of the vertical line on cumulative delivery schedule = Review point + Lead time for the step '10'.

Position of the vertical line on cumulative delivery schedule = 6 weeks + 2 days.

To determine planned number of units for the step '10', draw vertical line from the position determined above until cumulative schedule curve is reached. The length of the vertical line (69 units) gives planned number of units. Draw the horizontal line from the point of intersection to the 10th step in the progress chart (55 units). This line is above the bar chart of 10th step. So step 10 is 14 units behind the planned schedule (Fig. 20.11).

Determination of planned number of parts for the step 9:

Position of the vertical line on cumulative delivery schedule = Review point + Lead time for the step '9'.

Position of the vertical line on cumulative delivery schedule = 6 weeks + 2 days.

To determine planned number of units for the step '9', draw vertical line from the position determined above until cumulative schedule curve is reached. The length of the vertical line gives planned number of units (69 units). Draw the horizontal line from the point of intersection to the 9th step in the progress chart (55 units). This line is above the bar chart of 9th step. So step 9 is 14 units behind the planned schedule (Fig. 20.11).

Determination of planned number of parts for the step 3:

Position of the vertical line on cumulative delivery schedule = Review point + Lead time for the step '3'.

Position of the vertical line on cumulative delivery schedule = 6 weeks + 1 week = 7.

To determine planned number of units for the step '3', draw vertical line from the position determined above (7th week) until cumulative schedule curve is reached. The length of the vertical line gives planned number of units (79 units). Draw the horizontal line from the point of intersection to the 3rd step in the progress chart (67 units). This line is above the bar chart of 3rd step. So step 3 is 12 units behind the planned schedule (Fig. 20.11).

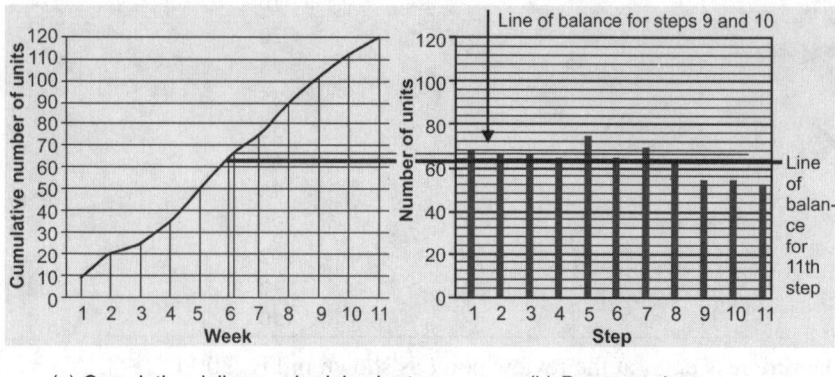

(a) Cumulative delivery schedule chart   (b) Progress chart

Fig. 20.11 Line of balance

Determination of planned number of parts for the step 4:

Position of the vertical line on cumulative delivery schedule = Review point+ Lead time for the step '4'.

Position of the vertical line on cumulative delivery schedule = 6 weeks + 1 week = 7 weeks.

To determine planned number of units for the step '4', draw vertical line from the position determined above (7th week) until cumulative schedule curve is reached. The length of the vertical line (79 units) gives planned number of units. Draw the horizontal line from the point of intersection to the 4th step in the progress chart (65 units). This line is above the bar chart of step 4. So step 4 is 14 units behind the planned schedule (Fig. 20.11).

Repeat the procedure for all steps.

Table 20.28 is generated using Fig. 20.11.

Table 20.28 Difference between planned schedule and actual schedule

Step	Planned number of units to be processed in each step (at the beginning of 6th week)	Actual number of units to be processed in each step (at the beginning of 6th week)	Behind/Ahead of schedule (Number of units)
1	85	68	17 Behind schedule
2	81	67	14 Behind schedule
3	79	67	12 Behind schedule
4	79	65	14 Behind schedule
5	77	75	02 Behind schedule
6	75	65	10 Behind schedule
7	73	70	03 Behind schedule
8	73	62	11 Behind schedule
9	69	55	14 Behind schedule
10	69	55	14 Behind schedule
11	65	52	13 Behind schedule

20.4 QUESTIONS AND PROBLEMS

1. What are the objectives of MRP I?
2. Define bill of materials.
3. What are the inputs to MRP I?
4. What is capacity?
5. Why is capacity plan needed?
6. Define the term aggregate production planning.
7. Explain master production schedule.
8. Explain MRP I with flowchart.
9. Explain capacity planning.
10. What are the outputs of MRP I?
11. Explain shop floor control with flowchart.
12. Explain line of balance.
13. Explain the following:
    a. Order release   b. Order progress   c. Order schedule
14. What do you understand by aggregate plan?
15. Describe the relevant costs involved in aggregate planning decision.
16. Explain various strategies in aggregate production planning.
17. Explain various aggregate production planning methods.
18. Given the product structure tree as shown in the figure and inventories shown in Table, compute the net requirements for A, B, C, D and E to produce 100 units of X.

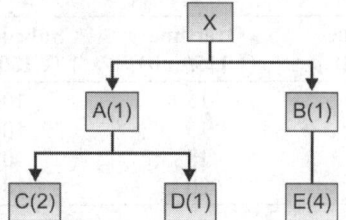

Fig. Product structure

Table Inventory details

Component	A	B	C	D	E
Inventory on hand	20	10	15	30	100

Answer: A = 80, B = 90, C = 145, D = 50, E = 260

19. Given the product structure tree shown in the figure, compute the net requirements to produce 200 units of assembly X. No stock is on hand or on order.

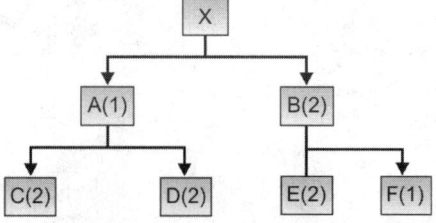

Fig. Product structure

Answer: A = 200, B = 400, C = 400, D = 400, E = 800, F = 400

20. Determine the net requirements for the three items shown in Table below.

	Switches	Microprocessors	Keyboards
Gross requirements	110	28	56
On hand inventory	18	2	7
Inventory on order (Schedule receipts)	12	12	10

Answer: Switches = 80, Microprocessors =14, Keyboards = 39

21. A company has developed a forecast for an item that has the following demand pattern.

Table Forecast for quarters

Quarter	1	2	3	4
Forecast demand (units)	500	900	700	300

a. Plot the demand as a histogram. Determine average demand. Determine the production rate required to meet average demand.
b. Plot the actual cumulative forecast required and cumulative production over planning horizon. Indicate the excess inventory and backorder on the graph.

22. Supply, demand, cost and inventory data for a firm that has a constant workforce is given in Table. Company wishes to satisfy demand with no back orders. Allocate production capacity to satisfy the demand at minimum cost.

Table Supply, demand, cost and inventory data for a firm

Period	Regular time (₹ 100/unit)	Overtime (₹ 125/unit)	Subcontract (₹ 130/unit)	Demand (units)
1	60	18	1000	100
2	50	15	1000	50
3	60	18	1000	70
4	65	20	1000	80

Initial inventory = 20
Final inventory = 25
Carrying cost = ₹ 2/unit/period

21

MRP II, ERP and Supply Chain Management

21.1 MANUFACTURING RESOURCE PLANNING (MRP II)

MRP II stands for Manufacturing Resource Planning. MRP I is concerned primarily with material requirements planning while MRP II is concerned with the coordination of the entire manufacturing including materials, finance, and human relations (Fig. 21.1). The goal of MRP II is to provide consistent data to all areas of manufacturing (Fig. 21.1). MRP I is one of the modules in MRP II. MRP II includes feedback from the shop floor on how the work has progressed to all levels of the schedule so that the next run can be updated on a regular basis. For this reason, it is sometimes called 'Closed Loop MRP'.

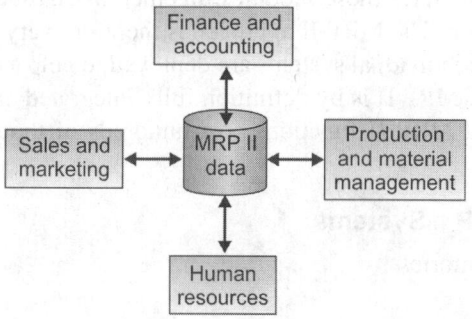

Fig. 21.1 MRP II

21.1.1 Basic Modules in MRP II System

MRP II includes software for many areas of manufacturing such as forecasting, purchasing, inventory, material requirements planning, shop floor control, performance measurement, capacity planning, financial management, customer order entry and accounting. Almost every MRP II system is modular in construction. Basic modules in MRP II system are:

- Master production schedule (MPS)
- Bill of materials (BOM)
- Production resources data (manufacturing technical data)
- Inventories and orders (inventory control)

- Purchasing management
- Material requirements planning (MRP)
- Shop floor control (SFC)
- Capacity planning or capacity requirements planning (CRP)
- Standard costing (cost control)
- Cost reporting/management (cost control)
- Business planning
- Lot traceability
- Contract management
- Tool management
- Engineering change control
- Configuration management
- Shop floor data collection
- Sales analysis and forecasting
- Finite capacity scheduling (FCS)
- General ledger
- Accounts payable (purchase ledger)
- Accounts receivable (sales ledger)
- Sales order management
- Distribution requirements planning (DRP)
- Automated warehouse management
- Project management
- Computer-aided design/computer-aided manufacturing (CAD/CAM)
- CAPP

The MRP II system integrates these modules together so that they use common data and freely exchange information. The MRP II approach is therefore very different from the "point solution" approach, where individual systems are deployed to help a company plan, control or manage a specific activity. MRP II is by definition fully integrated or at least fully interfaced. Systems that perform real MRP II functions cost hundreds of thousands, often millions of dollars.

21.1.2 Benefits of MRP II Systems

- Better control of inventories
- Improved scheduling
- Productive relationships with suppliers
- Improved design control
- Better quality and quality control
- Reduced working capital for inventory
- Improved cash flow through quicker deliveries
- Accurate inventory records
- Standardization and automation of business processes
- Improvements in cost control and revenue

21.1.3 Disadvantages of MRP II Systems

- MRP II implementation requires accurate information. If poor quantity information is used in any module, errors in automated planning processes will result.
- It assumes static nature of an enterprise and fits the system to it.

- Over the years, other tools like CAD, CAM, and CIM, etc. had evolved to automate design and manufacturing processes.

21.2 ENTERPRISE RESOURCE PLANNING (ERP)

ERP systems are extensions of MRP II systems. An ERP system integrates areas such as planning, purchasing, inventory, sales, marketing, finance, human resources, etc. ERP facilitates information sharing, business planning, and decision making on an enterprise. Initial ERP systems are used by large enterprises. Now smaller enterprises are also using ERP systems.

21.2.1 ERP and Internet

Using the web to access a single ERP system at a central location, companies can reduce their IT investments on hardware and personnel. Running ERP systems on a host computer relieves small businesses from the need to purchase a mainframe computer. In addition, this arrangement allows client companies to save money by paying only for the ERP applications they use rather than having to buy a certain number of modules. Leading vendors in the field are SAP of Germany; Oracle; and PeopleSoft.

21.2.2 Characteristics of ERP Systems

Enterprise Resource Planning systems typically include the following characteristics:

- An integrated system that operates in real time
- A common database, which supports all applications
- ERP systems can run on variety of computer hardware and network configurations.
- Installation of the system without elaborating application/data integration by the Information Technology (IT) department.

21.2.3 ERP Modules

An ERP system covers many functional areas. In many ERP systems these are grouped together as ERP modules. A typical ERP system manages functions and activities using 60 modules. Some common modules such as finance and accounting are adopted by nearly all users; others such as human resource management are not. Generally speaking, the greater the number of modules selected, the greater the integration benefits, but also the greater the costs, risks, and changes involved.

An ERP system covers the following common modules:

1. Financial accounting: General ledger, fixed assets, payables, receivables, cash management
2. Management accounting: Budgeting, costing, cost management, activity based costing
3. Human resources: Recruiting, training, payroll, benefits
4. Manufacturing: Engineering, bill of materials, work orders, scheduling, capacity, work-flow management, quality control, manufacturing process, manufacturing projects, manufacturing flow
5. Supply chain management: Supplier chain planning, supplier scheduling, cash order, purchasing, inventory
6. Project management: Project planning, resource planning, project costing, work break down structure, billing, time and expense, performance units, activity management
7. Customer relationship management: Sales and marketing, commissions, service, customer contact, call center support
8. Data services: Various "self-service" interfaces for customers, suppliers and/or employees

21.2.4 ERP Implementation

ERP systems are expensive and not a practical method for medium or small business owners. To address this issue, some software firms develop simpler, cheaper information processing tools specifically for smaller companies.

21.2.5 Benefits of ERP

- The same system could be used to forecast demand for a product, order the necessary raw materials, establish production schedules, track inventory, allocate costs, and project key financial measures.
- ERP improves the quality and efficiency of a business.
- ERP can lead to better outputs that benefit the company such as customer service, and manufacturing.
- ERP provides decision making information to upper level management.
- ERP also creates a more agile company that better adapts to change.
- ERP makes a company more flexible.

21.2.6 Disadvantages of ERP

- Implementation requires substantial time.
- Integration of truly independent businesses can create unnecessary dependencies.
- Implementation of ERP requires new procedures and employees' training.
- Extensive training requirements take resources from daily operations.

21.3 SUPPLY CHAIN MANAGEMENT (SCM)

A supply chain consists of all parties involved, directly or indirectly, in fulfilling a customer's request. Supply chain involves suppliers of raw materials, manufacturers, transporters, wholesalers, retailers and consumers.

A supply chain can be classified as a stage 1, 2 or 3 networks. In a stage 1 type supply chain, systems such as production, storage, distribution, and material control are not linked and are independent of each other. In a stage 2 supply chain, these are integrated under one plan. A stage 3 supply chain is one that achieves vertical integration with upstream suppliers and downstream customers. An example of this kind of supply chain is TISCO. Successful SCM requires a change from managing individual functions to integrating activities. Purchasing department places orders only after receiving the information about the requirement of materials from material planning department. The marketing department, responding to customer demand, communicates with several distributors and retailers as it attempts to determine ways to satisfy this demand. Information shared between supply chains partners can only be fully leveraged through process integration.

Figure 21.2 is an example of Supply chain. Consider a customer walking into a retail store to purchase detergent. The customer transfers funds to retail store. Retail store transfers products to customer. Retail store receives stock of detergent from a distributor. Retail store transfers funds to the distributor after the replenishment. Retail store also conveys sales data to distributor. The distributor receives stock of detergent from the manufacturer of detergent (say Hindustan Unilever limited in this case). Hindustan Unilever limited provides the product as well as pricing to distributor. Distributor transfers funds and sales data to the Hindustan Unilever limited after the replenishment. The manufacturing plant Hindustan Unilever Limited receives raw material from a variety of suppliers who may themselves have been supplied by lower tier suppliers. For example, packaging material may come from packaging company while packaging company receives raw materials to manufacture the packaging from other suppliers. A supply chain is dynamic and involves the constant flow of information, product, and funds between different stages.

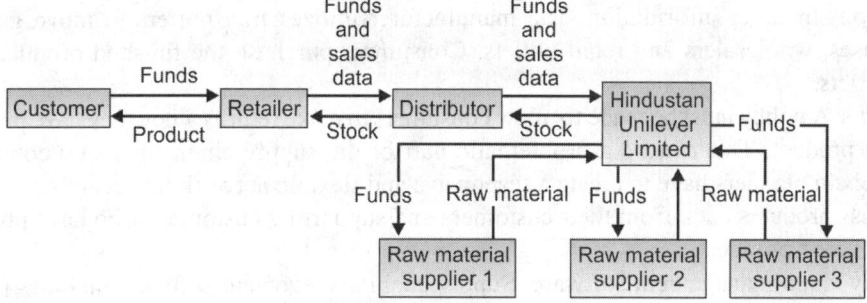

Fig. 21.2 Supply chain

The above example illustrates that the customer is an integral part of the supply chain. In reality, a manufacturer may receive material from several suppliers and then supply finished product to several distributors. Most supply chains are networks. The primary purpose of any supply chain is to visualize information, funds, and product flows along both directions of the supply chain in order to satisfy customer needs and generating profits for company. Supply chain activities begin with a customer order and end when a satisfied customer has paid for his or her purchase. A typical supply chain may involve a variety of stages. These supply chain stages include:

- Customers
- Retailers
- Wholesalers/Distributors
- Manufacturers
- Component/Raw material suppliers

21.3.1 Components of Supply Chain Management

A supply chain management (SCM) system is one in which there is a network of businesses that are interconnected to provide services and products required by a customer. A SCM system controls the planning, execution, design, monitoring and control of these activities to create value and improve performance.

The following are five basic components of supply chain management:

Planning: This is the strategic portion of supply chain management. Companies need a strategy for managing all the resources. Supply chain management planning is developing a set of metrics/standards to monitor the supply chain so that it is efficient, costless and delivers high quality and value to customers.

Suppliers: Next, companies must choose suppliers of raw materials, components and parts which are needed to create their product. Therefore, supply chain managers must develop a set of pricing, delivery and payment processes with suppliers and create metrics for monitoring and improving the relationships. SCM managers must design the methods for receiving and verifying shipments, transferring them to the manufacturing facilities and authorizing supplier payments.

Production: This is the manufacturing step. The actual manufacturing process must produce quality goods economically. Safety procedures, quality control, and personnel management come into play. Supply chain managers schedule the activities necessary for production, testing, packaging and preparation for delivery. This is the portion of the supply chain where companies are able to measure quality levels, production output and worker productivity.

Distribution: Finished products must be packaged properly, and may require printed assembly instructions or other information. The manufacturer utilizes transporters to move goods to warehouses, wholesalers and retail outlets. Consumers purchase the finished products from retail outlets.

Returns: A policy must be made to allow consumers to make returns when they have problems with the product. This can be a problematic part of the supply chain for many companies. Supply chain planners have to create a responsive and flexible network for receiving defective and excess products back from their customers and supporting customers who have problems with delivered products.

Supply chain management software: Supply chain management software includes tools or modules used to execute supply chain transactions, manage supplier relationships and control associated business processes. SCM system software allows you to know the position of your raw materials and your finished products by tracking both your suppliers and your distributors.

Supply chain execution: Supply chain execution means managing and coordinating the movement of materials, information and funds across the supply chain. The flow is bidirectional. Supply chain management applications provide real time analytical systems that manage the flow of products and information throughout the supply chain network.

21.3.2 Reverse Supply Chain

It is also referred to as "aftermarket customer services". Any time money is taken from a company's warranty reserve or service logistics budget. This is a reverse logistics operation.

21.3.3 Supply Chain Structure of Industry

Figure 21.3 is an example of supply chain structure in the aerospace industry. The aerospace industry is dominated by a few large companies. Important aircraft manufactures are Boeing and Airbus. These companies are supported by number of suppliers such as General Electric Aircraft Engines (GEAE), Rolls-Royce, Honeywell and Pratt & Whitney. They are referred to as tier 1 suppliers and play a significant role in the aerospace industry. Tier 1 suppliers are further supplied by a large base of tier 2 and tier 3 suppliers. Tier 2 and tier 3 suppliers serve multiple industries such as industrial manufacturing or automotive.

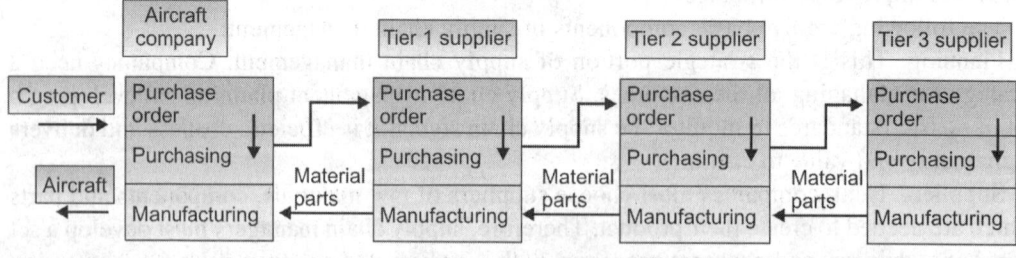

Fig. 21.3 Supply chain structure in the aerospace industry

21.3.4 Application of Supply Chain Management

A company's production operation contains material input components, each of which incurs a cost which is recovered in the price of the finished product. Much of market competition is based on keeping the prices of finished products as low as possible without sacrificing quality. For this reason, supply chain management (SCM) attempts to bridge this gap effectively by closely monitoring the cost, a company pays for materials and from whom the materials are being procured. In addition, SCM monitors the operational procedures from start to finish in order for identifying costly and unnecessary procedural steps. The purpose of supply chain management is to thus improving inventory visibility and the velocity of inventory movement.

21.3.5 Benefits of Supply Chain Management

- By adding an effective SCM system to a business can lead to lower costs of raw materials.
- SCM system can improve your company's relationship with vendors so that there are opportunities to cut costs like through a volume discount.
- Improves trust and collaboration among supply chain partners
- Improves inventory management
- Delivers high quality and value to customers

21.4 A COMPARISON OF ERP AND SCM

Comparison of ERP and supply chain management is given in Table 21.1.

Table 21.1 Comparison of ERP and SCM

ERP	SCM
• Primary function of ERP is to generate data.	• SCM provides capability to the enterprise so that it can make sense out of data to help to make decisions.
• ERP is the body of the enterprise.	• SCM is the brain of the enterprise.
• ERP systems are linear and interactive.	• SCM is constraint-based and optimized.
• ERP generates data.	• Data generated in ERP are used in the best possible way by optimizing the system in a SCM.
• ERP excels in the transaction management.	• SCM affords forecasting and the decision-support.
• ERP links processes only within the organization.	• SCM goes beyond the conventional boundaries of the organization and spans in the entire supply chain.

21.5 QUESTIONS

1. What is MRP II?
2. Explain MRP II.
3. Explain ERP.
4. Explain supply chain management.
5. What are the benefits of MRP II system?
6. What are MRP II modules?
7. What are the benefits of ERP system?
8. What are ERP modules?
9. Explain components of supply chain management.
10. Explain supply chain using an example.
11. What are the benefits of supply chain management?
12. Compare ERP and supply chain management.

22

Loading and Scheduling

22.1 TERMS USED IN SCHEDULING

Job shop: A work location in which a number of general purpose workstations exist and are used to perform a variety of jobs. For each job, sequence is not fixed.

Flow shop: In flow shop, sequence of operations performed on each job is fixed.

Capacity: Capacity is defined as the time available for work at work centers expressed in machine hours or in man hours.

Loading: Loading may be defined as assignment of work to facility. The facility may be people or equipment. Loading consists of assigning work to facilities without specifying when the work is to be done. Machine loading charts can be used to show the availability of machine. If the machine is underutilized, he can plan to put more work, while the machine is overloaded, he can plan for overtime or providing additional machine. Loading is useful to study the relationship between load and capacity at places where work is done. Loading of orders on various machines in a machine shop is shown in Table 22.1.

Table 22.1 Loading of orders on various machines in a machine shop

Machines	Daily machine capacities (hours)	Assigned orders (hours) January					
		1	2	3	4	5	6
CNC lathes	96	80	32	48	40	64	0
CNC milling machines	64	64	56	64	32	0	0
CNC drilling machines	32	24	16	16	0	0	0

22.2 SCHEDULING

The production schedule is a plan that authorizes the production department to produce a certain quantity of an item within a specified time frame. Scheduling gives information when to start the work and when to complete the work on a particular machine. In a large firm, the production schedule is drawn in the production planning department. Scheduling may be called the time phase of loading. Scheduling in practice needs to be dynamic, and so the scheduling rules need to be revised to adjust changes.

Scheduling involves:

- Assigning different jobs to different facilities
- Sequencing the jobs at a facility or on a machine using priority rules
- Specifying the start and the end time for each job
- Getting quick feedback from the shops regarding the delays and the various interruptions

22.3 FACTORS AFFECTING SCHEDULING

External factors:

- Customers' demand
- Delivery dates
- Stock of goods already with the dealers and retailers

Internal factors:

- Stock of goods available in the firm
- Time interval to convert raw materials into finished goods
- Availability of equipment and machinery
- Availability of material
- Availability of man power

22.4 OBJECTIVES OF PRODUCTION SCHEDULING

- Avoiding late completion of jobs
- Minimize the time, a job spends in the system
- Full utilization of work centers, equipment and personnel
- Making efficient use of the labor
- Making best possible use of the equipments
- Increasing the profit
- Increasing the output
- Improving the service level
- Maximizing the delivery performance, i.e. meeting the delivery dates
- Minimizing the inventory
- Reducing the manufacturing time
- Minimizing the production costs
- Minimizing the worker costs

22.5 METHODS USED IN SCHEDULING

Methods used in scheduling are:

- Forward scheduling
- Backward scheduling
- Gantt chart
- Johnson's algorithm
- Index method
- Critical ratio scheduling

22.5.1 Forward Scheduling

Steps:

- All the activities are scheduled from the date of the planned order release.
- First task of the job is scheduled first.

- Subsequent tasks are scheduled on the scheduled completion of the first task.
- Like this all the tasks of the job are scheduled.

22.5.2 Backward Scheduling

Backward scheduling begins with the due date of the final operation. Schedule jobs in reverse order.

Steps:

- Backward scheduling begins with the due date of the final operation.
- The last activity is scheduled first.
- Time of the start of the last task is considered as the time for the completion of the previous activity.

Used in many manufacturing environments, catering, scheduling surgery, etc.

22.5.3 Solved Problem

Problem 22.1: A job is due at the end of period 12 requires a 2 period lead time for material procurement, 1 period of run time for operation 1, 2 periods for operation 2 and 1 period for final assembly. Allow 1 period of transit time prior to each operation. Illustrate the complete schedule under:

a. Forward scheduling approach

b. Backward scheduling approach

Solution: Given data is shown in Table 22.2

Table 22.2 Given data

Activity	Description of activity	Time (Number of periods)
A	Obtain Raw material	2
B	Operation 1	1
C	Operation 2	2
D	Final Assembly	1
	Transit time prior to each operation	1

Schedule of each activity using forward and backward scheduling approaches is given in Fig. 22.1.

Fig. 22.1 Schedule of each activity using forward and backward scheduling approaches

22.5.4 Gantt Chart

A Gantt chart is commonly used in project management. Gantt chart is a simple graphical technique for displaying schedule. On the left of the chart is a list of the activities and along the top is a suitable time scale. Each activity is represented by a bar; the position and length of the bar reflects the start date, duration and end date of the activity. Gantt chart is suitable for less complex situations. Example of Gantt chart is shown in Fig. 22.2.

Gantt chart depicts the following:

- Various activities of the project
- Beginning and ending of each activity
- Duration of each activity
- Activities which are scheduled simultaneously
- The start and end date of the whole project

Limitations of Gantt chart:

- If the project consists of large number of activities, it is difficult to draw Gantt chart.
- The Gantt chart does not exhibit inter-relationship between the activities.
- Gantt charts are not useful in those projects where the activity time cannot be estimated exactly.

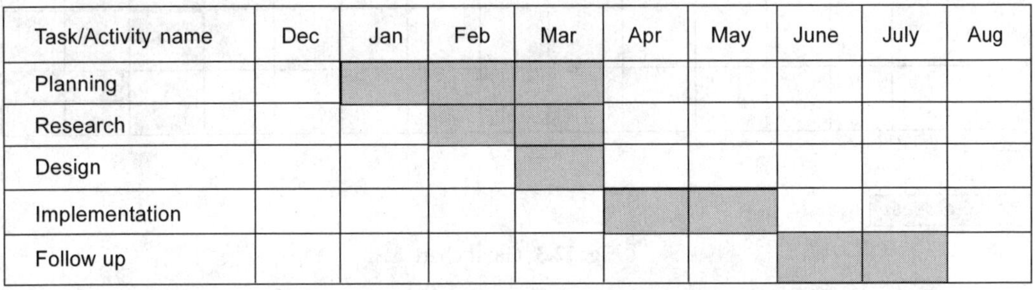

Task/Activity name	Dec	Jan	Feb	Mar	Apr	May	June	July	Aug
Planning		░░	░░	░░					
Research			░░	░░					
Design				░░					
Implementation					░░	░░			
Follow up							░░	░░	

Fig. 22.2 A simple Gantt chart

22.5.5 Solved Problem

Problem 22.2: The following activities are required to manufacture cast iron pulleys. The number of weeks required to complete each activity and actual progress of each activity in weeks is given in Table 22.3. Draw Gantt chart.

Table 22.3 The number of weeks required to complete each activity and actual progress of each activity in weeks

Activity	Description of activity	Time in weeks	Actual progress
1	Designing a drawing	2	2
2	Preparing pattern	2	1
3	Preparing moulds	1	–
4	Casting and cleaning	2	–
5	Inspection	1	–
6	Annealing	1	–
7	Machining	1	–
8	Making key ways	1	–
9	Grinding	2	–

Solution:

Gantt chart for the given data is shown in Fig. 22.3.

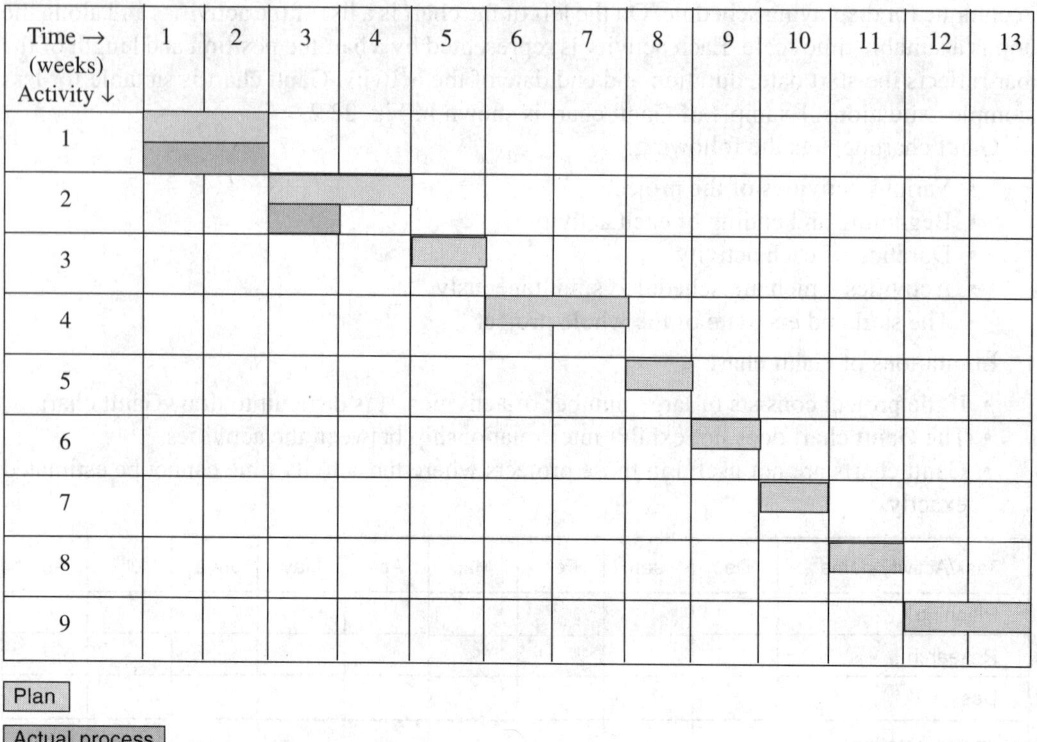

Time → (weeks) Activity ↓	1	2	3	4	5	6	7	8	9	10	11	12	13
1													
2													
3													
4													
5													
6													
7													
8													
9													

Plan

Actual process

Fig. 22.3 Gantt chart

22.5.6 Johnson's Algorithm

Processing N jobs through two machines: This sequencing problem is completely described as follows:

   i. Only two machines A and B are involved.

  ii. Each job is processed in order AB.

 iii. The actual processing times are known.

Johnson's algorithm can be used to solve N jobs and 2 machines problem Steps in Johnson algorithm:

1. Examine processing times on machine A and machine B and find smallest time. If there is a tie select any one of minimum value.
2. If the minimum processing time occurs on machine A, do that job first. If the minimum processing time occurs on machine B, do that job last.
3. Remove the job assigned and continue steps 1 and 2 until all the jobs are assigned.
4. Determine total elapsed time for optimal sequence.

22.5.7 Index Method

Used for the purpose of the loading and also for allocating the different jobs to the different machines.

Steps:
1. Calculate index for each job on different machines.

$$\text{Index for job '}i\text{'} = \frac{\text{Actual processing time for job '}i\text{'}}{\text{Minimum processing time in that row}}$$

2. Assign lowest index job to the work centers without exceeding the capacities.
3. Remove assigned jobs.
4. Repeat the steps 2 to 3 until all jobs are assigned.

22.5.8 Solved Problem

Problem 22.3: Assign the jobs given in Table 22.4 to work centers using index method.

Table 22.4 Time taken by the jobs at work centers

Jobs	Time taken by the jobs at work centers			
	1	2	3	4
A	10	9	8	12
B	3	4	5	2
C	25	20	14	16
D	7	9	10	9
E	18	14	16	25
Number of days available	20	20	20	20

Solution:
1. Calculate index for each job on different work centers.

$$\text{Index for job '}i\text{'} = \frac{\text{Actual processing time for job '}i\text{'}}{\text{Minimum processing time in that row}}$$

The index for each job on different work centers is given in Table 22.5.

Table 22.5 Index for each job on different work centers

Jobs	Work centers			
	1	2	3	4
A	1.25	1.13	1.00	1.50
B	1.50	2.00	2.50	1.00
C	1.78	1.42	1.00	1.14
D	1.00	1.28	1.42	1.28
E	1.28	1.00	1.14	1.78
Number of days available	20	20	20	20

Assign lowest index job to the work centers without exceeding the capacities. Assignment of jobs to the work centers is given in Table 22.6.

Table 22.6 Assignment of jobs to the work centers

Job	Work center	Days assigned	Remaining days available at work center
A	3	8	12
B	4	2	18
D	1	7	13
E	2	14	06
C	4	16	02

22.5.9 Critical Path Analysis

Network analysis is a system, which plans projects both large and small by analyzing the project activities. Project activities are arranged in logical sequence in network diagram. Time, costs and other resources are allocated to different activities. Network helps in designing, planning, coordinating, controlling and decision making in order to accomplish the project economically in the minimum available time with the limited available resources.

A number of network techniques have been developed and two important techniques are:

PERT: Program evaluation and review technique

CPM: Critical path method

22.5.10 Critical Ratio Scheduling

Critical ratio scheduling is a dynamic scheduling technique. Jobs are ranked according to critical ratio.

$$\text{Critical ratio} = \text{CR} = \frac{\text{Time remaining}}{\text{Work remaining}}$$

$$\text{Time remaining} = \text{Due date} - \text{Date now}$$

If $CR < 1$, the job is behind the schedule.

If $CR = 1$, the job is on schedule.

If $CR > 1$, the job is ahead of schedule.

22.5.11 Solved Problem

Problem 22.4: Today is day 22 on the production calendar, and four jobs are on order as shown in Table 22.7. Determine the critical ratio for each job and assign priority ranks.

Table 22.7 Due date and work remaining for each job

Job	Due date	Work remaining in days
A	28	8
B	26	2
C	24	2
D	30	12

Solution: Date now = 22

Computation of critical ratio for each job is given in Table 22.8. Scheduling of jobs is given in Table 22.9.

Table 22.8 Computation of critical ratio

Job	Due date	Work days remaining	Time remaining (Due date – Date now)	$CR = \dfrac{\text{Time remaining}}{\text{Work remaining}}$	Priority
A	28	8	6	0.75	2
B	26	2	4	2.00	4
C	24	2	2	1.00	3
D	30	12	8	0.67	1

Table 22.9 Job schedule

Job	Due date	Work days remaining	Job schedule In	Out
D	30	12	22	34
A	28	8	34	42
C	24	2	42	44
B	26	2	44	46

22.6 QUESTIONS AND PROBLEMS

1. What is Gantt chart? Explain with the help of a suitable example, the method of preparing a Gantt chart.
2. Explain the following:
   a. Job shop
   b. Flow shop
   c. Capacity
   d. Loading
   e. Scheduling
3. What are the factors affecting scheduling?
4. What are the objectives of scheduling?
5. Explain different methods of scheduling.
6. Explain forward and backward schedulings.
7. Explain index methods of loading and scheduling.
8. Explain critical ratio scheduling technique.
9. What are the limitations of Gantt chart?
10. Draw the Gantt chart for the schedule of installing of bus stand involving the following activities.

Activity	Activity description	Duration (days)	Preceding activities
A	Digging foundation	2	–
B	Pouring concrete	4	A
C	Curing concrete	10	B
D	Fabricating shed	6	–
E	Transport	2	D
F	Mounting and fixing shed	4	E
G	Procuring fitting	2	–
H	Fixing sign port	2	F

11. Draw The Gantt chart for a simple project involving following activities.
    Activities A and B can start simultaneously and proceed parallel.
    Activity A is to be completed in 25 days and B to complete in 20 days.
    Activity C start 5 days after the start of activity B and takes 15 days.
    Activity D starts at the end of activity C and takes 10 days.
    What is the total project duration?

Sequencing

Priority sequencing is a systematic procedure for assigning priority to waiting jobs thereby determining the sequence in which jobs will be performed.

23.1 PRIORITY SEQUENCING RULES

1. **First come first serve:** Gives top priority to the waiting jobs that arrived earliest in the production system.
2. **Earliest due date:** Gives top priority to the waiting jobs whose due date is earliest. The objective of this rule is to reduce the lateness of the job.
3. **Shortest processing time (SPT):** Gives top priority to the waiting jobs whose operation time at work center is shortest. The objective of this rule is to reduce waiting line in the queue.
4. **Least slack:** Gives top priority to the job, which is having least slack.
   Slack for job 'i' = Earliest due date of job 'i' – Processing time of job 'i'
5. **Longest processing time:** Gives top priority to the waiting job whose processing time at the work center is longest.
6. **Preferred customer:** Gives top priority based on importance of customer or order.

23.2 PARAMETERS FOR COMPARING PERFORMANCE OF VARIOUS SEQUENCING RULES

Parameters for comparing the performance of various sequencing rules are given below.

1. **Time of completing all jobs:** Total time of completing all jobs
2. **Average job flow time:** It is the total flow time of all jobs divided by number of jobs.

$$\text{Average flow time} = \frac{\text{Total flow time}}{\text{Number of jobs}}$$

Flow time for job 'i' = Waiting time of job 'i' + Processing time of job 'i'

3. **Average number of jobs:** It is the average number of jobs flowing in the system from the beginning of sequence through the time when the last job is finished.

$$\text{Average number of jobs} = \frac{\text{Total flow time}}{\text{Time of completing all jobs}}$$

4. **Average job lateness:** Job lateness is the difference between the actual completion time of job and its due date. Average lateness is the sum of lateness of all the jobs divided by the number of jobs in the system. This is also known as average tardiness.

$$\text{Average job lateness} = \frac{\text{Sum of lateness of all the jobs}}{\text{Number of jobs in the system}}$$

5. **Average job earliness:** Average earliness of job is the sum of earliness of all the jobs divided by the number of jobs in the system.

$$\text{Average job earliness} = \frac{\text{Sum of earliness of all the jobs}}{\text{Number of jobs in the system}}$$

6. **Number of tardy jobs:** It is the number of jobs which are completed after due date.

23. 3 SOLVED PROBLEMS

Problem 23.1: Processing time for jobs A, B, C, D and E and due date for these jobs in a machine shop are given in Table 23.1.

Table **23.1** Processing time and due date for jobs

Job	A	B	C	D	E
Processing time (days)	9	7	5	8	6
Due date	16	20	25	20	40

The job may be sequenced as: i. FCFS ii. Shortest process time iii. Least slack iv. Earliest due date. Determine: (a) Total completion time of all jobs (b) Average flow time (c) Average number of jobs in the machine shop (d) Average job lateness (e) Average job earliness (f) Number of tardy jobs.

Solution:

i. Sequence of jobs to be performed based on FCFS rule and calculations of various parameters are given in Table 23.2.

Table **23.2** Calculations of various parameters using FCFS rule

Job sequence	Processing time (days)	Job schedule (Start–End)	Flow time (days)	Due date	Job lateness	Job earliness
A	9	0–9	9	16	–	7
B	7	9–16	16	20	–	4
C	5	16–21	21	25	–	4
D	8	21–29	29	20	9	–
E	6	29–35	35	40	–	5
	Total		110		9	20

a. Time of completing all jobs = 35 days

b. Average job flow time $= \dfrac{110}{5} = 22$ days

c. Average number of jobs $= \dfrac{110}{35} = 3.14$ jobs

d. Average job lateness $= \dfrac{9}{5} = 1.8$ days

e. Average job earliness $= \dfrac{20}{5} = 4$ days

f. Number of tardy jobs $= 1$

ii. Sequence of jobs to be performed based on SPT rule and calculations of various parameters are given in Table 23.3.

Table 23.3 Calculations of various parameters using SPT rule

Job sequence	Processing time (days)	Job schedule (Start–End)	Flow time (days)	Due date	Job lateness	Job earliness
C	5	0–5	5	25	–	20
E	6	5–11	11	40	–	29
B	7	11–18	18	20	–	2
D	8	18–26	26	20	6	–
A	9	26–35	35	16	19	–
Total			95		25	51

a. Time of completing all jobs $= 35$ days

b. Average job flow time $= \dfrac{95}{5} = 19$ days

c. Average number of jobs $= \dfrac{95}{35} = 2.7$ jobs

d. Average job lateness $= \dfrac{25}{5} = 5$ days

e. Average job earliness $= \dfrac{51}{5} = 10.2$ days

f. Number of tardy jobs $= 2$

iii. Determining Slack is given in the Table 23.4.

Table 23.4 Determination of slack

Job	A	B	C	D	E
Processing time (days)	9	7	5	8	6
Due date	16	20	25	20	40
Slack (Due date – Processing time)	7	13	20	12	34

From Table 23.4, job sequence based on least slack rule and calculation of various parameters are given in Table 23.5.

Table 23.5 Calculations of various parameters using least slack rule

Job sequence	Processing time (days)	Job schedule (Start–End)	Flow time (days)	Due date	Job lateness	Job earliness
A	9	0–9	9	16	–	7
D	8	9–17	17	20	–	3
B	7	17–24	24	20	4	–
C	5	24–29	29	25	4	–
E	6	29–35	35	40	–	5
Total			114		8	15

a. Time of completing all jobs $= 35$ days

b. Average job flow time $= \dfrac{114}{5} = 22.8$ days

c. Average number of jobs $= \dfrac{114}{35} = 3.26$ jobs

d. Average job lateness $= \dfrac{8}{5} = 1.6$ days

e. Average job earliness $= \dfrac{15}{5} = 3$ days

f. Number of tardy jobs $= 2$ jobs

iv. Sequence of jobs to be performed based on earliest due date rule and calculation of various parameters given in Table 23.6.

Table 23.6 Calculations of various parameters using earliest due date

Job sequence	Processing time (days)	Job schedule (Start–End)	Flow time (days)	Due date	Job lateness	Job earliness
A	9	0–9	9	16	–	7
B	7	9–16	16	20	–	4
D	8	16–24	24	20	4	–
C	5	24–29	29	25	4	–
E	6	29–35	35	40	–	5
	Total		113		8	16

a. Time of completing all jobs = 35 days

b. Average job flow time $= \dfrac{113}{5} = 22.6$ days

c. Average number of jobs $= \dfrac{113}{35} = 3.23$ jobs

d. Average job lateness $= \dfrac{8}{5} = 1.6$ days

e. Average job earliness $= \dfrac{16}{5} = 3.2$ days

f. Number of tardy jobs $= 2$ jobs

Problem 23.2: Table 23.7 gives jobs that are waiting to be processed at a small machine center.

Table 23.7 Processing time and due date for jobs

Job	1	2	3	4	5
Due date	260	258	260	270	275
Duration (days)	30	16	8	20	10

In what sequence would the jobs to be ranked according to: i. FCFS and ii. LPT. All dates are specified as manufacturing calendar day, Assume that all jobs arrive on day 210. Which is the best decision rule?

Solution: Sequencing rule: Sequence of jobs to be performed based on FCFS rule and calculations of various parameters are given in Table 23.8.

Table 23.8 Calculations of various parameters using FCFS

Job sequence	Processing time (days)	Job schedule (Start–End)	Flow time (days)	Due date	Job lateness	Job earliness
1	30	210–240	30	260	–	20
2	16	240–256	46	258	–	2
3	8	256–264	54	260	4	–
4	20	264–284	74	270	14	–
5	10	284–294	84	275	19	–
Total			288		37	22

a. Time of completing all jobs = 294 − 210 = 84 days

b. Average job flow time $= \dfrac{288}{5} = 57.6$ days

c. Average number of jobs $= \dfrac{288}{84} = 3.42$ jobs

d. Average job lateness $= \dfrac{37}{5} = 7.4$ days

e. Average job earliness $= \dfrac{22}{5} = 4.4$ days

f. Number of tardy jobs $= 3$ jobs

ii. Sequence of jobs to be performed based on SPT rule and calculation of various parameters is given in Table 23.9.

Table 23.9 Calculations of various parameters using least processing time

Job sequence	Processing time (days)	Job schedule (Start–End)	Flow time (days)	Due date	Job lateness	Job earliness
3	8	210–218	8	260	–	42
5	10	218–228	18	275	–	47
2	16	228–244	34	258	–	14
4	20	244–264	54	270	–	6
1	30	264–294	84	260	34	–
Total			198		34	109

a. Time of completing all jobs: 294 − 210 = 84 days

b. Average job flow time: $\dfrac{198}{5} = 39.6$ days

c. Average number of jobs: $\dfrac{198}{84} = 2.35$ jobs

d. Average job lateness: $\dfrac{34}{5} = 6.8$ days

e. Average job earliness: $\dfrac{109}{5} = 21.8$ days

f. Number of tardy jobs: 1 job

23.4 JOB SHOP AND FLOW SHOP SEQUENCING PROBLEMS

The selection of appropriate order in which waiting jobs may be served is called sequencing. Here the jobs are different. But the sequence of operations performed on each job is same. That is why these problems are called job shop and flow shop sequencing problems.

Analytical methods have been developed for solving the following five cases:

1. N jobs and two machines A and B and all the jobs are processed in the order AB.
2. N jobs and three machines A, B and C and all jobs are processed in the order ABC.
3. N jobs and M machines and all jobs are processed in the order $M_1 M_2 \dots M_m$.
4. Two jobs and M machines and each job are to be processed through the machines in the prescribed order.
5. Traveling salesman problem.

Assumptions made in sequencing problem:

1. Only one operation is carried out on a machine at a particular time.
2. Each operation once started on a job must be completed.
3. Only one machine of each type is available.
4. A job is processed as soon as possible, but only in the order specified.
5. Processing times are independent of the order in which jobs are performed.
6. The time required to transport jobs from one machine to another is negligible.
7. All the jobs are available at the beginning.

Important terms used in sequence:

i. Optimal sequence: The sequence, which minimizes the total time of completing all jobs. There may be more than one optimal sequence.
ii. Total elapsed time: Minimum time of completing all jobs.

23.5 GANTT CHART

It is a tool used for both loading and scheduling of jobs on machines. Gantt chart (Fig. 23.2) is used to represent scheduling of jobs on each machine. Gantt chart consists of number of rectangular areas. These rectangular areas are used to represent work centers or machines for each job. The Gantt chart must be updated periodically to account for new jobs.

23.6 MODEL 1: PROCESSING N JOBS THROUGH TWO MACHINES

This sequencing problem is completely described as follows:

i. Only two machines A and B are involved.
ii. Each job is processed in order AB.
iii. The actual processing times are known.

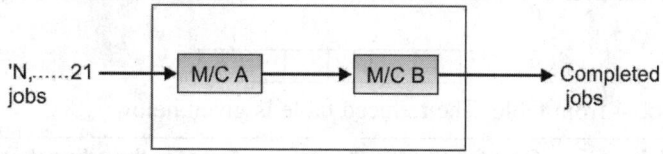

Fig. 23.1 Production system consists of 2 machines and N jobs

Johnson algorithm can be used to solve N jobs and 2 machines problem.

23.7 JOHNSON'S ALGORITHM

Steps in Johnson's algorithm:

1. Examine processing times on machine A and machine B and find smallest time. If there is a tie, select any one of minimum value.
2. If the minimum processing time occurs on machine 'A', do that job first. If the minimum processing time occurs on machine 'B', do that job last.

3. Remove the job assigned and continue steps 1 and 2 until all the jobs are assigned.
4. Determine total elapsed time for optimal sequence.

23.8 SOLVED PROBLEMS

Problem 23.3: A machine operator has to perform two operations turning and threading on a number of different jobs. The time required to perform these operations for each job is given in Table 23.10. Determine the order in which the jobs should be processed in order to minimize the total time required to complete all the jobs.

Table 23.10 The time required to perform the operations for each job

Job	Time for turning (hours) (Machine A)	Time for threading (hours) (Machine B)
1	3	8
2	12	10
3	5	9
4	2	6
5	9	3
6	11	1

Solution: There are six jobs and two machines.

Determining optimal sequence using Johnson's algorithm

Step 1:

The smallest time is 1 hour for job 6. It is occurring on machine B, therefore, do the job 6 last.

					6

Remove the job 6 from table. The reduced table is:

Job	Time for turning (hours) (Machine A)	Time for threading (hours) (Machine B)
1	3	8
2	12	10
3	5	9
4	2	6
5	9	3

Now the smallest time is 2 hours for job 4. It is occurring on machine A, therefore, do the job 4 first.

4					6

Remove the job 4 from table. The reduced table is given below.

Job	Time for turning (hours) (Machine A)	Time for threading (hours) (Machine B)
1	3	8
2	12	10
3	5	9
5	9	3

Now the smallest time is 3 hours. It is occurring on machine A on job 1 and also occurring on machine B on job 5. Here there is a tie for minimum time 3 hours. If there is tie then select any job (i.e. job 1 or job 5). If we select the job 5, minimum processing time occurs on machine B. Do the job 5 last.

4			5	6

Remove the job 5 from the above table. The reduced table is given below.

Job	Time for turning (hours) (Machine A)	Time for threading (hours) (Machine B)
1	3	8
2	12	10
3	5	9

Now the smallest time is 3 hours for job 1. It is occurring on machine A, therefore, do the job 1 first.

4	1			5	6

Remove the job 1 from table. The reduced table is:

Job	Time for turning (hours) (Machine A)	Time for threading (hours) (Machine B)
2	12	10
3	5	9

Now the smallest time is 5 hours for job 3. It is occurring on machine A, therefore, do the job 3 first.

4	1	3		5	6

Remove the job 3 from the table. Now the only job is 2. So optimal sequence is:

4	1	3	2	5	6

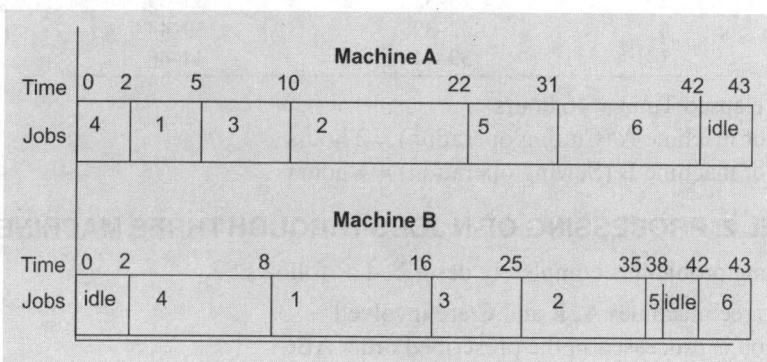

Fig. 23.2 Gantt chart

Table 23.11 To find total elapsed time

Job sequence	Turning operation (Time in – Time out)	Threading operation (Time in – Time out)
4	0–2	2–8
1	2–5	8–16
3	5–10	16–25
2	10–22	25–35
5	22–31	35–38
6	31–42	42–43

Minimum elapsed time = 43 hours
Idle time for machine A (Turning operation) = 1 hour
Idle time for machine B (Threading operation) = 6 hours

Problem 23.4: A readymade garment manufacturer has to process 7 items, through 2 stages of production, cutting and sewing. The time taken for each of these items is given in Table 23.12.

Table 23.12 The time required to perform the operations for each job

Item	1	2	3	4	5	6	7
Cutting time	5	7	3	4	6	7	12
Sewing time	2	6	7	5	9	5	8

Find an order in which these items are to be processed through these stages so as to minimize the total processing time.

Solution: There are six jobs and two machines.

Optimal sequence using Johnson's Algorithm

3	4	5	7	2	6	1

Table 23.13 To find total elapsed time

Job sequence	Cutting operation (Time in – Time out)	Sewing operation (Time in – Time out)
3	0–3	3–10
4	3–7	10–15
5	7–13	15–24
7	13–25	25–33
2	25–32	33–39
6	32–39	39–44
1	39–44	44–46

Minimum elapsed Time = 46 hours
Idle time for machine A (Cutting operation) = 2 hours
Idle time for machine B (Sewing operation) = 4 hours

23.9 MODEL 2: PROCESSING OF N JOBS THROUGH THREE MACHINES

This sequencing problem is completely described as follows:

   i. Only three machines A, B and C are involved.
  ii. Each job is processed in the prescribed order ABC.
 iii. No passing of jobs is permitted.
  iv. The actual processing times are known.

Johnson's algorithm can be extended to cover the special cases where either one or both of the following conditions hold good:

   i. The minimum time on machine A $\geq$ Maximum time on machine B
  ii. The minimum time on machine C $\geq$ Maximum time on machine B

If either one or both conditions are satisfied then convert N jobs-3 machines problem into N jobs-2 machines problem. Here G and H are two fictitious machines and their corresponding processing times are given by:

$$G_i = A_i, + B_i, i = 1, 2, 3, ..., N$$
$$H_i = B_i, + C_i, i = 1, 2, 3, ..., N$$

23.10 SOLVED PROBLEMS

Problem 23.5: Find the sequence that minimizes the total time required in performing the 6 jobs on three machines in the order ABC. Processing times of each job on each machine is given in Table 23.14. Also find total elapsed time.

Table 23.14 The time required to perform the operations for each job

Job → Machine ↓	1	2	3	4	5	6
A	8	3	7	2	5	1
B	3	4	5	2	1	6
C	8	7	6	9	10	9

Solution:

Minimum of A = 1

Maximum of B = 6

Minimum of C = 6

The condition (ii), i.e. the minimum time on machine C ≥ Maximum time on machine B is satisfied. So Johnson's rule is applicable to N jobs-3 machines problem. Convert N jobs-3 machines problem into N jobs-2 machines problem.

Table 23.15 Conversion of N jobs-3 machines problem into N jobs-2 machines problem

Job → Machine ↓	1	2	3	4	5	6
G = (A + B)	11	7	12	4	6	7
H = (B + C)	11	11	11	11	11	15

Determining optimal sequence using Johnson's rule. (This problem has two optimal sequences)

Optimal solution 1:

4	5	6	2	1	3

Optimal solution 2:

4	5	2	6	1	3

Table 23.16 To determine total elapsed time for optimal sequence

Job	Machine A (Time in – Time out)	Machine B (Time in – Time out)	Machine C (Time in – Time out)
4	0–2	2–4	4–13
5	2–7	7–8	13–23
6	7–8	8–14	23–32
2	8–11	14–18	32–39
1	11–19	19–22	39–47
3	19–26	26–31	47–53

Minimum elapsed time = 53 hours

Idle time for machine A = 27 hours

Idle time for machine B = 32 hours

Idle time for machine C = 4 hours

Problem 23.6: Find the sequence for the following 8 jobs that will minimize the total elapsed time for the completion of all jobs. Each job is processed in the order of CAB. Calculate idle time.

Table 23.17 The time required to perform the operations for each job

Job → Machine ↓	1	2	3	4	5	6	7	8
A	4	6	3	4	5	3	6	2
B	8	10	7	8	11	8	9	13
C	5	6	2	3	4	9	15	11

Solution: Here the sequence of machines is CAB.

Minimum of C = 2

Maximum of A = 6

Minimum of B = 7

The condition, i.e. the minimum time on machine B ≥ maximum time on machine A is satisfied. So Johnson's rule is applicable to N jobs-3 machines problem.

Table 23.18 Converting N jobs-3 machines problem into N jobs-2 machines problem

Job → Machine ↓	1	2	3	4	5	6	7	8
G = (C+A)	9	12	5	7	9	12	21	13
H = (A+B)	12	16	10	12	16	11	15	15

Optimal sequence of jobs using Johnson's rule

3	4	1	5	2	8	7	6

Table 23.19 To determine total elapsed time for optimal sequence

Job sequence	Machine C (Time in – Time out)	Machine A (Time in – Time out)	Machine B (Time in – Time out)
3	0–2	2–5	5–12
4	2–5	5–9	12–20
1	5–10	10–14	20–28
5	10–14	14–19	28–39
2	14–20	20–26	39–49
8	20–28	28–30	49–62
7	28–43	43–49	62–71
6	43–52	52–59	71–79

Minimum elapsed time = 79 hours

Idle time for machine A = 42 hours

Idle time for machine B = 5 hours

Idle time for machine C = 27 hours

23.11 MODEL 3: PROCESSING OF N JOBS THROUGH M MACHINES

This sequencing problem is completely described as follows:

    i. There are N jobs to be performed.

    ii. There are M machines denoted by M_1, M_2, ..., M_m

iii. Each job is processed in the order $M_1 M_2 ... M_m$

iv. No passing of jobs is permitted.

v. Actual processing times are known.

Johnson's algorithm can be extended to cover the special cases where either one or both of the following conditions hold well.

i. The minimum time on machine $M_1 \geq$ Maximum time on machines $M_2, M_3, ..., M_{m-1}$

ii. The minimum time on machine $M_m \geq$ Maximum time on machines $M_2, M_3, ..., M_{m-1}$

If either one or both conditions are satisfied, convert N jobs-M machines problem into N jobs-2 machines problem. Here G and H are two fictitious machines and their corresponding processing times are given by:

$G_i = Mi_1 + Mi_2 + ... + Mi_{m-1}$ $i = 1, 2, ..., N$

$H_i = Mi_2 + Mi_3 + ... + Mi_m$ $i = 1, 2, ..., N$

23.12 SOLVED PROBLEMS

Problem 23.7: Four jobs 1, 2, 3 and 4 are to be processed on four machines A, B, C and D in the order ABCD. The processing times in hours are given in Table 23.20. For no passing find the minimum elapsed time.

Table 23.20 The processing times in hours

Machine → Job ↓	A	B	C	D
1	58	14	14	48
2	30	10	18	32
3	28	12	16	44
4	64	16	12	42

Solution:

Minimum of A = 28

Maximum of B and C = 18

Minimum of D = 32

Condition (i), i.e. Minimum of A ≥ Maximum of B and C and Condition (ii), i.e. Minimum of D ≥ Maximum of B and C are satisfied. So Johnson's rule is applicable to N jobs-4 machines problem. Convert N jobs-4 machines problem into N jobs-2 machines problem.

Table 23.21 Conversion of N jobs-4 machines problem into N jobs-2 machines problem

Machine → Job ↓	G = A + B + C	H = B + C + D
1	86	76
2	58	60
3	56	72
4	92	70

Optimum sequence of jobs using Johnson's rule

3	2	1	4

Table 23.22 To determine total elapsed time for optimal sequence

Job	Machine A (Time in – Time out)	Machine B (Time in – Time out)	Machine C (Time in – Time out)	Machine D (Time in – Time out)
3	0–28	28–40	40–56	56–100
2	28–58	58–68	68–86	100–132
1	58–116	116–130	130–144	144–192
4	116–180	180–196	196–208	208–250

Minimum elapsed time = 250 hours
Idle time for Machine A = 70 hours
Idle time for Machine B = 198 hours
Idle time for Machine C = 190 hours
Idle time for Machine D = 84 hours

Problem 23.8: Four jobs 1, 2, 3 and 4 are to be processed on each of the five machines A, B, C, D and E in the order ABCDE. Find the total minimum elapsed time, if no passing of jobs is permitted.

Table 23.23 The processing times in hours

Machine → Job ↓	A	B	C	D	E
1	7	5	2	3	9
2	6	6	4	5	10
3	5	4	5	6	8
4	8	3	3	2	6

Solution: Minimum of A = 5
Maximum of B, C and D = 6
Minimum of E = 6
Condition (ii), i.e. Minimum of E ≥ Maximum of B, C and D is satisfied. So Johnson's rule is applicable to N jobs-5 machines problem. Convert N jobs-4 machines problem into N jobs-2 machines problem.

Table 23.24 Conversion of N jobs-4 machines problem into N jobs-2 machines problem

Machine → Job ↓	G = A + B + C + D	H = B + C + D + E
1	17	19
2	21	25
3	20	23
4	16	14

Optimum sequence of jobs using Johnson's rule

1	3	2	4

Table 23.25 Total elapsed time for optimal sequence

Job	Machine A (In–Out)	Machine B (In–Out)	Machine C (In–Out)	Machine D (In–Out)	Machine E (In–Out)
1	0–7	7–12	12–24	14–17	17–26
3	7–12	12–16	16–21	21–27	27–35
2	12–18	18–24	24–28	28–33	35–45
4	18–26	26–29	29–32	33–35	45–51

Minimum elapsed time = 51 hours
Idle time for machine A = 25 hours
Idle time for machine B = 33 hours
Idle time for machine C = 37 hours
Idle time for machine D = 35 hours
Idle time for machine E = 18 hours

23.13 MODEL 4: TWO JOBS THROUGH M MACHINES

Processing two jobs though M machines:
   i. Only two jobs are to be performed.
   ii. Sequence of machines for each job is known.
   iii. Sequence of the machines for each job may be same or different.
   iv. Processing times are known.
   v. Each machine can work only one job at a time.

The problem is to minimize the total elapsed time, i.e. minimum time of completing both jobs. Here the objective is to determine scheduling of jobs on each machine so that time of completing both jobs is minimized. Use graphical method to solve 2 jobs through M machines problem. The graphical procedure is given below.

Graphical method

   1. Draw two axes at right angles to each other. Represent processing times of job 1 along the horizontal axis and processing times of job 2 along vertical axis.
   2. Layout the machine times for two jobs on corresponding axis in the given order of machines. Rectangular areas represent both jobs are competing the same machine at same time. Only one job is processed on each machine at a time. So passing through these rectangular areas is not permitted.
   3. Determine optimal path starting from origin to the final point F.
       i. Line inclined to 45° on X-axis or Y-axis represents both jobs are processing.
       ii. Line parallel to X-axis represents job 1 is processing and job 2 is idle.
       iii. Line parallel to Y-axis represents job 2 is processing and job 1 is idle.
   4.   Determine total elapsed time from graph using the equation below.

Total elapsed time = Processing time of job 1 + Idle time of job 1

Or

Processing time of job 2 + Idle time of job 2

23.14 SOLVED PROBLEMS

Problem 23.9: Use graphical method to calculate the minimum time needed to process jobs 1 and 2 on five machines A, B, C, D and E, i.e. for each machine find the job, which should be done first.

Table 23.26 Given data

Job 1	Sequence	A	B	C	D	E
	Time (hours)	1	2	3	5	1
Job 2	Sequence	C	A	D	E	B
	Time (hours)	3	4	2	1	5

Solution: Draw the graph using the procedure presented in graphical method. The resultant graph is shown in Fig. 23.4.

From the graph (Fig. 23.4),

Total elapsed time = Processing time of job 1 + Idle time of job 1

Or

Processing time of job 2 + Idle time of job 2

= 12 + 3 = 15 hours

Or

= 15 + 0 = 15 hours

To determine which job should be done first: From Fig. 23.4, the schedulings of job 1 and job 2 are determined. The resultant schedule of each job on each machine is given in Fig. 23. 3.

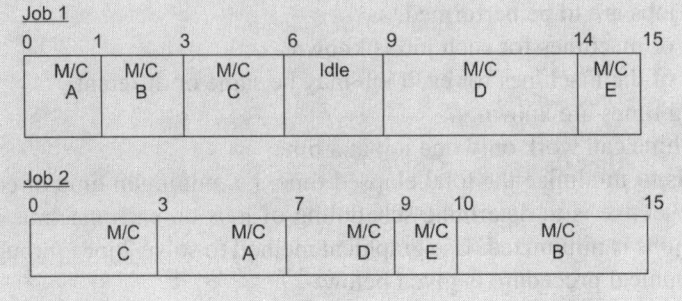

Fig. 23.3 Gantt chart

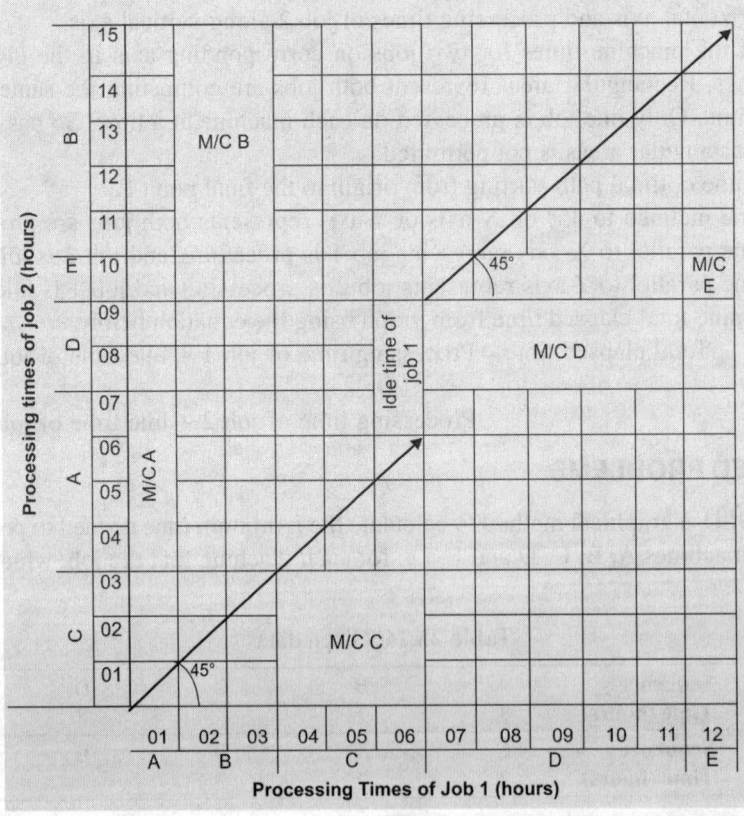

Fig. 23.4 Determining optimal path

From Fig. 23.3,
On machine A, job 1 should be done first.
On machine B, job 1 should be done first.
On machine C, job 2 should be done first.
On machine D, job 2 should be done first.
On machine E, job 2 should be done first.

Problem 23.10: Use graphical method to minimize the time required to process the following jobs on machines, i.e. for each machine specify the job which should be done first. Also calculate the total elapsed time to complete both jobs.

Table 23.27 Given data

Job 1	Sequence	A	B	C	D	E
	Time (hours)	6	8	4	12	4
Job 2	Sequence	B	C	A	D	E
	Time (hours)	10	8	6	4	12

Solution: Draw the graph using the procedure presented in graphical method. The resultant graph is shown in Fig. 23.5.

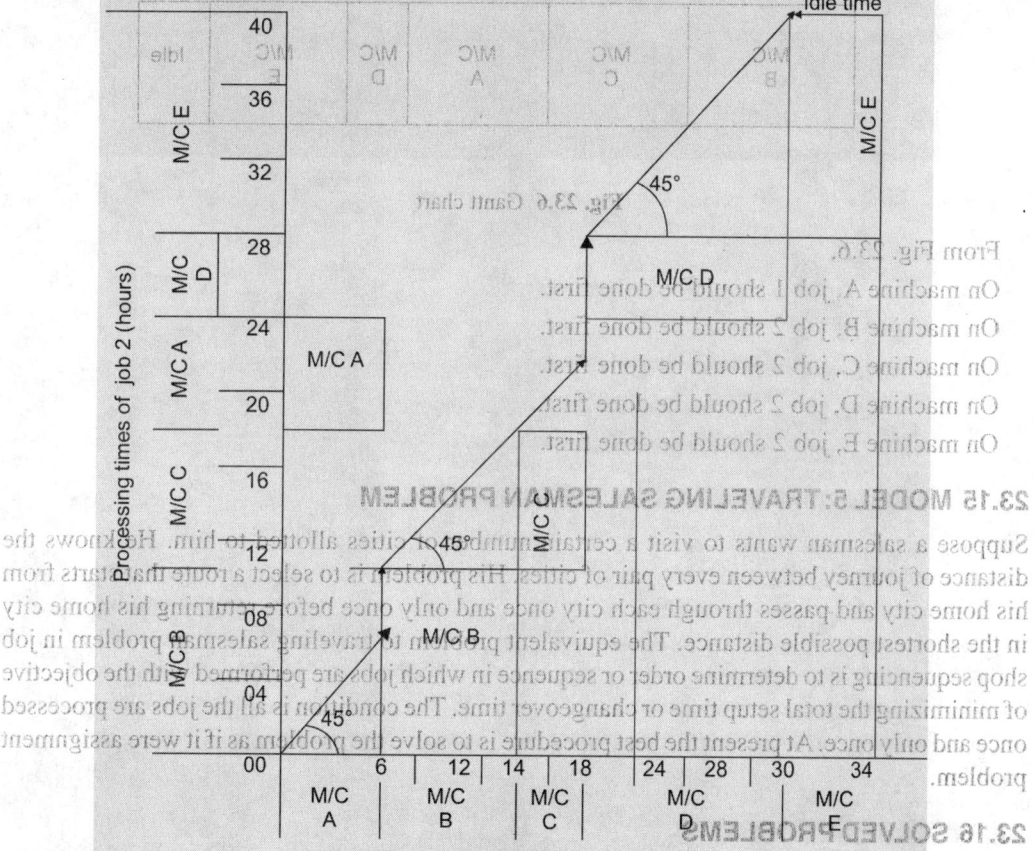

Fig. 23.5 Determining optimal path

From Fig. 23.5, Total elapsed time = Processing time of job 1 + Idle time of job 1

Or

Processing time of job 2 + Idle time of job 2

= 34 + (4 + 6) = 44 hours

Or

= 40 + 4 = 44 hours

To determine which job should be done first: From Fig. 23.5, the scheduling of job 1 and job 2 are determined. The resultant schedule of each job on each machine is given in Fig. 23.6.

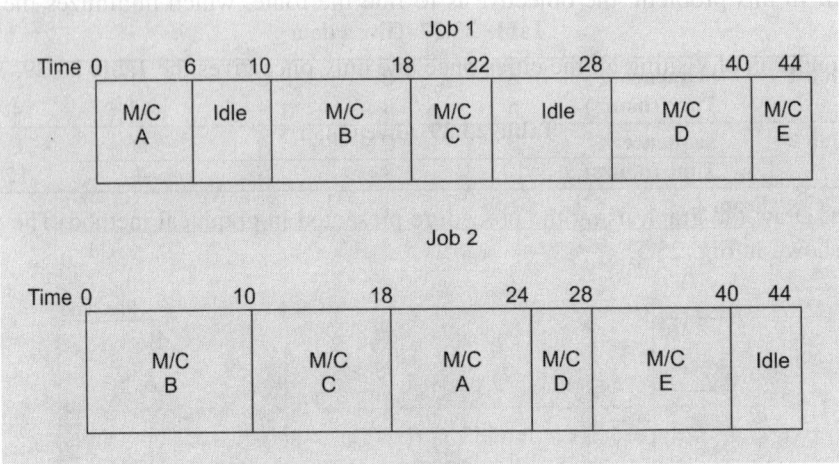

Fig. 23.6 Gantt chart

From Fig. 23.6,

On machine A, job 1 should be done first.

On machine B, job 2 should be done first.

On machine C, job 2 should be done first.

On machine D, job 2 should be done first.

On machine E, job 2 should be done first.

23.15 MODEL 5: TRAVELING SALESMAN PROBLEM

Suppose a salesman wants to visit a certain number of cities allotted to him. He knows the distance of journey between every pair of cities. His problem is to select a route that starts from his home city and passes through each city once and only once before returning his home city in the shortest possible distance. The equivalent problem to traveling salesman problem in job shop sequencing is to determine order or sequence in which jobs are performed with the objective of minimizing the total setup time or changeover time. The condition is all the jobs are processed once and only once. At present the best procedure is to solve the problem as if it were assignment problem.

23.16 SOLVED PROBLEMS

Problem 23.11: Solve the following traveling salesman problem. The distance matrix is given in Table 23.28.

Table 23.28 The distance matrix

To → From ↓	A₁	A₂	A₃	A₄	A₅
A₁	–	2	5	7	1
A₂	6	–	3	8	2
A₃	8	7	–	4	7
A₄	12	4	6	–	5
A₅	1	3	2	8	–

Solution: In this problem, the objective is to find the route, which minimizes the distance traveled.

The condition of visiting all the cities once and only once gives the Table 23.29.

Table 23.29 Given matrix

To → From ↓	A₁	A₂	A₃	A₄	A₅
A₁	∞	2	5	7	1
A₂	6	∞	3	8	2
A₃	8	7	∞	4	7
A₄	12	4	6	∞	5
A₅	1	3	2	8	∞

Apply assignment algorithm to above matrix. The resultant row opportunity matrix and total opportunity matrix are given in Table 23.30 and Table 23.31.

Table 23.30 Row opportunity matrix

To → From ↓	A₁	A₂	A₃	A₄	A₅
A₁	∞	1	4	6	0
A₂	4	∞	1	6	0
A₃	4	3	∞	0	3
A₄	8	0	2	∞	1
A₅	0	2	1	7	∞

Table 23.31 Total opportunity matrix

To → From ↓	A₁	A₂	A₃	A₄	A₅
A₁	∞	1	3	6	0
A₂	4	∞	0	6	0
A₃	4	3	∞	0	3
A₄	8	0	1	∞	1
A₅	0	2	0	7	∞

Number of lines to cover all zeros in total opportunity matrix is equal to number of rows or columns. So we obtain optimal solution matrix. Make assignment for the optimal solution matrix using assignment algorithm.

Table 23.32 Optimal solution matrix

To → From ↓	A_1	A_2	A_3	A_4	A_5
A_1	∞	1	3	6	⓪
A_2	4	∞	⓪	6	0
A_3	4	3	∞	⓪	3
A_4	8	⓪	1	∞	1
A_5	⓪	2	0	7	∞

Solution: In this problem, the objective is to find the route, which minimizes the distance

Route obtained from above Matrix is A_1-A_5-A_1. This route violates restriction of visiting all the cities once and only once. So select next higher element other than zero. There are four ones. Select any one and make assignment. If the route is satisfied, determine cost. First make assignment to A_1A_2 cell. Remove row A_1 and column A_2. Make the remaining assignments after searching rows and columns, which are having single zero using similar procedure in assignment problem. If there are no single zero rows are columns then make assignment to any zero.

Table 23.33 Optimal assignment

To → From ↓	A1	A2	A3	A4	A5
A1	∞	①	3	6	0
A2	4	∞	⓪	6	0
A3	4	3	∞	⓪	3
A4	8	0	1	∞	①
A5	⓪	2	0	7	∞

Repeat the above procedure for all ones. If the condition is satisfied, determine cost. The route, which minimizes the cost, is an optimal solution.

Optimal solution: A1-A2-A3-A4-A5 = 15

Problem 23.12: A machine operator processes four types of items on his machine each week and must choose a sequence for them. The setup cost per change depends on the items presently on the machines and setup to be made according to the following Table 23.34. Determine sequence, which minimizes total setup cost.

Table 23.34 Setup costs

To → From ↓	A	B	C	D
A	–	4	7	3
B	4	–	6	3
C	7	6	–	7
D	3	3	7	–

Solution:
The condition of producing four types of items once and only in a week gives the Table 23.35.

Table 23.35 Given matrix

To → From ↓	A	B	C	D
A	∞	4	7	3
B	4	∞	6	3
C	7	6	∞	7
D	3	3	7	∞

Table 23.36 Row opportunity matrix

To → From ↓	A	B	C	D
A	∞	1	4	0
B	1	∞	3	0
C	1	0	∞	1
D	0	0	4	∞

Solution: The condition of producing all the jobs once and only once in a production cycle gives the Table 23.41.

Table 23.37 Total opportunity matrix

To → From ↓	A	B	C	D
A	∞	1	1	0
B	1	∞	0	0
C	1	0	∞	1
D	0	0	1	∞

Number of lines to cover all zeros in total opportunity matrix is equal to number of rows or columns. So we obtain optimal solution. Make assignment for the optimal solution matrix using assignment algorithm. The resultant assignment is given in Table 23.38.

Table 23.38 Total opportunity matrix (Optimal solution matrix)

To → From ↓	A	B	C	D
A	∞	1	1	⓪
B	1	∞	⓪	0
C	1	⓪	∞	1
D	⓪	0	1	∞

From Table 23.38 of optimal solution matrix, the route is A-D-A. This route violates the condition of producing all the jobs once and only once. Select next higher value in the optimal solution matrix.

Table 23.39 Making assignment by selecting AC cell

To → From ↓	A	B	C	D
A	∞	1	①	0
B	1	∞	0	⓪
C	1	⓪	∞	1
D	⓪	0	1	∞

Minimum number of lines to cover all the zeros in total opportunity matrix is less than number of rows or columns so solution is not an optimal solution. So we have to revise the solution. The revised solution is given in Table 23.44.

From the Table 23.39, the route is A-C-B-D-A.

The cost = 7 + 6 + 3 + 3 = 19

Problem 23.13: The matrix of setup costs is given in Table 23.40. Determine sequence production so as to minimize setup cost per cycle.

Table 23.40 Setup costs

To → From ↓	A	B	C	D	E
A	–	3	6	2	3
B	3	–	5	2	3
C	6	5	–	6	4
D	2	2	6	–	6
E	3	3	4	6	–

Solution: The condition of producing all the jobs once and only once in a production cycle gives the Table 23.41.

Table 23.41 Given matrix

To → From ↓	A	B	C	D	E
A	∞	3	6	2	3
B	3	∞	5	2	3
C	6	5	∞	6	4
D	2	2	6	∞	6
E	3	3	4	6	∞

Table 23.42 Row opportunity matrix

To → From ↓	A	B	C	D	E
A	∞	1	4	0	1
B	1	∞	3	0	1
C	2	1	∞	2	0
D	0	0	4	∞	4
E	0	0	1	3	∞

Table 24.43 Total opportunity matrix

To → From ↓	A	B	C	D	E
A	∞	1	3	0	1
B	1	∞	2	0	1
C	2	1	∞	2	0
D	0	0	3	∞	4
E	0	0	0	3	∞

Minimum number of lines to cover all the zeros in total opportunity matrix is less than number of rows or columns so solution is not an optimal solution. So we have to revise the solution. The revised solution is given in Table 23.44.

Table 23.44 Revised solution (Optimal solution)

To → From ↓	A	B	C	D	E
A	∞	0	2	0	1
B	0	∞	1	0	1
C	1	0	∞	2	0
D	0	0	3	∞	5
E	0	0	0	4	∞

Minimum number of lines to cover all the zeros is equal to number of rows or columns. So solution is an optimal solution. So we have to make assignment. Assignment to the optimal solution using assignment algorithm is given in Table 23.45.

Table 23.45 Optimal solution (Assignment)

To → From ↓	A	B	C	D	E
A	∞	0	2	⓪	1
B	⓪	∞	1	0	1
C	1	0	∞	2	⓪
D	0	⓪	3	∞	5
E	0	0	⓪	4	∞

A-D-B-A is not satisfied. Next, select higher value after zero and make assignment.

Table 23.46 Assignment by selecting the cell AE

To → From ↓	A	B	C	D	E
A	∞	0	2	0	①
B	0	∞	1	⓪	1
C	1	⓪	∞	2	0
D	⓪	0	3	∞	5
E	0	0	⓪	4	∞

Sequence of jobs from Table 23.46 = A-E-C-B-D-A

Total setup cost = 3 + 4 + 5 + 2 + 2 = 16

23.17 QUESTIONS AND PROBLEMS

1. What is priority sequencing and what are the priority sequencing rules?
2. Explain parameters used for comparing performance of various sequencing rules.
3. What are the assumptions made in sequencing problem?
4. Describe models in job shop and flow shop sequencing problems.
5. Explain sequence of steps in Johnson algorithm.
6. Describe traveling sales man problem.
7. What is Gantt chart? Illustrate with an example.
8. There are five jobs, which are to be processed on work center sheet metal shop. The processing times are given below.

Job	A	B	C	D	E
Processing time (days)	4	17	14	9	11
Due date	6	12	12	18	20

Determine the sequence using SPT rule.

Answer: Sequence according to SPT rule: A-D-E-C-B

9. The following jobs are waiting to be processed at the same machine canter. Jobs are logged as they arrive.

Job	A	B	C	D	E
Due date	313	312	325	314	314
Duration (days)	8	16	40	5	3

All the dates are specified as manufacturing calendar days. Assume that all jobs arrive on the day 275. In what sequence would the jobs be ranked according to the following decision rules?
a. Earliest due date
b. Shortest processing time
What is the best decision rule?

Answer: Sequence according to earliest due date: B-A-D-E-C
Sequence according to SPT rule: E-D-A-B-C

10. There are five jobs, which are to be processed on work center sheet metal shop. The processing times are given below.

Job	A	B	C	D	E
Processing time (days)	4	17	14	9	11
Due date	6	12	12	18	20

Determine the sequence using SPT rule.

11. Find the sequence that minimizes the total elapsed time for 10 jobs through 2 machines M_1 and M_2 shown below. Also find total elapsed time.

Job	A	B	C	D	E	F	G	H	I	J
Time on M_1	7	3	10	8	13	9	5	11	7	10
Time on M_2	6	5	15	7	12	12	2	8	5	11

Answer: Minimum time of completing all jobs = 90 hours

12. Four jobs 1, 2, 3 and 4 are to be processed on each of the five machines A, B, C, D and E in the order ABCDE. Find the total minimum elapsed time, if no passing of jobs is permitted.

Machine → Job ↓	A	B	C	D	E
1	7	5	2	3	9
2	6	6	4	5	10
3	5	4	5	6	8
4	8	3	3	2	6

Answer: Optimal sequence of jobs: 1324; Minimum elapsed time = 51 hours

13. Use graphical method to minimize the time needed to process the following jobs on machines A, B, C, D and E. Find the total elapsed time to complete both the jobs. Also find for each job the machine on which it should be processed first.

23.17 QUESTIONS AND PROBLEMS

1. What is priority sequencing and what are the priority sequencing rules?
2. Explain parameters used for comparing performance of various sequencing rules.
3. What are the assumptions made in sequencing problem?
4. Describe models in job shop and flow shop sequencing problems.

Job 1	Sequence	A	B	C	D	E
	Time (hours)	2	3	5	2	1
Job 2	Sequence	D	C	A	B	E
	Time (hours)	6	2	3	1	3

Answer: Minimum elapsed time = 16 hours
Job 1 idle time = 3 hours; Job 2 idle time = 1 hour

14. A machine shop has six machines A, B, C, D, E and F. Two jobs must be processed through each of these machines. The time spent on each machine and the necessary sequence of the jobs through the shops is given below. In what sequence should the job be done on each of the machines, in order to minimize total time necessary to finish the jobs. Also determine total elapsed time.

Order	1	2	3	4	5	6
Job 1	A-25	C-15	D-15	B-35	E-30	F-30
Job 2	B-20	C-35	A-15	D-15	F-20	E-25

Answer: Minimum elapsed time = 155 hours
Job 1 idle time = 5 hours; Job 2 idle time = 25 hours

15. Solve the following traveling sales man problem with the given distance matrix.

From	To				
	1	2	3	4	5
1	–	14	10	24	41
2	6	–	10	12	10
3	7	13	–	8	15
4	11	14	30	–	17
5	6	8	12	16	–

Answer: Optimal Route: 1-3-4-2-5-1; Minimum distance = 48

16. A shoe manufacturer has to process 6 items through three stages of production, i.e. cutting, pasting and curing. The time taken for each of these items at the different stages is given below in hours.

Item	A	B	C	D	E	F
Cutting	4	1.5	3.5	1	2.5	0,5
Pasting	1.5	2	2.5	1	0.5	3
Curing	4	3.5	3	4.5	5	4.5

Find an order in which these items can be processed in minimum time.

Answer: Optimal sequence: DEBFAC; Total elapsed time = 26.5 hours

17. Use Johnson's rule to determine the best sequence for five jobs and use Gantt chart to show an optimum schedule for the machines. Each job is processed in the order ABC.

Job	1	2	3	4	5
Machine A	4	9	8	6	5
Machine B	5	6	2	3	4
Machine C	8	10	6	7	11

Answer: Optimal sequence:1-4-5-2-3; Total elapsed time = 51 hours

18. Use Johnson's rule to determine the best sequence for six jobs and use Gantt chart to show an optimum schedule for the machines. Each job is processed in the order ACB.

Job	1	2	3	4	5	6
Machine A	12	8	7	11	10	5
Machine B	7	10	9	6	10	4
Machine C	3	4	2	5	1.5	4

Answer: Optimal sequence: 3-5-2-4-1-6; Total elapsed time = 62 hours

19. Use Johnson's rule to determine the best sequence for six jobs and use Gantt chart to show an optimum schedule for the machines. Each job is processed in the order M_1M_2.

Job	A	B	C	D	E	F
M_1	4	8	3	6	7	5
M_2	6	3	7	2	8	4

Answer: Optimal sequence: C-A-E-F-B-D; Total elapsed time = 35 hours

20. Four jobs are to be processed on each of the five machines in the order $M_1M_2M_3M_4M_5$. Find the total minimum elapsed time, if no passing of jobs is permitted.

Machine → Job ↓	M_1	M_2	M_3	M_4	M_5
A	9	7	4	5	11
B	8	8	6	7	12
C	7	6	7	8	10
D	10	5	5	4	8

Answer: Optimal sequence of jobs: A-C-B-D; Minimum elapsed time = 67 hours

21. Four jobs are to be processed on each of the five machines in the order $M_1M_2M_3M_4$. Find the total minimum elapsed time, if no passing of jobs is permitted.

Machine → Job ↓	M_1	M_2	M_3	M_4
A	13	8	7	14
B	12	6	8	19
C	9	7	5	15
D	8	5	6	15

Answer: Optimal sequence of jobs: D-C-B-A; Minimum elapsed time = 82 hours

Part VII

Quality Management

Quality and Taguchi Methods

24.1 DEFINITION OF QUALITY

Quality is a much more complicated term than it appears. Quality is, "meeting or exceeding customer expectations".

According to experts, the word quality can be defined as:
- Fitness for use or purpose
- To do a right thing at first time
- Features that meet consumer needs and give customer satisfaction
- Freedom from deficiencies or defects
- Conformance to standards

24.2 FACTORS AFFECTING PRODUCT QUALITY

Product quality mainly depends on important factors like:
- The type of raw materials used for making a product.
- Various production technologies used in production.
- Skill and experience of manpower involved in the production process.
- Availability of production related overheads like power and water supply, transport, etc.

24.3 PRODUCT QUALITY

Product quality means to incorporate features that have a capacity to meet consumer needs (wants) and gives customer satisfaction by improving products (goods) and making them free from any deficiencies or defects".

24.4 DIMENSIONS IN PRODUCT QUALITY

A number of scholars in the quality field have developed lists of dimensions that define quality for a product and/or a service. David Garvin developed a list of 8 dimensions of product quality.

The eight dimensions of product quality according to Garvin:

1. Performance
2. Features
3. Reliability
4. Conformance
5. Durability
6. Serviceability
7. Aesthetics
8. Perceived Quality

1. Performance: The primary operating characteristics of a product or service. For a car, it is speed and acceleration. For a restaurant, it is good food.
2. Features: The secondary characteristics of a product or service. For a restaurant, it is table cloths and napkins. For cell phone, it is a camera.
3. Conformance: The match with specifications or pre-established standards. For a part, it is whether this part is in the right size.
4. Durability or product life. For a light bulb, it is how long it works before the filament burns out. For capital items, it is a life of the machine tool, automobile, etc.
5. Reliability: The frequency with which a product or service fails. For a car, it is how often it needs repair.
6. Serviceability: The speed, courtesy and competence of repair. For a car, it is how quickly and easily it can be repaired and how long it stays repaired.
7. Appearance: Aesthetics or fits and finishes. For a product or service, it is look, feel, sound taste or smell.
8. Image: Perceived quality or reputation. For a product or service, it is the positive or negative feelings people attach to any new offerings, based on their past experiences with the company.

24.5 QUALITY COSTS

1. Prevention costs
2. Appraisal costs
3. Failure costs

Prevention costs: The costs of all activities specifically designed to prevent poor quality in products or services. Examples are the costs of new product review, quality planning, supplier capability surveys, process capability evaluations, quality improvement team meetings, quality improvement projects, quality education and training.

Appraisal costs: The costs associated with measuring, evaluating or auditing products or services to assure conformance to quality standards and performance requirements. These include the costs of inspection of purchased material and final inspection/test product, calibration of measuring and test equipment.

Failure costs: The costs resulting from products which are not confirming to requirements or customer needs. Failure costs are divided into internal and external failure categories.

- Internal failure costs: Failure costs occurring prior to delivery or shipment of the product to the customer. Examples are the costs of scrap, rework, re-testing, etc.
- External failure costs: Failure costs occurring after delivery or shipment of the product to the customer. Examples are the costs of processing customer complaints, customer returns, warranty claims, product recalls
- Total quality costs: The sum of the above costs.

24.6 QUALITY ASSURANCE AND QUALITY CONTROL

Quality assurance and quality control are interdependent.

Quality assurance: Quality assurance (QA) is a broad concept that focuses on the entire quality system including suppliers and ultimate consumers of the product or service. It includes all activities designed to produce products and services of appropriate quality. The role of a quality assurance department is to establish procedures and systems in collaboration with other departments that ensure all deliverables are constantly of good quality. Quality assurance aims to prevent defects or problems from occurring.

Quality control: Quality control (QC) is the collection of methods and techniques for ensuring that a product or service is produced and delivered according to given requirements. This includes the development of specifications and standards, performance measures, and tracking procedures, and corrective actions to maintain control. The data collection and analysis functions for quality control involve statistical sampling, estimation of parameters, and construction of various control charts for monitoring the processes in making products. This area of quality control is formally known as statistical process control (SPC) and, along with acceptance sampling, represents the traditional perception of quality management. Statistical process control focuses primarily on the conformance element of quality, and to somewhat less extent on operating performance and durability. Quality control focuses on the process of producing the product or service with the intent of eliminating problems that might result in defects. Concurrent engineering, Taguchi methods, six sigma, just in time manufacturing and total quality management (TQM) are modern management approaches for improving quality through effective planning and integration of design, manufacturing, and materials management functions throughout an organization. Quality improvement programs typically include goals for reducing warranty claims and associated costs because warranty data directly or indirectly impact on most of the product quality. The quality assurance department relies on feedback from quality control to identify areas where the preventive process needs changes. For example, the quality assurance department may investigate the causes of defects as reported by quality control, and subsequently establish a new procedure to prevent them from happening again. After the new procedures are established, the quality control department checks that the goods meet the new quality standards.

In some organizations, particularly service-oriented ones, it may be more difficult to tell the difference between the two functions, and in fact, the same department may be responsible for both quality assurance and quality control.

24.7 TAGUCHI METHODS

Genichi Taguchi, a Japanese engineer and statistician, began formulating the Taguchi method while developing a telephone-switching system for Electrical Communication Laboratory, a Japanese company, in the 1950s. As a result of his success, he eventually became well-known in both Japan and United States, with companies such as Toyota, Ford, Boeing and Xerox adopting his methods.

Genichi Taguchi proposed several approaches to experimental designs that are sometimes called Taguchi methods. Taguchi methods emphasize the effective application of engineering strategies in reducing the occurrence of defects and failures in products rather than application of advanced statistical techniques.

The quality system is divided into two basic functions:
- Offline quality control
- Online quality control

24.7.1 Offline Quality Control

Offline quality control is concerned with both product and process design. Product design is concerned with development of new product or a new model of an existing product. The process design is concerned with the processes and equipment, setting work standards, determining route sheets, etc. Offline quality control method uses small scale experiments to reduce variability and to find cost-effective robust designs. Robustness means high quality product. Produces products that are robust against noises. Noise factors cause variation in performance of the functioning of the product. Build designs (product and process) that can perform low variability in the presence of noise. Selects operating parameters which are least sensitive to noises. Thus robust products are insensitive to changes in conditions. Noise factors fall into three types.

1. **Outer noise factors/Environmental factors:** Variations in performance of the finished product due to the factors such as temperature, humidity, input voltage, etc. External factors are generally more difficult to control.
2. **Internal noise factors:** Variations in performance of the finished product due to the factors such as wear of mechanical components, improper setting of the product, etc.
3. **Piece-to-piece noise factors:** Variations in performance of the finished product due to the factors such as variations in input materials, machine input, labor input, etc.

24.7.1.1 Robust product design

Robust product design is Taguchi's important and useful contribution. The aim here is to make a product or process less variable (more robust) in the face of variation over which we have little or no control.

Some examples of robust design:

A software company's power operating room that maintains same supply of power when the electrical power to the company is interrupted.

A machine tool produces same surface finish to all products in specified range of cutting parameters such as speed, feed and depth of cut.

A robust car is a car that not only starts and runs well under ideal environmental conditions, but performs well in all conditions.

A Spark plug ignites fuel air mixture of petrol engine in all climatic conditions.

Taguchi considered three levels of design. The Taguchi method contains system design, parameter design, and tolerance design procedures to achieve a robust process and result for the best product quality.

Level 1/System design: System design involves the application of new ideas, techniques, philosophies, science and engineering knowledge to develop a prototype design that will meet customer needs.

Level 2/Parameter design: Taguchi's method for making products more robust is called parameter design. This method is based on the idea that the design parameters of a product can be manipulated by experimental methods with the objective of finding product design configurations that are less sensitive to variation.

Level 3/ Tolerance design: The objective is to specify appropriate tolerance which maximizes the quality and also minimizes the cost of production. A tolerance is allowable variation that is permitted about the nominal value.

24.7.2 Online Quality Control

This is concerned with production operations and relations with the customer after shipment. "Online" quality control (manufacturing process control) can achieve a more cost-effective process control. Every product or process consists of a large number of components. It is a natural belief that the quality and performance of any item can easily be improved by merely tightening tolerances. We believe that we can obtain better performance by specifying machining to ± 0.5 micron than specifying machining to ± 1 micron. Specifying machining to ± 0.5 micron is an expensive production process. Sometimes tightening of tolerances is not guaranteed much better performance. Therefore, it is recommended that only after extensive parameter design, studies have been completed then tolerance calculation is to be performed. Tight tolerances are last resorts to improve quality and productivity.

Applications of Taguchi methods have centered around two main areas:
- Improving an existing product
- Improving process for a specific process

24.8 TAGUCHI'S LOSS FUNCTION

Taguchi's loss function is used to measure financial loss to society resulting from poor quality. Traditionally, companies measure quality by the number of defects or the defect rate. In this system, defects are identified through inspections of the materials and products. Upper and lower quality limits are established. Everything that does not fall within the limits is considered a defect. Taguchi methods in a manufacturing context are statistical methods to improve the quality of products. It is vital to produce a product on target. The variation around the target causes poor manufactured quality. Taguchi quantified costs over the lifetime of the product. This includes costs associated with poor performance, operating costs (which changes as a product ages) and any added expenses due to harmful side effects of the product in use. Long-term costs to the manufacturer include brand reputation and loss of customer satisfaction leading to declining market share. Other costs to the consumer include costs from low durability, difficulty interfacing with other parts, or the need to build in safety margins. There are many types of quality loss functions. However, in all types, the loss is determined by evaluating variation from a specific target. Deviation is the difference between the output value (y) and the target value (m). The loss (L) is proportional to the square of deviation (equation). Therefore, loss (L) function is a quadratic function of square of the deviation (equation). Quality is defined in terms of deviation from the target value. Products meeting tolerance also inflict a quality loss. Best quality when performance is on target. Quality loss of product increases with increase in deviation.

The equation used to describe the loss function of one unit of product:

$$L = k(y - m)^2$$

Where

   L = Loss in rupees
   y = Actual value (diameter, length, weight, heat loss, power, speed, voltage, current, etc.)
  m = Target Value
   k = Constant (defined below)

Let A is cost spent by the customer to account for a defective product at either end of the specification range. It is found by substituting $y = m + \Delta$ into the loss function, where Δ is the deviation from target value.

$$A = k(y - m)^2$$

$$A = k((m + \Delta) - m)^2$$

$$A = k(\Delta)^2$$

$$k = \frac{A}{(\Delta)^2}$$

A graphical representation of the loss in rupees is shown in the Fig. 24.1. If deviation of output value (y) from the target value (m) increases then the loss (L) increases by the mean square of deviation. There is no loss when the output value is equal to the target value ($y = m$).

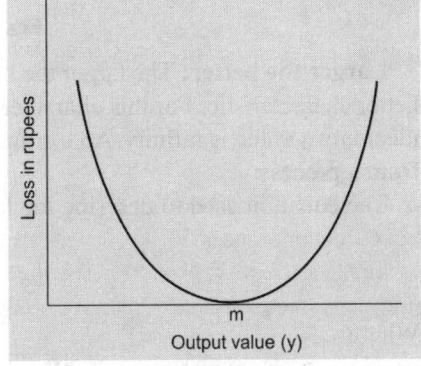

Fig. 24.1 Loss (L) function as a function of square of the deviation

24.8.1 Quality Loss Functions for Various Quality Characteristics

There are three characteristics used to define the quality loss function:

1. Smaller the better
2. Larger the better
3. Nominal the best

Each of these types is defined by a different set of equations, which is different from the general form of the loss function equation.

Smaller the better: In the case of smaller the better characteristic, the ideal target value is defined as zero. As the value gets larger, the loss incurred grows as shown in Fig. 24.2. An example of this characteristic is minimization of heat losses in a heat exchanger. Another example is minimizing carbon dioxide emissions. Minimizing this characteristic (y) as much as possible would produce a more desirable product.

The equation used to describe the loss function of one unit of product:

$$L = k(y - m)^2$$
$$m = 0$$
$$L = k(y)^2$$

Where,

k = Proportionality constant

y = Output value

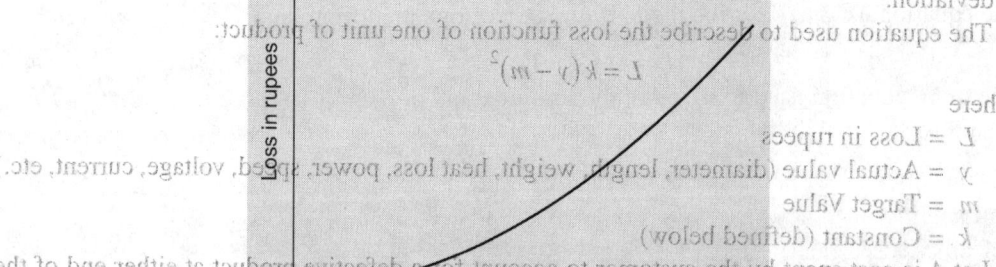

Fig. 24.2 Smaller the better

Larger the better: The larger the better characteristic is just the opposite of the smaller the better characteristic. For this characteristic type, it is preferred to maximize the result, and the ideal target value is infinity. An example of this characteristic is maximizing the product yield from a process.

The equation used to describe the loss function of one unit of product:

$$L = \frac{k}{y^2}$$

Where

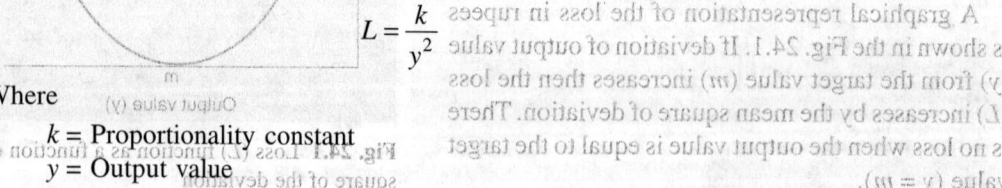

k = Proportionality constant

y = Output value

A graphical representation of the larger the better characteristic is shown in Fig. 24.3. This characteristic is the opposite of the smaller the better characteristic, as the loss is minimized as the output value is maximized.

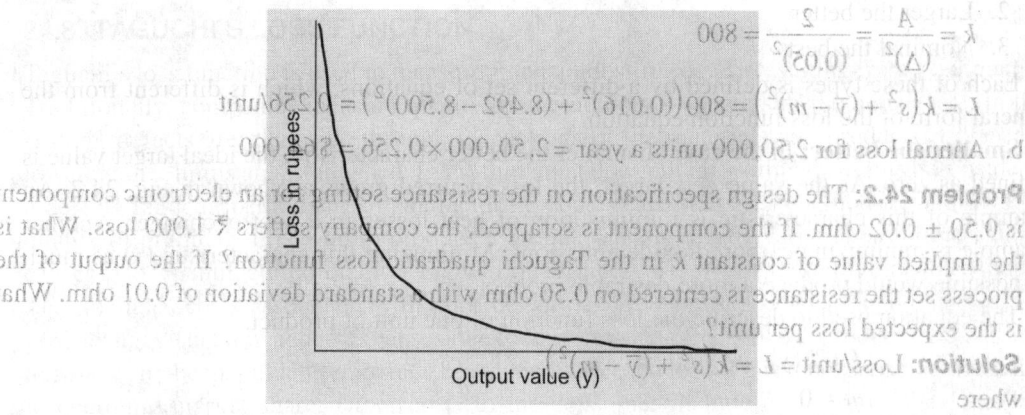

Fig. 24.3 Larger the better characteristic

Nominal the best: In this approach, the closer to the target value, the better. It does not matter whether the deviation is above or below the target value. Under this approach the deviation is quadratic. Quality is in this case is defined in terms of deviation from the target value. Products meeting tolerance also inflict a quality loss. Best quality when performance is on target as shown in Fig. 24.1. An example of this characteristic is the thickness of a windshield in a car. Another example is mating part in an assembly.

The equation used to describe the loss function of one unit of product:

$$L = k(y - m)^2$$

One more equation used to describe the loss function of one unit of product using process mean and process deviation is:

$$L = k\left(s^2 + (\bar{y} - m)^2\right)$$

Where

s is process (product) standard deviation

$\bar{y}$ is process (product) mean

24.9 SOLVED PROBLEMS

Problem 24.1: The cost spent by the customer to account of defective product when a part is exactly 8.55 or 8.45 units is $2. Process specification is: 8.5 + 0.05 units and $\bar{y} = 8.492$ and $s = 0.016$. Determine:

a. Average loss per unit
b. Average loss for 2,50,000 units a year

Solution: Loss/unit = $L = k\left(s^2 + (\bar{y} - m)^2\right)$

L = Loss in rupees

A = Cost spent by the customer to account for a defective product when a part is exactly 8.55 or 8.45 units = $2

Process mean = $\bar{y} = 8.492$

Standard deviation = $s = 0.016$

Target value $= m = 8.5$

$\quad \Delta = 0.05$ units

a. $A = k(\Delta)^2$

$$k = \frac{A}{(\Delta)^2} = \frac{2}{(0.05)^2} = 800$$

$$L = k\left(s^2 + (\overline{y} - m)^2\right) = 800\left((0.016)^2 + (8.492 - 8.500)^2\right) = 0.256/\text{unit}$$

b. Annual loss for 2,50,000 units a year $= 2,50,000 \times 0.256 = \$64,000$

Problem 24.2: The design specification on the resistance setting for an electronic component is 0.50 ± 0.02 ohm. If the component is scrapped, the company suffers ₹ 1,000 loss. What is the implied value of constant k in the Taguchi quadratic loss function? If the output of the process set the resistance is centered on 0.50 ohm with a standard deviation of 0.01 ohm. What is the expected loss per unit?

Solution: Loss/unit $= L = k\left(s^2 + (\overline{y} - m)^2\right)$

where

$\quad L =$ Loss in rupees

$\quad A =$ Cost spent by the customer to account for a defective product when a part is exactly

$\qquad$ 0.52 or 0.48 ohms $=$ ₹ 1,000

Process mean $= \overline{y} = 0.5$ ohm

Standard deviation $= s = 0.01$ ohm

Target value $= m = 0.5$

$\Delta = 0.02$ ohms

a. $\quad A = k(\Delta)^2$

$$k = \frac{A}{(\Delta)^2} = \frac{1000}{(0.02)^2} = 25,00,000$$

Loss/unit $= k\left(s^2 + (\overline{y} - m)^2\right) = 25,00,000\left((0.01)^2 + (0.5 - 0.5)^2\right) = 250/\text{unit}$

Problem 24.3: Dimension of a part is specified as 100 ± 0.20 mm. To investigate the impact of this tolerance on product performance, the company has studied its repair records to discover that, if the ± 0.20 tolerance is exceeded there is a 60% chance that the product will be returned for repair at cost of ₹ 5,000 to the company (during warrant period). Estimate Taguchi loss function constant k for the data.

Solution: The equation used to describe the loss function of one unit is:

$$L = k(y - m)^2$$

Where

$\quad L =$ Loss in rupees

$\quad A =$ The average repair cost/unit, if the dimensions of part exceed 100.20 or 99.80

$\quad A =$ 0.6 (5,000) + 0.4 (0) = ₹ 3,000

$\quad \Delta =$ 0.2 mm

$\quad A = k(\Delta)^2$

$$k = \frac{A}{(\Delta)^2} = \frac{3,000}{(0.2)^2} = 75,000$$

The loss function: $L = k(y - m)^2 = 75,000(y - 100)^2$

Problem 24.4: Dimension of a part is specified as 100 ± 0.20 mm. To investigate the impact of this tolerance on product performance, the company has studied its repair records to discover that, if the ± 0.20 tolerance is exceeded there is a 60% chance that the product will be returned for repair at cost of ₹ 5,000 to the company (during warrant period). Estimate Taguchi loss function constant k for the data. Determine cost for tolerances:

    a. ± 0.10 mm

    b. ± 0.05 mm

Solution: The equation used to describe the loss function of one unit is:

$$L = k(y - m)^2$$

Where

    $L =$ Loss in rupees

    $A =$ The repair cost per unit, if the dimensions of part exceed 100.20 mm or 99.80 mm

    $A = 0.6\,(5,000) + 0.4\,(0) = ₹\ 3,000$

    $\Delta = 0.2$ mm

    $A = k(\Delta)^2$

    $k = \dfrac{A}{(\Delta)^2} = \dfrac{3,000}{(0.2)^2} = 75,000$

The loss function $L = k(y - m)^2 = 75,000(y - 100)^2$

a. $L = k(y - m)^2 = 75,000\,(0.1)^2 = ₹\ 750$

b. $L = k(y - m)^2 = 75,000\,(0.05)^2 = ₹\ 187.5$

Problem 24.5: For manufacturing green paint target is 200 grams of pigment in each gallon of paint. The average cost to the consumer is $500 per gallon from adjusting the pigment when the pigment in a gallon of paint equal to 215 or 185. The paint becomes unsatisfactory, if it is out of the range (200 ± 15) grams. Calculate the loss imparted to society from a gallon of paint with only 190 grams of pigment.

Solution: The equation used to describe the loss function of one gallon of paint:

$$L = k(y - m)^2$$

Where

    $L =$ Loss in rupees

    $A =$ The average cost per gallon spent by the customer for adjusting pigments when the pigment in gallon of paint equal to 215 or 185 = ₹ 500

    $\Delta = 15$ grams

    $A = k(\Delta)^2$

    $k = \dfrac{A}{(\Delta)^2} = \dfrac{500}{(15)^2} = 2.22/(\text{gram})^2$

The loss imparted to society from a gallon of paint with only 190 grams of pigment.

$$L = k(y - m)^2 = 2.22(190 - 200)^2 = ₹\ 222$$

24.10 SPECIFYING TOLERANCES FOR A PROCESS

A manufacturer is responsible for only shipping products that meet certain specifications. Products that do not meet these determined specifications are defective and cannot be shipped for sale, resulting in a loss to the company. In aiming to meet these specifications, manufacturers have a determined level of tolerance for deviation from the desired target specification. More stringent tolerances would result in fewer products failing on customers, reducing losses in the market, but they would also result in increased costs to manufacturers. Before Taguchi, there was no set method for determining optimal tolerances for a given process. In quality engineering, tolerance is defined as the deviation from the target. Taguchi's method determines tolerances that aim for a balance between losses to the manufacturer and the customer. To determine these tolerances, the quality loss function can be used to determine how much it costs the manufacturer to fix the defective product before shipment, and compare that value to the cost that the defective product would have on the customer (society).

24.11 USES OF QUALITY LOSS FUNCTION

1. Reduces costs: There are three ways that managers can use quality loss function to reduce costs.
   a. Move the average of the actual distribution closer to the target value
   b. Reduce variability
   c. Do a combination of both
2. Setting specific limits: The data from the quality loss function can be used to determine where limits should be set to help minimize losses.

24.12 EIGHT STEPS IN TAGUCHI METHODOLOGY

The main trust of Taguchi's techniques is the use of parameter design, which is an engineering method for product or process design that focuses on determining the parameter (factor) settings producing the best levels of a quality characteristic (performance measure) with minimum variation. Taguchi's approach to design experiments is easy to be adopted and applied for users with limited knowledge of statistics; hence it has gained a wide popularity in the engineering and scientific communities.

Step 1: Identify the main function, side effects and failure mode of finished product.

Step 2: Identify the noise factors, testing conditions, and quality characteristics.

Step 3: Identify the objective function to be optimized (Smaller the better, Larger the better, Nominal the best).

Step 4: Identify the parameters and their levels.

For example, machining parameters in a machining process are cutting tools of different materials, depth of cut, cutting speed, feed rate, working temperature and power.

For example, the range of the spindle speed is between 500 and 1,000 rpm.

Step 5: Select the orthogonal array matrix experiment.

A well-planned set of experiments, in which all parameters of interest are varied over a specified range, is a much better approach to obtain systematic output data (performance of an output characteristic). Mathematically speaking, such a complete set of experiments give desired results. Usually the number of experiments and resources (materials and time) required are large for conducting experiments. The method of investigating all possible combinations and conditions in an experiment (involving multiple factors) is traditionally known as factorial design.

Orthogonal arrays (OA) are a special set of Latin squares, constructed by Taguchi to layout the product design experiments. By using this table, an orthogonal array of standard procedure can be used for a number of experimental situations.

The OA facilitates the experimental design process by assigning factors to the appropriate columns. In this case, referring to Table 24.1, there are seven 2-level factors, these are arbitrarily assigned factors A, B, C, D, E, F and G to columns 1, 2, 3, 4, 5, 6 and 7 respectively, for a L8 array. From the Table 24.1, 8 trials of experiments are needed, with the level of each factor for each trial-run as indicated on the array. The experimental descriptions are reflected through the condition level. For example, 0 may indicates the factor is not applied, and 1 represents the factor that is fully applied. The factors may be variation in chemical concentration, material purity, and mechanical pressure and so on. The experimenter may use different designators for the columns. The OA also ensures that factors influencing the end product's quality are properly investigated and controlled during the initial design stage.

Table 24.1 An orthogonal array of L8

Trial number	Factors							Performance of an output characteristic
	A	B	C	D	E	F	G	
1	0	0	0	0	0	0	0	
2	0	0	0	1	1	1	1	
3	0	1	1	0	0	1	1	
4	0	1	1	1	1	0	0	
5	1	0	1	0	1	0	1	
6	1	0	1	1	0	1	0	
7	1	1	0	0	1	1	0	
8	1	1	0	1	0	0	1	

Step 7: Analyze the data; predict the optimum levels and performance.

Sony corporation example

Sony uses the Taguchi model in managing the television sets it produces. The quality characteristic is the color density of the televisions. The Sony engineers set specific limits for color density at a plus or minus tolerance level. One of the Sony's plants uniformly distributed televisions that fell between the specification limits. The other plant followed a normal distribution with an average near the set target. A comparison of customer responses shows that a higher level of satisfaction was reported on televisions from the second plant. Also, the second plant's warranty expenses were lower. This case shows the problem with focusing on a defect rate rather than a variation from the target. The first plant shipped at a zero defect rate, however, the specification limits allowed for too much variation. In the second plant, the limits were smaller and the quality was more consistent. The Taguchi model provides a good way to analyze the costs associated with variability, even within the limits. In conclusion, if companies want to remain competitive, they have to provide quality products. To accomplish this, a company must focus on the reduction of variability of a product's characteristics around a specific target value. The traditional approach is not enough. To remain a world competitor, a company must consider the Taguchi quality loss function approach.

24.13 QUESTIONS AND PROBLEMS

1. Define the quality.
2. Define various factors which affect product quality.
3. What are the dimensions of quality?
4. Explain quality costs.
5. Compare quality assurance and quality control.
6. Why is Taguchi loss function parabolic in nature? Explain.
7. Explain quality loss functions for various quality characteristics.
8. What are the uses of Taguchi quality loss function?

9. Explain the steps in Taguchi methodology.

10. Dimension of a part is specified as 100 ± 0.20 mm. To investigate the impact of this tolerance on product performance, the company has studied its repair records to discover that, if the ± 0.20 tolerance is exceeded, there is a 60% chance that the product will be returned for repair at cost of $100 to the company (during warrant period). Estimate Taguchi loss function constant k for the data. **Answer:** $k = \$1,500$

11. Dimension of a part is specified as 100 ± 0.20 mm. To investigate the impact of this tolerance on product performance, the company has studied its repair records to discover that, if the ± 0.020 tolerance is exceeded, here is a 60% chance that the product will be returned for repair at cost of $100 to the company (during warrant periods). Estimate Taguchi loss function constant k for the data. Determine cost for tolerances.

    a. ± 0.10 mm

    b. ± 0.05 mm

    Answer: $k = \$1,500$, a. \$15, b. \$3.75

12. The design specification on the resistance setting for an electronic component is (0.50 ± 0.02) ohm. If the component is scrapped, the company suffers $ 200 loss. What is the implied value of constant k in the Taguchi quadratic loss function? If the output of the process set the resistance is centered on 0.50 ohm with a standard deviation of 0.01 ohm, what is the expected loss per unit?

    Answer: $k = \dfrac{A}{(\Delta)^2} = \dfrac{200}{(\Delta)^2} = 5,00,000,$

    $$\text{Loss/unit} = k\left(s^2 + (\bar{y} - m)^2\right) = 5,00,000 \left((0.01)^2 + (0.5 - 0.5)^2\right) = 50 \text{ units}$$

13. A certain part dimension on a power garden tool is specified as (25.50 ± 0.3) mm. Company repair records indicate that, if the ± 0.3 mm tolerance is exceeded, there is 75% chance that the product will be returned for replacement. The cost associated with replacing the product, which includes not only the product cost itself but also additional paper work and handling associated with replacement, is estimated to be $300. Determine the constant k in the Taguchi loss function for these data.

    Answer: $k = \dfrac{A}{(\Delta)^2} = \dfrac{300 \times 0.75 + 0 \times 0.25}{(0.3)^2} = ₹\, 2,500$

Inspection and Statistical Quality Control (SQC)

25.1 INSPECTION

Definition: Inspection means checking the acceptability of the manufactured product. Inspection measures the quality of a product or service in terms of predefined standards. Product quality may be specified by its strength, hardness, shape, surface finish, chemical composition and dimensions.

Purpose or objectives of inspection are:

- Inspection of raw material eliminates those materials which do not meet specifications and are likely to cause trouble during processing.
- After inspection of raw material right quality material is sent to processing or assembly line.
- Inspection separates defective components from no defective ones and thus ensures the adequate quality of product.
- Inspection locates defects in raw material.
- Inspection prevents further work being done on semi-finished product already detected as spoiled.
- Inspection makes sure that the product works without hurting anybody, i.e. its operation is safe.
- Inspection detects source of weakness and trouble in the finished products.
- Inspection reduces number of complaints from the customer.

25.2 STATISTICAL QUALITY CONTROL (SQC)

Statistical quality control (SQC) is the term used to describe the set of statistical tools used by quality professionals. Statistical quality control can be divided into three broad categories:

1. **Descriptive statistics:** Descriptive statistics are used to describe quality characteristics and relationships such as the mean, standard deviation, the range, etc.
2. **Statistical process control (SPC):** Statistical process control involves inspecting a random sample of the output from a process and deciding whether the process is producing products with characteristics that fall within a predetermined range. SPC answers the question of whether the process is functioning properly or not.

3. **Acceptance sampling:** Acceptance sampling is the process of randomly inspecting a sample of goods and deciding whether to accept the entire lot based on the results. Acceptance sampling determines whether a batch of goods should be accepted or rejected.

Statistical quality control (SQC) tools are used to identify quality problems during the production process.

25.3 SOURCES OF VARIATION

Variation in the production process leads to quality defects. If you look at bottles of a soft drink, you will notice that no two bottles are filled to exactly the same level. Some are filled slightly higher and some slightly lower. No two products are exactly same because of slight differences in materials, workers, machines, tools, and other factors. These are called random causes of variation. Common causes of variation are based on random causes that we cannot identify. These types of variation are unavoidable and are due to slight differences in processing. An important task in quality control is to find out the range of natural random variation in a process. The second type of variation that can be observed involves variations where the causes can be precisely identified and eliminated. These are called assignable causes of variation. Examples of this type of variation are poor quality in raw materials, an employee who needs more training, or a machine in need of repair. In each of these examples the problem can be identified and corrected.

25.4 DESCRIPTIVE STATISTICS

Descriptive statistics can be helpful in describing certain characteristics of a product and a process. The most important descriptive statistics are:

- Mean
- Range
- Standard deviation
- Distribution of data

25.4.1 Mean

The arithmetic average or the mean measures the central tendency of a set of data. To compute the mean we simply sum all the observations and divide by the total number of observations. The equation for computing the mean is

$$\bar{X} = \text{Mean} = \frac{\text{Sum of all the observations}}{\text{Number of observations}} = \frac{\Sigma X}{n}$$

25.4.2 Range and Standard Deviation

There are two measures that can be used to determine the amount of variation in the data. The first measure is the range, which is the difference between the largest and smallest observations. Another measure of variation is the standard deviation. The equation for computing the standard deviation is

$$\sigma = \sqrt{\frac{\Sigma (X_i - \bar{X})^2}{n-1}}$$

Where

σ = Standard deviation of a sample

$\bar{X}$ = Mean

X_i = Observation $i, i = 1, 2, \ldots, n$

n = Number of observations

25.4.3 Distribution of Data

A third descriptive statistic used to measure quality characteristics is the shape of the distribution of the observed data. When a distribution is symmetric, there are the same numbers of observations on left and right of the mean and we say that the data has a Normal distribution. Normal distribution is shown in the Fig. 25.1. When a disproportionate number of observations are present on either side of the mean, we say that the data has a skewed distribution. Figure 25.2 shows difference between symmetric and skewed distributions.

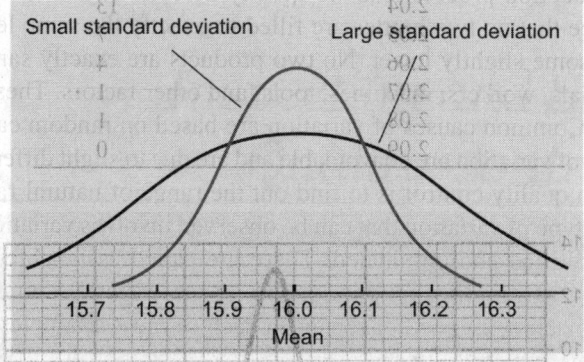

Fig. 25.1 Normal distribution with varying standard deviations

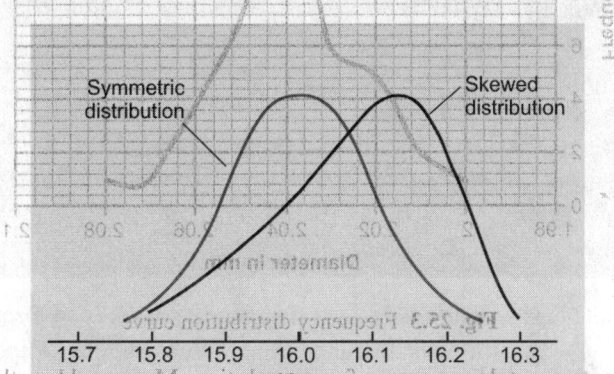

Fig. 25.2 Difference between symmetric and skewed distributions

25.4.4 Solved Problems

Problem 25.1: Forty spindles were manufactured on a machine and their diameters as measured in mm are given in Table 25.1.

Table 25.1 Spindle diameters in mm

Spindle number	1	2	3	4	5	6	7	8	9	10	11	12	13	14	15
Diameter	2.01	2.02	2.04	2.04	2.04	2.03	2.01	2.02	2.03	2.06	2.02	2.04	2.05	2.03	2.04
Spindle number	16	17	18	19	20	21	22	23	24	25	26	27	28	29	30
Diameter	2.05	2.02	2.03	2.04	2.06	2.05	2.04	2.04	2.02	2.03	2.05	2.06	2.03	2.04	2.05
Spindle number	31	32	33	34	35	36	37	38	39	40					
Diameter	2.04	2.05	2.08	2.05	2.04	2.07	2.00	2.04	2.06	2.04					

Determine frequency distribution and draw frequency distribution graph.

Solution: Frequency distribution of spindle diameters is shown in Table 25.2 and in Fig. 25.3.

Table 25.2 Frequency distribution of spindle diameters

Dimension (diameter in mm)	Frequency
2.00	1
2.01	2
2.02	5
2.03	6
2.04	13
2.05	7
2.06	4
2.07	1
2.08	1
2.09	0

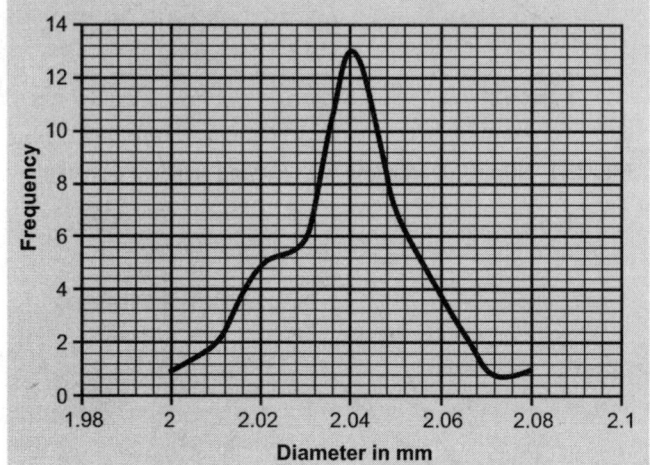

Fig. 25.3 Frequency distribution curve

Problem 25.2: Twenty crystals are grown from a solution. Measured lengths of each crystal in millimetres are given below:

9, 2, 5, 4, 12, 7, 8, 11, 9, 3, 7, 4, 12, 5, 4, 10, 9, 6, 9, 4.

Calculate the mean and standard deviation of the length of the crystals.

Solution:

Calculating the mean of the data:

$$\overline{X} = \text{Mean} = \frac{\text{Sum of all the observations}}{\text{Total number of observations}} = \frac{\Sigma X}{N}$$

$$\overline{X} = \text{Mean} = \frac{(9+2+5+4+12+7+8+11+9+3+7+4+12+5+4+10+9+6+9+4)}{20}$$

$$= \frac{140}{20} = 7$$

Calculation of standard deviation is shown in Table 25.3.

Table 25.3 Calculation of standard deviation

X_i	$(X_i - \bar{X})^2$			
9	$(9-7)^2$	=	$(2)^2$	= 4
2	$(2-7)^2$	=	$(-5)^2$	= 25
5	$(5-7)^2$	=	$(-2)^2$	= 4
4	$(4-7)^2$	=	$(-3)^2$	= 9
12	$(12-7)^2$	=	$(5)^2$	= 25
7	$(7-7)^2$	=	$(0)^2$	= 0
8	$(8-7)^2$	=	$(1)^2$	= 1
11	$(11-7)^2$	=	$(4)^2$	= 16
9	$(9-7)^2$	=	$(2)^2$	= 4
3	$(3-7)^2$	=	$(-4)^2$	= 16
7	$(7-7)^2$	=	$(0)^2$	= 0
4	$(4-7)^2$	=	$(-3)^2$	= 9
12	$(12-7)^2$	=	$(5)^2$	= 25
5	$(5-7)^2$	=	$(-2)^2$	= 4
4	$(4-7)^2$	=	$(-3)^2$	= 9
10	$(10-7)^2$	=	$(3)^2$	= 9
9	$(9-7)^2$	=	$(2)^2$	= 4
6	$(6-7)^2$	=	$(-1)^2$	= 1
9	$(9-7)^2$	=	$(2)^2$	= 4
4	$(4-7)^2$	=	$(-3)^2$	= 9

$$\Sigma(X_i - \bar{X})^2 = 178$$

$$\sigma = \sqrt{\frac{\Sigma(X_i - \bar{X})^2}{n-1}} = \sqrt{\frac{178}{20-1}} = \sqrt{9.368} = 3.061$$

1. The sample variance is 9.368.
2. The sample standard deviation is 3.061.

25.5 CONTROL CHARTS

The most commonly used tool for monitoring the production process is a control chart. Different types of control charts are used to monitor different aspects of the production process. A control chart is a graph that shows whether a sample of data falls within upper and lower control limits. A control chart has upper and lower control limits. We say that a process is out of control when a plot of data reveals that one or more samples fall outside the control limits.

The upper and lower control limits on a control chart are usually set at 3 standard deviations from the mean. If we assume that the data exhibit a normal distribution, these control limits will capture 99.74 percent of the normal variation. Control limits can be set at 2 standard deviations from the mean. In that case, control limits would capture 95.44 percent of the values.

The different characteristics that can be measured by control charts can be divided into two groups:

1. A control chart for variables
2. A control chart for attributes

25.5.1 Control Chart for Variables

A control chart for variables is used to monitor characteristics that can be measured and have a continuum of values, such as height, weight, or volume. Examples are the weight of a bag of sugar, the temperature of a baking oven and the diameter of a shaft.

25.5.1.1 $\bar{X}$ charts

A mean control chart is often referred to as a $\bar{X}$ chart. It is used to monitor changes in the mean of a process. To construct a mean chart, we first need to construct the centre line of the chart. Steps required in constructing a $\bar{X}$ chart:

1. Take multiple samples. Usually these samples are small, with about four or five observations.
2. Compute their means. Each sample has its own mean $\bar{X}$.
3. The centre line of the chart is then computed as the mean of all 'K' sample means, where 'K' is the number of samples.

$$\bar{\bar{X}} = \frac{\bar{X}_1 + \bar{X}_2 + \ldots + \bar{X}_K}{K}$$

Construct the upper and lower control limits of the chart, we use the following formulas:

$$\text{Upper control limit (UCL)} = \bar{\bar{X}} + z\sigma_{\bar{x}}$$

$$\text{Lower control limit (LCL)} = \bar{\bar{X}} - z\sigma_{\bar{x}}$$

Where

$\bar{\bar{X}}$ = The average of the sample means

z = Standard normal variable (2 for 95.44% confidence, 3 for 99.74% confidence)

$\sigma_{\bar{x}}$ = Standard deviation of the distribution of sample means $= \dfrac{\sigma}{n}$

σ = Population standard deviation

n = Number of observations for sample

Another way to construct the control limits is to use the sample range as an estimate of the variability of the process. Remember that the range is simply the difference between the largest and smallest values in the sample. The spread of the range can tell us about the variability of the data. In this case control limits are:

$$\text{Upper control limit (UCL)} = \bar{\bar{X}} + A_2\bar{R}$$

$$\text{Lower control limit (LCL)} = \bar{\bar{X}} - A_2\bar{R}$$

Where values for A_2 are obtained from Table 25.4.

$$\bar{R} = \frac{\Sigma R}{K}$$

25.5.1.2 Range (R) charts

Range (R) charts are another type of control chart for variables. Whereas $\bar{X}$ charts measure shift in the central tendency of the process, range charts monitor the dispersion or variability of the process. The method for developing and using R charts is the same as that for $\bar{X}$. The centre line of the control chart is the average range, and the upper and lower control limits are computed as follows:

$$\text{CL} = \bar{R}$$

$$\text{UCL} = D_4\bar{R}$$

$$\text{LCL} = D_3\bar{R}$$

Where values for D_4 and D_3 are obtained from Table 25.4.

Table 25.4 Factors for three sigma control limits of $\bar{X}$ and R charts

Sample size (n)	Factor for $\bar{X}$ chart A_2	Factor for R chart D_3	D_4
2	1.88	0	3.27
3	1.02	0	2.57
4	0.73	0	2.28
5	0.58	0	2.11
6	0.48	0	2.00
7	0.42	0.08	1.92
8	0.37	0.14	1.86
9	0.34	0.18	1.82
10	0.31	0.22	1.78
11	0.29	0.26	1.74
12	0.27	0.28	1.72
13	0.25	0.31	1.69
14	0.24	0.33	1.67
15	0.22	0.35	1.65
16	0.21	0.36	1.64
17	0.20	0.38	1.62
18	0.19	0.39	1.61
19	0.19	0.40	1.60
20	0.18	0.41	1.59
21	0.17	0.43	1.58
22	0.17	0.43	1.57
23	0.16	0.44	1.56
24	0.16	0.45	1.55
25	0.15	0.46	1.54

25.5.1.3 Process capability

A critical aspect of statistical quality control is evaluating the ability of a production process to meet or exceed preset specifications. This is called process capability. Product specifications are called tolerances. For a product to be considered acceptable, its characteristics must fall within this preset range. Otherwise, the product is not acceptable. Product specifications, or tolerance limits, are usually established by design engineers or product design specialists. For example, the specifications for the width of a machine part may be specified as 10 mm ± 0.03. This means that the width of the part should be 10 mm, though it is acceptable, if it falls within the limits of 9.97 mm and 10.03 mm. The spread of the process or process capability is less than the difference between the upper specification limit and lower specification limit.

$$\text{Process capability} = \pm 3\sigma' = 6\sigma'$$

$$\sigma' = \frac{\bar{R}}{d_2}$$

Process capability is measured by the process capability index, C_p, which is computed as the ratio of the specification width to the width of the process variability.

$$C_p = \frac{\text{Specification width}}{\text{Process width}} = \frac{\text{Upper specification limit} - \text{Lower specification limit}}{6\sigma}$$

There are three possible range of values for C_p.

$C_p = 1$: A value of C_p equal to 1 means that the process variability just meets specifications, as in Fig. 25.4a. We would then say that the process is minimally capable.

$C_p \le 1$: A value of C_p below 1 means that the process variability is outside the range of specification, as in Fig. 25.4b. This means that the process is not capable of producing within specification and the process must be improved.

$C_p \ge 1$: A value of C_p above 1 means that the process variability is tighter than specifications and the process exceeds minimal capability, as in Fig. 25.4c.

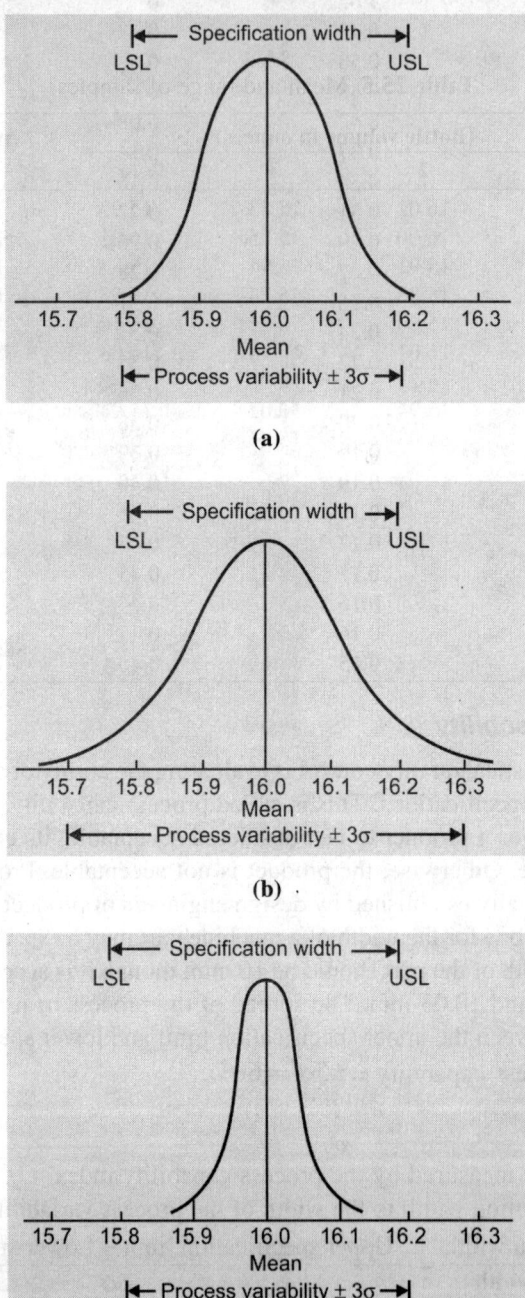

(a)

(b)

(c)

Fig. 25.4 Relationship between process variability and specification width

25.5.1.4 Solved problems

Problem 25.3: A quality control inspector at the Cocoa Fizz soft drink company has taken twenty-five samples with four observations each of the volume of bottles filled. The data and the computed means are shown in Table 25.5. If the standard deviation of the bottling operation is 0.14 ounces, use this information to develop control limits of three standard deviations for the bottling operation.

Table 25.5 Mean and range of samples

Sample	(Bottle volume in ounces)				Average $\bar{X}$	Range R
	1	**2**	**3**	**4**		
1	15.85	16.02	15.83	15.93	15.91	0.19
2	16.12	16.00	15.85	16.01	15.99	0.27
3	16.00	15.91	15.94	15.83	15.92	0.17
4	16.20	15.85	15.74	15.93	15.93	0.46
5	15.74	15.86	16.21	16.10	15.98	0.47
6	15.94	16.01	16.14	16.03	16.03	0.20
7	15.75	16.21	16.01	15.86	15.96	0.46
8	15.82	15.94	16.02	15.94	15.93	0.20
9	16.04	15.98	15.83	15.98	15.96	0.21
10	15.64	15.86	15.94	15.89	15.83	0.30
11	16.11	16.00	16.01	15.82	15.99	0.29
12	15.72	15.85	16.12	16.15	15.96	0.43
13	15.85	15.76	15.74	15.98	15.83	0.24
14	15.73	15.84	15.96	16.10	15.91	0.37
15	16.20	16.01	16.10	15.89	16.05	0.31
16	16.12	16.08	15.83	15.94	15.99	0.29
17	16.01	15.93	15.81	15.68	15.86	0.33
18	15.78	16.04	16.11	16.12	16.01	0.34
19	15.84	15.92	16.05	16.12	15.98	0.28
20	15.92	16.09	16.12	15.93	16.02	0.20
21	16.11	16.02	16.00	15.88	16.00	0.23
2	15.98	15.82	15.89	15.89	15.90	0.16
23	16.05	15.73	15.73	15.93	15.86	0.32
24	16.01	16.01	15.89	15.86	15.94	0.15
25	16.08	15.78	15.92	15.98	15.94	0.30
Total					398.75	7.17

Solution:

σ = Population standard deviation = 0.14 ounce

n = Number of observations for sample = 4

The centre line of the control data is the average of samples

$$\bar{\bar{X}} = \frac{398.75}{25} = 15.95$$

Upper control limit $(\text{UCL}) = \bar{\bar{X}} + z\sigma_{\bar{X}} = 15.95 + 3\frac{0.14}{\sqrt{4}} = 16.16$

Lower control limit $(\text{LCL}) = \bar{\bar{X}} - z\sigma_{\bar{X}} = 15.95 - 3\frac{0.14}{\sqrt{4}} = 15.74$

The above control limits for $\bar{X}$ charts are shown in Fig. 25.5. The sample $\bar{X}$ values are also plotted on Fig. 25.5.

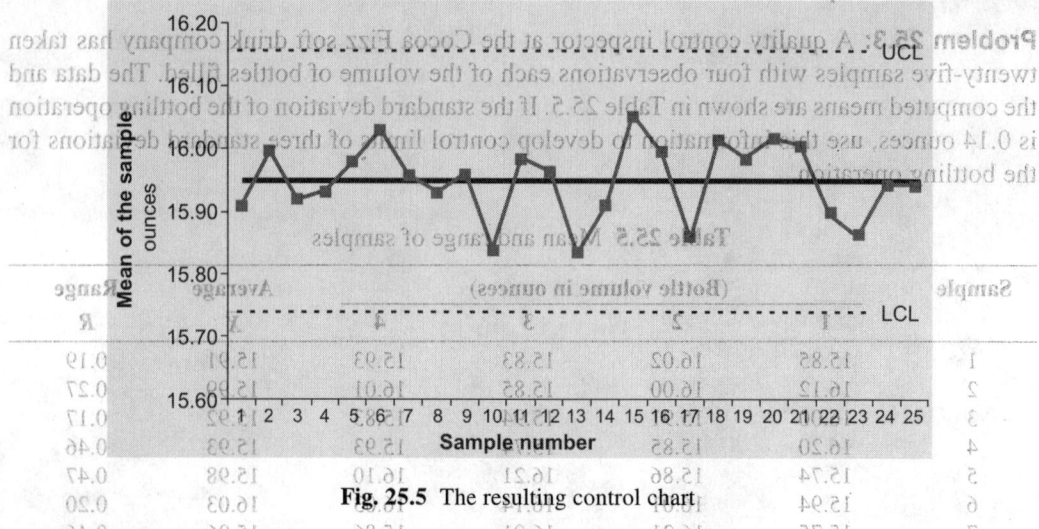

Fig. 25.5 The resulting control chart

Figure 25.5 shows mean of all observations are within the control limits.

Problem 25.4: Assembly times at different times are given in Table 25.6. Number of observations at each time is 3.

Table 25.6 Assembly times at different times

Hour	Assembly time (minutes)		
	X_1	X_2	X_3
9:00 AM	5	12	4
12:00 Noon	6	8	10
3:00 PM	9	4	2

Compute the following:

i. $\bar{X}$

ii. R

iii. $\bar{\bar{X}}$

iv. $\bar{R}$

Solution: Computations of mean and range are given in Table 25.7.

Table 25.7 To determine mean and range

Hour	Assembly time (minutes)			Sample mean	Sample range
	X_1	X_2	X_3	$\bar{X}$	R
9:00 AM	5	12	4	21/3 = 7	12 − 4 = 8
12 Noon	6	8	10	24/3 = 8	10 − 6 = 4
3:00 PM	9	4	2	15/3 = 5	9 − 2 = 7
				$\bar{\bar{X}} = 20/3 = 6.6$	$\bar{R} = 19/3 = 6.3$

Problem 25.5: The data in Table 25.8 were obtained over a five days period. The sample size was 5. Two samples were taken per day. Comment on the process using $\bar{X}$ and R charts.

Table 25.8 Five days period data

Sample number	Observations				
	1	2	3	4	5
1	10	12	13	8	9
2	7	10	8	11	9
3	11	12	9	12	10
4	10	9	8	13	11
5	8	11	11	7	7
6	11	8	8	11	10
7	10	12	13	13	9
8	10	12	12	10	12
9	12	13	11	12	10
10	10	13	7	9	12

Solution: Computation of $\bar{X}$ and R is shown in the Table 25.9.

Table 25.9 To determine $\bar{X}$ and R

Sample number	Observations					$\bar{X}$	R
	1	2	3	4	5		
1	10	12	13	8	9	10.4	5
2	7	10	8	11	9	9.0	4
3	11	12	9	12	10	10.8	3
4	10	9	8	13	11	10.2	5
5	8	11	11	7	7	8.8	4
6	11	8	8	11	10	9.6	3
7	10	12	13	13	9	11.4	4
8	10	12	12	10	12	11.2	2
9	12	13	11	12	10	11.6	3
10	10	13	7	9	12	10.2	6

$$\Sigma\bar{X} = 103.2$$

$$\Sigma R = 39$$

$$\bar{\bar{X}} = \frac{\Sigma\bar{X}}{K} = \frac{103.2}{10} = 10.32$$

$$\bar{R} = \frac{\Sigma R}{K} = \frac{39}{10} = 3.9$$

From Table 25.5, factors for the sample size of 5 are:

A_2 factor for $\bar{X}$ chart = 0.58

D_4 factor for R chart = 2.11

D_3 factor for R chart = 0.00

Upper control limit (UCL) $= \bar{\bar{X}} + A_2\bar{R} = 10.32 + 0.58 \times 3.9 = 10.32 + 2.262 = 12.582$

Lower control limit (LCL) $= \bar{\bar{X}} - A_2\bar{R} = 10.32 - 0.58 \times 3.9 = 10.32 - 2.262 = 8.058$

Control limits for R:

$$CL = \bar{R} = 3.9$$

$$UCL = D_4\bar{R} = 2.11 \times 3.9 = 8.229$$

$$LCL = D_3\bar{R} = 0 \times 3.9 = 0$$

$\overline{X}$ and R charts are shown in Fig. 25.6 and Fig. 25.7 respectively. The sample $\overline{X}$ values are plotted on the Fig. 25.6. The sample R values are also plotted on the Fig. 25.7.

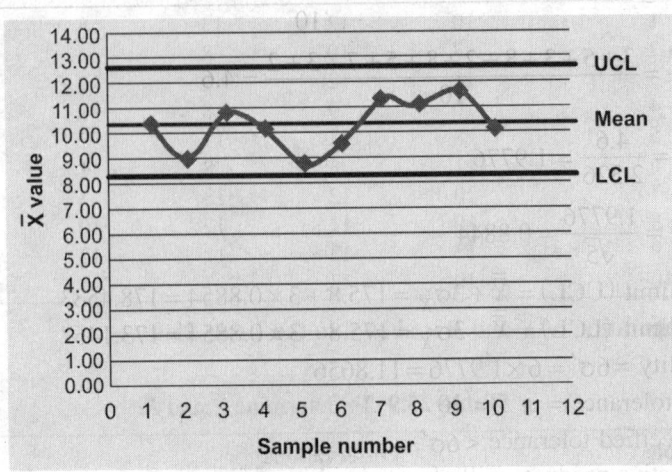

Fig. 25.6 $\overline{X}$ chart

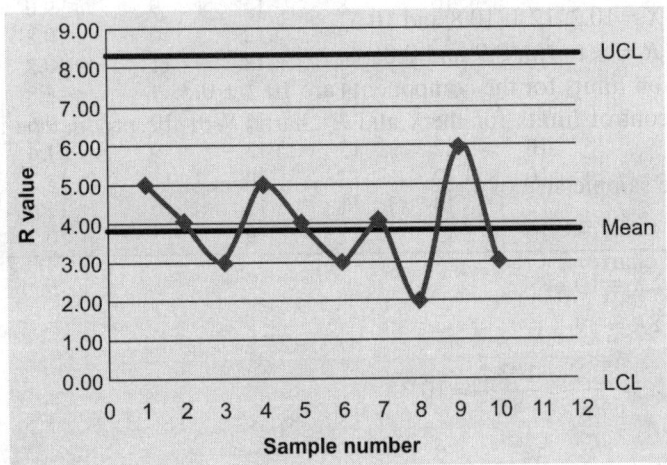

Fig. 25.7 R chart

Problem 25.6: In an automatic filling, 175 grams of certain chemical is to be packed in a certain container. The permissible variation is ± 5 grams. To investigate the capacity of a process, a sample of 5 each were drawn from 10 successive batches and data were recorded in Table 25.10.

Table 25.10 $\overline{X}$ and R for 10 successive batches

Batch	1	2	3	4	5	6	7	8	9	10
Mean ($\overline{X}$)	177	177	176	176	174	177	175	176	176	174
Range (R)	3	5	3	8	2	8	5	7	3	2

Assuming the process to be within control, establish the capability of the process and compare with the stipulated specifications. Take $d_2 = 2.326$.

Solution: Given $d_2 = 2.326$

Sample size $n = 5$

$$\bar{\bar{X}} = \frac{\Sigma \bar{X}}{K} = \frac{177+177+176+176+174+177+175+176+176+174}{10} = 175.8$$

$$\bar{R} = \frac{\Sigma R}{K} = \frac{3+5+3+8+2+8+5+7+3+2}{10} = 4.6$$

$$\sigma' = \frac{\bar{R}}{d_2} = \frac{4.6}{2.326} = 1.9776$$

$$\sigma_{\bar{X}} = \frac{\sigma'}{\sqrt{n}} = \frac{1.9776}{\sqrt{5}} = 0.8844$$

Upper control limit (UCL) $= \bar{\bar{X}} + 3\sigma_{\bar{X}} = 175.8 + 3 \times 0.8854 = 178.4533$

Lower control limit (LCL) $= \bar{\bar{X}} - 3\sigma_{\bar{X}} = 175.8 - 3 \times 0.8854 = 173.1467$

Process capability $= 6\sigma' = 6 \times 1.9776 = 11.8656$

Now specified tolerance $= \pm 5 = 10$

In this case, specified tolerance $< 6\sigma'$

Therefore, defective products will always be there.

Problem 25.7: The following are the $\bar{X}$ and R values of 4 subgroups of 5 readings

$$\bar{X} = 10.2, 12.1, 10.8 \text{ and } 10.5$$
$$R = 1.1, 1.3, 0.9 \text{ and } 0.8$$

The specification limits for the components are 10.7 ± 0.2.

Establish the control limits for the $\bar{X}$ and R charts. Will the product be able to meet its specifications?

Solution: For the sample size of 5,

A_2 factor for $\bar{X}$ chart $= 0.58$

D_4 factor for R chart $= 2.11$

D_3 factor for R chart $= 0.00$

$$\bar{\bar{X}} = \frac{\bar{X}_1 + \bar{X}_2 + \bar{X}_3 + \bar{X}_4}{4} = \frac{10.2 + 12.1 + 10.8 + 10.5}{4} = 10.9$$

$$\bar{R} = \frac{\Sigma R}{K} = \frac{1.1+1.3+0.9+0.8}{4} = 1.025$$

Upper control limit (UCL) $= \bar{\bar{X}} = A_2\bar{R} = 10.9 + 0.58 \times 1.025 = 11.4945$

Lower control limit (LCL) $= \bar{\bar{X}} = A_2\bar{R} = 10.9 - 0.58 \times 1.025 = 9.305$

The specification limits for the components are 10.7 ± 0.2

The upper specification limits for the components $= 10.7 + 0.2 = 10.90$

The lower specification limits for the components $= 10.7 - 0.2 = 10.50$

Since the lower limit is less than the lower specification limit, the process will meet lower limit specifications. Since the upper limit is greater than the upper specification limit, the process does not meet upper specification limit. Therefore, defective products will always be there.

Control limits for R:

$$CL = \bar{R} = 1.025$$
$$UCL = D_4\bar{R} = 2.11 \times 1.025 = 2.1627$$
$$LCL = D_3\bar{R} = 0 \times 1.025 = 0$$

Problem 25.8: Three machines producing the same part are being evaluated for their capability. Bottling machine's standard deviation is given in Table 25.11.

Table 25.11 Bottling machine's standard deviation

Bottling machine	Standard deviation
A	0.05
B	0.1
C	0.2

If specifications are set between 15.8 mm and 16.2 mm, determine which of the machines are capable of producing within specifications.

Solution: C_p of each machine is determined in Table 25.12.

Table 25.12 To determine the capability of each machine

Bottling machine	Standard deviation	USL − LSL	6σ	$C_p = \dfrac{\text{Upper specification limit − Lower specification limit}}{6\sigma}$
A	0.05	0.4	0.3	1.33
B	0.1	0.4	0.6	0.67
C	0.2	0.4	1.2	0.33

Looking at the C_p values, only machine A is capable of filling bottles within specifications, because it is the only machine that has a C_p value at or above 1.

Problem 25.9: Three bagging machines at Potato Chip Company are being evaluated for their capability. Bagging machine's standard deviation is given in Table 25.13.

Table 25.13 Bagging machine's standard deviation

Bagging machine	Standard deviation
A	0.2
B	0.3
C	0.05

If specifications are set between 12.35 and 12.65 ounces, determine which of the machines are capable of producing within specification.

Solution: C_p of each machine is given in Table 25.14.

Table 25.14 To determine the capability of each machine

Bagging machine	Standard deviation	USL − LSL	6σ	$C_p = \dfrac{\text{Upper specification limit − Lower specification limit}}{6\sigma}$
A	0.2	0.3	1.2	0.25
B	0.3	0.3	1.8	0.17
C	0.05	0.3	3	1.00

Looking at the C_p values, only machine C is capable of bagging the potato chips within specifications, because it is the only machine that has a C_p value at or above 1.

25.6 CONTROL CHARTS FOR ATTRIBUTES

Control charts for attributes are used to measure quality characteristics that are counted rather than measured. Attributes are discrete in nature. For example, this could be the number of nonfunctioning light bulbs, the proportion of broken eggs in a carton, the number of rotten apples, the number of scratches on a tile, or the number of complaints issued. Two of the most common types of control charts for attributes are:

- p charts
- c charts

25.6.1 p Charts

p charts are used to measure the proportion of items in a sample that are defective. Each item is classified as good (non-defective) or bad (defective). p charts are used when both the number of defectives and the size of the total sample can be counted.

To construct the upper and lower control limits for a p chart, we use the following formulas:

$$\text{The centre line of the chart} = \text{CL} = \bar{p} = \frac{\text{Total number of defective items}}{\text{Total number of observations}}$$

$$\text{Upper control limit} = \text{UCL} = \bar{p} + z\sigma_p$$

$$\text{Lower control limit} = \text{LCL} = \bar{p} - z\sigma_p$$

Where

z = Standard normal variable

$\bar{p}$ = The sample proportion defective

σ_p = The standard deviation of the average proportion defective

As with the other charts, z is selected to be either 2 or 3 standard deviations, depending on the amount of data we wish to capture in our control limits. Usually, however, they are set at 3.

The sample standard deviation is computed as follows:

$$\sigma_p = \sqrt{\frac{\bar{p}(1-\bar{p})}{n}}$$

Where

n is the sample size.

25.6.2 c Charts

c charts count the actual number of defects. For example, we can count the number of complaints from customers in a month or the number of trucks that exceed their weight limit in a month. However, we cannot compute the proportion of complaints from customers, the proportion of trucks that exceed their weight limit in a month.

The average number of defects is the centre line of the control chart. The upper and lower control limits are computed as follows:

$$\text{UCL} = \bar{c} + z\sqrt{\bar{c}}$$

$$\text{LCL} = \bar{c} - z\sqrt{\bar{c}}$$

25.6.3 Solved Problems

Problem 25.10: A production manager at a tile manufacturing plant has inspected the number of defective tiles in 20 random samples with 20 observations each. The number of defective tiles found in each sample is given in Table 25.15.

Table 25.15 The number of defective tiles found in each sample

Sample number	Number of defective tiles	Sample size	Fraction defective
1	3	20	0.15
2	2	20	0.10
3	1	20	0.05
4	2	20	0.10
5	1	20	0.05
6	3	20	0.15
7	3	20	0.15
8	2	20	0.10
9	1	20	0.05
10	2	20	0.10
11	3	20	0.15
12	2	20	0.10
13	2	20	0.10
14	1	20	0.05
15	1	20	0.05
16	2	20	0.10
17	4	20	0.20
18	3	20	0.15
19	1	20	0.05
20	1	20	0.05
Total	40	400	

Construct a three sigma control chart ($z = 3$) with this information.

Solution:

The centre line of the chart $= \text{CL} = \bar{p} = \dfrac{\text{Total number of defective tiles}}{\text{Total number of observations}} = \dfrac{40}{400} = 0.10$

$$\sigma_p = \sqrt{\frac{\bar{p}(1-\bar{p})}{n}} = \sqrt{\frac{0.10 \times 0.90}{20}} = 0.067$$

Upper control limit $= \text{UCL} = \bar{p} + z\sigma_p = 0.10 + 3 \times (0.67) = 0.301$

Lower control limit $= \text{LCL} = \bar{p} - z\sigma_p = 0.10 - 3 \times (0.67) = -0.101$

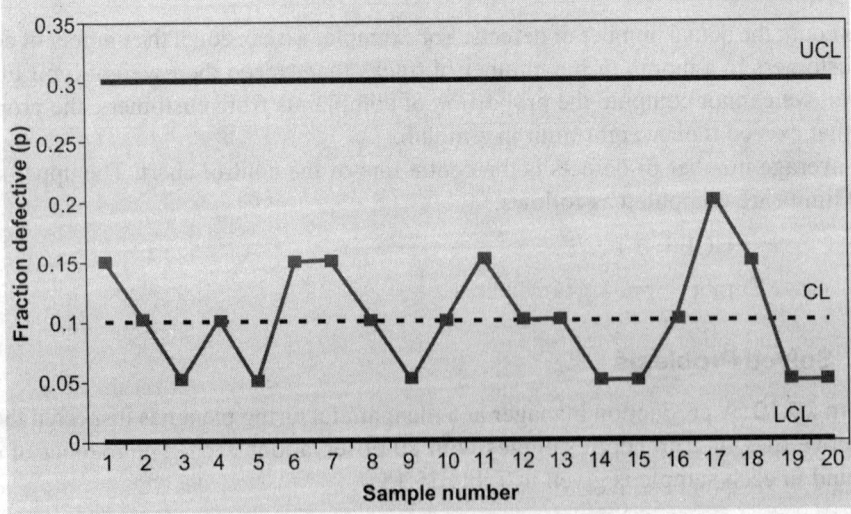

Fig. 25.8 Control chart

In this example the lower control limit is negative, which sometimes occurs because the computation is an approximation of the binomial distribution. When this occurs, the LCL is rounded up to zero because we cannot have a negative control limit. The resulting control chart is shown in Fig. 25.8.

Problem 25.11: The following are the inspection results of 20 lots of magnets, each having 750 magnets. Numbers of defective magnets in each lot are 48, 83, 70, 85, 45, 56, 48, 67, 37, 52, 47, 50, 47, 57, 51, 71, 53, 34, 29 and 30. Calculate the average fraction defective and sigma control limits for p chart.

Solution: $\text{CL} = \overline{p} = \dfrac{\text{Total number of defectives}}{\text{Total number inspected}}$

$$= \frac{\begin{aligned}&48+83+70+85+45+56+48+67+37+52+47+50+47+57\\&+51+71+53+34+29+30\end{aligned}}{750 \times 20}$$

$$= \frac{1060}{750 \times 20} = 0.07067$$

$$\sigma_p = \sqrt{\frac{\overline{p}(1-\overline{p})}{n}} = \sqrt{\frac{0.07067 \times (1-0.07067)}{750 \times 20}} = 0.000262$$

Upper control limit = UCL = $\overline{p} + z\sigma_p$ = $0.07067 + 3 \times (0.000209) = 0.0769$

Lower control limit = LCL = $\overline{p} - z\sigma_p$ = $0.07067 - 3 \times (0.000209) = 0.0643$

Problem 25.12: The number of weekly customer complaints are monitored at a large hotel using a c chart. Complaints have been recorded over the past 20 weeks and are shown in Table 25.16. Develop three sigma control limits using the data.

Table 25.16 Complaints over the past 20 weeks

Week	1	2	3	4	5	6	7	8	9	10	11	12	13	14	15	16	17	18	19	20
Number of complaints	3	2	3	1	3	3	2	1	3	1	3	4	2	1	1	1	3	2	2	3

Solution:

Table 25.16(a) Determining total number of complaints

Week	1	2	3	4	5	6	7	8	9	10	11	12	13	14	15	16	17	18	19	20	Total
Number of complaints	3	2	3	1	3	3	2	1	3	1	3	4	2	1	1	1	3	2	2	3	44

The average number of complaints per week = $\overline{c} = \dfrac{44}{20} = 2.2$

The upper and lower control limits are computed as follows:

UCL = $\overline{c} + z\sqrt{\overline{c}} = 2.2 + 3\sqrt{2.2} = 6.65$

LCL = $\overline{c} - z\sqrt{\overline{c}} = 2.2 - 3\sqrt{2.2} = -2.25$

The LCL is negative and rounded up to zero. Control chart is shown in Fig. 25.9.

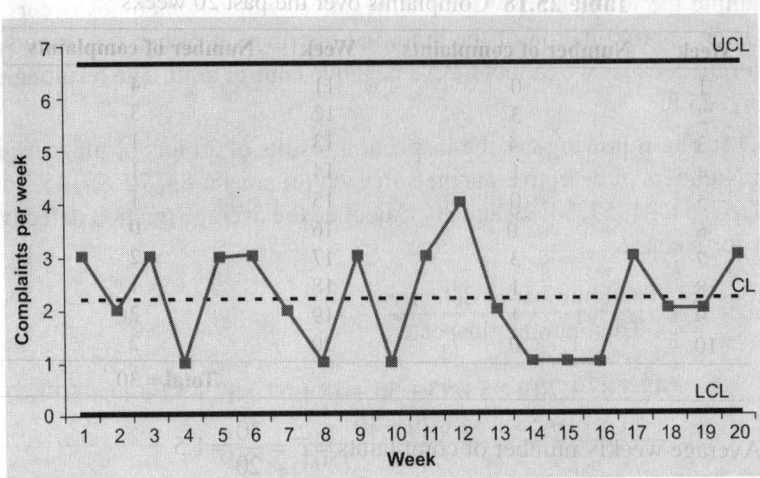

Fig. 25.9 Control chart

Problem 25.13: Table 25.17 shows the number of defects observed in 25 similar castings.

Table 25.17 Number of defects observed in 25 similar castings

Casting	Number of defects	Casting	Number of defects	Casting	Number of defects
1	7	11	22	21	9
2	14	12	15	22	11
3	14	13	8	23	7
4	18	14	24	24	26
5	8	15	14	25	8
6	14	16	9		
7	8	17	9		
8	11	18	11		
9	20	19	10		
10	12	20	8		

i. Find $\bar{c}$.
ii. Compute control limits.

Solution: Total number of defects from Table 25.17 = 317

The average number of defects = $\bar{c}$ = $\dfrac{317}{25}$ = 12.68

The upper and lower control limits are computed as follows:

$$UCL = \bar{c} + z\sqrt{\bar{c}} = 12.68 + 3\sqrt{12.68} = 23.3627$$

$$LCL = \bar{c} - z\sqrt{\bar{c}} = 12.68 - 3\sqrt{12.68} = 1.9973$$

Casting number 24 is above the upper control limit.

Problem 25.14: A service company uses a c chart to monitor the number of customer complaints per week. Complaints have been recorded over the past 20 weeks and are shown in Table 25.18. Develop a control chart with three sigma control limits using the following data:

Table 25.18 Complaints over the past 20 weeks

Week	Number of complaints	Week	Number of complaints
1	0	11	4
2	3	12	3
3	4	13	1
4	1	14	1
5	0	15	1
6	0	16	0
7	3	17	2
8	1	18	1
9	1	19	2
10	0	20	2
			Total = 30

Solution: Average weekly number of complaints $= \bar{c} = \dfrac{30}{20} = 1.5$

The upper and lower control limits are computed as follows:

$$\text{UCL} = \bar{c} + z\sqrt{\bar{c}} = 1.5 + 3\sqrt{1.5} = 5.17$$
$$\text{LCL} = \bar{c} - z\sqrt{\bar{c}} = 1.5 - 3\sqrt{1.5} = -2.17 = 0$$

Control chart is shown in Fig. 25.10.

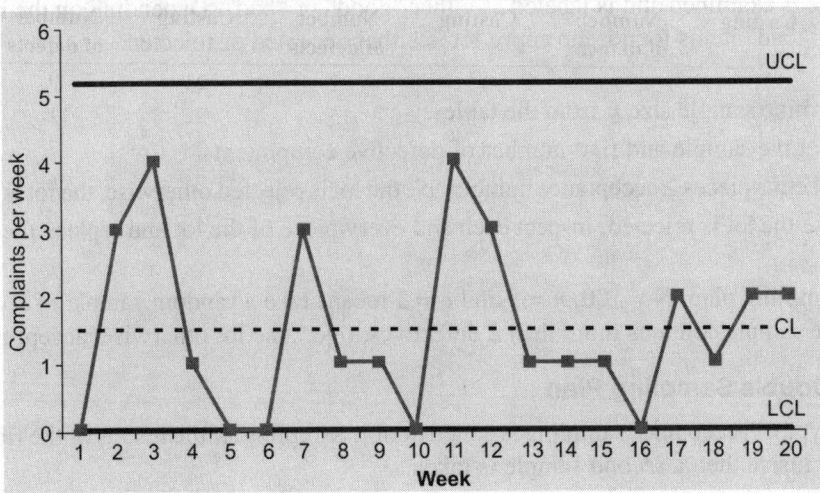

Fig. 25.10 Control chart

25.7 SAMPLING PLANS

Accepting sampling is the process of evaluation of a sample in a lot for the purpose of accepting or rejecting a lot. Inspection for acceptance is carried out in many stages in manufacturing. There are two ways in which inspection is carried out.

1. 100% inspection
2. Sampling inspection

In 100% inspection, all the parts or products are subjected to inspection. In sampling inspection, only a sample drawn from the lot is inspected. If the sample is good, entire lot is accepted otherwise, the entire lot is rejected.

Advantages of sampling inspection:

* The items which are subjected to destructive tests must be inspected by sampling inspection only.
* The cost and time required for sampling inspection is less compared to 100% inspection.
* Smaller inspection staff is required.
* Less damage to products because only few items are subjected to handling during inspection.
* Rejection of entire lot on the basis of sample brings stronger pressure on quality improvement.

Sampling inspection: The variables in a sampling plan are the size of the lot (N), the size of the sample inspected from the lot (n), the number of defects above which a lot is rejected (c), and the number of samples that will be taken.

There are different types of sampling plans.

* Single sampling plan
* Double sampling plan
* Multiple sampling plans

25.7.1 Single Sampling Plan

Single sampling plan is a plan in which a random sample is drawn from every lot. Each item in the sample is examined and is labeled as either "good" or "bad". Depending on the number of defects or "bad" items found, the entire lot is either accepted or rejected.

Steps:

* Determine sample size n from the tables.
* Inspect the sample and find number of defective components.
* If defective pieces $\geq$ acceptance number 'c', the lot is rejected otherwise, the lot is accepted.
* In case the lot is rejected, inspect each and every piece of the lot and replace the defective parts.

In a sampling plan, $N = 100$, $n = 5$ and $c = 2$ means take a random sample of 5 from a lot 100. If the sample contains more than 2 defectives, reject the lot otherwise, accept the lot.

25.7.2 Double Sampling Plan

Another type of acceptance sampling is called double sampling. If the results of the first sample are inconclusive then a second sample is taken.

Parameters of sampling plan:

n_1 = Number of pieces in the first sample
c_1 = Acceptance number for the first sample
n_2 = Number of pieces in the second sample
c_2 = Acceptance number for the second sample
$n_1 + n_2$ = Number of pieces in the two samples combined

Steps:

1. Inspect the first sample n_1 and find number of defective components.
2. If defective pieces > acceptance number 'c_2', the lot is rejected.
3. If defective pieces $\leq$ acceptance number 'c_1', the lot is accepted.
4. If defective pieces $\leq$ acceptance number 'c_2' and if defective pieces > acceptance number 'c_1' then take second sample of n_2 pieces from the lot.

5. If defective pieces in the first and second combined $\leq$ acceptance number 'c_2', the lot is accepted otherwise, the lot is rejected.

6. In case the lot is rejected, inspect each and every piece of the lot and replace the defective parts.

25.7.3 Multiple Sampling Plans

A lot is accepted or rejected based upon the results obtained from several samples from the lot.

25.8 OPERATING CHARACTERISTIC (OC) CURVES

Different sampling plans have different capabilities for discriminating between good and bad lots. At one extreme is 100 percent inspection, which has perfect discriminating power. However, as the size of the sample inspected decreases, the chance of accepting a defective lot increases. Operating characteristic (OC) (Fig. 25.11) curve shows the probability or chance of accepting a lot given various proportions of defects in the lot.

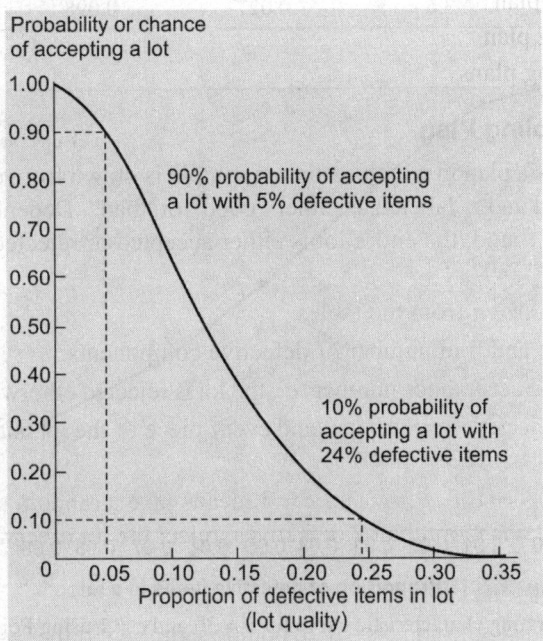

Fig. 25.11 Operating characteristic (OC) curve

Figure 25.11 shows a typical OC curve. The *X*-axis shows the percentage of items that are defective in a lot. This is called "lot quality". The *Y*-axis shows the probability or chance of accepting a lot. When the proportion of defects in the lot increases, probability of accepting the lot decreases. For example, we have a 90 percent probability of accepting a lot with 5 percent defects and an 80 percent probability of accepting a lot with 8 percent defects.

An OC curve using the Poisson's distribution

The Poisson's formula $P_x = \dfrac{e^{-\mu}\mu^x}{x!}$ is used to compute the probabilities of acceptance for $n = 30$ and $c = 1$ in Table 25.19. Figure 25.12 is the operating characteristic curve for $n = 30$ and $c = 1$ using the data of Table 25.19.

Table 25.19 Probabilities of acceptance for $n = 30$ and $c = 1$ using Poisson's distribution

$p = AIQ$	$\mu = np$	$P(0)$	$P(1)$	P_a
0.01	0.30	0.741	0.222	0.963
0.02	0.60	0.549	0.329	0.878
0.03	0.90	0.407	0.366	0.772
0.04	1.2	0.301	0.361	0.663
0.05	1.5	0.223	0.335	0.558
0.06	1.8	0.165	0.298	0.463
0.07	2.1	0.122	0.257	0.380
0.08	2.4	0.091	0.218	0.308
0.09	2.7	0.067	0.181	0.249
0.10	3.0	0.050	0.149	0.199
0.11	3.3	0.037	0.122	0.159
0.12	3.6	0.027	0.098	0.126

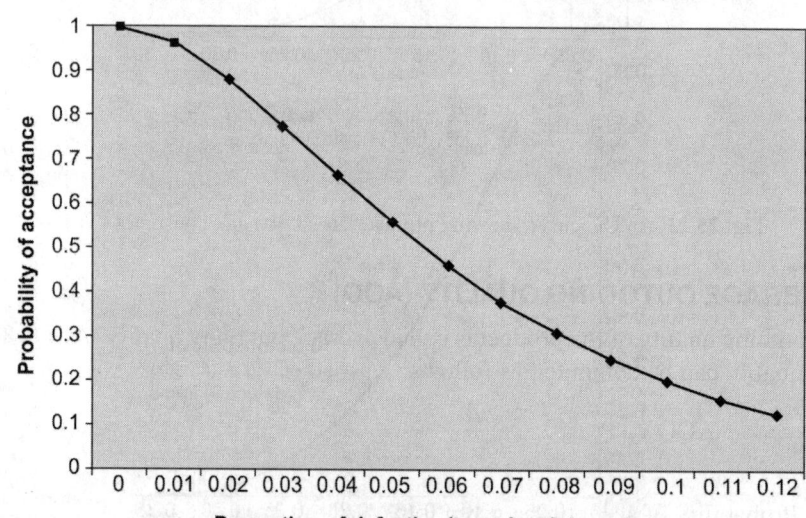

Fig. 25.12 The operating characteristic curve for $n = 30$ and $c = 1$ using Poisson's distribution

Acceptable quality level (AQL): There are a small percentage of defects that consumers are willing to accept. This is called the acceptable quality level (AQL) and is generally in the order of 1 to 2 percent. Acceptable quality level (AQL) is shown in Fig. 25.13.

Lot tolerance percent defective (LTPD): Consumers will usually tolerate defects up to some point. Beyond which consumers will not tolerate them. This threshold level is called the lot tolerance percent defective (LTPD). Lot tolerance percent defective (LTPD) is shown in Fig. 25.13.

Consumer's risk: Consumer's risk is the chance or probability that a lot will be accepted that contains a greater number of defects than the LTPD limit. Consumer's risk is generally denoted by beta (β). Consumer's risk is shown in Fig. 25.13.

Producer's risk: Producer's risk is the chance or probability that a lot containing an acceptable quality level will be rejected. It is generally denoted by alpha (α). Producer's risk is also shown in Fig. 25.13.

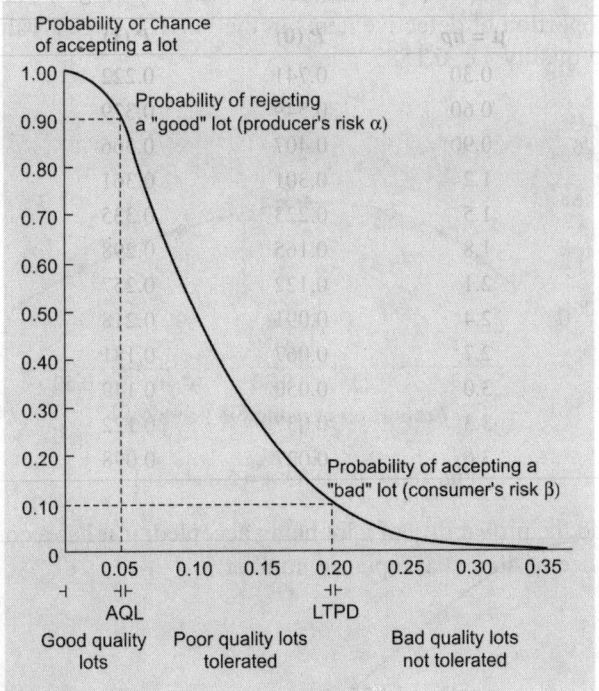

Fig. 25.13 An OC curve showing producer's risk (α) and consumer's risk (β)

25.9 AVERAGE OUTGOING QUALITY (AOQ)

Overall outgoing quality of the product is called average outgoing quality (AOQ). The average outgoing quality can be computed as follows:

$$AOQ = (P_{ac})\, p\, \frac{(N-n)}{n}$$

Where

P_{ac} = Probability of accepting a given lot
p = Proportion of defective items in a lot
N = The size of the lot
n = The sample size chosen for inspection

Usually we assume $\dfrac{(N-n)}{n} = 1$. Then

$$AOQ = (P_{ac})\, p$$

For the parameters $N = 1,000$, $n = 5$ and $c = 1$, P_{ac} is given in Table 25.20. Then compute the value of AOQ as $AOQ = (P_{ac})\, p$. The resultant data is shown in Table 25.20.

Table 25.20 Computing the value of AOQ as $AOQ = (P_{ac})\, p$

p	0.05	0.10	0.15	0.20	0.25	0.30	0.35	0.40	0.45	0.50
P_{ac}	0.9974	0.9185	0.8352	0.7373	0.6328	0.5282	0.4284	0.3370	0.2562	0.1875
AOQ	0.0499	0.0919	0.1253	0.1475	0.1582	0.1585	0.1499	0.1348	0.1153	0.0938

Figure 25.14 shows a graphical representation of the AOQ values. The AOQ value varies, depending on the proportion of defective items in the lot. The largest value of AOQ is called the average outgoing quality, i.e. 0.1585.

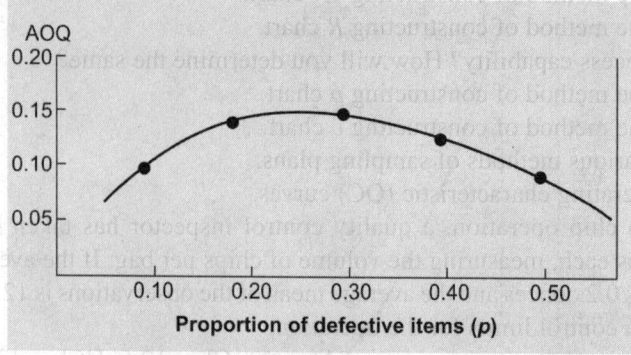

Fig. 25.14 AOQ for $n = 5$ and $c = 1$

Problem 25.15: Find the probability of a lot being accepted, if it has a coming quality, of 5% defective, a sample size of 40 and acceptance number 1.

Solution: Given

$$p = 0.05$$
$$n = 40$$
$$a = 1$$

Probability of acceptance $= \sum_{r=0}^{1} n_{c_r} p^r q^{n-r} = \sum_{r=0}^{1} 40_{c_r} (0.05)^r (0.95)^{40-r}$

Probability of acceptance $= 40_{c_0} (0.05)^0 (0.95)^{40-0} + 40_{c_1} (0.05)^1 (0.95)^{40-1} = 0.39905$

25.10 COMPARISON BETWEEN ATTRIBUTE CHARTS AND VARIABLE CHARTS

Comparison between attribute charts and variable charts is given in Table 25.21.

Table 25.21 Comparison between attribute charts and variable charts

Variable charts	Attribute charts
• Examples are $\overline{X}$, R and σ charts.	• Examples are p and c charts.
• Type of the data required is measured the value of characteristics.	• Attribute data using Go and NO Go gauges
• Control of variables	• Control of proportion of defectives

25.11 QUESTIONS AND PROBLEMS

1. Define inspection.
2. What are the objectives of inspection?
3. What is the mean of the data?
4. Explain two measures that can be used to determine the amount of variation in the data.
5. Define quality control.
6. What is statistical quality control?
7. Define variation.
8. Explain sources of variation.

9. Explain control charts for variables.
10. Explain control charts for attributes.
11. What are the differences between attribute chats and variable charts?
12. Describe the method of constructing $\bar{X}$ chart.
13. Describe the method of constructing R chart.
14. What is process capability? How will you determine the same?
15. Describe the method of constructing p chart.
16. Describe the method of constructing c chart.
17. Describe various methods of sampling plans.
18. Explain operating characteristic (OC) curves.
19. In a potato chip operation, a quality control inspector has taken 4 samples with 5 observations each, measuring the volume of chips per bag. If the average range for the 4 samples is 0.2 ounces and the average mean of the observations is 12.5 ounces, develop three-sigma control limits for the operation.

    Answer: CL = 12.5, UCL = 12.62, LCL = 12.38
20. Ten samples with 5 observations each have been taken Potato Chip Company plant in order to test for volume dispersion in the bagging process. The average sample range was found to be 0.3 ounces. Develop control limits for the sample range.

    Answer: UCL = 0.633, LCL = 0
21. A production manager at a light bulb plant has inspected the number of defective light bulbs in 10 random samples with 30 observations each. Following are the numbers of defective light bulbs found:

Sample	Number of defectives	Number of observations in sample
1	1	30
2	3	30
3	3	30
4	1	30
5	0	30
6	5	30
7	1	30
8	1	30
9	1	30
10	1	30
Total	17	300

Construct a three-sigma control chart ($z = 3$) with this information.

Answer: CL = 0.057, UCL = 0.183, LCL = −0.069 = 0
22. In 400 observations of a computer operator, an analyst found him idle for 32 times. Find:
    a. The sample proportion   b. Standard error of proportion

    Answer: a. 0.08 b. 0.014
23. In a study to find the time to service customers, a bank teller worked 60 minutes and served 36 customers. A record of individual service times showed that
$$\Sigma(X - \bar{X})^2 = (0.79)^2$$
Find:
   a. Sample mean
   b. Standard deviation

   Answer: a. 1.67 minutes/customer b. 0.15 minute

24. Problem: Plot $\bar{X}$ and R charts using the following data

Sample number (Sample size = 5)	$\bar{X}$	R
1	7	2
2	7.5	3
3	8.0	2
4	10.0	2
5	9.5	3
6	11.0	4
7	11.5	3
8	4.0	2
9	3.5	3
10	4.0	2

Answer: i. For $\bar{X}$ chart, UCL = 9.11, LCL = 6.09

ii. For R chart, UCL = 5.48, LCL = 0

25. Problem: Plot 'p' chart using the following data

Number of pieces inspected	Number of defective pieces found
300	25
300	30
300	35
300	40
300	45
300	35
300	40
300	30
300	20
300	50

Answer: i. For p chart; UCL = 0.1723, LCL = 0.0611.

Normal Distribution and Six Sigma

26.1 INTRODUCTION

Data can be "distributed" (spread out) in different ways. It can be spread out more on the left as shown in Fig. 26.1a or more on the right as shown in Fig. 26.1b or it can be all jumbled up as shown in Fig. 26.1c.

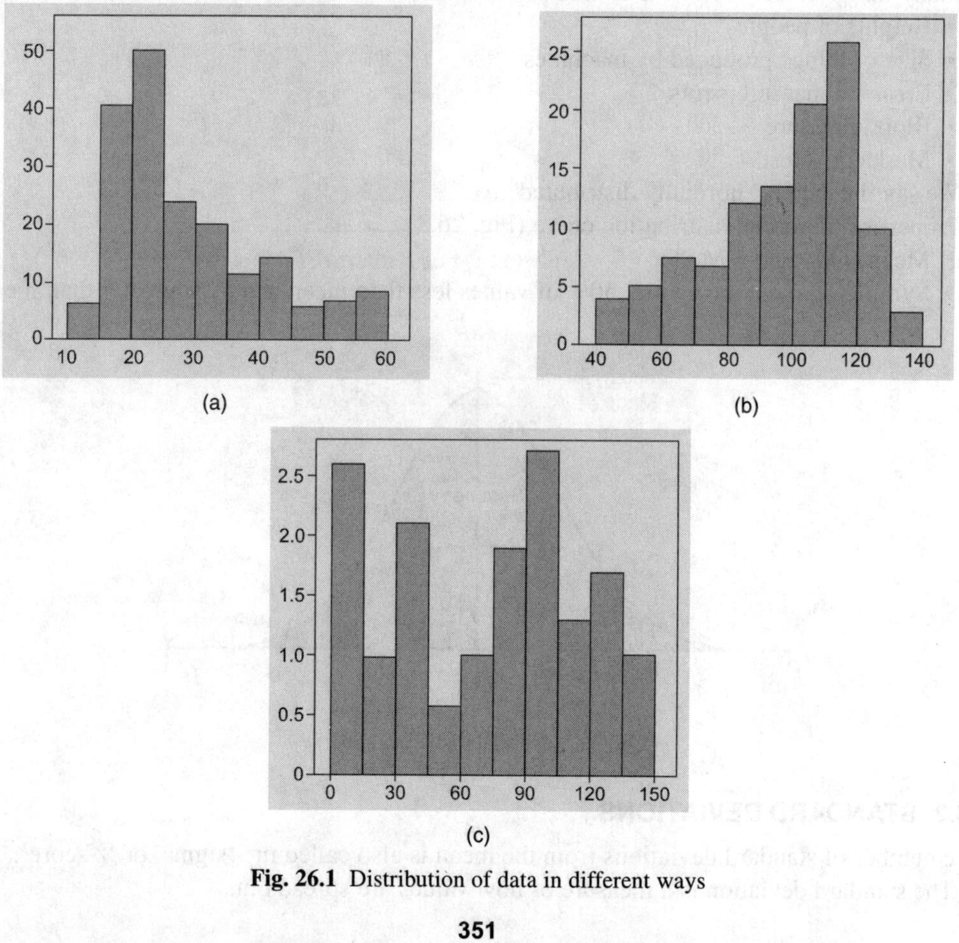

(a)

(b)

(c)

Fig. 26.1 Distribution of data in different ways

But there are many cases where the data tends to be around a central value with no bias left or right, and it gets close to a "normal distribution" as shown in Fig. 26.2. For example, you are weighing bags of potatoes. The bags are supposed to weigh 10 kg, but the actual weights will vary.

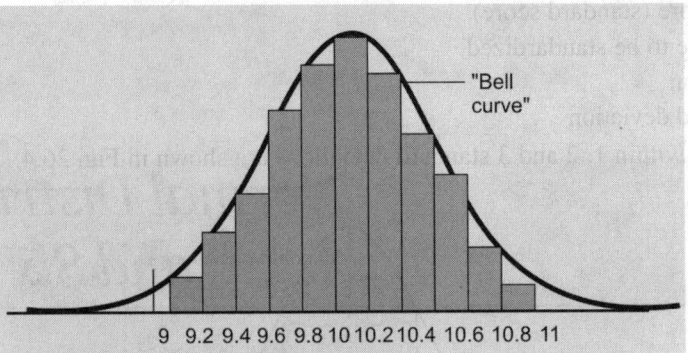

Fig. 26.2 Distribution of actual weights in kilograms

The "bell curve" is a normal distribution. And the histogram in Fig. 26.2 shows some data that follows it closely, but not perfectly (which is usual).

Many things closely follow a normal distribution:
- Heights of people
- Size of things produced by machines
- Errors in measurements
- Blood pressure
- Marks on a test

We say the data is "normally distributed".

Properties of normal distribution curve (Fig. 26.3):
- Mean = Median = Mode
- Symmetry about the centre: 50% of values less than mean and 50% greater than mean.

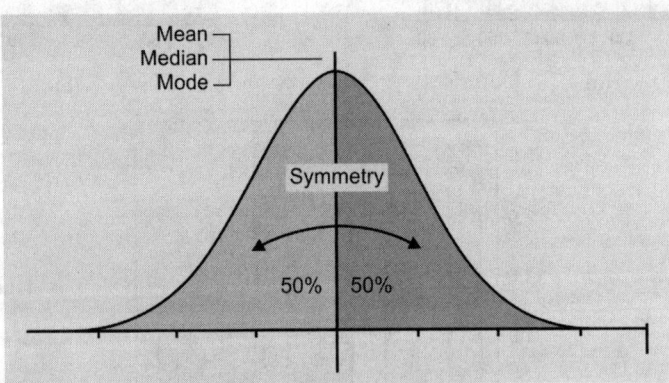

Fig. 26.3 Normal distribution curve

26.2 STANDARD DEVIATIONS

The number of standard deviations from the mean is also called the 'sigma' or 'z-score'.

The standard deviation is a measure of how values are spread out.

Here is the formula to convert value to a standard or z-score:

$$z = \frac{x - \mu}{\sigma}$$

Where

z is the z-score (standard score)

x is the value to be standardized

μ is the mean

σ = Standard deviation

% of values within 1, 2 and 3 standard deviations are shown in Fig. 26.4.

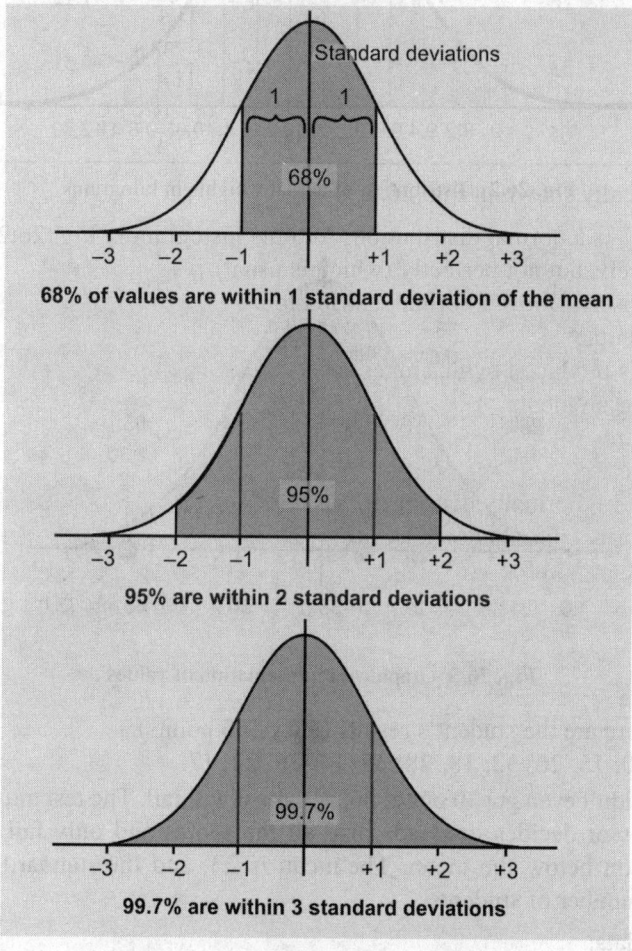

Fig. 26.4 % of values within 1, 2 and 3 standard deviations

26.3 SOLVED PROBLEMS

Problem 26.1: A survey of daily travel time had these results (in minutes):

26, 33, 65, 28, 34, 55, 25, 44, 50, 36, 26, 37, 43, 62, 35, 38, 45, 32, 28, 34.

The mean is 38.8 minutes, and the standard deviation is 11.4 minutes. Convert the values 26, 33 and 65 to z-scores ("standard scores").

Solution: Given data:

Mean = μ = 38.8 minutes

Standard deviation = σ = 11.4 minutes

Conversion of the values 26, 33 and 65 to z-scores is shown in Table 26.1.

Table 26.1 Conversion of the values 26, 33 and 65 to z-scores

Original value	Standard score (z-score) $z = \dfrac{x - \mu}{\sigma}$
26	$z = \dfrac{26 - 38.8}{11.4} = -1.12$
33	$z = \dfrac{33 - 38.8}{11.4} = -0.51$
65	$z = \dfrac{65 - 38.8}{11.4} = +2.30$

They are graphically shown in Fig. 26.5.

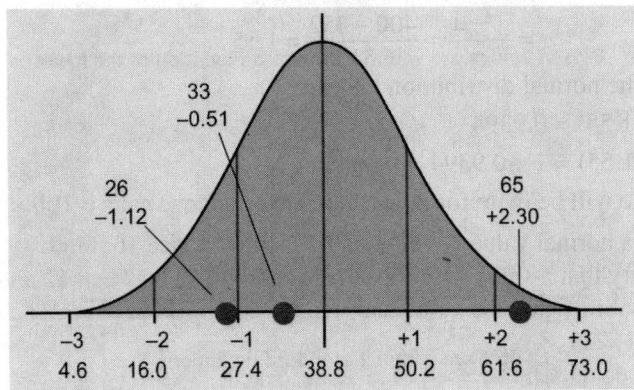

Fig. 26.5 Graphical representation of values

Problem 26.2: Here are the student's results (out of 60 points):

20, 15, 26, 32, 18, 28, 35, 14, 26, 22, 17.

Most students didn't even get 30 out of 60, and most will fail. The test must have been really hard, so the Professor decides to standardize all the scores and only fail people who have 1 standard deviation below the mean. The mean is 23, and the standard deviation is 6.6. Determine failed number of students.

Solution: Given data:

Mean = μ = 23 points

Standard deviation = σ = 6.6 points

Actual scores = x = 20, 15, 26, 32, 18, 28, 35, 14, 26, 22, 17

Standard scores (z-scores) using the formula $z = \dfrac{x - \mu}{\sigma}$ are:

−0.45, −1.21, 0.45, 1.36, −0.76, 0.76, 1.82, −1.36, 0.45, −0.15, −0.91

Number of students who got below −1 is 2.

Here only 2 students will fail (the ones who scored 15 and 14 on the test).

Problem 26.3: 95% of students at school are between 1.1 m and 1.7 m tall. Assuming this data is normally distributed. Calculate the mean and standard deviation.

Solution: The mean is halfway between 1.1 m and 1.7 m:

$$\text{Mean} = \frac{(1.1+1.7)\,m}{2} = 1.4\,m$$

95% is 2 standard deviations either side of the mean (a total of 4 standard deviations) so:

$$1 \text{ standard deviation} = \frac{(1.7-1.1)\,m}{4} = 0.15\,m$$

Problem 26.4: It is known from the past experience that the number of telephone calls made daily in a certain community between 3 pm and 4 pm have a mean of 352 and standard deviation of 31. What percentage of time will be there for more than 400 telephone calls made in this community between 3 pm and 4 pm.

Solution: Given data:

Mean = $\mu = 352$

Standard deviation = $\sigma = 31$

Standard score (z-score) for $x = 400$ using the formula

$$z = \frac{x-\mu}{\sigma} = \frac{400-352}{31} = 1.55$$

From the normal distribution tables:

$P\,(z < 1.55) = 0.9394$

$P\,(z \ge 1.55) = 1 - 0.9394 = 0.0606$

Percentage of time will be there for more than 400 telephone calls = $100 \times 0.0606 = 6.06\%$

Problem 26.5: For a normal value x with mean 20 and variance 16. Find:

a. The probability that a value of x chosen at random lies between 12 and 24.
b. Find the probability that the value chosen at random lies between 16 and 18.

Solution: Mean = $\mu = 20$

Standard deviation = $\sigma = 4$

a. Standard score (z-score) for $x = 12$ using the formula

$$z = \frac{x-\mu}{\sigma} = \frac{12-20}{4} = -2$$

Standard score (z-score) for $x = 24$ using the formula

$$z = \frac{x-\mu}{\sigma} = \frac{24-20}{4} = 1$$

From the normal distribution tables:

$P\,(z \le -2) = 0.0228$

$P\,(z \le 1) = 0.8413$

$P\,(-2 \le z \le 1) = 0.8413 - 0.0228 = 0.8185$

The probability that a value of x chosen at random lies between 12 and 24 = 0.8185

b. Standard score (z-score) for $x = 16$ using the formula

$$z = \frac{x-\mu}{\sigma} = \frac{16-20}{4} = -1$$

Standard score (z-score) for $x = 18$ using the formula

$$z = \frac{x-\mu}{\sigma} = \frac{18-20}{4} = -0.5$$

From the normal distribution tables:

$P(z \leq -1) = 0.1587$

$P(z \leq -0.5) = 0.3085$

$P(-0.5 \leq z \leq -1) = 0.3085 - 0.1587 = 0.1498$

Problem 26.6: The life time of certain kinds of electronic devices have a mean of 300 hours and a standard deviation of 25 hours. Assuming the distribution of those lifetimes approximately close with a normal curve,

    a. Find the probability that any one of these electronic devices will have a lifetime of more than 350 hours.

    b. What percentage will have lifetimes of 300 hours or less?

    c. What percentage will have lifetimes from 220 to 260 hours?

Solution: Mean = μ = 300 hours

               Standard deviation = σ = 25 hours

    a. Standard score (z-score) for $x = 350$ using the formula

$$z = \frac{x - \mu}{\sigma} = \frac{350 - 300}{25} = 2$$

From the normal distribution tables:

$$P(z > 2) = 1 - P(z < 2) = 1 - 0.9772 = 0.0228$$

The probability that any one of these electronic devices will have a lifetime of more than 350 hours = 0.0228

    b. Standard score (z-score) for $x = 300$ using the formula

$$z = \frac{x - \mu}{\sigma} = \frac{300 - 300}{25} = 0$$

From the normal distribution tables:

$$P(z < 0) = 0.5$$

Percentage will have lifetimes 300 hours or less = $0.5 \times 100 = 50\%$

    c. Standard score (z-score) for $x = 220$ using the formula

$$z = \frac{x - \mu}{\sigma} = \frac{220 - 300}{25} = -3.2$$

Standard score (z-score) for $x = 260$ using the formula

$$z = \frac{x - \mu}{\sigma} = \frac{260 - 300}{25} = -1.6$$

From the normal distribution tables:

$P(z < -3.2) = 0.0007$

$P(z < -1.6) = 0.0548$

$P(-3.2 < z < -1.6) = 0.0548 - 0.0007 = 0.0541$

Percentage will have failures from 220 to 260 hours = $0.0541 \times 100 = 5.41\%$

26.4 SIX SIGMA

Six sigma methodology is not a new concept. It combines elements of statistical quality control, breakthrough thinking and management science. Six sigma means a measure of quality for near perfection. Six sigma is a disciplined, data-driven approach and methodology for eliminating defects in any process from manufacturing to service. Six sigma is now increasingly being used by industries such as healthcare, information technology (IT) and tele-communications. Quality has now become vital for a company's success. When 6 sigma is integrated with all the functional departments such as sales, purchase, production, inventory, and others, it makes it easier for the business to increase efficiency, improve quality and reduce operational costs, developing competencies and maximizing profits. The level of quality in an organization that does not have proper quality control systems is usually 1 sigma, 2 sigma, or 3 sigma. The main objective of 6 sigma quality control systems is to take the organization to the highest possible quality level (i.e. six sigma) within the specified time.

26.5 COMPARISON OF ± 3 SIGMA WITH ± 6 SIGMA

Sigma (σ) stands for the number of standard deviations of the process. ± 3 sigma (σ) means that 2,600 defects per million. The level of defects associated with ± 6 sigma is approximately 3.4 defects per million. Figure 26.6 shows a process distribution with quality levels of 3 sigma and 6 sigma. You can see the difference in the number of defects produced.

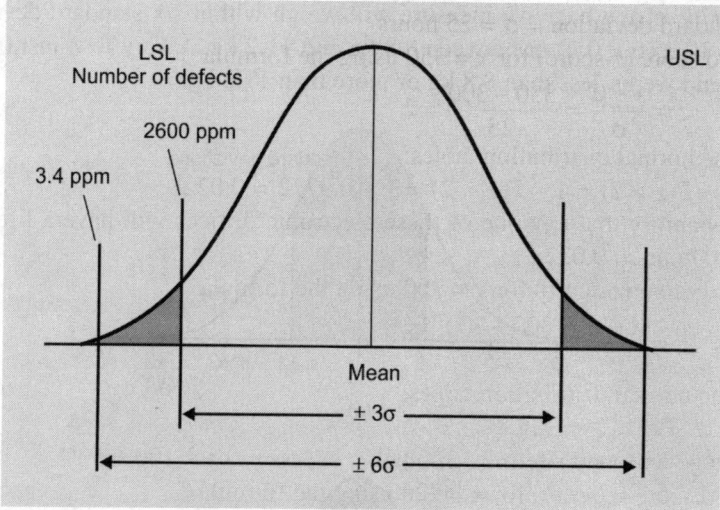

Fig. 26.6 3 σ quality versus 6 σ quality

Statistical quality control tools are used to implement the six sigma concept.

26.6 STATISTICAL REPRESENTATION OF SIX SIGMA

The statistical representation of six sigma describes quantitatively how a process is performing. To achieve six sigma, a process must not produce more than 3.4 defects per million opportunities. A six sigma defect is defined as anything outside of customer specifications.

The steps required to determine sigma limits:

For example, you are weighing bags of Onions. The bags are supposed to weigh 10 kg, but the actual weights will vary. If they are overweight, you are giving away Onions. If they are underweight, customers are losing Onions. The steps required to determine sigma limits are given below:

  a. Determine the weight for each bag.
  b. Construct a histogram of the distribution weight of the bags.
  c. Suppose the data follows normal distribution then determine mean and the standard deviation.

$$\text{Mean weight} = \frac{\text{Total weight of all bags}}{\text{Number of bags}}$$

d. Standard deviation gives you an idea of how much variation is there around the mean. If the standard deviation is high, that means you have a lot of variations in the process. Greek letter sigma is usually used to symbolize the standard deviation in statistical equations.

Suppose for the data, mean is 10 kg and the standard deviation is 0.2 kg, then:

- 68% of the bags will weigh within one standard deviation, or one sigma or $10 \pm (1 \times 0.2)$ or between 9.8 kg and 10.2 kg.
- 95.5% of the bags will weigh within two standard deviations, or two sigma or $10 \pm (2 \times 0.2)$ or between 9.6 kg and 10.4 kg.
- 99.7% of the bags we measure will weigh within three standard deviations, or three sigma or $10 \pm (3 \times 0.2)$ or between 9.4 kg and 10.6 kg. A very few, just 0.3% of all bags, would weigh less than 9.4 kg or more than 10.6 kg.
- 99.999997% of the bags we measure will weigh within six standard deviations, or six sigma or $10 \pm (6 \times 0.2)$ or between 8.8 kg and 11.2 kg. A very few, just 0.0003% of all bags, would weigh less than 8.8 kg or more than 11.2 kg.

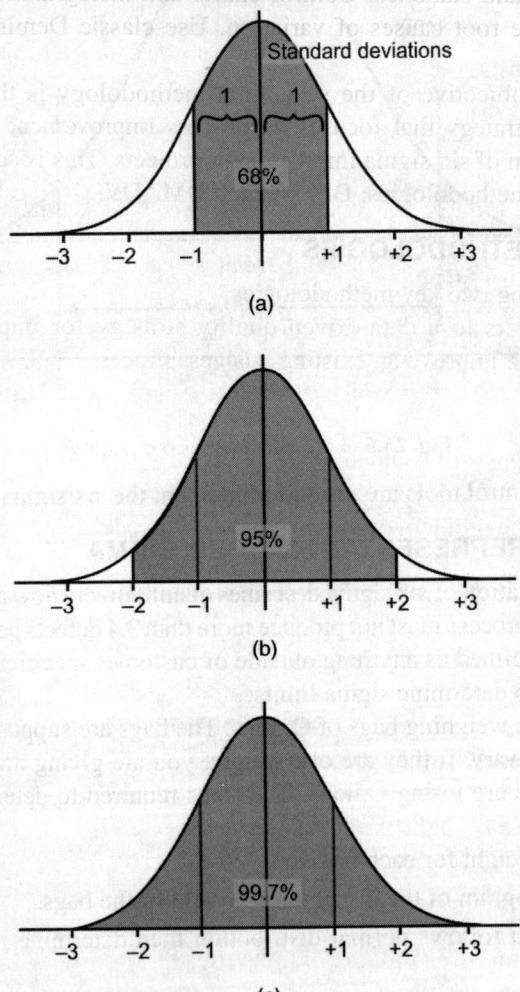

Fig. 26.7 (a) 1 Standard deviation (b) 2 Standard deviations (c) 3 Standard deviations

Number of defects/million at different sigma levels is given in Table 26.2.

Table 26.2 Number of defects/million

Sigma level	Number of defects/million
σ	$(1-0.68) \times 10,00,000 = 3,20,000$
2σ	$(1-0.95) \times 10,00,000 = 50,000$
3σ	$(1-0.9974) \times 10,00,000 = 2,600$
6σ	$(1-0.999997) \times 10,00,000 = 3$

Six sigma advocates believe that for many processes, there should be six sigmas between the mean and the specification limits, so that the process is only making a few bad "parts" in every million. You can, of course, do that by relaxing the specifications, but that isn't usually the way to please customers. Instead, the variation in the process needs to be driven towards zero, so that the histogram gets narrower, and fits more comfortably inside the specification limits. Clearly, to get an accurate view of your critical processes, you need to have people who understand variation and statistics. Control charts and histograms are the tools of quality improvement. Explore root causes of variation. Use classic Deming PDCA cycle to plan improvements.

The fundamental objective of the six sigma methodology is the implementation of a measurement-based strategy that focuses on process improvement and variation reduction through the application of six sigma improvement projects. This is accomplished through the use of two six sigma methodologies: DMAIC and DMADV.

26.7 SIX SIGMA METHODOLOGIES

Six sigma has following two key methodologies:
DMAIC: DMAIC refers to a data-driven quality strategy for improving processes. This methodology is used to improve an existing business process.

DMAIC methodology

This methodology consists of following five steps:
Define → Measure → Analyze → Improve → Control

- Define: Define the problem or project goals that needs to be addressed.
- Measure: Measure the problem and process from which it was produced.
- Analyze: Analyze data and process to determine root causes of defects and opportunities.
- Improve: Improve the process by finding solutions to fix, diminish, and prevent future problems.
- Control: Implement, control, and sustain the improvements solutions to keep the process on the new course.

DMAIC methodology is shown in Fig. 26.8.
DMADV: DMADV refers to a data-driven quality strategy for designing products and processes. This methodology is used to create new product designs or process designs in such a way that it results in a more predictable, mature and defect-free performance.

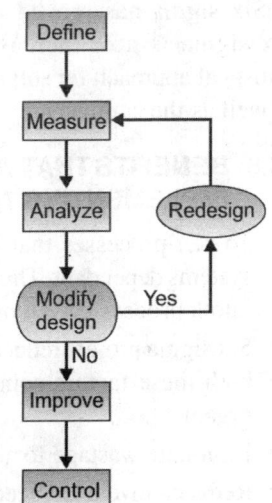

Fig. 26.8 DMAIC methodology

DMADV methodology

This methodology consists of following five steps:

Define → Measure → Analyze → Design → Verify

- Define: Define the problem or project goals that needs to be addressed.
- Measure: Measure and determine customers' needs and specifications.
- Analyze: Analyze the process to meet the customers' needs.
- Design: Design a process that will meet customers' needs.
- Verify: Verify the design performance and ability to meet customers' needs.

Both six sigma processes are executed by six sigma green belts and six sigma black belts.

Six sigma black belts: Six sigma black belts are well trained experts in quality, process improvement, and statistical process control who work within companies as "problem-solvers for hire". They lead process improvement projects, and focus on areas which will have the highest impact on the bottom line.

Jack Welch, the energetic chairman of GE, has been six sigma's most influential advocate. Other companies, notably Motorola and Allied Signal, have been incubators and proponents of the movement. Mikel Harry is its most colorful champion. The consulting and training firm he founded, Six Sigma Academy, has, become the most well-known educator of Black Belts. Many other traditional quality consultancies have been quick to follow suit, including Six Sigma Qualtec, the Juran Institute, and Oriel.

According to the Six Sigma Academy, black belts save companies approximately $230,000 per project and can complete 4 to 6 projects per year. (Given that the average Black Belt salary is $80,000 in the United States that is a fantastic return on investment.) General Electric, one of the most successful companies implementing six sigma, has estimated benefits on the order of $10 billion during the first 5 years of implementation. GE first began six sigma in 1995 after Motorola and Allied Signal blazed the six sigma trail. Since then, thousands of companies around the world have discovered the far reaching benefits of six sigma.

Many frameworks exist for implementing the six sigma methodology. Six sigma consultants all over the world have developed proprietary methodologies for implementing six sigma quality, based on the similar change management philosophies and applications of tools.

Six sigma has proved itself the best quality control program, available in the market. Six sigma is now increasingly used in many fields and sectors. Six sigma provides a statistical approach for solving any problem and thereby improves the quality level of the product as well as the company.

26.8 BENEFITS THAT AN ORGANIZATION DERIVES AS THE RESULT OF IMPLEMENTING A SIX SIGMA PROJECT

- To run processes that have less than 3.4 defects per million, organizations need to be systems dependant. Thus, six sigma mind set helps to transform a people-driven organization into a process-driven one.
- Six sigma project reduces requirement of labor and the requirement of skilled labor. Hence, both these factors combined have an effect of drastically reducing the labor bill of the organization.
- Eliminate wastage to a large extent.
- Reduces inventory needs
- Reduces reworks and defects.

- Organizations are plagued with defective processes which result in the manufacturing of defective products. Each defect has costs attached. The costs include material, time, overheads and loss of reputation for the firm. Implementing Six Sigma projects often pays for itself in the long run by providing the financial benefits of near zero defects to the firm that implements it.
- Increases customer satisfaction
- Many companies that have implemented Six Sigma have not only found their costs reduced but their market share increased considerably. Hence, Six Sigma is also capable of positively impacting on the marketing of the firm.

26.9 QUESTIONS AND PROBLEMS

1. Explain normal distribution curve.
2. What does six sigma black belts mean?
3. What is six sigma?
4. What are the benefits of six sigma?
5. Explain the method of determining defects in million for one sigma, 2 sigma, 3 sigma and 6 sigma.
6. Explain six sigma methodologies.
7. Compare 3 sigma with 6 sigma.
8. The marks obtained by a large group of students in a final examination in physics have a mean of 58 and standard deviation of 8.5. Assuming that these marks are approximately normally distributed, what percentage of students can be expected to have obtained marks from 60 to 69 both inclusive? **Answer: 30.67%**
9. In an intelligence test administered to 1,000 students, the average was 42 and standard deviation was 24. Find:
   a. The number of students exceeding a score 50.
   b. The number of children lying between 30 and 54.
   Answer: a. 371 b. 383
10. The local authorities in a certain city installed 2,000 electrical lamps in street of the city. If the lamps have an average life of 1,000 burning hours with a standard deviation of 200 hours, what number of lamps might be expected to fail in the first 700 burning hours? **Answer:** 134 lamps
11. A research organization recently conducted a market survey to estimate demand for a certain product. The demand followed a normal distribution with a mean of 150 and a standard deviation of 25. Determine the probability that demand for the product in a given geographical area will exceed 200 units. **Answer:** 0.0228
12. The local authorities in a certain city installed 10,000 electrical lamps in street of the city. These lamps have an average life of 1,000 burning hours with a standard deviation of 200 hours. Assuming normal distribution i. what number of lamps might be expected to fail in the first 800 burning hours ii. between 800 and 1,200 burning hours. **Answer:** i. 1587 ii. 6826

26.10 NORMAL DISTRIBUTION TABLES

Standard Normal Cumulative Probability Table

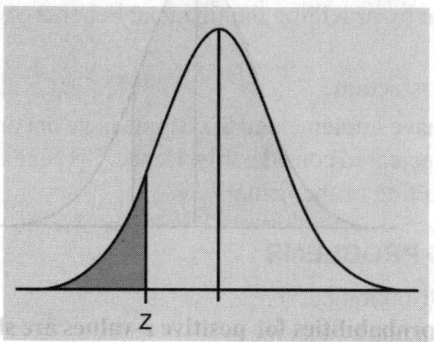

Table 26.3 Cumulative probabilities for negative z-values are shown in the following table:

z	0.00	0.01	0.02	0.03	0.04	0.05	0.06	0.07	0.08	0.09
−3.4	0.0003	0.0003	0.0003	0.0003	0.0003	0.0003	0.0003	0.0003	0.0003	0.0002
−3.3	0.0005	0.0005	0.0005	0.0004	0.0004	0.0004	0.0004	0.0004	0.0004	0.0003
−3.2	0.0007	0.0007	0.0006	0.0006	0.0006	0.0006	0.0006	0.0005	0.0005	0.0005
−3.1	0.0010	0.0009	0.0009	0.0009	0.0008	0.0008	0.0008	0.0008	0.0007	0.0007
−3.0	0.0013	0.0013	0.0013	0.0012	0.0012	0.0011	0.0011	0.0011	0.0010	0.0010
−2.9	0.0019	0.0018	0.0018	0.0017	0.0016	0.0016	0.0015	0.0015	0.0014	0.0014
−2.8	0.0026	0.0025	0.0024	0.0023	0.0023	0.0022	0.0021	0.0021	0.0020	0.0019
−2.7	0.0035	0.0034	0.0033	0.0032	0.0031	0.0030	0.0029	0.0028	0.0027	0.0026
−2.6	0.0047	0.0045	0.0044	0.0043	0.0041	0.0040	0.0039	0.0038	0.0037	0.0036
−2.5	0.0062	0.0060	0.0059	0.0057	0.0055	0.0054	0.0052	0.0051	0.0049	0.0048
−2.4	0.0082	0.0080	0.0078	0.0075	0.0073	0.0071	0.0069	0.0068	0.0066	0.0064
−2.3	0.0107	0.0104	0.0102	0.0099	0.0096	0.0094	0.0091	0.0089	0.0087	0.0084
−2.2	0.0139	0.0136	0.0132	0.0129	0.0125	0.0122	0.0119	0.0116	0.0113	0.0110
−2.1	0.0179	0.0174	0.0170	0.0166	0.0162	0.0158	0.0154	0.0150	0.0146	0.0143
−2.0	0.0228	0.0222	0.0217	0.0212	0.0207	0.0202	0.0197	0.0192	0.0188	0.0183
−1.9	0.0287	0.0281	0.0274	0.0268	0.0262	0.0256	0.0250	0.0244	0.0239	0.0233
−1.8	0.0359	0.0351	0.0344	0.0336	0.0329	0.0322	0.0314	0.0307	0.0301	0.0294
−1.7	0.0446	0.0436	0.0427	0.0418	0.0409	0.0401	0.0392	0.0384	0.0375	0.0367
−1.6	0.0548	0.0537	0.0526	0.0516	0.0505	0.0495	0.0485	0.0475	0.0465	0.0455
−1.5	0.0668	0.0655	0.0643	0.0630	0.0618	0.0606	0.0594	0.0582	0.0571	0.0559
−1.4	0.0808	0.0793	0.0778	0.0764	0.0749	0.0735	0.0721	0.0708	0.0694	0.0681
−1.3	0.0968	0.0951	0.0934	0.0918	0.0901	0.0885	0.0869	0.0853	0.0838	0.0823
−1.2	0.1151	0.1131	0.1112	0.1093	0.1075	0.1056	0.1038	0.1020	0.1003	0.0985
−1.1	0.1357	0.1335	0.1314	0.1292	0.1271	0.1251	0.1230	0.1210	0.1190	0.1170
−1.0	0.1587	0.1562	0.1539	0.1515	0.1492	0.1469	0.1446	0.1423	0.1401	0.1379
−0.9	0.1841	0.1814	0.1788	0.1762	0.1736	0.1711	0.1685	0.1660	0.1635	0.1611
−0.8	0.2119	0.2090	0.2061	0.2033	0.2005	0.1977	0.1949	0.1922	0.1894	0.1867
−0.7	0.2420	0.2389	0.2358	0.2327	0.2296	0.2266	0.2236	0.2206	0.2177	0.2148
−0.6	0.2743	0.2709	0.2676	0.2643	0.2611	0.2578	0.2546	0.2514	0.2483	0.2451
−0.5	0.3085	0.3050	0.3015	0.2981	0.2946	0.2912	0.2877	0.2843	0.2810	0.2776
−0.4	0.3446	0.3409	0.3372	0.3336	0.3300	0.3264	0.3228	0.3192	0.3156	0.3121
−0.3	0.3821	0.3783	0.3745	0.3707	0.3669	0.3632	0.3594	0.3557	0.3520	0.3483
−0.2	0.4207	0.4168	0.4129	0.4090	0.4052	0.4013	0.3974	0.3936	0.3897	0.3859
−0.1	0.4602	0.4562	0.4522	0.4483	0.4443	0.4404	0.4364	0.4325	0.4286	0.4247
0.0	0.5000	0.4960	0.4920	0.4880	0.4840	0.4801	0.4761	0.4721	0.4681	0.4641

Standard Normal Cumulative Probability Table

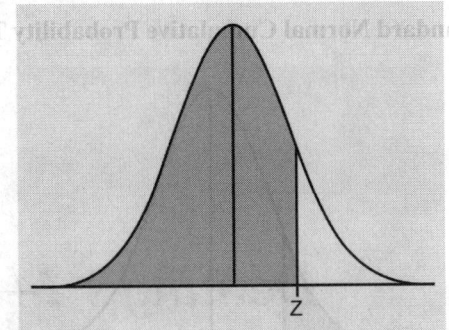

Table 26.4: Cumulative probabilities for positive z-values are shown in the following table:

z	0.00	0.01	0.02	0.03	0.04	0.05	0.06	0.07	0.08	0.09
0.0	0.5000	0.5040	0.5080	0.5120	0.5160	0.5199	0.5239	0.5279	0.5319	0.5359
0.1	0.5398	0.5438	0.5478	0.5517	0.5557	0.5596	0.5636	0.5675	0.5714	0.5753
0.2	0.5793	0.5832	0.5871	0.5910	0.5948	0.5987	0.6026	0.6064	0.6103	0.6141
0.3	0.6179	0.6217	0.6255	0.6293	0.6331	0.6368	0.6406	0.6443	0.6480	0.6517
0.4	0.6554	0.6591	0.6628	0.6664	0.6700	0.6736	0.6772	0.6808	0.6844	0.6879
0.5	0.6915	0.6950	0.6985	0.7019	0.7054	0.7088	0.7123	0.7157	0.7190	0.7224
0.6	0.7257	0.7291	0.7324	0.7357	0.7389	0.7422	0.7454	0.7486	0.7517	0.7549
0.7	0.7580	0.7611	0.7642	0.7673	0.7704	0.7734	0.7764	0.7794	0.7823	0.7852
0.8	0.7881	0.7910	0.7939	0.7967	0.7995	0.8023	0.8051	0.8078	0.8106	0.8133
0.9	0.8159	0.8186	0.8212	0.8238	0.8264	0.8289	0.8315	0.8340	0.8365	0.8389
1.0	0.8413	0.8438	0.8461	0.8485	0.8508	0.8531	0.8554	0.8577	0.8599	0.8621
1.1	0.8643	0.8665	0.8686	0.8708	0.8729	0.8749	0.8770	0.8790	0.8810	0.8830
1.2	0.8849	0.8869	0.8888	0.8907	0.8925	0.8944	0.8962	0.8980	0.8997	0.9015
1.3	0.9032	0.9049	0.9066	0.9082	0.9099	0.9115	0.9131	0.9147	0.9162	0.9177
1.4	0.9192	0.9207	0.9222	0.9236	0.9251	0.9265	0.9279	0.9292	0.9306	0.9319
1.5	0.9332	0.9345	0.9357	0.9370	0.9382	0.9394	0.9406	0.9418	0.9429	0.9441
1.6	0.9452	0.9463	0.9474	0.9484	0.9495	0.9505	0.9515	0.9525	0.9535	0.9545
1.7	0.9554	0.9564	0.9573	0.9582	0.9591	0.9599	0.9608	0.9616	0.9625	0.9633
1.8	0.9641	0.9649	0.9656	0.9664	0.9671	0.9678	0.9686	0.9693	0.9699	0.9706
1.9	0.9713	0.9719	0.9726	0.9732	0.9738	0.9744	0.9750	0.9756	0.9761	0.9767
2.0	0.9772	0.9778	0.9783	0.9788	0.9793	0.9798	0.9803	0.9808	0.9812	0.9817
2.1	0.9821	0.9826	0.9830	0.9834	0.9838	0.9842	0.9846	0.9850	0.9854	0.9857
2.2	0.9861	0.9864	0.9868	0.9871	0.9875	0.9878	0.9881	0.9884	0.9887	0.9890
2.3	0.9893	0.9896	0.9898	0.9901	0.9904	0.9906	0.9909	0.9911	0.9913	0.9916
2.4	0.9918	0.9920	0.9922	0.9925	0.9927	0.9929	0.9931	0.9932	0.9934	0.9936
2.5	0.9938	0.9940	0.9941	0.9943	0.9945	0.9946	0.9948	0.9949	0.9951	0.9952
2.6	0.9953	0.9955	0.9956	0.9957	0.9959	0.9960	0.9961	0.9962	0.9963	0.9964
2.7	0.9965	0.9966	0.9967	0.9968	0.9969	0.9970	0.9971	0.9972	0.9973	0.9974
2.8	0.9974	0.9975	0.9976	0.9977	0.9977	0.9978	0.9979	0.9979	0.9980	0.9981
2.9	0.9981	0.9982	0.9982	0.9983	0.9984	0.9984	0.9985	0.9985	0.9986	0.9986
3.0	0.9987	0.9987	0.9987	0.9988	0.9988	0.9989	0.9989	0.9989	0.9990	0.9990
3.1	0.9990	0.9991	0.9991	0.9991	0.9992	0.9992	0.9992	0.9992	0.9993	0.9993
3.2	0.9993	0.9993	0.9994	0.9994	0.9994	0.9994	0.9994	0.9995	0.9995	0.9995
3.3	0.9995	0.9995	0.9995	0.9996	0.9996	0.9996	0.9996	0.9996	0.9996	0.9997
3.4	0.9997	0.9997	0.9997	0.9997	0.9997	0.9997	0.9997	0.9997	0.9997	0.9998

Deming's 14 Points and Total Quality Management

William Edwards Deming (October 14, 1900–December 20, 1993) was an American statistician, professor, author, lecturer and consultant. He is perhaps best known for the *"Plan-Do-Check-Act"* cycle popularly named after him. In Japan, from 1950 onwards, he taught top management how to improve design, product quality, testing, and sales through various methods including the application of statistical methods.

27.1 THE DEMING CYCLE (or Shewhart Cycle)

As a repetitive process to determine the next action, the Deming cycle describes a simple method to test information before making a major decision. The 4 steps in the Deming Cycle are: Plan-Do-Check-Act (PDCA), also known as Plan-Do-Study-Act or PDSA. Deming called the cycle the Shewhart Cycle, after Walter A. Shewhart. The cycle can be used in various ways, such as running an experiment; PLAN (design) the experiment; DO the experiment by performing the steps; CHECK the results by testing information; and ACT on the decisions based on those results. Figure 27.1 shows the Deming cycle.

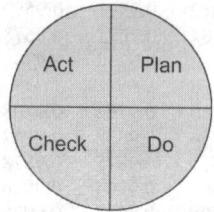

Fig. 27.1 The Deming cycle

27.2 DEMING'S 14 POINTS

Deming produced his 14 points for management, in order to help people understand and implement the necessary transformation. Deming said that adoption of, and action on, the 14 points are the signals that management intend to stay in business. They apply to small or large organizations, and to service industries as well as to manufacturing.

1. Constancy of purpose: Create constancy of purpose for continual improvement of products and service to society, allocating resources to provide for long range needs

rather than only short-term profitability, with a plan to become competitive, to stay in business, and to provide jobs.

2. The new philosophy: Adopt the new philosophy. We are in a new economic age. We can no longer live with commonly accepted levels of delays, mistakes, defective materials and defective workmanship. Transformation is necessary to halt the continued decline of business and industry.

3. Cease dependence on mass inspection: Eliminate the need for mass inspection as the way of life to achieve quality by building quality into the product in the first place. Require statistical evidence of built in quality in both manufacturing and purchasing functions.

4. End lowest tender contracts: End the practice of awarding business solely on the basis of price tag. Instead, require meaningful measures of quality along with price. Reduce the number of suppliers for the same item by eliminating those that do not qualify with statistical and other evidence of quality. The aim is to minimize total cost, not merely initial cost, by minimizing variation. This may be achieved by moving towards a single supplier for any one item, on a long-term relationship of loyalty and trust. Purchasing managers have a new job, and must learn it.

5. Improve every process: Improve constantly and forever every process for planning, production, and service. Search continually for problems in order to improve every activity in the company, to improve quality and productivity, and thus to constantly decrease costs. Institute innovation and constant improvement of product, service, and process. It is management's job to work continually on the system (design, incoming materials, maintenance, improvement of machines, supervision, training, retraining).

6. Institute training on the job: Institute modern methods of training on the job for all, including management, to make better use of every employee. New skills are required to keep up with changes in materials, methods, product and service design, machinery, techniques, and service.

7. Institute leadership: Adopt and institute leadership aimed at helping people do a better job. The responsibility of managers and supervisors must be changed from sheer numbers to quality. Improvement of quality will automatically improve productivity. Management must ensure that immediate action is taken on reports of inherited defects, maintenance requirements, poor tools, fuzzy operational definitions, and all conditions detrimental to quality.

8. Drive out fear: Encourage effective two ways communication and other means to drive out fear throughout the organization so that everybody may work effectively and more productively for the company.

9. Break down barriers: Break down barriers between departments and staff areas. People in different areas, such as leasing, maintenance, administration, must work in teams to tackle problems that may be encountered with products or service.

10. Eliminate exhortations: Eliminate the use of slogans, posters and exhortations for the workforce, demanding zero defects and new levels of productivity, without providing methods. Such exhortations only create adversarial relationships; the bulk of the causes of low quality and low productivity belong to the system, and thus lie beyond the power of the work force.

11. Eliminate arbitrary numerical targets: Eliminate work standards that prescribe quotas for the work force and numerical goals for people in management. Substitute aids and helpful leadership in order to achieve continual improvement of quality and productivity.

12. Permit pride of workmanship: Remove the barriers that rob hourly workers, and people in management, of their right to pride of workmanship. This implies, among other things, abolition of the annual merit rating (appraisal of performance) and of management by objective. Again, the responsibility of managers, supervisors, foremen must be changed from sheer numbers to quality.

13. Encourage education: Institute a vigorous program of education, and encourage self improvement for everyone. What an organization needs is not just good people; it needs people that are improving with education. Advances in competitive position will have their roots in knowledge.

14. Top management commitment and action: Clearly define top management's permanent commitment to ever improving quality and productivity, and their obligation to implement all of these principles. Indeed, it is not enough that top management commit themselves for life to quality and productivity. They must know what it is that they are committed to—that is, what they must do. Create a structure in top management that will push every day on the preceding 13 points, and take action in order to accomplish the transformation.

15. Support is not enough: Action is required.

27.3 TOTAL QUALITY MANAGEMENT

Total quality management (TQM) is a management approach that originated in the 1950s and has steadily become more popular since the early 1980s. Total quality management (TQM) refers to management methods used to enhance quality and productivity in organizations, particularly in businesses. TQM is a comprehensive system approach that works horizontally across an organization, involving all departments and employees and extending backward and forward to include both suppliers and clients/customers. Total quality management measures its success on the basis of customer satisfaction with regard to all aspects of the product (quality, price and availability). Like most quality management concepts, TQM views "quality" entirely from the point of view of "the customer". All businesses have many types of customer. A customer can be someone "internal" to the business (e.g. a production employee working at the end of the production line is the "customer" of the employees involved earlier in the production process). A customer can also be "external" to the business. When you fly with an airline you are their customer. When you buy products from food manufacturers you are their customer.

27.4 THE IMPORTANCE OF CUSTOMER-SUPPLIER RELATIONSHIPS (QUALITY CHAINS)

TQM focuses strongly on the importance of the relationship between customers (internal and external) and supplier. These are known as the **"quality chains"** and they can be broken at any point by one person or one piece of equipment not meeting the requirements of the customer. Failure to meet the requirements in any part of a quality chain has a way of multiplying, and failure in one part of the system creates problems elsewhere, leading to yet more failure and problems, and so the situation is exacerbated. The ability to meet customers' (external and internal) requirements is vital. To achieve quality throughout a business, every person in the quality chain must be trained.

27.5 MAIN PRINCIPLES OF TQM

The main principles that underline TQM are summarized below:

Prevention: Prevention is better than cure. In the long run, it is cheaper to stop products' defects than trying to find them.

Zero defects: The ultimate aim is no (zero) defects or exceptionally low defect levels, if a product or service is complicated.

Getting things right first time: Better not to produce at all than produce something defective

Quality involves everyone: Quality is not just the concern of the production or operations department. It involves everyone, including marketing, finance and human resources.

Continuous improvement: Continues improvement of all aspects of the firm through two "learning" processes:

- Single loop learning: Improving the existing product and production processes.
- Second order learning: Changing products and production processes.

Single loop learning is often the result of innovations from the work floor, while second order learning is a result from R & D, marketing studies and design studies.

Employee involvement: Those involved in production and operations have a vital role to play in spotting improvement opportunities for quality and in identifying quality problems.

27.6 INTRODUCING TQM INTO A BUSINESS

TQM is not an easy concept to introduce into businesses, particularly those that have not traditionally concerned with understanding customer needs and business processes. TQM also focuses on the activities of the business that are closest to the customer, e.g. the production department, the employees facing the customer. To be successful in implementing TQM, an organization must concentrate on the eight key elements:

1. Ethics
2. Integrity
3. Trust
4. Training
5. Teamwork
6. Leadership
7. Recognition
8. Communication

These elements can be divided into four groups according to their function.

The groups are:

i. Foundation: It includes ethics, integrity and trust.
ii. Building bricks: It includes training, teamwork and leadership.
iii. Binding mortar: It includes communication.
iv. Roof: It includes recognition.

Figure 27.2 shows the total quality management house.

Fig. 27.2 Total quality management house

I. Foundation

TQM is built on a foundation of ethics, integrity and trust. These three elements move together, however, each element offers something different to the TQM concept.

Ethics: Ethics are the disciplines concerned with good and bad in any situation. It is represented by organizational and individual ethics. Organizational ethics establish a business code of ethics that outlines guidelines that all employees are to adhere to in the performance of their work. Individual ethics include personal rights or wrongs.

Integrity: Integrity implies honesty, morals, values, fairness, and adherence to the facts and sincerity. The characteristic is what customers (internal or external) expect and deserve to receive. People see the opposite of integrity as duplicity. TQM will not work in an atmosphere of duplicity.

Trust: Trust is a byproduct of integrity and ethical conduct. Without trust, the framework of TQM cannot be built. Trust fosters full participation of all members. It allows empowerment that encourages pride ownership and it encourages commitment. It allows decision making at appropriate levels in the organization, fosters individual risk-taking for continuous improvement and helps to ensure that measurements focus on improvement of process and are not used to contend people. Trust is essential to ensure customer satisfaction. So, trust builds the cooperative environment essential for TQM.

II. Bricks

Basing on the strong foundation of trust, ethics and integrity, bricks are placed to reach the roof of recognition. It includes:

Training: Training is very important for employees to be highly productive. Supervisors are responsible for implementing TQM within their departments, and teaching their employees about the philosophies of TQM. Training that employees require are interpersonal skills, the ability to function within teams, problem solving, decision making, job management performance analysis and improvement, business economics and technical skills. During the creation and formation of TQM, employees are trained so that they can become effective employees for the company.

Teamwork: To become successful in business, teamwork is also a key element of TQM. With the use of teams, the business will receive quicker and better solutions to problems. Teams also provide more permanent improvements in processes and operations. In teams, people feel more comfortable bringing up problems that may occur, and can get help from other workers to find a solution and put into place. There are mainly three types of teams that TQM organizations adopt:

- Quality improvement teams (QITs) are temporary teams with the purpose of dealing with specific problems that often recur. These teams are set up for a period of three to twelve months.
- Problem solving teams (PSTs) are temporary teams to solve certain problems and also to identify and overcome causes of problems. They generally last from one week to three months.
- Natural work teams consist of small groups of skilled workers who share tasks and responsibilities. These teams use concepts such as employee involvement teams, self-managing teams and quality circles. These teams generally work for one to two hours a week.

Leadership: It is possibly the most important element in TQM. It appears everywhere in organization. Leadership in TQM requires the manager to provide an inspiring vision, make strategic directions that are understood by all and to install values that guide subordinates. For TQM to be successful in the business, the supervisor must be committed in leading his employees. A supervisor must understand TQM, believe in it and then demonstrate their belief and commitment through their daily practices of TQM. The supervisor makes sure that the strategies, philosophies, values and goals are transmitted down throughout the organization to provide focus, clarity and direction. A key point is that TQM has to be introduced and led by top management. Commitment and personal involvement is required from top management in creating and deploying clear quality values and goals consistent with the objectives of the company and in creating and deploying well defined systems, methods and performance measures for achieving those goals.

III. Binding mortar

Communication: It binds everything together. Starting from foundation to roof of the TQM house, everything is bounded by strong mortar of communication. It acts as a vital link between all elements of TQM. Communication means a common understanding of ideas between the sender and the receiver. The success of TQM demands communication with and among all the organizations' members, suppliers and customers. Supervisors must keep open airways where employees can send and receive information about the TQM process. Communication coupled with the sharing of correct information is vital. For communication to be credible the message must be clear and receiver must interpret in the way the sender intended. There are different ways of communication such as presentations and discussions basically do it. By this the supervisors are able to make the employees clear about TQM.

Upward communication: By this the lower level of employees are able to provide suggestions to upper management of the affects of TQM. As employees provide insight and constructive criticism, supervisors must listen effectively to correct the situation that comes about through the use of TQM. This forms a level of trust between supervisors and employees. This is also similar to empowering communication, where supervisors keep open ears and listen to others.

Sideways communication: This type of communication is important because it breaks down barriers between departments. It also allows dealing with customers and suppliers in a more professional manner.

IV. Roof

Recognition: Recognition is the last and final element in the entire system. It should be provided for both suggestions and achievements for teams as well as individuals. Employees strive to receive recognition for themselves and their teams. Detecting and recognizing contributors is the most important job of a supervisor. As people are recognized, there can be huge changes in self-esteem, productivity, quality and the amount of effort exhorted to the task at hand. Recognition comes in its best form when it is immediately following an action that an employee has performed. Recognition comes in different ways, such as places and time,

Ways: It can be by way of personal letter from top management. Also by award banquets, plaques, trophies, etc.

Places: Good performers can be recognized in front of departments, on performance boards and also in front of top management.

Time: Recognition can be given at any time like in staff meeting, annual award banquets, etc.

These eight elements are important in ensuring the success of TQM in an organization and that the supervisor is a huge part in developing these elements in the work place.

Since 1970, total quality management (TQM) has become an important philosophy in many companies all over the world. Prior 1950, Japan produced consumer products of poor quality. The management of Japanese companies realized that quality is a vital element. As a result in many industrial sectors, Japanese companies were become the market leaders by the end of the 20th century. Japanese companies are successful because of its strong cooperative culture, where the employees and management have similar values and goals and are also strongly committed to achieve these goals. Thus total quality management is a mix of statistical process control, design engineering, marketing and organizational culture strengthening techniques.

The TQM is applied to many stages of industrial cycle which are listed below:

1. Marketing
2. Engineering
3. Purchasing
4. Manufacturing

27.7 BENEFITS OF TQM

- Improvement in product quality
- Improvement in product design
- Improvement in production flow
- Improvement in employee morale and quality consciousness
- Improvement in product service
- Improvement in market place acceptance
- Reduction in operating costs
- Reduction in operating losses
- Reduction in field service costs

27.8 QUALITY IN MANUFACTURING AND SERVICE ORGANIZATIONS

Quality in manufacturing organizations is often different from that of services. Manufacturing organizations produce a tangible product that can be seen, touched and directly measured. Examples include cars, CD players, clothes, computers, and food items. Therefore, quality in manufacturing usually focuses on tangible product features. The most common quality definitions in manufacturing are:

Conformance: Conformance which is the degree to which a product characteristic meets preset standards. Other common definitions of quality in manufacturing include:

Performance: Performance such as acceleration of a vehicle

Reliability: Probability that the product will function as expected without failure.

Features: The extras that are included beyond the basic characteristics.

Durability: Expected operational life of the product

Serviceability: How readily a product can be repaired.

The relative importance of these definitions is based on the preferences of each individual customer.

In contrast to manufacturing, service organizations produce a product that is intangible. Usually, the complete product cannot be seen or touched. Rather, it is experienced. Examples include delivery of health care, experience of staying at a vacation resort, and learning at a university. The intangible nature of the product makes defining quality difficult. Also, since a service is experienced, perceptions can be highly subjective. In addition to tangible factors, quality of services is often defined by perceptual factors. These include responsiveness to customer needs, courtesy and friendliness of staff, promptness in resolving complaints, and atmosphere.

Other definitions of quality in services include:

Time: The amount of time a customer has to wait for the service; and

Consistency: The degree to which the service is same at each time.

27.9 QUALITY COST

Quality cost is the extra cost incurred due to poor or bad quality of the product or service. Categories of quality cost: Many companies summarize quality costs into four broad categories:

a. Internal failure costs: The cost associated with defects that are found prior to transfer of the product to the customer.

b. External failure costs: The cost associated with defects that are found after product is shipped to the customer.

c. Appraisal costs: The cost incurred in determining the degree of conformance to quality requirement.

d. Prevention costs: The cost incurred in keeping failure and appraisal costs to a minimum.

27.10 QUESTIONS

1. State Deming's 14 principles on total quality management.
2. Explain Deming's cycle.
3. Elaborate your approach on TQM implementation in an organization.
4. State the benefits of TQM.
5. Define TQM.
6. State the principles of TQM.
7. Describe various elements of TQM.
8. Explain quality chains.

28

PERT and CPM

28.1 INTRODUCTION

Network analysis: Network analysis is a system, which plans projects both large and small by analyzing the project activities. Project activities are arranged in logical sequence in network diagram. Time, costs and other resources are allocated to different activities. Network helps in designing, planning, coordinating, controlling and decision making in order to accomplish the project economically in the minimum available time with the limited available resources.

Examples of projects:

1. To launch the satellite:

   Activities: Production of rocket, production of satellite, launching of satellite, etc. Each major activity consists of number of subactivities.

2. Construction of building:

   Activities: Purchase of land, foundation, slab, flooring, pipelines, sanitation works, electrification, etc.

3. Construction of industry:

   Activities: Land procurement, building construction, purchasing of machinery, clearances from various departments, installation of machinery, procurement of working capital, etc.

4. Overhauling of boilers, turbines, etc.

28.2 NETWORK TECHNIQUES

A number of network techniques have been developed and two important techniques are:

1. PERT: Program evaluation and review technique
2. CPM: Critical path method

28.3 TERMS USED IN PERT AND CPM

1. Event: An event is a specific instant of time which marks the start and end events of an activity. Event consumes neither time nor resources. It is represented by a circle and the event number is written within the circle.

   Examples of events are starting the motor, loan approved, starting of foundation work, etc.

2. Activity: Every project consists of number of job operations or tasks, which are called activities. An activity is shown by an arrow and it begins and ends with an event. Unlike

event an activity consumes time and resources. An individual or a group of individuals may perform an activity.

Activities are classified as:

i. Critical activities: In a network diagram critical activities are those which if consume more than their estimated time the project will be delayed. Slack is zero for critical activities. Critical activities are represented by double arrow

ii. Non-critical activities: Non critical activities are those, even if they consume a specified time over and above the estimated time, the project will not be delayed.

iii. Dummy activities: Dummy activities are used to show logical relationships. They are shown by dotted arrow. A dummy activity does not consume time and other resources. A dummy activity may be non-critical or critical.

3. Critical path: Critical path is formed by critical activities. It is the longest path and consumes maximum time. Critical path consists of critical activities. Even one critical activity is delayed, the project is delayed. A project may have more than one critical paths.

4. Duration of an activity: Duration is the estimated or actual time required to complete an activity.

5. Total project time: It is the time to complete the project.

6. Earliest start time (EST): It is the earliest possible time at which an activity can start and EST is calculated by moving from first event to last event in the network diagram.

7. Earliest finish time (EFT): It is the earliest possible time at which an activity can finish. EFT of an activity i = EST of an activity i + Duration of an activity i

8. Latest finish time (LFT): It is calculated by moving from last event to the first event in the network diagram.

9. Latest start time (LST): It is the latest possible time by which an activity can start without project delay.
LST of an activity i = LFT of an activity i – Duration of an activity i

10. Float or slack: It is the extra time over and above its duration, which a non-critical activity can consume without delaying the project. There are three types of floats for each activity.

i. Total float: It is the additional time on which non-critical activity can consume without increasing the project duration. However, total float may affect the floats in previous and subsequent activities.

Total float of an activity i = LST of an activity i – EST of an activity i

Or

Total float of an activity i = LFT of an activity i – EFT of an activity i

ii. Free float: If all the non-critical activities start as early as possible, the surplus time is the free float. Free float, if used, does not change the float in later activities.
Free float of an activity i = EST of end event of an activity i – EST of starting event of an activity i – Duration of activity i

iii. Independent float: If independent float is negative, take it as zero.
Independent float of an activity i = EST of end event of an activity i – LFT of starting event of an activity i – Duration of activity i

In simple network problems, it is assumed that resources are unlimited. Here the problem is only to determine schedule of each activity and slacks or floats of each activity such that project will be completed in minimum time.

28.4 CRITICAL PATH METHOD (CPM)

Critical path method is a deterministic problem. Critical path method is used when duration of each activity or time is known exactly. Critical path method is used for repetitive projects. Here the problem is only to determine schedule of each activity and slacks or floats of each activity such that project will be completed in minimum time.

Data or inputs required for determining schedule of each activity is given below:

1. Number of activities
2. Precedence relationship between these activities
3. The time required to complete each activity

28.5 STEPS REQUIRED FOR DETERMINING VARIOUS PARAMETERS OF A PROJECT USING CPM

1. Draw the network diagram for the precedence relationship given in the problem using the rules of drawing network diagram.
2. Calculation of Earliest start time of an activity (EST): Calculate EST of activities by moving from first event to last event in the network diagram.
3. Determination of project duration: EST of end event is the project duration. All the remaining parameters are calculated based on project duration.
4. Determination of critical path: Critical path is calculated by moving from last event to the first event in the network diagram using EST of events. A project may have more than one critical path.
5. Calculation of Earliest finish time (EFT) of an activity:
   EFT of an activity i = EST of activity i + Duration of activity i.
6. Calculation of Latest finish time (LFT): It is calculated by moving from last event to the first event in the network diagram.
7. Calculation of Latest start time of an activity (LST): LST of an activity i = LFT of an activity i – Duration of an activity i.
8. Calculation of Total float of activity: Total float of an activity i = LST of an activity i – EST of an activity i or Total float of an activity i = LFT of an activity i – EFT of an activity i.
9. Calculation of Free float: Free float of an activity i = EST of end event of an activity i – EST of starting event of an activity i – Duration of activity i.
10. Calculation of Independent float: If independent float is negative, take it as zero.
   Independent float of an activity i = EST of end event of an activity i – LFT of starting event of an activity i – Duration of activity i.

28.6 RULES FOR DRAWING NETWORK DIAGRAM

1. No two activities should have the same starting and ending events.
2. Activities don't cross each other.
3. Use dummy activities to show logical relationship. Dummy activities do not consume any time and other resources. Dummy activities are represented by dotted arrow.
4. Each activity is represented once and only once in the network diagram.
5. There is only one starting and ending event in the network diagram.

28.7 SOLVED PROBLEMS

Problem 28.1: A small engineering project consists of 6 activities namely A, B, C, D, E and F with duration of 4, 6, 5, 4, 3 and 3 days. Details are shown in the network diagram (Fig. 28.1).

Calculate EST, LST, EFT, LFT and floats. Mark the critical path and find total project duration.

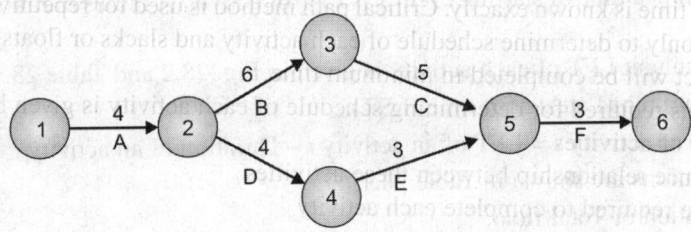

Fig. 28.1 Network diagram

Solution:

Step 1: Network diagram is given in the problem.

Step 2: Calculation of Earliest start time (EST): Calculate EST of activities by moving from first event to last event in the network diagram. Procedure of calculating EST for each activity is given below.

Earliest start time of an activity A is zero.

Earliest start time of activities B and D is EST of activity A plus duration of activity A, i.e. 4.

Earliest start time of an activity C is EST of activity B plus duration of activity B, i.e. 10.

Earliest start time of an activity E is EST of D plus duration of activity D, i.e. 8.

Earliest start time of an activity F is bigger of (EST of activity C plus duration of activity C, EST of activity E plus duration of activity E), i.e. bigger of (15, 11), i.e. 15.

EST of end event 6 is EST of activity F plus duration of activity F, i.e. 18.

The resultant EST of each activity is given in Fig. 28.2 and Table 28.1.

Step 3: Determination of project duration: EST of end event is the project duration. EST of end event 6 is 18. So project duration is 18 days.

Step 4: Critical path is calculated by moving from last event to the first event in the network diagram using EST of events.

EST of end event 6 is 18 days. 18 days is coming from event 5. So activity 5-6 or activity F is a critical activity 15 is from event 3. So activity 3-5 or Activity C is critical activity.

EST of 10 for the event 3 is from event 2. So activity 2-3 is a critical activity.

EST of 4 for the event 2 is from 1. So Activity 1-2 or activity A is critical activity.

So critical path is A-B-C-F or 1-2-3-5-6.

Step 5: Calculation of Earliest finish time (EFT):

EFT of an activity i = EST of activity i + Duration of activity i

Calculations are shown in Table 28.1.

Step 6: Latest finish time (LFT): It is calculated by moving from last event to the first event in the network diagram.

Latest finish time of an activity F is 18 days.

Latest finish time of activities C and E is LFT of activity F minus duration of F, i.e. 15.

Latest finish time of an activity D is LFT of activity E minus duration of E, i.e. 12.

Latest finish time of an activity B is LFT of activity C minus duration of C, i.e. 10.

Latest finish time of an activity A is smaller of (LFT of activity B minus duration of B, LFT of an activity D minus duration of D), i.e. minimum of (4,8), i.e. 4.

Latest finish time of starting event 1 is LFT of activity A minus duration of activity A, i.e. 0.

The resultant LFT of each activity is given in Fig. 28.2 and Table 28.1.

Step 7: Calculation of Latest start time (LST):

LST of an activity i = LFT of an activity i – Duration of an activity i

Calculations are shown in Table 28.1.

Step 8: Calculation of Total float:

Total float of an activity i = LST of an activity i – EST of an activity i

Or

Total float of an activity i = LFT of an activity i – EFT of an activity i

Calculations are shown in Table 28.1.

Step 9: Calculation of Free float:

Free float of an activity i = EST of end event of an activity i – EST of starting event of an activity i – Duration of activity i.

For example,

Free float of an activity A = EST of end event of an activity A – EST of starting event of an activity A – Duration of activity A.

Free float of an activity A = 4 – 0 – 4 = 0

Free float of each activity is calculated from Fig. 28.2 and the resultant free float of each activity is shown in Table 28.1.

Step 10: Calculation of Independent Float: If independent float is negative, take it as zero.

Independent float of an activity i = EST of end event of an activity i – LFT of starting event of an activity i – Duration of activity i.

For example,

Independent float of an activity A = EST of end event of an activity A – LFT of starting event of an activity A – Duration of activity A.

Independent float of an activity A = 4 – 0 – 4 = 0

Similarly, Independent float of each activity is calculated from Fig. 28.2 and the resultant independent float of each activity is shown in Table 28.1.

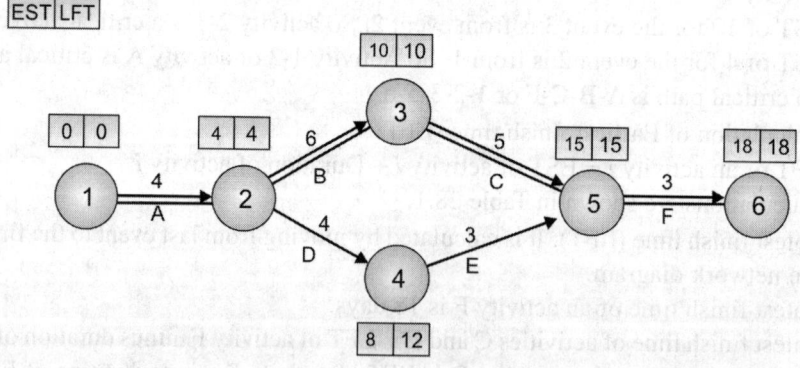

Fig. 28.2 Network diagram and calculations of EST and LFT for each activity

Table 28.1 Calculation of floats for each activity

Activity i	Duration of activity i (days)	EST_i	LST_i	EFT_i	LFT_i	Total float of activity i	Free float of activity i	Independent float of activity i
A	4	0	0	4	4	0	0	0
B	6	4	4	10	10	0	0	0
C	5	10	10	15	15	0	0	0
D	4	4	8	8	12	4	0	0
E	3	8	12	11	15	4	4	0
F	3	15	15	18	18	0	0	0

Problem 28.2: A small project consists of 3 activities namely A, B and C with duration of 10, 15 and 20 days. Information is given in Table 28.2. Draw the network diagram. Calculate EST, LST, EFT, LFT and floats. Mark the critical path and find total project duration.

Table 28.2 Information about the project

Activity	A	B	C
Preceding activities	–	–	A, B
Duration (days)	10	15	20

Solution:

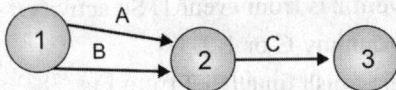

Fig. 28.3 Network diagram

The diagram (Fig. 28.3) is not correct. Because the activities A and B have same starting and end events.

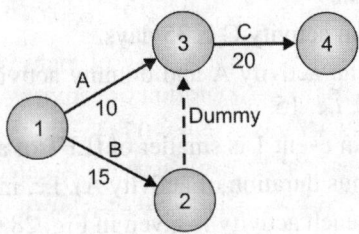

Fig. 28.4 Network diagram

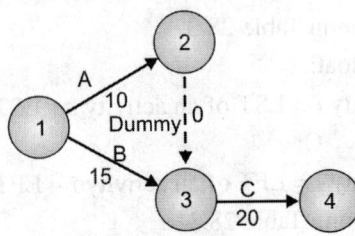

Fig. 28.5 Network diagram

Figures 28.4 and 28.5 are correct network diagrams for the given data. Either of Fig. 28.4 or 28.5 is selected. Remaining calculations are done using Fig. 28.4.

Step 2: Calculation of Earliest start time (EST): Calculate EST of activities by moving from first event to last event in the network diagram. Procedure of calculating EST for each activity is given below:

Earliest start time of activities A and B is zero.

Earliest start time of dummy activity is EST of activity B plus duration of activity B, i.e. 15.

Earliest start time of an activity C is bigger of (EST of activity A plus duration of activity A, EST of dummy activity plus duration of dummy activity)

Earliest start time of an activity C is bigger of (10, 15), i.e.15.

Earliest start time of an activity C is 15.

Step 3: Determination of project duration: EST of end event 4 is the project duration. EST of end event is 35. So project duration is 35 days.

Step 4: Critical path is calculated by moving from last event to the first event in the network diagram using EST of events.

EST of end event 4 is 35 days. 35 days is coming from event 3. So activity 3-4 or activity C is a critical activity.

EST of 15 for the event 3 is from event 2. So activity 2-3 or dummy activity is critical activity.

EST of 15 for the event 2 is from event 1. So activity 1-2 is a critical activity.

So critical path is B-dummy-C or 1-2-3-4.

Step 5: Calculation of earliest finish time (EFT):

EFT of an activity i = EST of activity i – Duration of activity i

Calculations are shown in Table 28.3.

Step 6: Latest finish time (LFT): It is calculated by moving from last event to the first event in the network diagram.

Latest finish time of an activity C is 35 days.

Latest finish time of an activity A and dummy activity is LFT of activity C minus duration of activity C, i.e. 15.

Latest finish time of an event 1 is smaller of (LFT of activity B minus duration of B, LFT of activity A minus duration of activity A), i.e. minimum of (5, 0), i.e. 0.

The resultant LFT of each activity is given in Fig. 28.6.

Step 7: Calculation of latest start time (LST):

LST of an activity i = LFT of an activity i – Duration of an activity i

Calculations are shown in Table 28.3.

Step 8: Calculation of Total float:

Total float of an activity i = LST of an activity i – EST of an activity i

$$Or$$

Total float of an activity i = LFT of an activity i – EFT of an activity i

Calculations are shown in Table 28.3.

Step 9: Calculation of free float:

Free float of an activity i = EST of end event of an activity i – EST of starting event of an activity i – Duration of activity i.

For example,

Free float of an activity A = EST of end event of an activity A – EST of starting event of an activity A – Duration of activity A.

Free float of an activity A = 15 – 0 – 10 = 5

Free float of each activity is calculated from Fig. 28.6 and the resultant free float of each activity is shown in Table 28.3.

Step 10: Calculation of Independent Float: If independent float is negative, take it as zero.

Independent float of an activity i = EST of end event of an activity i – LFT of starting event of an activity i – Duration of activity i

For example,

Independent float of an activity A = EST of end event of an activity A – LFT of starting event of an activity A – Duration of activity A.

Independent float of an activity A = 15 – 0 – 10 = 5

Similarly Independent float of each activity is calculated from Fig. 28.6 and the resultant independent float of each activity is shown in Table 28.3.

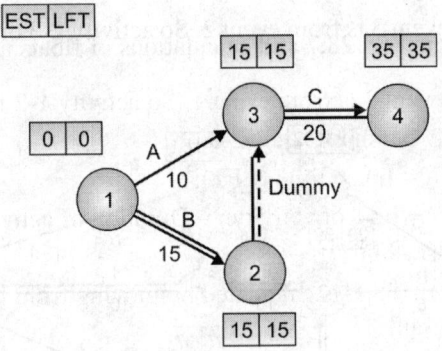

Fig. 28.6 Calculations of EST and LFT

Table 28.3 Calculation of floats for each activity

Activity i	Duration of activity i (days)	EST_i	EFT_i	LFT_i	LST_i	Total float of activity i	Free float of activity i	Independent float of activity i
A	10	0	10	15	5	5	5	5
B	15	0	15	15	0	0	0	0
C	20	15	35	35	15	0	0	0
Dummy	0	15	15	15	15	0	0	0

Problem 28.3: A project consists of a series of tasks labeled A, B, C, D, E, F, G, H and I with the following relationship.

A   < D, E

B, D < F

C   < G

G   < H

F, G < I

The duration of performing each task is given in Table 28.4.

Table 28.4 The duration of performing each task

Task	A	B	C	D	E	F	G	H	I
Time (days)	23	8	20	16	24	18	19	4	10

i. Construct the network diagram.

ii. Find critical path and floats.

Solution: i. Obtain the precedence relationship (Table 28.5) from the given logical relationship. A < D, E means, the activities D and E cannot be started without completing activity A.

Table 28.5 Precedence relationship among the activities

Task	A	B	C	D	E	F	G	H	I
Time (days)	23	8	20	16	24	18	19	4	10
Immediate predecessors				A	A	B, D	C	G	F, G

Draw the network diagram using the data in Table 28.5. The resultant network diagram is shown in Fig. 28.7.

ii. Critical path is given in Fig. 28.7 and calculations of floats are given in Table 28.6.

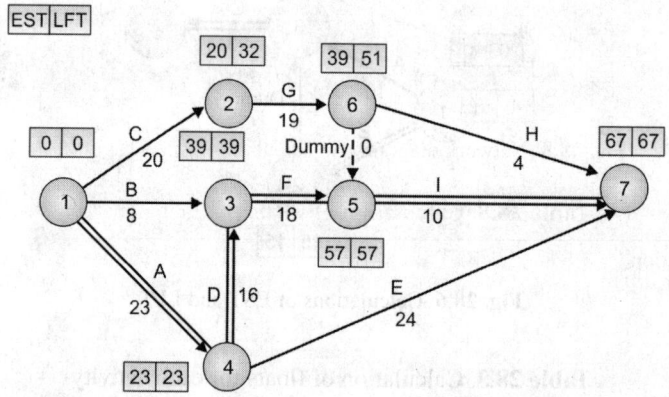

Fig. 28.7 Network diagram and calculations of EST and LFT

Table 28.6 Calculation of floats for each activity

Activity i	Duration of activity i (days)	EST_i	EFT_i	LFT_i	$LSFT_i$	Total float of activity i	Free float of activity i	Independent float of activity i
A	23	0	23	23	0	0	0	0
B	8	0	8	39	31	31	31	31
C	20	0	20	32	12	12	0	0
Dummy 1	0	39	39	39	39	0	0	0
D	16	23	39	39	23	0	0	0
E	24	23	47	67	43	20	20	20
F	18	39	57	57	39	0	0	0
G	19	20	39	51	32	12	0	0
H	4	39	43	67	63	24	14	12
I	10	57	67	67	57	0	0	0

Critical path: A-D-F-I
Project duration: 67 days

Problem 28.4: For the information given in Table 28.7 draw network and determine slacks or floats.

Table 28.7 Given information about the project

Activity	A	B	C	D	E	F	G	H
Predecessors	–	–	A	A	C	D	B	E, F, G
Time (weeks)	3	4	5	6	7	8	9	3

Solution:

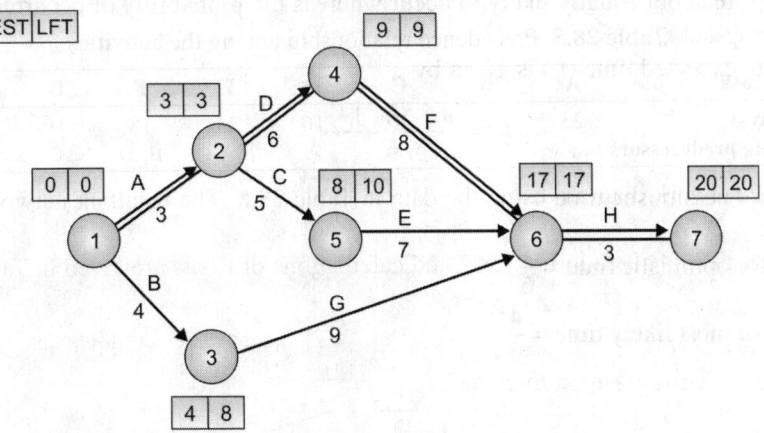

Fig. 28.8 Network diagram and calculations of EST and LFT

Table 28.8 Calculation of floats for each activity

Activity i	Duration of activity i (days)	EST_i	EFT_i	LFT_i	LST_i	Total float of activity i	Free float of activity i	Independent float of activity i
A	3	0	3	3	0	0	0	0
B	4	0	4	8	4	4	0	0
C	5	3	8	10	5	2	0	0
D	6	3	9	9	3	0	0	0
E	7	8	15	17	10	2	2	0
F	8	9	17	17	9	0	0	0
G	9	4	13	17	8	4	4	0
H	3	17	20	20	17	0	0	0

Critical path: A-D-F-H
Project duration: 20 weeks

28.8 PROGRAM EVALUATION REVIEW TECHNIQUE (PERT)

PERT can be employed at those places where a project cannot be easily defined in terms of time and resources. PERT technique proves very much advantageous when used for non-repetitive types of projects such as research and development, prototype production, irrigation projects, defense projects, etc. Because of uncertainty of activity timing, PERT acquires the shape of probabilistic model. Probability concept helps in estimating activity times.

Estimation of activity time: For dealing with uncertainties associated with different activities, PERT approach computes expected time for each activity from the following three time estimates:

1. Optimistic time (t_o): It is the shortest possible time in which an activity can be completed, if everything goes actually well.
2. Most likely time (t_m): It is the time in which the activity will take to complete under normal conditions.
3. Pessimistic time (t_p): It is the time in which an activity will take to complete in case of difficulty, i.e. if mostly the things go wrong. It is the longest of all the three time estimates.

The t_o, t_m and t_p are combined statistically to develop the expected time (t_e). The fundamental assumption in PERT is that three time estimates form Beta distribution. It is further assumed that t_p and t_o are about equally likely to occur whereas the probability of occurrence of t_m is 4 times that of t_p and t_o.

Therefore, expected time (t_e) is given by

$$t_e = \frac{t_o + 4t_m + t_p}{6}$$

Weight for pessimistic time $= \dfrac{1}{6}$

Weight for optimistic time $= \dfrac{1}{6}$

Weight for most likely time $= \dfrac{4}{6}$

Estimation of variance of activity time

$$v_t = \left(\frac{t_p - t_o}{6}\right)^2$$

Standard deviation of activity time

$$s_t = \left(\frac{t_p - t_o}{6}\right)$$

The variance of project = Sum of the variances of activities in critical path

Standard deviation of project $= \sqrt{\text{The variance of project}}$

Probability that the project is completed in D

$$z = \frac{D - T_e}{\text{Standard deviation of project}}$$

For the value of z, determine the corresponding value of probability from the normal distribution table (Appendix A).

28.9 SOLVED PROBLEMS

Problem 28.5: The Table 28.9 gives the values of t_o, t_m and t_p for each activity. Calculate expected time, standard deviation and variance for each activity. Also calculate the probability that the project will meet the schedule or due date of 38 days.

Table 28.9 Values of t_o, t_m and t_p for each activity

Activity	1-2	1-6	2-3	2-4	3-5	4-5	6-7	5-8	7-8
Optimistic time (t_o)	2	2	5	1	5	2	3	2	7
Most likely time (t_m)	5	5	11	4	11	5	9	2	13
Pessimistic time (t_p)	14	8	29	7	17	14	27	8	31

Solution:

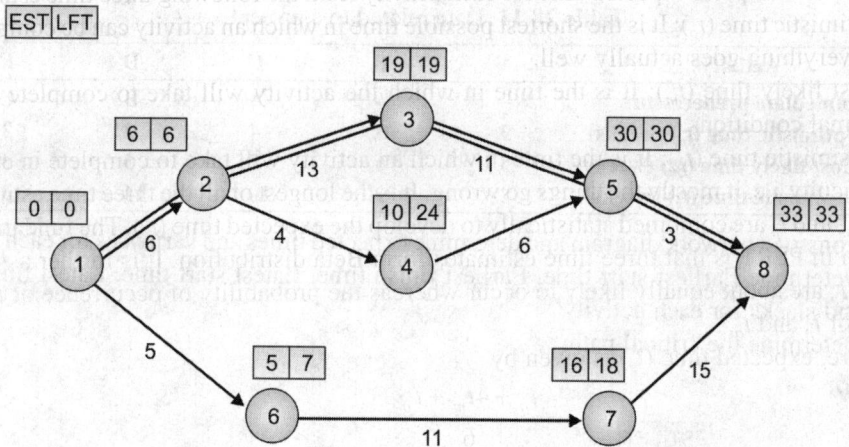

Fig. 28.9 Network diagram and calculations of EST and LFT

Table 28.10 Calculation of floats for each activity

Activity$_i$	1-2	1-6	2-3	2-4	3-5	4-5	6-7	5-8	7-8
Optimistic time (t_o) (days)	2	2	5	1	5	2	3	2	7
Most likely time (t_m) (days)	5	5	11	4	11	5	9	2	13
Pessimistic time (t_p) (days)	14	8	29	7	17	14	27	8	31
$t_e = \dfrac{t_o + 4t_m + t_p}{6}$ (days)	6	5	13	4	11	6	11	3	15
$s_t = \left(\dfrac{t_p - t_o}{6}\right)$ (days)	2	1	4	1	2	2	4	1	4
$v_t = \left(\dfrac{t_p - t_o}{6}\right)^2$ (days)	4	1	16	1	4	4	16	1	6
EST$_i$	0	0	6	6	19	10	5	30	16
EFT$_i$	6	5	19	10	30	16	16	33	31
LFT$_i$	6	7	19	24	30	18	18	33	33
LST$_i$	0	2	6	20	19	12	7	30	18
Total slack of activity$_i$	0	2	0	14	0	2	2	0	2
Free slack of activity$_i$	0	0	0	0	0	14	0	0	2
Independent slack of activity$_i$	0	0	0	0	0	0	0	0	0

Critical path = 1-2-3-5-8

The length of critical path = 6 + 13 + 11 + 3 = 33 days

The variance of project = Sum of the variances of activities in critical path.

The variance of project = 4 + 16 + 4 + 1 = 25 days

Standard deviation of project = $\sqrt{25}$ = 5

$$z = \frac{D - T_e}{\text{Standard deviation of project}} = \frac{38 - 33}{5} = 1$$

For the value of $z = 1$, the corresponding value of probability from the normal distribution table is 0.841.

Problem 28.6: Table 28.11 gives the data related to a project.

Table 28.11 Data related to a project

Activity	A	B	C	D	E
Immediate predecessors	–	–	A	B	C
Optimistic time (t_o) (weeks)	2	3	4	2	3
Most likely time (t_m) (weeks)	5	6	7	5	3
Pessimistic time (t_p) (weeks)	8	9	10	14	3

i. Construct network diagram and determine expected times and variances for each activity.
ii. Determine Earliest start time, Earliest finish time, Latest start time, Latest finish time and slacks for each activity.
iii. Determine the critical path.

Solution:

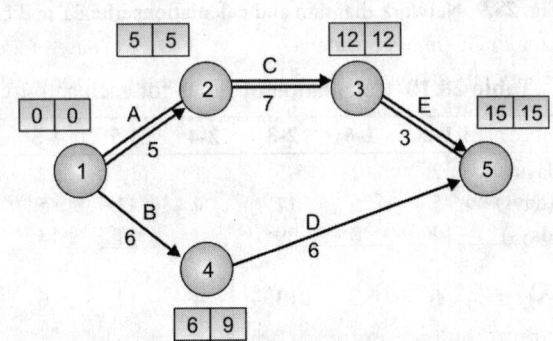

Fig. 28.10 Network diagram and calculations of EST and LFT

Table 28.12 Calculation of floats for each activity

Activity$_i$	A	B	C	D	E
Optimistic time (t_o) (weeks)	2	3	4	2	3
Most likely time (t_m) (weeks)	5	6	7	5	3
Pessimistic time (t_p) (weeks)	8	9	10	14	3
$t_e = \dfrac{t_o + 4t_m + t_p}{6}$ (weeks)	5	6	7	6	3
$s_t = \left(\dfrac{t_p - t_o}{6}\right)$ (weeks)	1	1	1	2	0
$v_t = \left(\dfrac{t_p - t_o}{6}\right)^2$ (weeks)	1	1	1	4	0
EST	0	0	5	6	12
EFT	5	6	12	12	15
LFT	5	9	12	15	15
LST	0	3	5	9	12
Total slack	0	3	0	3	0
Free slack	0	0	0	3	0
Independent slack	0	0	0	0	0

Critical path = A-C-E

The length of critical path = 5 + 7 + 3 = 15 weeks

The variance of project = Sum of the variances of activities in critical path

The variance of project = 1 + 1 + 0 = 2 weeks

Standard deviation of project = $\sqrt{2}$ = 1.414

28.10 DIFFERENCES BETWEEN PERT AND CPM

Differences between PERT and CPM are given in Table 28.13.

Table 28.13 Differences between PERT and CPM

PERT	CPM
1. A probabilistic model with uncertainty in activity duration.	1. A deterministic model with certainty in project duration.
2. Expected time is calculated from t_o, t_p and t_m.	2. Single activity based on past experience.
3. PERT is used for single time project.	3. CPM is used for repetitive projects.
4. Suitable for defense, irrigation and R and D projects where activity times cannot be reliably predicted.	4. Suitable for problems in industrial setting, plant maintenance and civil construction projects.
5. Variance and standard deviation of activities are estimated.	5. Only one time estimate, therefore, there is no variance and standard deviation.

28.11 CRASHING OF NETWORK

To reduce the project duration by employing additional resources is known as crashing. There are two types of activities. They are critical activities and non-critical activities. Crashing of non-critical activities do not serve any purpose, therefore, crashing is centered on critical activities. Reduction of timing of critical activities only reduces the project duration.

Terms related to the crashing of network:

Normal cost (N_c): It is the lowest cost of completing an activity by using normal resources.

Normal time (N_t): It is the time required to complete the activity with normal resources.

Crash cost (C_c): It is the cost of completing an activity by using additional resources.

Crash time (C_t): It is the time required to complete the activity with additional resources.

$$\text{Slope of an activity} = \theta = \frac{C_c - N_c}{N_t - C_t}$$

Steps required for crashing network:

1. Draw the network diagram.
2. Determine slope of each activity.
3. Determine EST of each activity.
4. Determine project duration and corresponding total cost.
5. Identify critical activities.
6. Crash the critical activity (reduce one unit of time), which has least cost slope.
7. Redraw the new network diagram.
8. Repeat the steps 3 to 7 till the project duration cannot be reduced further or the cost of crashing is more than the amount of saving in return.

28.12 SOLVED PROBLEMS

Problem 28.7: Determine the optimum duration for Table 28.14.

Table 28.14 Data about the project

Activity	Normal time (N_t)(days)	Normal cost (N_c) (₹)	Crash time (C_t) (days)	Crash cost (C_c) (₹)
1-2	1	20	1	20
2-3	3	70	2	120
1-3	5	110	3	180
3-4	3	80	2	140
4-5	1	30	1	30
1-5	12	250	9	370
5-6	3	90	2	160
6-7	1	20	1	20

Indirect cost per day = ₹ 50

Solution:

Iteration 1:

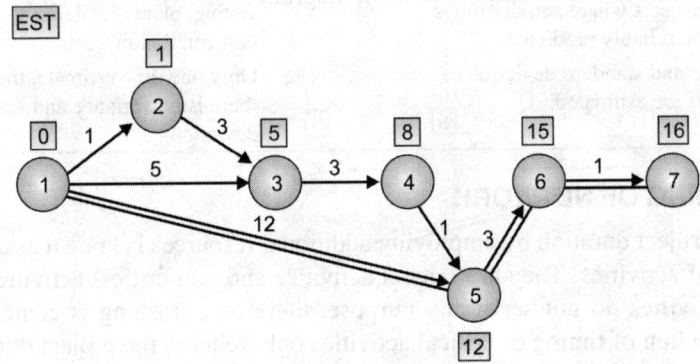

Fig. 28.11 Network diagram and calculation of project duration

Critical path: 1-5-6-7

Normal duration = 16 days

Table 28.15 To determine slope of each activity and critical activities

Activity i	Normal time (N_t)(days)	Normal cost (N_c)(₹)	Crash time (C_t)(days)	Crash cost (C_c)(₹)	$\theta = \dfrac{C_c - N_c}{N_t - C_t}$ ₹/day
1-2	1	20	1	20	–
2-3	3	70	2	120	50
1-3	5	110	3	180	35
3-4	3	80	2	140	60
4-5	1	30	1	30	–
1-5 (critical)	12	250	9	370	40
5-6 (critical)	3	90	2	160	70
6-7 (critical)	1	20	1	20	–
Total normal cost		670			

Iteration 2: Reduce time of activity 1-5 (which has least cost slope among critical activities) from 12 days to 11 days. The resultant network diagram is shown in Fig. 28.12.

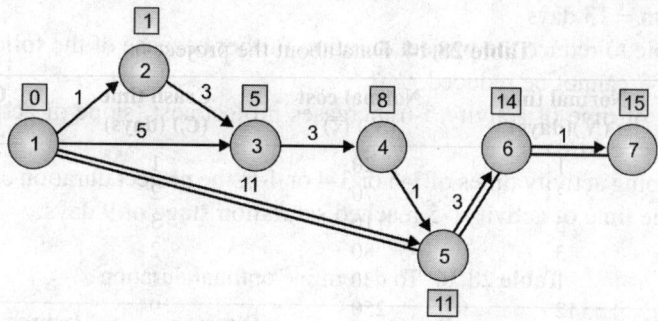

Fig. 28.12 Network diagram and calculation of project duration

From Fig. 28.12,
Critical path: 1-5-6-7
Project duration = 15 days

Iteration 3: Reduce time of activity 1-5 (which has least cost slope among critical activities) from 11 days to 10 days. The resultant network diagram is shown in Fig. 28.13.

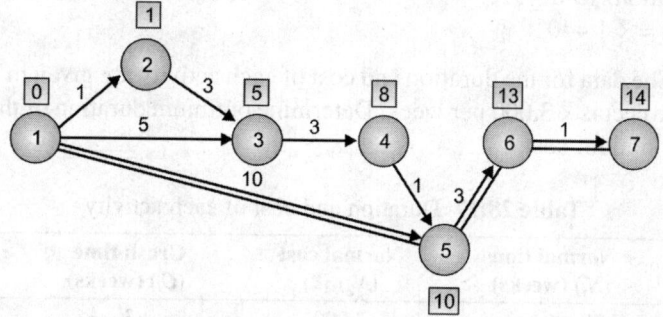

Fig. 28.13 Network diagram and calculation of project duration

From Fig. 28.13,
Critical path: 1-5-6-7
Project duration = 14 days

Iteration 4: Reduce time of activity 1-5 (which has least cost slope among critical activities) from 10 days to 9 days. The resultant network diagram is shown in Fig. 28.14.

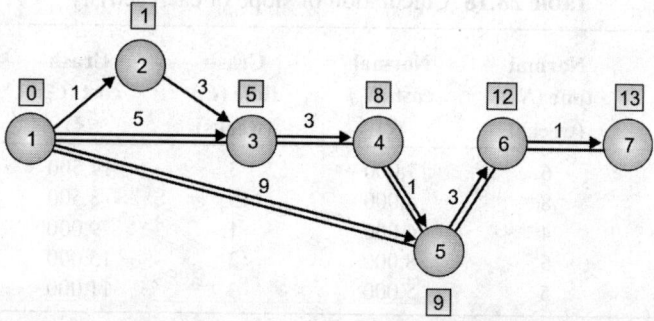

Fig. 28.14 Network diagram and calculation of project duration

From Fig. 28.14,

Critical paths: 1-5-6-7 and 1-3-4-5-6-7

Project duration = 13 days

It is not possible to reduce the project duration further because of the following reasons:

i. Activity 6-7 cannot be reduced.

ii. Reduction of time of activity 5-6 increases project cost (slope of activity 5-6 is more than indirect cost/day).

iii. Even reducing activity times of 1-3 or 3-4 or 4-5, the project duration cannot be reduced because the time of activity1-5 reached saturation stage of 9 days.

Table 28.16 To determine optimal duration

Iteration	Activity or Activities crashed	Project duration (days)	Direct cost (₹)	Indirect cost (₹)	Total cost (₹)
1	Normal duration	16	670	800	1,470
2	Activity 1-5 by 1 day	15	670 + 40 = 710	750	1,460
3	Activity 1-5 by 1 day	14	710 + 40 = 750	700	1,450
4	Activity 1-5 by 1 day	13	750 + 40 = 790	650	1,440

Optimum duration: 13 days

Optimum cost = ₹ 1,440

Problem 28.8: The data for the duration and cost of each activity are given in Table 28.17. The indirect cost of project is ₹ 3,000 per week. Determine optimum duration of the project and the corresponding minimum cost.

Table 28.17 Duration and cost of each activity

Activity	Normal time (N_t) (weeks)	Normal cost (N_c) (₹)	Crash time (C_t) (weeks)	Crash cost (C_c) (₹)
1-2	6	7,000	3	14,500
1-3	8	4,000	5	8,500
2-3	4	6,000	1	9,000
2-4	5	8,000	3	15,000
3-4	5	5,000	3	11,000

Solution:

Table 28.18 Calculation of slope of each activity

Activity	Normal time (N_T) (weeks)	Normal cost (N_c) (₹)	Crash time (C_t) (weeks)	Crash cost (C_c) (₹)	$\theta = \dfrac{C_c - N_c}{N_t - C_t}$
1-2 (critical)	6	7,000	3	14,500	2,500
1-3	8	4,000	5	8,500	1,500
2-3 (critical)	4	6,000	1	9,000	1,000
2-4	5	8,000	3	15,000	3,500
3-4 (critical)	5	5,000	3	11,000	3,000
Total normal cost		30,000			

Iteration 1:

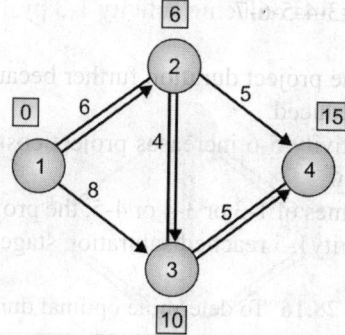

Fig. 28.15 Network diagram

Critical path: 1-2-3-4
Normal duration = 15 weeks

Iteration 2:

Reducing activity 2-3 by 1 week (which has least cost slope among critical activities)

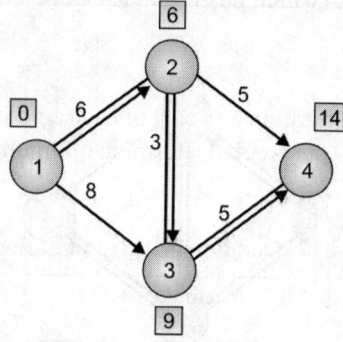

Critical path: 1-2-3-4
Project duration = 14 weeks

Iteration 3:

Reducing activity 2-3 by 1 week (which has least cost slope among critical activities)

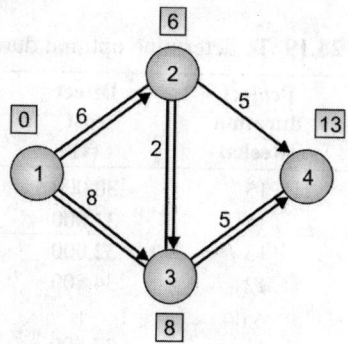

Critical paths: 1-2-3-4 and 1-3-4
Now 1-3 is also critical activity
Project duration = 13 weeks

Iteration 4:
Reducing activity 2-3 by 1 week and reducing activity 1-3 by 1 week simultaneously

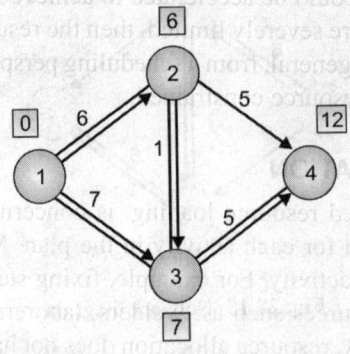

Critical paths: 1-2-3-4 and 1-3-4
Project duration = 12 weeks

Iteration 5:
Reducing activity 3-4 by 1 week (which has least cost slope among critical activities)

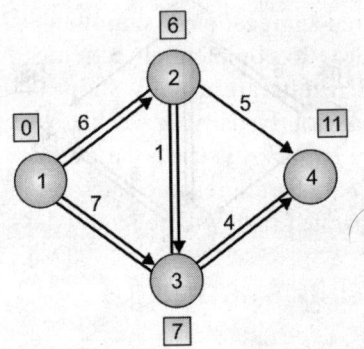

Critical paths: 1-2-3-4 , 1-3-4 and 1-2-4.
Now all are critical activities.
Project duration = 11 weeks

Table 28.19 To determine optimal duration

Iteration	Activity or Activities crashed	Project duration (weeks)	Direct cost (₹)	Indirect cost (₹)	Total cost (₹)
1	Normal duration	15	30,000	45,000	75,000
2	Reducing 2-3 by 1 day	14	31,000	42,000	73,000
3	Reducing 2-3 by 1 day	13	32,000	39,000	71,000
4	Reducing 2-3 by 1 day Reducing 1-3 by 1 day	12	34,500	36,000	70,500
5	Reducing 3-4 by 1 day	11	37,500	33,000	70,500

Optimum project duration: 11 weeks
Optimum cost = ₹ 70,500

28.13 RESOURCE MANAGEMENT

The most important resources that project managers have to plan and manage on day-to-day basis are people, machines, materials, and money. Obviously, if these resources are available in abundance then the project could be accelerated to achieve shorter project duration. On the other hand, if these resources are severely limited, then the result more likely will be a delay in the project completion time. In general, from a scheduling perspective, projects can be classified as either time constrained or resource constrained.

28.14 RESOURCE ALLOCATION

Resource allocation, also called resource loading, is concerned with assigning the required number of resources identified for each activity in the plan. More than one type of resource may be assigned to a specific activity. For example, fixing steel plates on a bridge deck may require different types of resources such as: welders, laborers and a certain type of welding machine. From a practical view, resource allocation does not have to follow a constant pattern; some activities may initially require fewer resources but may require more of the same resources during the later stages of the project.

28.15 RESOURCE AGGREGATION (LOADING)

After each activity has been assigned its resources, the next step is to aggregate the resources used by all activities. Resource aggregation is simply the summation, on a period-by-period basis, of the resources required to complete all activities based on the resource allocation carried out previously. The results are usually shown graphically as a histogram. Such aggregation may be done on an hourly, daily, or weekly basis, depending on the time unit used to allocate resources. When a bar chart is used, the resource aggregation is fairly simple and straightforward. For a given bar chart, a resource aggregation chart can be drawn underneath the bar chart. However, a separate graph will be required for each resource type.

28.16 RESOURCE LEVELING (SMOOTHING)

A project is classified as time constraint in situations where the project completion time cannot be delayed. In this case, resources are not limited and project duration is not allowed to be delayed. In network scheduling, the basic inputs to critical path analysis are the project activities, their durations, and their dependency relationships. The forward and backward calculations determine the start and finish times of the activities. The CPM algorithm, therefore, is duration-driven. The CPM formulation, therefore, assumes that all the resources needed for the schedule are available. This assumption, however, is not always true for all projects. Smoothing of resources of a network implies scheduling the activities within the limits of their total floats such that fluctuations of resource requirements are minimized. In other words, peaks and valleys in the load chart are leveled in order to keep constant and stable optimum workforce throughout the project duration. Resource leveling (smoothing) improves resource utilization. This process is called resource leveling or smoothing. Resource leveling (smoothing) reduces the hiring and firing of resources and to smooth the fluctuation in the daily demand of a resource, as shown in Fig. 28.21. The objective in this case is to shift non-critical activities of the original schedule, within their float times so that a better resource profile is achieved.

Problem 28.9: Duration and resource required for each activity is given in Table 28.20. Determine the schedule which minimizes resource fluctuations.

Table 28.20 Duration and resource required for each activity

Activity	1	2	3	4	5	6	7
Duration	3	3	1	2	2	3	3
Resource required	2	3	2	3	4	2	1

Entry in ( ) represents duration of activity.

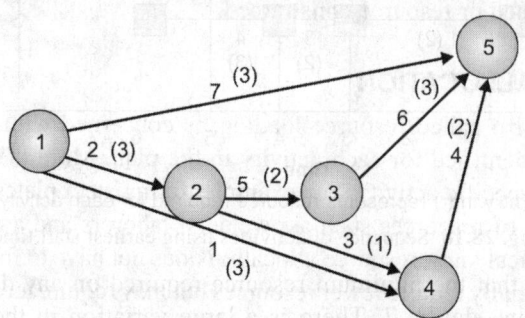

Entry in ( ) represents duration of activity

Fig. 28.16 Network diagram

Solution:

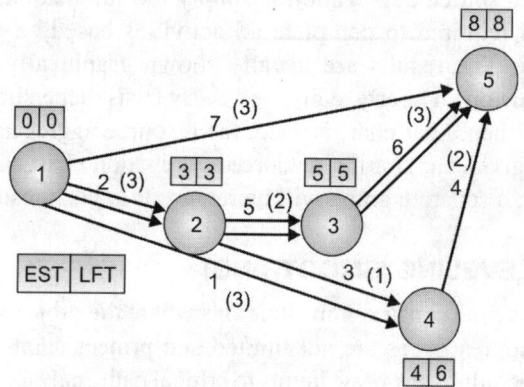

Entry in ( ) represents duration of activity

Fig. 28.17 Project scheduling under no resource constraints

Table 28.21 Determining float for each activity

Activity	1	2	3	4	5	6	7
Duration	3	3	1	2	2	3	3
Resource required	2	3	2	3	4	2	1
EST	0	0	3	4	3	5	0
EFT	3	3	4	6	5	8	3
LFT	6	3	6	8	5	8	8
LST	3	0	5	6	3	5	5
Total float	3	0	1	2	1	0	5

Schedule the activities using earliest start time. The resultant schedule is given in Fig. 28.18.

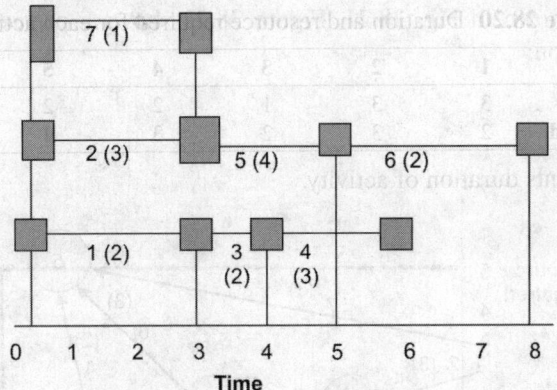

Entry in ( ) represents resource required for each activity

Fig. 28.18 Sequence of activities using earliest start time

Figure 28.19 shows that the minimum resource required on any day is 2 and maximum resource required on any day is 7. There is a large variation in the day-to-day resource distribution. Figure 28.20 shows the modified schedule graph. The resource requirement chart

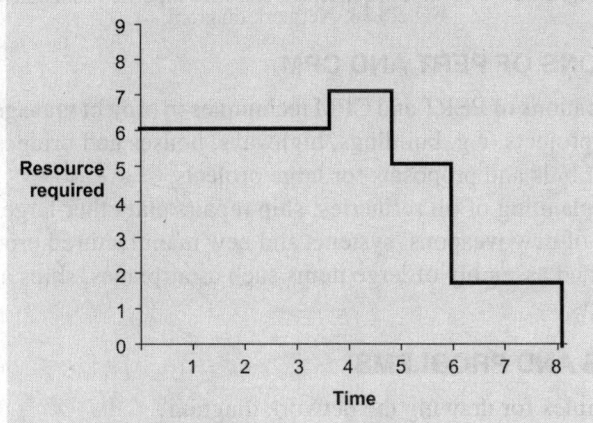

Fig. 28.19 Resource requirement chart

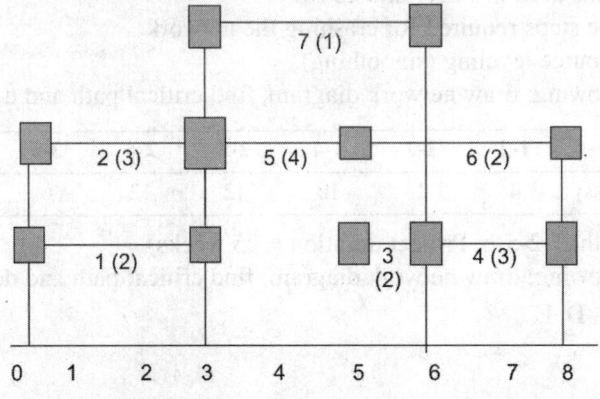

Entry in ( ) represents resource required for each activity

Fig. 28.20 Modified schedule chart

for the schedule is shown in Fig. 28.21. Figure 28.21 shows a perfectly constant resource throughout the duration.

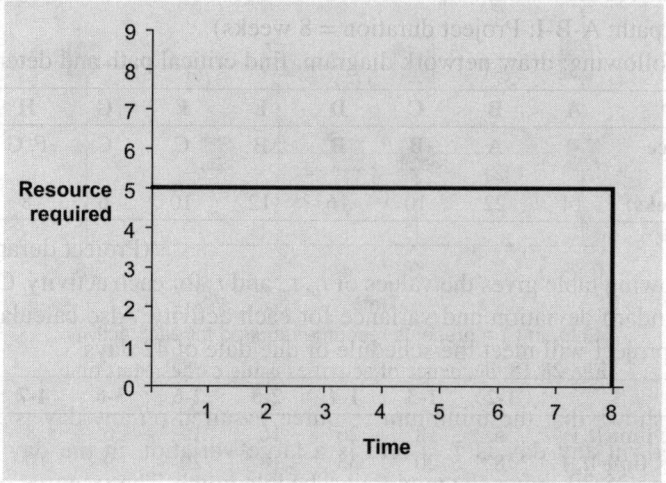

Fig. 28.21 Resource requirement chart for modified schedule

28.17 APPLICATIONS OF PERT AND CPM

A partial list of applications of PERT and CPM techniques in project management is as follows:
1. Construction projects (e.g. buildings, highways, houses and bridges)
2. Preparation of bids and proposals for large projects.
3. Maintenance planning of oil refineries, ship repairs and other large operations.
4. Development of new weapons' systems and new manufactured products.
5. Manufacture and assembly of large items such as airplanes, ships and computers.
6. Simple projects.

28.18 QUESTIONS AND PROBLEMS

1. What are the rules for drawing the network diagram?
2. Explain the differences between PERT and CPM.
3. Explain terms used in PERT and CPM.
4. Describe the steps required for crashing the network.
5. Explain resource leveling (smoothing).
6. For the following; draw network diagram, find critical path and determine slacks.

Activity	1-2	1-3	1-4	2-5	2-6	3-5	4-6	5-6
Time (weeks)	4	2	10	12	12	6	8	9

(Critical path: 1-2-5-6; Project duration = 25 weeks)
7. For the following; draw network diagram; find critical path and determine slacks.

A   < B, C, D, E
E   < F
D   < G
G, F < H
B   < I

Activity	A	B	C	D	E	F	G	H	I
Time (weeks)	1	4	2	3	2	3	2	1	3

(Critical path: A-B-I; Project duration = 8 weeks)

8. For the following; draw network diagram, find critical path and determine slacks.

Activity	A	B	C	D	E	F	G	H	I	J
Precedence activities	–	A	B	B	B	C	C	F, G	D, E, H	I
Time (weeks)	14	22	10	16	12	10	6	8	24	16

(Project duration = 104 weeks)

9. The following table gives the values of t_o, t_m and t_p for each activity. Calculate expected time, standard deviation and variance for each activity. Also calculate the probability that the project will meet the schedule or due date of 42 days.

Activity	1-2	1-3	1-4	2-5	2-6	3-6	4-7	5-7	6-7
Optimistic time (t_o)	6	18	26	16	15	6	8	7	3
Most likely time (t_m)	8	20	33	18	20	9	10	8	4
Pessimistic time (t_p)	10	22	40	20	25	12	12	9	5

(**Answer:** Critical path: 1-4-7; Project duration = 43 days; Probability = 0.34)

10. Determine the optimum cost for the data given in following Table.

Activity	Normal time (N_t) (days)	Normal cost (N_c) (₹)	Crash time (C_t) (days)	Crash cost (C_c) (₹)
1-2	4	60	3	90
1-3	2	38	1	60
1-4	6	150	4	250
2-4	5	150	3	250
3-4	2	100	2	100
2-5	7	115	5	175
4-5	4	100	2	240

Indirect cost per day = ₹ 60

(Normal duration = 13 days; Optimal duration = 8 days; Optimal cost = ₹ 1,523)

11. Determine the optimum cost for the following data and Table.

A < D

B < F

C < E

D < G

E < F

Activity	Normal time (N_t) (days)	Normal cost (N_c) (₹)	Crash time (C_t) (days)	Crash cost (C_c) (₹)
A	2	100	1	150
B	6	200	4	400
C	3	200	1	300
D	4	500	2	700
E	2	400	1	550
F	8	200	5	450
G	5	300	3	500

Indirect cost per day = ₹ 200

12. Duration and resource required for each activity is given in the Table. Determine the schedule which minimizes resource fluctuations.

Table Duration and resource required for each activity

Activity	1	2	3	4	5	6	7
Duration	3	3	1	2	2	3	3
Resource required	2	3	2	4	4	2	3

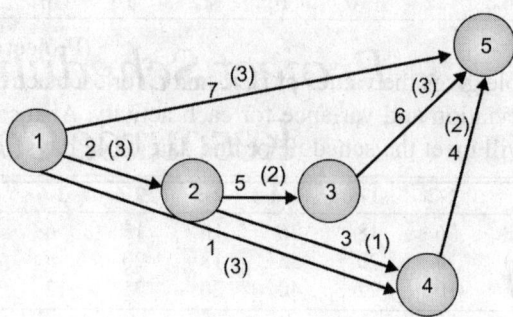

Entry in ( ) represents duration of activity

Fig. 28.22 Network diagram

29 Project Scheduling under Resource Constraints

29.1 INTRODUCTION

CPM/PERT procedures are considered on the basis for project management techniques. A network consists of number of activities arranged in a logical sequence. If in a particular project resources are released in desired amounts then critical path is based on activity times. However, in many practical situations resources are limited. However, when the resources are limited many other possible paths may become critical. It is then difficult to find the sequence which gives minimum time for the project. Scheduling of projects with limited resources is a large combinatorial problem. Unfortunately, analytical techniques like 0-1 integer programing are computationally impractical. Therefore, there is need to turn into heuristic approaches.

29.2 HEURISTIC METHODS OF PROJECT SCHEDULING UNDER RESOURCE CONSTRAINTS

1. ACTIM: ACTIM is the sum of the times of all the activities that each activity controls through the network on any path. The ACTIM has the following steps:
   - Develop the project network taking into consideration of precedence requirements.
   - Determine time and resources required for each activity.
   - Determine maximum resource available for each of the resource.
   - Estimate ACTIM for each activity for all paths.
   - Determine for each activity the maximum ACTIM it controls through the network on any path.
   - Rank these activities in decreasing order as per maximum ACTIM.

2. ACTRES: ACTRES is the sum of the resources of all the activities that each activity controls through the network on any path. For each activity there may be more than one path. The ACTRES has the following steps:
   - Develop the project network taking into consideration of precedence requirements.
   - Determine time and resources required for each activity.
   - Determine maximum resource available for each of the resource.
   - Determine for each activity the maximum ACTRES it controls through the network on any path.
   - Rank these activities in decreasing order as per maximum ACTRES.

3. TIMRES: The TIMRES has the following steps:
   - Determine TIMRES for each activity 'i'.
   - TIMRES for each activity 'i' = ACTIM for each activity 'i' + ACTRES for each activity 'i'
   - Rank these activities in decreasing order as per TIMRES.

4. GENRES: Weighted average of ACTIM and ACTRES
   The GENRES has the following steps:
   - Determine GENRES.
   - GENRES for each activity 'i' = $w \times$ ACTIM for each activity 'i' + $(1 - w)$ ACTRES for each activity 'i', where $0 \leq w \geq 1$.
   - Rank these activities in decreasing order as per GENRES.

5. ROT (Resource over time): ROT is the sum of the product of the time and resources of all the activities that each activity controls through the network on any path.
   The ROT has the following steps:
   - Develop the project network taking into consideration of precedence requirements.
   - Determine time and resources required for each activity.
   - Determine maximum resource available for each of the resource.
   - For each activity, determine 'Time for activity × Resource required'.
   - Determine for each activity the maximum ROT it controls through the network on any path.
   - Rank these activities in decreasing order as per ROT.

6. ROT-ACTIM:
   - Determine ROT-ACTIM for each activity 'i'.
   - ROT-ACTIM for each activity 'i' = $w \times$ ROT for each activity 'i' + $(1 - w)$ ACTIM for each activity 'i', where $0 \leq w \geq 1$.
   - Rank these activities in decreasing order as per maximum ROT-ACTIM.

7. ROT-ACTRES:
   - Determine ROT-ACTRES.
   - ROT-ACTRES for each activity 'i' = $w \times$ ROT for each activity 'i' + $(1 - w)$ ACTRES for each activity 'i', where $0 \leq w \geq 1$.
   - Rank these activities in decreasing order as per maximum ROT-ACTRES.

29.3 SOLVED PROBLEM

Problem 29.1: As per the information given in Table 29.1 and in Fig. 29.1, determine sequence of activities using:

a. Float/slack	e. GENRES
b. ACTIM	f. ROT
. ACTRES	g. ROT-ACTIM
. TIMRES	h. ROT-ACTRES

Table 29.1 Time and resources required for each activity

Activity	1	2	3	4	5	6	7
Duration	3	3	1	2	2	3	3
Resource required	2	3	2	3	4	2	1

Take $w = 0.2$

Maximum resource available = 4

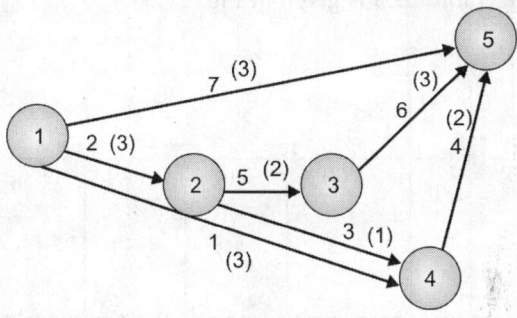

Entry in ( ) represents duration of activity

Fig. 29.1 Network diagram

Solution:

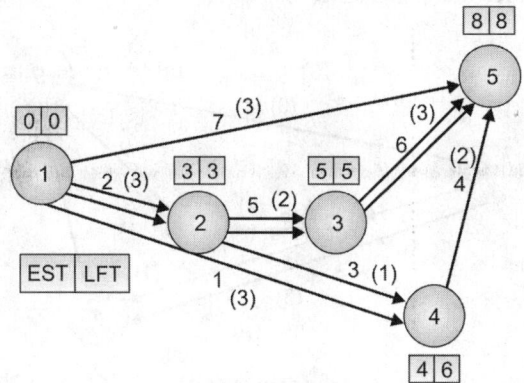

Fig. 29.2 Project scheduling under no resource constraints

Table 29.2 Project schedule under no resource constraints

Activity	1	2	3	4	5	6	7
Duration	3	3	1	2	2	3	3
Resource required	2	3	2	3	4	2	1
EST	0	0	3	4	3	5	0
EFT	3	3	4	6	5	8	3
LFT	6	3	6	8	5	8	8
LST	3	0	5	6	3	5	5
Total float	3	0	1	2	1	0	5

Under no resource constraints shortest possible time of completing the project is 8.

If activities are done one at a time, the time of completion of project is 17.

a. Sequence of activities using float/slack:

Arranging the tasks in descending order according to availability of float is given in Table 29.3.

Table 29.3 Arranging the tasks in descending order according to availability of float

Activity	2	6	3	5	4	1	7
Duration	3	3	1	2	2	3	3
Resource	3	2	2	4	3	2	1

Project schedule using Table 29.3 is given in Fig. 29.3.

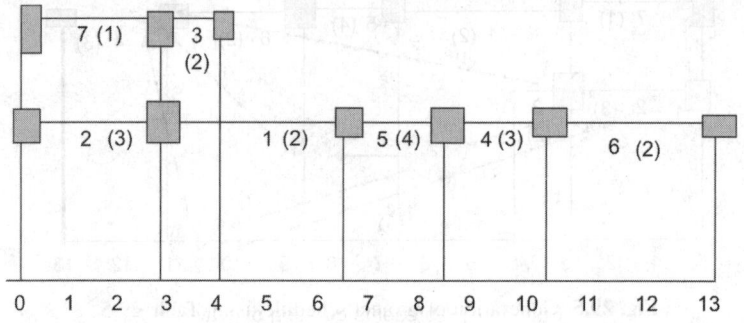

Fig. 29.3 Project schedule using Table 29.3

b. Sequence of activities using ACTIM:

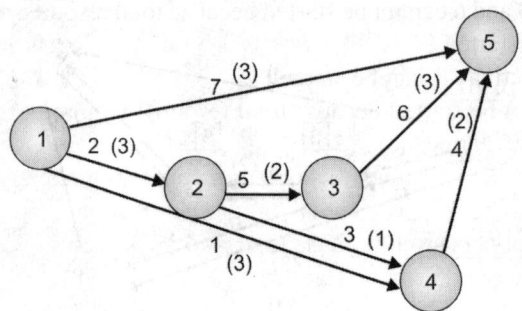

Entry in ( ) represents duration of activity

Fig. 29.4 Network diagram

Table 29.4 Determining ACTIM for each Activity

Activity	1	2	3	4	5	6	7
Duration	3	3	1	2	2	3	3
Resource required	2	3	2	3	4	2	1
Maximum ACTIM	5	8	3	2	5	3	3

For example, Activity '2' has two paths. Path 1 and Path 2 are 2-5-6 and 2-3-4. ACTIM for the Path 1 and Path 2 are 8 and 6 respectively. Maximum ACTIM for the Activity '2' is 8. Sequence of activities using ACTIM is given in Table 29.5.

Table 29.5 Sequence of activities using ACTIM

Activity	2	1	5	3	6	7	4
Duration	3	3	2	1	3	3	2
Resource	3	2	4	2	2	1	3

If activities are done one at a time, the time of completion of project is 17.
Generation of project schedule using Table 29.5 is given in Fig. 29.5.

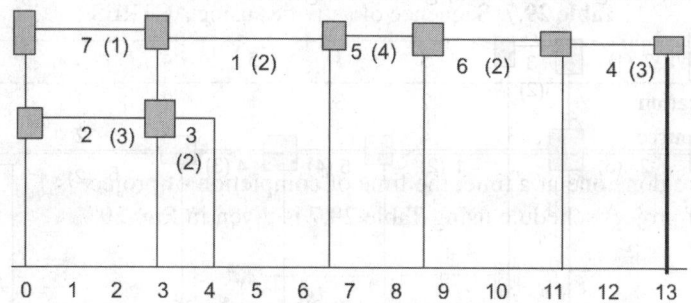

Fig. 29.5 Generation of project schedule using Table 29.5

Scheduling procedure: Sequence of steps to generate the schedule as shown in Fig. 29.5 is given below.

Step 1: Activity 2 is started at duration 0.

Activities 1, 5, 3 and 6 cannot be started because total resource requirement exceeds 4. So Activity 7 is started.

Step 2: At duration 3 Activity 1 can be started.

Activity 5 cannot be started because total resource required exceeds 4.

Activity 3 can be started.

Step 3: At duration 4 activity 5 cannot be started because resource required exceeds 4.

Activity 6 cannot be started without completion of activity 5.

Activity 4 cannot be started because total resource required exceeds 4.

Step 4: At duration 6 Activity 5 can be started.

Step 5: At duration 8 activity 6 can be started.

At duration 8 activity 4 cannot be started because resource requirement exceeds 4.

Step 6: At duration 11 activity 4 can be started.

c. Sequence of activities using ACTRES:

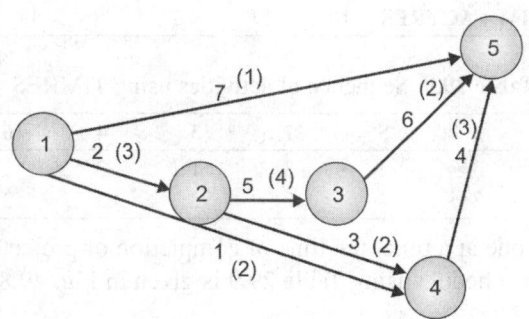

Entry in ( ) represents resource required for each activity

Fig. 29.6 Network diagram

Table 29.6 Determining ACTRES for each Activity

Activity	1	2	3	4	5	6	7
Duration	3	3	1	2	2	3	3
Resource required	2	3	2	3	4	2	1
Maximum ACTRES	5	9	5	3	6	2	1

Table 29.7 Sequence of activities using ACTRES

Activity	2	5	1	3	4	6	7
Duration	3	2	3	1	2	3	3
Resource	3	4	2	2	3	2	1

If activities are done one at a time, the time of completion of project is 17. Generation of project schedule using Table 29.7 is given in Fig. 29.7.

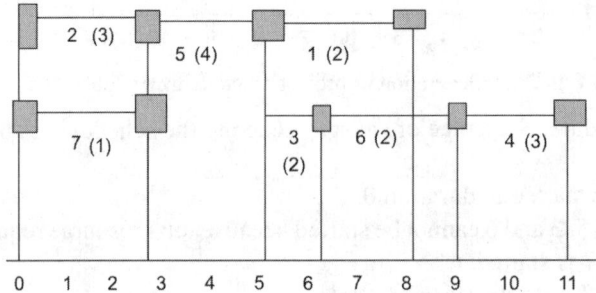

Fig. 29.7 Generation of project schedule using Table 29.7

Project duration is 11 days.

d. Sequence of activities using TIMRES:

Table 29.8 Determining TIMRES for each Activity

Activity	1	2	3	4	5	6	7
Duration	3	3	1	2	2	3	3
Resource required	2	3	2	3	4	2	1
ACTIM	5	8	3	2	5	3	3
ACTRES	5	9	5	3	6	2	1
TIMRES = ACTIM + ACTRES	10	17	8	5	11	5	4

Table 29.9 Sequence of activities using TIMRES

Activity	2	5	1	3	4	6	7
Duration	3	2	3	1	2	3	3
Resource	3	4	2	2	3	2	1

If activities are done one at a time, the time of completion of project is 17. Generation of project schedule using Table 29.9 is given in Fig. 29.8.

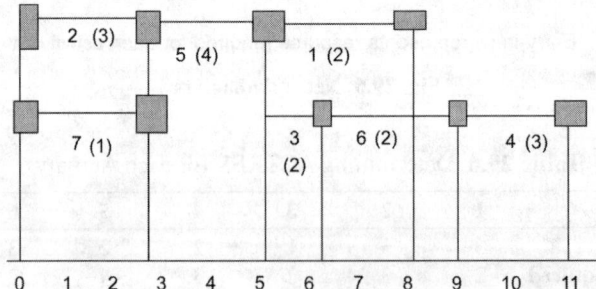

Fig. 29.8 Generation of project schedule using Table 29.9

Project duration is 11 days.

e. Sequence of activities using GENRES:

Table 29.10 Determining GENRES for each Activity

Activity	1	2	3	4	5	6	7
Duration	3	3	1	2	2	3	3
Resource required	2	3	2	3	4	2	1
ACTIM	5	8	3	2	5	3	3
ACTRES	5	9	5	3	6	2	1
GENRES = 0.2 × ACTIM + 0.8 × ACTRESS	5	8.8	4.6	2.8	5.8	2.2	1.4

Table 29.11 Sequence of activities using GENRES

Activity	2	5	1	3	4	6	7
Duration	3	2	3	1	2	3	3
Resource	3	4	2	2	3	2	1

If activities are done one at a time, the time of completion of project is 17. Generation of project schedule using Table 29.11 is given in Fig. 29.9.

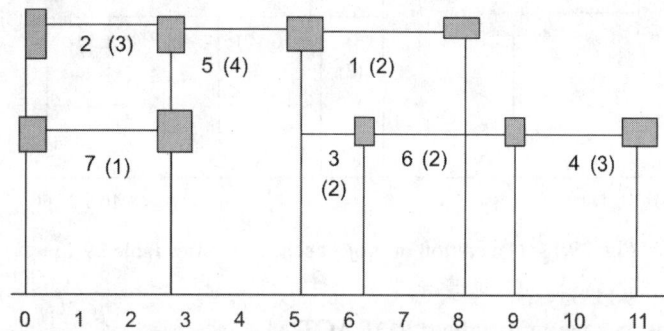

Fig. 29.9 Generation of project schedule using Table 29.11

Project duration is 11 days.

f. Sequence of activities using ROT:

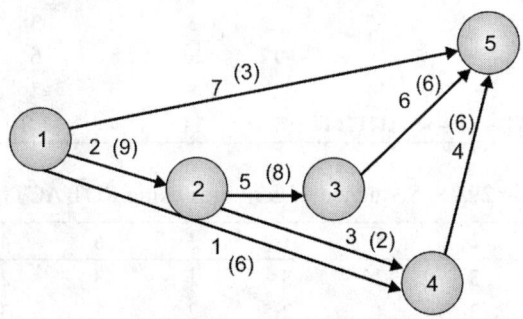

Entry in ( ) represents 'Duration x Resource' of activity

Fig. 29.10 Network diagram

Table 29.12 Determining ROT for each Activity

Activity	1	2	3	4	5	6	7
Duration	3	3	1	2	2	3	3
Resource required	2	3	2	3	4	2	1
Duration × Resource required	6	9	2	6	8	6	3
Maximum ROT	12	23	8	6	14	6	3

Table 29.13 Sequence of activities using ROT

Activity	2	5	1	3	4	6	7
Duration	3	2	3	1	2	3	3
Resource	3	4	2	2	3	2	1

If activities are done one at a time, the time of completion of project is 17.
Generation of project schedule using Table 29.13 is given in Fig. 29.11.

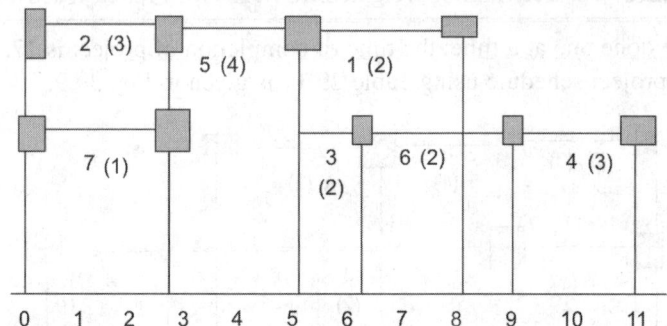

Fig. 29.11 Generation of project sche..le using Table 29.13

Project duration is 11 days.
f. Sequence of activities using weighted ROT-ACTIM:

Table 29.14 Determining weighted ROT-ACTIM for each Activity

Activity	1	2	3	4	5	6	7
Duration	3	3	1	2	2	3	3
Resource required	2	3	2	3	4	2	1
ROT	12	23	8	6	14	6	3
ACTIM	5	8	3	2	5	3	3
ROT-ACTIM = $w \times$ ROT + $(1 - w)$ ACTIM	6.4	11	4	2.8	6.8	3.6	3

Table 29.15 Sequence of activities using ROT-ACTIM

Activity	2	5	1	3	6	4	7
Duration	3	2	3	1	3	2	3
Resource	3	4	2	2	2	3	1

If activities are done one at a time, the time of completion of project is 17.
Generation of project schedule using Table 29.15 is given in Fig. 29.12.

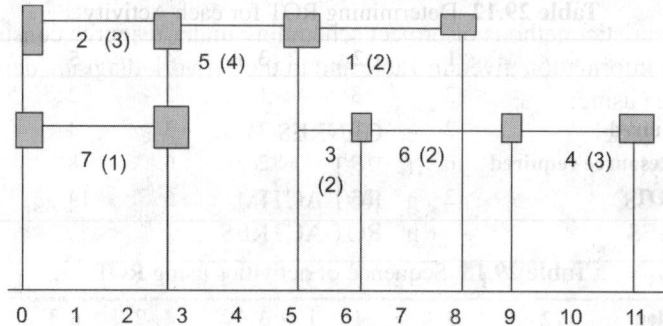

Fig. 29.12 Generation of project schedule using Table 29.15

Project duration is 11 days.

g. Sequence of activities using weighted ROT-ACTRES:

Table 29.16 Determining weighted ROT-ACTRES for each activity

Activity	1	2	3	4	5	6	7
Duration	3	3	1	2	2	3	3
Resource required	2	3	2	3	4	2	1
ROT	12	23	8	6	14	6	3
ACTRES	5	9	5	3	6	2	1
ROT-ACTRES = $w \times$ ROT + $(1 - w)$ ACTRES	6.4	11.8	5.6	3.6	7.6	2.8	1.4

Table 29.17 Sequence of activities using ROT-ACTRES

Activity	2	5	1	3	4	6	7
Duration	3	2	3	1	2	3	3
Resource	3	4	2	2	3	2	1

If activities are done one at a time, the time of completion of project is 17.
Generation of project schedule using Table 29.17 is given in Fig. 29.13.

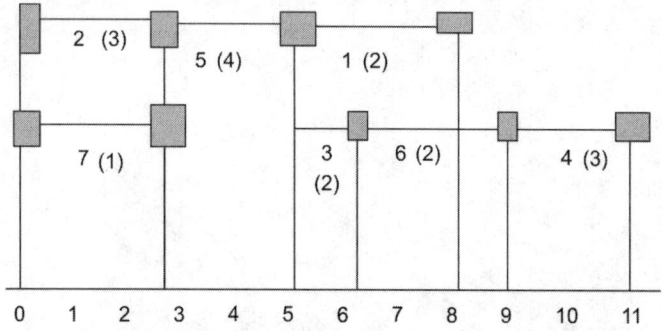

Fig. 29.13 Generation of project schedule using Table 29.17

Project duration is 11 days.

29.4 QUESTIONS AND PROBLEMS

1. Explain heuristic methods of project scheduling under resource constraints.
2. As per the information given in Table and in the network diagram, determine sequence of activities using:

   a. Float/slack   e. GENRES
   b. ACTIM   f. ROT
   c. ACTRES   g. ROT-ACTIM
   d. TIMRES   h. ROT-ACTRES

Activity	1	2	3	4	5	6	7
Duration	3	3	1	2	2	3	3
Resource required	2	3	2	3	4	2	1

Take $w = 0.3$
Maximum resource available = 5

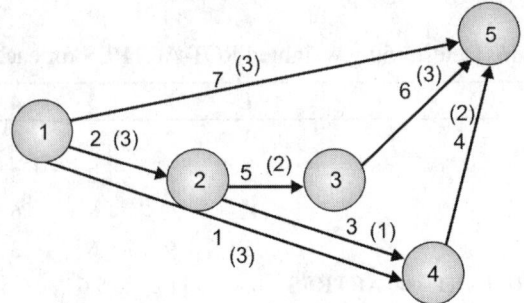

Part IX

Maintenance Management

30

Maintenance

30.1 INTRODUCTION

Maintenance encompasses all those activities that relate to keeping facilities and equipment in good working order and making necessary repairs when breakdowns occur, so that the system can perform as intended. Equipment maintenance is responsible for maintaining machinery and equipment in good working condition and making all necessary repairs.

30.2 EQUIPMENT FAILURE RATE OVER TIME

If we graph the failure rate of equipment versus time, it is likely that graph would follow the bathtub shape as shown in Fig. 30.1. In Fig. 30.1, the Y-axis represents the failure rate and the X-axis is time. From its shape, the curve can be divided into three distinct periods:
- Infant mortality
- Useful life
- Wear-out periods

The initial infant mortality period of bathtub curve is characterized by high failure rate followed by a period of decreasing failure. Many of the failures associated with this region are linked to poor design, poor installation, or misapplication. The infant mortality period is followed

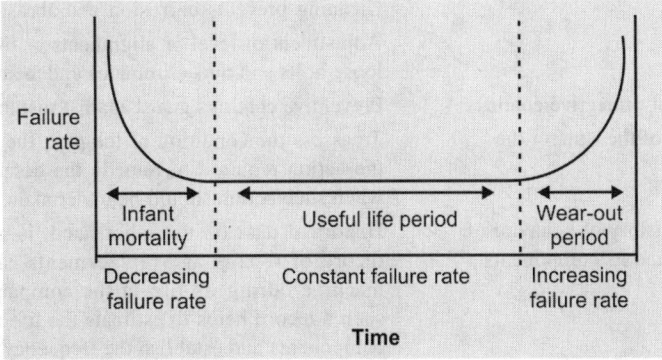

Fig. 30.1 Equipment failure rate over time

by a nearly constant failure rate period known as useful life. There are many theories on why components fail in this region. Poor maintenance often plays significant role. It is also generally agreed that maintenance practices such as preventive and predictive elements can extend this period. The wear-out period is characterized by a rapid increasing failure rate with time. The design life of most equipment requires periodic maintenance. Belts need adjustment, alignment needs to be maintained, and proper lubrication on rotating equipment is required, and so on. In some cases, certain components need replacement (e.g. a wheel bearing on a motor vehicle) to ensure the main piece of equipment (in this case a car), last for its design life. Anytime we fail to perform maintenance activities intended by the equipment's designer, we shorten the operating life of the equipment. There are different approaches of maintenance. In addition to waiting for a piece of equipment to fail (reactive maintenance), we can utilize preventive maintenance, predictive maintenance, or reliability centered maintenance.

30.3 MAINTENANCE OBJECTIVES

The goal of maintenance is to keep the production system in good working order at minimum cost. Other objectives are:
- Avoid production disruptions
- Maintain high quality
- Avoid missed delivery dates

30.4 CONSEQUENCES OF BREAKDOWNS

When breakdowns occur, there are a number of adverse consequences arised:
- Production capacity is reduced, and orders are delayed.
- There is no production, but overhead continues, increasing the cost per unit.
- There are quality issues; product may be damaged.
- There are safety issues; employees or customers may be injured.

30.5 MAINTENANCE FUNCTIONS

The maintenance objectives are attained by taking certain action illustrated in Table 30.1:

Table 30.1 Maintenance actions

Action	Purpose
1. Lubrication	Lubrication reduces friction.
2. Cleaning	Cleaning prevents corrosion and abrasion due to dust.
3. Adjustments	Adjustment of level or alignments or by tightening the loose bolts and nuts eliminates undue stresses.
4. Application of protective coatings	Preventive coatings guard against rust and corrosion.
5. Examination of the state of the components	To assess the condition of the part, the extent of wear, the action required to remedy the defect and the time when such action should be undertaken.
6. Analysis of history of behavior of the machine and its components	Historical data on the other hand, is a chronological record of repairs and replacements carried out on a machine during its life in the company. Analysis of such a record helps to estimate the life span of various components and establish the frequency of inspections, repairs, and replacements.

(Contd...)

(Contd...)

7. Replacement of worn out components	Replacement and repair of components can be undertaken on the basis of inspection reports, analysis of history or complaints of operating personnel.
8. Repair of cracks or other repairable damage	Restore the original operational capacity of the machine and prevent further damage.
9. Modification of design of the components or location of the equipment.	Affect improvements to reduce the frequency of attention or to reduce cost of maintaining the equipment.
10. Capital replacement	Replace the item when increase in maintenance cost and availability of better technology.

30.6 MAINTENANCE STRATEGIES

Maintenance strategies are:
- Corrective maintenance
- Preventive maintenance
- Predictive maintenance
- Reliability centered maintenance
- Total productive maintenance

In practice, all these types are used in maintaining engineering systems. The challenge is to optimize the balance between these types for maximum profitability.

30.6.1 Reactive Maintenance (Breakdown Maintenance)

Basic philosophy:
- Allow machinery to run to failure.
- Repair or replace damaged equipment when obvious problems occur.

The task of the maintenance team is usually to repair a system or component as soon as possible. Costs associated with corrective maintenance include repair costs (replacement of components, labor, and consumables), loss of production during breakdown. To minimize breakdown time, actions such as increasing the size of maintenance teams, the use of back-up systems and implementation of emergency procedures can be considered. Unfortunately, such measures are relatively costly and/or only effective in the short-term.

Advantages:
- Low cost
- Less staff

Disadvantages:
- Increased cost due to unplanned downtime of equipment.
- Increased labor cost, especially, if overtime is needed.
- Cost involved with repair or replacement of equipment.
- Possible secondary equipment or process damage from equipment failure.
- Inefficient use of staff resources.

Reactive element applications:
- Small parts and equipment
- Non-critical equipment
- Equipment unlikely to fail
- Redundant systems.

30.6.2 Preventive Maintenance

Basic philosophy:
- Schedule maintenance activities at predetermined time intervals.
- Repair or replace damaged equipment before obvious problems occur.

In preventive maintenance, equipment is repaired and serviced before failures occur. The frequency of maintenance activities is pre-determined by schedules. The higher the failure consequences, then greater the level of preventive maintenance is justified. Inspection assumes a crucial role in preventive maintenance strategies. Components are essentially inspected for corrosion and other damage at planned intervals, in order to identify corrective action before failures actually occur. Preventive maintenance performed at regular intervals will usually results in reduced failure rates. There is loss of production during the preventive maintenance period. Managers usually schedule preventive maintenance using some combination of the following:
- The result of planned inspections that reveal a need for maintenance
- According to the calendar (passage of time)
- After a predetermined number of operating hours

An important issue in preventive maintenance is the frequency of preventive maintenance. As the time between periodic maintenance increases, the cost of preventive maintenance decreases while the risk (and cost) of breakdowns increases.

Advantages:

Preventive maintenance helps to:
- Protect assets and prolong the useful life of production equipment
- Improve system reliability
- Decrease cost of replacement
- Decreases system downtime
- Reduce injury

Disadvantages:
- Catastrophic failures still likely to occur
- Labor intensive
- Includes performance of unneeded maintenance
- Potential for incidental damage to components in conducting unneeded maintenance

Preventive maintenance applications:
- Equipment subject to wear
- Consumable equipment
- Equipment with known failure patterns

30.6.3 Predictive Maintenance (Condition-based Maintenance)

Basic philosophy:
- Schedule maintenance activities when mechanical or operational conditions warrant.
- Repair or replace damaged equipment before obvious problems occur.

Predictive maintenance refers to maintenance based on the actual condition of a component. This philosophy consists of scheduling maintenance activities only if and when mechanical or operational conditions warrant by periodically monitoring the machinery for excessive vibration, temperature and/or lubrication degradation, or by observing any other unhealthy trends that occur over time. When the condition gets to a predetermined unacceptable level, the equipment is shut down to repair or replace damaged components so as to prevent a more costly failure from occurring. Predictive maintenance refers to maintenance based on a certain change in characteristics. Changing the oil every 5,000 km to prolong engine life, irrespective of whether the oil change is really needed or not, is a preventive maintenance strategy. Predictive maintenance of changing the oil is based on changes in its properties, such as the build-up of wear debris. Advantages of this approach are that it works very well if personnel have adequate knowledge, skills, and time to perform the predictive maintenance work, and that it allows equipment repairs to be scheduled in an orderly fashion. It also provides some lead-time to purchase materials for the necessary repairs, reducing the need for a high parts inventory. Since maintenance work is only performed when it is needed, there is likely to be an increase in production capacity. A good predictive maintenance effort relies on complete records for each piece of equipment. Records must include information such as date of installation, operating hours, and dates and types of repairs.

Advantages:

- Increased component's operational life/availability
- Allows for preemptive corrective actions
- Decrease in equipment or process downtime
- Decrease in costs for parts and labor
- Better product quality
- Improved worker and environmental safety
- Improved worker morale
- Energy savings

Disadvantages:

- Increased investment in diagnostic equipment
- Increased investment in staff training
- Savings potential not readily seen by management

Predictive maintenance applications:

- Equipment with random failure patterns
- Critical equipment
- Equipment not subject to wear
- Systems which failure may be induced by incorrect preventive maintenance

30.6.4 Reliability Centered Maintenance (RCM)

This philosophy utilizes all of the previously discussed predictive/preventive maintenance techniques. It recognizes that all equipments in a facility are not of equal importance. It utilizes approach that is based on the consequences of failure. It emphasizes the functional importance of system components and their failure/maintenance history. It recognizes that maintenance activities on equipment that is inexpensive and unimportant to facility reliability may best be left to a reactive maintenance approach.

For example,

<10% Reactive

25% to 35% Preventive

45% to 55% Predictive

RCM is heavily weighted in utilization of predictive maintenance technologies. A major concern of airlines was that existing time-based preventive maintenance programs would threaten the economic viability of larger, more complex aircraft. The experience of airlines with the

RCM approach was that maintenance costs remained roughly constant but that the availability and reliability of their planes improved. RCM is now standard practice for most of the world's airlines.

Advantages:

- Lower costs by eliminating unnecessary maintenance or overhauls
- Minimize frequency of overhauls
- Reduced probability of sudden equipment failures
- Able to focus maintenance activities on critical components
- Increased component reliability
- Incorporates root cause analysis

Disadvantages:

- Training costs
- Savings potential not readily seen by management

30.6.5 Total Productive Maintenance

Some Japanese companies' workers perform preventive and predictive maintenance on the machines they operate, rather than use separate maintenance personnel for that task. This approach is called total productive maintenance. This approach is consistent with JIT systems, where employees are given greater responsibility for quality, productivity, and the general functioning of the system. Employees are trained in proper operating procedures and in how to keep equipment in good operating order and providing the incentive to do so are major part of this. More and more, North American organizations are taking a cue from the Japanese and transferring routine maintenance (e.g. cleaning, adjusting, inspecting) to the users of equipment, in an effort to give them a sense of responsibility and awareness of the equipment they use and to cut down on abuse and misuse of the equipment.

30.6.6 Determining Optimum Period between Periodic Maintenance

The machines can be used without any interruption, if machines are replaced frequently or if a stand by machine is maintained. Stand by machine can be put into operation as and when the original unit is stopped for checks, repairs and component replacements. But, no industrial unit can possibly afford to throw away its capital resources by replacing machines frequently, nor can a company block its money in equipment that will be only partially utilized by having stand bys. An organization has to accept a certain loss in productive capacity of its investments to perform maintenance in order to reduce capital costs. Management of maintenance has therefore, to concern with the balancing of costs against gains so as to evolve the most suitable maintenance policies. Maintenance procedures can vary from industry to industry. Decision makers have two basic options with respect to maintenance. One option is breakdown maintenance. The other option is preventive maintenance. Decision makers try to make a trade-off between these two basic options that will minimize their combined cost for the period T.

Cost of breakdown maintenance for the period T = Repair cost + Loss of revenue in repair period + Cost of wages while equipment is not in service + Cost of injuries or damage to other equipment and facilities + Cost of sub-standard performance.

Cost of preventive maintenance for the period T = Wages for maintenance personnel + Loss of revenue in maintenance period + Cost of replacement of parts + Cost of wages while equipment is not in service.

Total cost for period T = Cost of breakdown maintenance for the period T + Cost of Preventive maintenance for the period T.

With no preventive maintenance, breakdown and repair costs would be tremendous. However, beyond a certain point, the cost of preventive maintenance activities exceeds the benefit. The best approach is to seek a balance between preventive maintenance and breakdown maintenance. An important issue in preventive maintenance is the frequency of preventive maintenance. As the time between periodic maintenance increases, the cost of preventive maintenance decreases while the risk (and cost) of breakdowns increases. The goal is to strike a balance between the two costs (i.e. to minimize total cost). This concept is illustrated in Fig. 30.2.

T = optimum period between periodic maintenance

Frequency of preventive maintenance = $1/T$

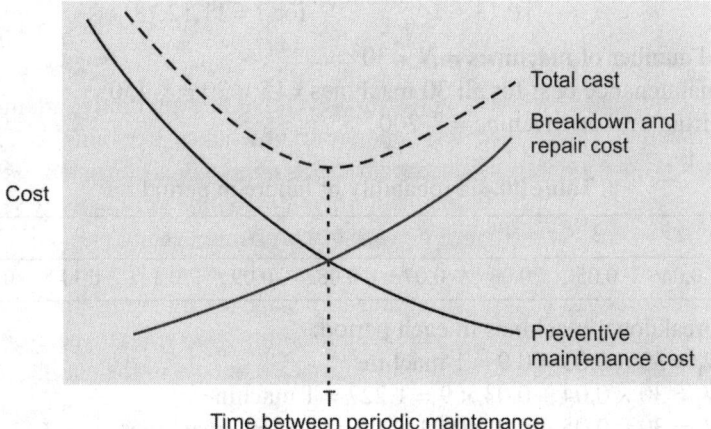

Fig. 30.2 To determine optimum period of maintenance

30.7 SOLVED PROBLEMS

Problem 30.1: There are 5 identical parallel centrifugal pump sets. They are all running at 100% capacity and have a stressful service. The data of failures using various maintenance strategies is given in Table 30.2. Estimated costs per unit per hour of downtime is ₹ 2,100. Calculate production loss cost per year in downtime using different maintenance strategies.

Table 30.2 The data of failures using various maintenance strategies

Strategy	Run to failure	Time-based	Condition-based
Failures per year	4	1	0
Downtime hours per failure	16	16	0
Overhauls per year	0	5	3
Downtime hours per overhaul	0	3	3

Solution: Calculation of Production loss cost per year (₹) is given in Table 30.3.

Table 30.3 Production loss cost per year in downtime (₹)

Strategy	Run to failure	Time-based	Condition-based
Failures per year	4	1	0
Downtime hours per failure	16	16	0
Overhauls per year	0	5	3
Downtime hours per overhaul	0	3	3
Total down time per year (hours)	$4 \times 16 = 64$	$1 \times 16 + 5 \times 3 = 31$	$0 \times 0 + 3 \times 3 = 9$
Production loss cost per year (₹)	$64 \times 2{,}100 = 1{,}34{,}400$	$31 \times 2{,}100 = 65{,}100$	$9 \times 2{,}100 = 18{,}900$

Problem 30.2: Let $p(t)$ be the probability that a machine in a group of 30 machines would breakdown in period 't'. The cost of repairing broken machine is ₹ 200. Preventive maintenance is performed by servicing all the 30 machines at the end of 'T' units of time. Preventive maintenance cost is ₹ 15 per machine. Find optimal 'T' that will minimizes the expected total cost per period of servicing given that,

$$p(t) = \begin{cases} 0.03 & \text{for } t = 1 \\ p(t-1) + 0.01 & \text{for } t = 2, 3, ..., 10 \\ 0.13 & \text{for } t = 11, 12, 13 \end{cases}$$

Solution: Total number of machines = $N = 30$

Preventive maintenance cost for all 30 machines $= 15 \times 30 = ₹ 450$

Cost of repairing broken machine = ₹ 200

Table 30.4 Probability of failure in period t

t	1	2	3	4	5	6	7	8	9	10	11
$P(t)$	0.03	0.04	0.05	0.06	0.07	0.08	0.09	0.1	0.11	0.12	0.12

Number of breakdown machines in each period:

Period 1 $= N_1 = 30 \times 0.03 = 0.9 = 1$ machine

Period 2 $= N_2 = 30 \times 0.04 + 0.03 \times 9 = 1.227 = 1$ machine

Period 3 $= N_3 = 30 \times 0.05 + 0.9 \times 0.04 + 1.2 \times 0.03 = 2$ machines

Similarly,

Period 4 $= N_4 = 2$ machines

Period 5 $= N_5 = 2$ machines

Period 6 $= N_6 = 3$ machines

Calculation of period at which all the machines are serviced is given in Table 30.5.

Table 30.5 To determine period at which all the machines are serviced

Period	Number of breakdown machines	Cumulative number at the end of period	Individual repair cost (₹)	Group cost (₹)	Total cost (₹)	Average cost $= \dfrac{\text{Total cost}}{\text{Period}}$ (₹)
1	1	1	200	450	650	650
2	1	2	400	450	850	425
3	2	4	800	450	1,250	416
4	2	6	1,200	450	1,650	412
5	2	8	1,600	450	2,050	410
6	3	11	2,200	450	2,650	442

Group preventive maintenance at the end of 5th period.

Problem 30.3: The historical frequency of breakdowns of a machine per month is shown in Table 30.6. The cost of a breakdown repair is ₹ 1,000 and the cost of preventive maintenance is ₹ 1,250 per month (daily inspections, lubrication, and possible parts changes). If preventive maintenance is performed, the probability of machine breakdown is negligible. Should the manager use preventive maintenance, or would it be cheaper to repair the machine when it breaks down?

Table 30.6 The historical frequency of breakdowns of a machine per month

Number of breakdowns	0	1	2	3
Frequency of occurrence	0.2	0.3	0.4	0.1

Solution: The expected number of breakdowns without preventive maintenance per month

$$= 0 \times 0.2 + 1 \times 0.3 + 2 \times 0.4 + 3 \times 0.1 = 1.4$$

The expected number of breakdowns without preventive maintenance is 1.40 per month.

Expected cost using repair maintenance policy $= 1.4 \times 1,000 = ₹ 1,400$/month

Preventive maintenance cost $= ₹ 1,250$ per month.

Therefore, preventive maintenance would yield a savings of $150/month.

30.8 QUESTIONS AND PROBLEMS

1. What is the goal of a maintenance program?
2. List the costs associated with equipment breakdown.
3. State the various functions of maintenance.
4. How does corrective maintenance differ from preventive maintenance?
5. Describe preventive maintenance.
6. Describe reactive maintenance.
7. Explain various maintenance strategies.
8. What are three different ways preventive maintenance is scheduled?
9. Explain the term predictive maintenance.
10. List the major approaches organizations use to deal with breakdowns.
11. Explain the term reliability centered maintenance.
12. What advantages does preventive maintenance have over breakdown maintenance?
13. The probability that equipment used in a hospital lab will need a number of recalibrations per month is given in the following table. A service firm is willing to provide any number of necessary calibrations for a fixed fee of $650 per month. Without this plan, recalibration costs $500 per occurrence. Which approach would be most economical, recalibration as needed or the fixed fee service contract?

Number of recalibrations	0	1	2	3	4
Probability of occurrence	0.15	0.25	0.30	0.20	0.10

Hint: Expected number of recalibrations/month = 1.85
Preventive maintenance cost/month = $650
Breakdown maintenance cost = 1.85 × 500 = $925

14. The frequency of breakdown of a machine that issues lottery tickets is given in the following table. Repairs cost an average of $240 per repair. A service firm is willing to provide service contracts under either of two options. # 1 is $500 and covers all necessary repairs, and # 2 is $350 and covers any repairs after the first one (which will cost $240).

Which option would have the lowest expected cost: Pay for all repairs as they arise, service option # 1, or service option # 2?

Number of breakdowns/month	0	1	2	3	4
Probability of occurrence	0.10	0.30	0.30	0.20	0.10

Hint: Expected number of breakdowns/month = 1.9, Option # 1 cost = ₹ 950, Option # 2 cost = ₹ 566, Option # 2 is preferred

15. Determine the optimum preventive maintenance frequency for each of the equipment below, if breakdown times are normally distributed.

Equipment	Average time between breakdowns (days)	Standard deviation (days)	Preventive maintenance cost (₹)	Breakdown cost (₹)
201	20	2	300	2,300
400	30	3	200	3,500
850	40	4	530	4,800

Replacement and Capital Investment Decisions

31.1 INTRODUCTION

Replacement problems are concerned with replacement of men, machines, bulbs, etc. due to their decreased efficiency, failures or breakdowns. Failures may be sudden or gradual. Replacement problem arises because of the following factors.

1. Increase in maintenance cost
2. Decrease in efficiency
3. Failure of the item due to accident
4. Obsolescence of item

31.2 DECISION TO BE MADE IN REPLACEMENT PROBLEMS

The decision to be made in the replacement model is when to replace an item or determine the age at which an item is replaced.

31.3 REPLACEMENT MODELS

Model 1: Replacement policy of items whose maintenance cost increases with time and money value is constant.

Model 2: Replacement policy of items whose maintenance cost increases with time and money value changes with a constant rate.

Model 3: Replacement policy of items that fail completely.

31.4 MODEL 1

Replacement policy of items whose maintenance cost increases with time and money value is constant.

C = Initial cost of item

R_i = Maintenance cost of the item at the beginning of the year i

S_n = Scrap value or final value or end value of the item at the end of nth year

Total cost of item for n years = TC

$$TC = C + \sum_{i=1}^{n} R_i - S_n$$

$$\text{Average cost} = AC = \frac{TC}{n} = \frac{C + \Sigma R_i - S_n}{n}$$

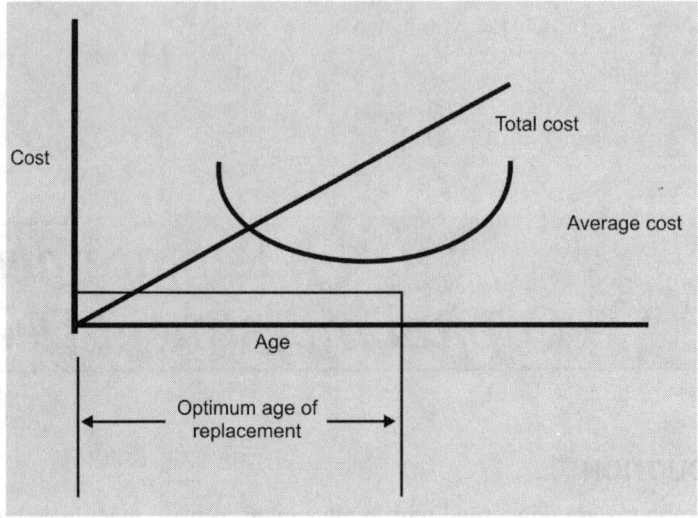

Fig. 31.1 Relationship between cost and age

Optimum replacement policy is to replace the item at the end of period when average cost is least.

31.5 SOLVED PROBLEMS

Problem 31.1: The cost of a machine is ₹ 6,100 and its scrap value is only ₹ 100. The maintenance costs are given in Table 31.1.

Table 31.1 Maintenance costs

Year	1	2	3	4	5	6	7	8
Maintenance cost (₹)	100	250	400	600	900	1,250	1,600	2,000

When should machine be replaced?

Solution: C = Initial cost of machine = ₹ 6,100

Table 31.2 To determine the year at which machine is to be replaced

i or n	Maintenance cost (R_i) (₹)	Scrap value S_n (₹)	ΣR_i (₹)	$C - S_n$ (₹)	Total cost $C + \sum_{i=1}^{n} R_i - S_n$ (₹)	Average cost $\dfrac{TC}{n}$ (₹)
1	100	100	100	6,000	6,100	6,100
2	250	100	350	6,000	6,350	3,175
3	400	100	750	6,000	6,750	2,250
4	600	100	1,350	6,000	7,350	1,837
5	900	100	2,250	6,000	8,250	1,650
6	1,250	100	3,500	6,000	9,500	**1,583**
7	1,600	100	5,100	6,000	11,100	1,586
8	2,000	100	7,100	6,000	13,100	1,638

Minimum average cost is at the end of 6th year. So replace at the end of 6th year.

Problem 31.2: A fleet owner finds from his past records that the cost per year of running a vehicle whose purchasing price is ₹ 50,000 are given in Table 31.3.

Table 31.3 Maintenance cost and resale price

Year	1	2	3	4	5	6	7
Maintenance cost (₹)	5,000	6,000	7,000	9,000	11,500	16,000	18,000
Resale price (₹)	30,000	15,000	7,500	3,750	2,000	2,000	2,000

At what age is replacement due?

Solution: C = Initial cost of vehicle = ₹ 50,000

Table 31.4 To determine the year at which vehicle is to be replaced

i	Maintenance cost (R_i) (₹)	ΣR_i (₹)	Scrap value (₹) S_n	$C - S_n$ (₹)	Total cost $C + \sum_{i=1}^{n} R_i - S_n$ (₹)	Average cost $\dfrac{TC}{n}$ (₹)
1	5,000	5,000	30,000	20,000	25,000	25,000
2	6,000	11,000	15,000	35,000	46,000	23,000
3	7,000	18,000	7,500	42,500	60,500	20,166
4	9,000	27,000	3,750	46,250	73,250	18,312
5	11,500	38,500	2,000	48,000	86,500	17,300
6	16,000	54,500	2,000	48,000	1,02,500	17,083
7	18,000	72,500	2,000	48,000	1,20,500	17.214

Minimum average cost is at the end of 6th year. So replace at the end of 6th year.

31.6 MODEL 2

Replacement policy of items whose maintenance cost increases with time and money value changes with a constant rate.

V = Discount factor = $(1 + r)^{-1}$

r = Rate of interest in %

C = Initial cost of item

R_i = Maintenance cost of the item at the beginning of the year i

S_i = Scrap value or final value or end value of the item at the end of nth year

Total cost of the items for n years = TC

$$TC = C + \sum_{i=1}^{n} R_i v^{i-1} - S_i v^{i-1}$$

Average cost = AC

$$AC = \frac{TC}{\sum v^{i-1}}$$

Optimum replacement policy is to replace the item at the age when average cost of item is minimum.

31.7 SOLVED PROBLEMS

Problem 31.3: A machine is priced at ₹ 5,000 and running costs are estimated at ₹ 800 for each of the first five years and increasing by ₹ 200 per year in the sixth and subsequent years. If the money is worth 10% per year. Determine the year at which machine is replaced.

Solution:

$$C = \text{Initial cost of item} = ₹\ 5,000$$
$$r = 10\%$$
$$v = \frac{1}{1+r} = \frac{1}{1+0.1} = 0.9091$$

Table 31.5 To determine the year at which vehicle is to be replaced

Year i	R_i (₹)	v^{i-1}	$R_i\,v^{i-1}$ (₹)	Total cost = $C + \sum_{i=1}^{n} R_i v^{i-1}$ (₹)	$\sum v^{i-1}$ (₹)	Average cost = $AC = \dfrac{TC}{\sum v^{i-1}}$ (₹)
1	800	1.000	800	5,800	1.00	5,800
2	800	0.9091	727	6,527	1.9091	3,419
3	800	0.8264	661	7,188	2.7355	2,628
4	800	0.7513	601	7,889	3.4868	2,234
5	800	0.6830	546	8,335	4.1698	1,999
6	1,000	0.6209	621	8,956	4.7907	1,896
7	1,200	0.5645	677	9,633	5.3552	1,799
8	1,400	0.5132	718	10,351	5.8684	1,764
9	1,600	0.4665	746	11,097	6.3349	1,752
10	1,800	0.4241	763	11,860	6.7590	1,755

Minimum average cost is at the end of 9th year. So replace at the end of 9th year.

Problem 31.4: The cost of a new machine is ₹ 5,000. The maintenance cost of nth year is given by $R_n = 500\,(n-1)$ $n = 1, 2, 3, \ldots$. The discount rate per year is 0.5. After how many years it will be economical to replace the machine?

Solution: The cost of a new machine $= C = ₹\ 5,000$

$$\text{Discount rate} = v = \frac{1}{1+r} = 0.5$$

Table 31.6 To determine the year at which machine is to be replaced

n	R_i (₹)	v^{i-1}	$R_i\,v^{i-1}$ (₹)	$\sum R_i v^{i-1}$ (₹)	Total cost = $C + \sum_{i=1}^{n} R_i v^{i-1}$ (₹)	$\sum v^{i-1}$ (₹)	Average cost = $AC = \dfrac{TC}{\sum v^{i-1}}$ (₹)
1	0	1	0	0	5,000	1	5,000
2	500	0.5	250	250	5,250	1.5	3,500
3	1,000	0.25	250	500	5,500	1.75	3,142
4	1,500	0.125	187.5	687.5	5,687	1.875	3,033
5	2,000	0.0625	125	812.5	5,812	1.9375	3,000
6	2,500	0.0312	78.12	890.6	5,890	1.96	2,992
7	3,000	0.156	46.87	930	5,936	1.98	2,992.39

Minimum average cost is at the end of 6th year. So replace at the end of 6th year.

Problem 31.5: A truck has been purchased at a cost of ₹ 1,60,000. The value of the truck is depreciated in the first three years by ₹ 20,000 per year and 16,000 per year thereafter. If the maintenance and operating costs for the first three years are ₹ 16,000, ₹ 18,000 and ₹ 20,000 in that order and increase by ₹ 4,000 every year. Assume interest rate of 10%. Find the economic life of the truck.

Solution: Initial cost $= C = ₹\ 1,60,000$

$$r = 10\% = 0.1$$

$$v = \frac{1}{1+r} = \frac{1}{1+0.1} = 0.909$$

Table 31.7 To determine the year at which truck is to be replaced

| n | R_i | S_i | v^{i-1} | $R_i\,v^{i-1}$ | v^i | $S_i\,v^i$ | $(C + \Sigma R_i\,v^{i-1})$ | Total cost $= (C + \Sigma R_i\,v^{i-1} - S_i\,v^i)$ $(₹)$ | Σv^{i-1} | $AC = \dfrac{TC}{\Sigma v^{i-1}}$ |
	$(₹)$	$(₹)$		$(₹)$		$(₹)$	$(₹)$			$(₹)$
1	16,000	1,40,000	1.000	16,000	0.9091	1,27,260	1,76,000	48,740	1.00	48,740
2	18,000	1,20,000	0.909	16,362	0.8264	99,120	1,92,363	92,243	1.9091	48,841
3	20,000	1,00,000	0.826	16,520	0.7510	75,000	2,08,891	1,33,761	2.7355	48,907
4	24,000	84,000	0.751	18,031	0.6830	57,372	2,26,923	1,69,557	3.4868	**48,633**
5	28,000	68,000	0.683	19,124	0.6209	42,160	2,46,047	2,03,819	4.1698	48,885
6	32,000	52,000	0.620	19,872	0.5645	29,328	2,65,919	2,36,565	4.7907	49,490
7	36,000	36,000	0.564	20,322	0.5132	18,473	2,86,241	2,67,761	5.3552	50,048
8	40,000	20,000	0.513	20,528	0.4665	9,330	3,06,769	2,97,449	5.8684	50,690

Minimum average cost is at the end of 4th year. So replace at the end of 4th year.

Problem 31.6: A machine which requires an initial investment of $₹\ 12,000$ has its salvage value at the end of the year i is $₹\ (7,000 - 500\,(i - 1))$. The operating and maintenance costs are given in Table 31.8.

Table 31.8 The operating and maintenance costs

i	1	2	3	4	5	6	7	8	9
$R_i\,(₹)$	1,100	1,300	1,700	2,100	2,300	2,700	3,100	3,500	3,900

Determine optimal replacement year when money increases by 12% every year.

Solution: Initial cost $= C = ₹\ 12,000$

$$r = 12\% = 0.12$$

$$v = \frac{1}{1+r} = \frac{1}{1+0.12} = 0.892$$

Table 31.9 To determine the year at which machine is to be replaced

| i | R_i | S_i | v^{i-1} | $R_i\,v^{i-1}$ | $\Sigma R_i\,v^{i-1}$ | Σv^{i-1} | v^i | $S_i\,v^i$ | Total cost $= (C + \Sigma R_i\,v^{i-1} - S_i\,v^i)$ | $AC = \dfrac{TC}{\Sigma v^{i-1}}$ |
	$(₹)$	$(₹)$		$(₹)$	$(₹)$			$(₹)$		$(₹)$
1	1,100	7,000	1.000	1,100	1,100	1.00	0.892	6,244	6,856	6,856
2	1,300	6,500	0.892	1,160	2,260	1.892	0.797	5,180	9,080	4,799
3	1,700	6,000	0.797	1,355	3,615	2.689	0.711	4,266	11,349	4,221
4	2,100	5,500	0.711	1,493	5,108	3.4	0.635	3,493	13,615	4,004
5	2,300	5,000	0.635	1,461	6,569	4.035	0.567	2,835	15,734	3,899
6	2,700	4,500	0.567	1,531	8,100	4.602	0.506	2,277	17,823	**3,873**
7	3,100	4,000	0.506	1,569	9,669	5.108	0.452	1,808	19,861	3,888
8	3,500	3,500	0.452	1,582	11,251	5.56	0.404	1,414	21,837	3,928
9	3,900	3,000	0.404	1,570	12,821	5.964	0.360	1,081	23,746	3,981

Minimum average cost is at the end of 6th year. So replace at the end of 6th year.

31.8 MODEL 3

Replacement policy of items that fail completely.

Examples: Bulbs, Resistors, Transistors, etc.

Assumptions made in this model:

1. The system contains large number of identical items.
2. Individual replacement cost per item is greater than group replacement cost per item.

There are two types of replacement policies:

1. Individual replacement policy
2. Group replacement policy

Individual replacement policy:

Under this policy an item is replaced as soon as it fails. The number of items failed per unit time is determined using Mortality theorem.

According to mortality theorem,

$$\text{Number of failures/unit time} = \frac{\text{Total number of items}}{\text{Average age of the item}}$$

Group replacement policy:

Under this policy, items are replaced individually as and when it fails till optimum period T. After time T all the items are replaced irrespective of the fact that items have failed or items have not failed. Here the problem is to find optimum period T.

31.9 SOLVED PROBLEMS

Problem 31.7: A computer contains 10,000 resistors. When any resistor fails, it is replaced. The cost of replacing a resistor individually is ₹ 10 only. If all the resistors are replaced at a time, the cost of resistor would be reduced to ₹ 3.50. The percent surviving by the end of month t is given in Table 31.8.

Table 31.10 The percent surviving by the end of month t

Month (t)	0	1	2	3	4	5	6
% Surviving by the end of the month	100	97	90	70	30	15	0

What is the optimum replacement plan?

Solution:

Table 31.11 Probability of failure in each month

Month (t)	1	2	3	4	5	6
% Surviving by the end of the month	97	90	70	30	15	0
% failing at the end of month	3	10	30	70	85	100
% Failing in each month	3	7	20	40	15	15
Probability of failure in each month	0.03	0.07	0.2	0.4	0.15	0.15

Total number of resistors in a computer = 10,000
The cost of replacing single resistor = ₹ 10
For group replacement the cost per resistor = ₹ 3.50
Individual replacement policy:
Average age of the resistor = $0.03 \times 1 + 0.07 \times 2 + 0.2 \times 3 + 0.4 \times 4 + 0.15 \times 5 + 0.15 \times 6$
$= 4.02$ months

Average number of resistors failed in each month = $\dfrac{10,000}{4.02}$ = 2,488

Individual replacement cost = 2,488 × 10 = ₹ 24,880/month

Group replacement policy

Number of resistors failed in each month is given below:

1st Month = N_1 = 10,000 × 0.03 = 300

2nd Month = N_2 = 10,000 × 0.07 + 300 × 0.03 = 709

3rd Month = N_3 = 10,000 × 0.2 + 300 × 0.07 + 709 × 0.03 = 2,042

4th Month = N_4 = 10,000 × 0.4 + 300 × 0.2 + 709 × 0.07 + 2,042 × 0.03 = 4,171

5th Month = N_5 = 10,000 × 0.15 + 300 × 0.4 + 709 × 0.2 + 2,042 × 0.07 + 4,171 × 0.03
= 2,030

6th Month = N_6 = 10,000 × 0.15 + 300 × 0.15 + 709 × 0.4 + 2,042 × 0.2 + 4,171 × 0.07 +
2,030 × 0.03 = 2,590

To determine optimum group replacement period:

Table 31.12 To determine month at which all the resistors are to be replaced

Month	Number of failures in each month	Cumulative number of failures at the end of month	Individual replacement cost (₹)	Total cost = Individual replacement cost + Group replacement cost (₹)	Average cost = $\dfrac{\text{Total cost}}{\text{Month}}$ (₹)
1	300	300	3,000	35,000 + 3,000 = 38,000	38,000
2	709	1,009	10,090	45,090	22,545
3	2,042	3,051	30,510	65,510	**21,836**
4	4,171	7,222	72,220	1,07,220	26,805
5	2,030	9,252	92,520	1,27,520	25,504
6	2,590	11,842	1,18,420	1,53,420	25,570

Minimum group replacement cost/month = ₹ 21,836

From the above table, optimum group replacement period is at the end of 3rd month. Group replacement cost per month is less than individual replacement cost per month. So decision is followed group replacement policy.

Problem 31.8: Table 31.13 gives mortality rate for a certain type of light bulbs.

Table 31.13 Mortality rate for a certain type of light bulbs

Week (t)	1	2	3	4	5
% Failing at the end of week	10	25	50	80	100

There are 1,000 bulbs in use and it costs ₹ 10 to replace an individual bulb, which has burnt out. If all bulbs were replaced simultaneously, it would ₹ 4 per bulb. It is proposed to replace all the bulbs at fixed intervals, whether or not they have burnt out and continues to replace the bulbs as and when they fail. At what intervals all the bulbs should be replaced. At what group replacement, price per bulb would a policy of strictly group replacement become preferable to the individual replacement policy?

Solution: Number of bulbs = 1,000

The cost of replacing single bulb = ₹ 10

For group replacement cost per bulb = ₹ 4

Table 31.14 Probability of failure in each week

Week (t)	1	2	3	4	5
% Failing at the end of week	10	25	50	80	100
Number of bulbs failed in each week	10	15	25	30	20
Probability of failure in each week	0.1	0.15	0.25	0.3	0.2

Individual replacement policy:

Average age of the bulb $= 0.1 \times 1 + 0.15 \times 2 + 0.25 \times 3 + 0.3 \times 4 + 0.2 \times 5 = 3.35$ weeks

Average number of bulbs failed in each week $= \dfrac{1,000}{3.35} = 299$

Individual replacement cost $= 299 \times 10 = ₹\ 2,990$/week

Group replacement policy

Number of bulbs failed in each week is given below:

1st Week $= N_1 = 1,000 \times 0.10 = 100$

2nd Week $= N_2 = 1,000 \times 0.15 + 100 \times 0.1 = 160$

3rd Week $= N_3 = 1,000 \times 0.25 + 100 \times 0.15 + 160 \times 0.1 = 281$

4th Week $= N_4 = 1,000 \times 0.3 + 100 \times 0.25 + 160 \times 0.15 + 281 \times 0.1 = 377$

5th Week $= N_5 = 1,000 \times 0.2 + 100 \times 0.3 + 160 \times 0.25 + 281 \times 0.15 + 377 \times 0.1 = 350$

6th Week $= N_6 = 1,000 \times 0 + 100 \times 0.2 + 160 \times 0.3 + 281 \times 0.25 + 377 \times 0.15 + 350 \times 0.1$
$= 230$

Table 31.15 To determine optimum group replacement period

Week	Number of bulbs failed in each week	Cumulative number of bulbs failed	Individual replacement cost (₹)	Total cost = Individual replacement cost + Group replacement cost (₹)	Average cost = Total cost / Week (₹)
1	100	100	1,000	5,000	5,000
2	160	260	2,600	6,600	3,300
3	281	541	5,410	9,410	**3,136**
4	377	918	9,180	13,180	3,295
5	350	1,268	12,680	16,680	3,336

From the above table, optimum group replacement period is at the end of 3rd week. Group replacement cost per week is more than individual replacement cost per week. So optimum replacement policy is individual replacement policy.

Determining group replacement price per bulb at which group replacement is economical.

In order that group replacement is economical then average cost should be less than ₹ 2,990. Then we have to reduce group replacement cost per bulb.

$$\frac{1,000x + 541 \times 10}{3} = 2,990$$

$1,000x + 5,410 = 8,970$

$1,000x = 3,560$

$x = ₹\ 3.56$

Group replacement cost/bulb is less than ₹ 3.56 group replacement is preferred.

Application to maintenance problem

Problem 31.9: Let $p(t)$ be the probability that a machine in a group of 30 machines would breakdown in period 't'. The cost of repairing broken machine is ₹ 200. Preventive maintenance is performed by servicing all the 30 machines at the end of 'T' units of time. Preventive maintenance cost is ₹ 15 per machine. Find optimal 'T' that will minimize the expected total cost per period of servicing given that,

$$p(t) = \begin{cases} 0.03 & \text{for } t = 1 \\ p(t-1) + 0.01 & \text{for } t = 2, 3, ..., 10 \\ 0.13 & \text{for } t = 11, 12, 13 \end{cases}$$

Solution:

Total number of machines = $N = 30$

Preventive maintenance cost for all 30 machines = $15 \times 30 = ₹\ 450$

Cost of repairing broken machine = ₹ 200

Table 31.16 Probability of failure in period t

t	1	2	3	4	5	6	7	8	9	10	11
$p(t)$	0.03	0.04	0.05	0.06	0.07	0.08	0.09	0.1	0.11	0.12	0.12

Number of breakdown machines in each period:

Period 1 = N_1 = 30 × 0.03 = 0.9 = 1 machine

Period 2 = N_2 = 30 × 0.04 + 0.03 × 9 = 1.227 = 1 machine

Period 3 = N_3 = 30 × 0.05 + 0.9 × 0.04 + 1.2 × 0.03 = 2 machines

Similarly,

Period 4 = N_4 = 2 machines

Period 5 = N_5 = 2 machines

Period 6 = N_6 = 3 machines

Table 31.17 To determine period at which all the machines are serviced

Period	Number of breakdown machines	Cumulative number at the end of period	Individual repair cost (₹)	Group cost (₹)	Total cost (₹)	Average cost = $\dfrac{\text{Total cost}}{\text{Periods}}$ (₹)
1	1	1	200	450	650	650
2	1	2	400	450	850	425
3	2	4	800	450	1,250	416
4	2	6	1,200	450	1,650	412
5	2	8	1,600	450	2,050	410
6	3	11	2,200	450	2,650	442

Group preventive maintenance at the end of 5th period.

31.10 CAPITAL INVESTMENT DECISION USING NET PRESENT VALUE (NPV)

Terms used:

Money: Money has value overtime. Spending ₹ 100 today is not equal to spending ₹ 100 after one year.

Final value: Value of money after n years

Suppose money is worth $r\%$ per year

Then Final value = Present value $(1 + r)^n$

Present value or Net Present value:

Net Present value = Final value $(1 + r)^{-n}$

Present worth factor (pwf): $(1 + r)^{-n}$ is called present worth factor

Discount factor (v): $(1 + r)^{-1}$ is called discount factor

Capital investment decision: Purchase the item whose net present value is minimum.

31.11 SOLVED PROBLEMS

Problem 31.10: The cost pattern for two machines A and B, when money value is not considered is given in Table 31.18.

Table 31.18 The cost pattern for two machines A and B

Year	Cash inflows for machine A (₹)	Cash inflows for machine B (₹)
1	900	1,400
2	600	100
3	700	700

a. Find the machine to be purchased.

b. Find which machine to be purchased when the money is worth 10%.

Solution:

  a. Cost of machine A for 3 years = 900 + 600 + 700 = ₹ 2,200

    Cost of machine B for 3 years = 1,400 + 100 + 700 = ₹ 2,200

    Costs of machine A and machine B are same. So either purchase machine A or machine B.

  b. $r = 0.1$

Net Present value of machine A = $900 + 600 (1 + r)^{-1} + 700 (1 + r)^{-2}$

        = $900 + 600 (1 + 0.1)^{-1} + 700 (1 + 0.1)^{-2}$ = ₹ 2,023.97

Net Present value of machine B = $1,400 + 100 (1 + r)^{-1} + 700 (1 + r)^{-2}$

        = $1,400 + 100 (1 + 0.1)^{-1} + 700 (1 + 0.1)^{-2}$ = ₹ 2,069.43

Net Present value of machine A is less than Net Present value of machine B. So purchase machine A.

Problem 31.11: A manual stamping machine is currently valued at ₹ 1,000 is expected to last 2 years and costs ₹ 4,000 per year to operate. An automatic stamping machine which can be purchased for ₹ 3,000 will last 4 years and can be operated at an annual cost of ₹ 3,000. If money carries the rate of interest 10% per annum, determine which stamping machine should be purchased.

Solution: Life of automatic stamping machine is two times the life of manual stamping machine. Life of the manual machine is 2 years. So consider cash outflows for 4 years.

  Net Present value of manual stamping machine

      = $5,000 + 4,000 (1 + r)^{-1} + 5,000 (1 + r)^{-2} + 4,000 (1 + r)^{-3}$

      = $5,000 + 4,000 (1 + 0.1)^{-1} + 5,000 (1 + 0.1)^{-2} + 4,000 (1 + 0.1)^{-3}$ = ₹ 15,773.8

  Net Present value of automatic stamping machine

      = $6,000 + 3,000 (1 + r)^{-1} + 3,000 (1 + r)^{-2} + 3,000 (1 + r)^{-3}$

      = $6,000 + 3,000 (1 + 0.1)^{-1} + 3,000 (1 + 0.1)^{-2} + 3,000 (1 + 0.1)^{-3}$ = 13,458.9.

Net Present value of automatic stamping machine is less than Net Present value of manual stamping machine. So purchase automatic stamping machine.

Problem 31.12: A person is considering to purchase a machine for his own factory; relevant data about alternative machines are given in Table 31.19.

Table 31.19 Cost data about alternative machines

	Machine A	Machine B	Machine C
Present investment (₹)	10,000	12,000	15,000
Total Annual cost (₹)	2,000	1,500	1,200
(Investment at the beginning of year)			
Life (years)	10	10	10
Salvage value (₹)	500	1,000	1,200

Select the best machine considering 12% normal rate of return. You are given that:

a. Single payment present worth factor 12% for 10 years = 0.322

b. Annual series present worth factor 12% for 10 years = 5.65

Solution: $r = 0.12$

Net Present value of machine A

$$= 10,000 + 2,000 + 2,000\,(1 + r)^{-1} + 2,000\,(1 + r)^{-2} + ... + 2,000\,(1 + r)^{-9}$$
$$-500\,(1 + r)^{-10}$$

$$= 10,000 + 2,000\,(1 + (1 + r)^{-1} + (1 + r)^{-2} + ... + (1 + r)^{-9}) - 500\,(1 + r)^{-10}$$

$$= 10,000 + 2,000 \times 5.650 - 500 \times 0.322 = ₹ 21,139$$

Net Present value of machine B = $12,000 + 1,500 \times 5.650 - 1,000 \times 0.322 = ₹ 20,153$

Net Present value of machine C = $15,000 + 1,200 \times 5.650 - 1,200 \times 0.322 = ₹ 21,393$

Net Present value of machine B is less. So purchase machine B.

Problem 31.13: If you wish to have return 10% per annum on your investment, which of the plan in Table 31.20 would you prefer?

Table 31.20 Given data

	Plan A	Plan B
First cost (₹)	75,000	75,000
Scrap value for 20 years (₹)	37,500	6,000
Excess annual revenue over annual disbursement (₹)	7,500	9,000

Given:

a. Annual series present worth factor at 10% for 20 years = 8.514

b. Single payment present worth factor at 10% for 20 years = 0.2472

Solution:

Here the problem is return on investment. So objective is maximizing returns.

In flows = +

Out flows = −

Net Present value of Plan A = $-75,000 + 37,500 \times 0.2472 + 7,500 \times 8.514 = ₹ -1,875$

Net Present value of Plan B = $-75,000 + 6,000 \times 0.2472 + 9,000 \times 8.514 = ₹ 3,109$

Return on plan B is more. So invest on plan B.

31.12 QUESTIONS AND PROBLEMS

1. What are the important replacement models? Explain.

2. A machine owner finds from his past record that the costs per year of maintaining a machine whose purchase price is ₹ 6,000 are given below:

Year	1	2	3	4	5	6	7	8
Resale price	3,000	1,500	750	375	200	200	200	200
Maintenance cost (₹)	1,000	1,200	1,400	1,800	2,300	2,800	3,400	4,000

Determine at which age replacement is due.

Answer: Replace at the end of the 5th year

3. A fleet owner finds from his past records that the cost per year of running a vehicle whose purchasing price is ₹ 5000 are given below:

Year	1	2	3	4	5	6	7	8
Maintenance cost (₹)	1,500	1,600	1,800	2,100	2,500	2,900	3,400	4,000
Resale price (₹)	3,500	2,500	1,700	1,200	800	500	500	500

At what age is replacement due?

Answer: Replace at the end of the 5th year

4. A truck is priced at ₹ 60,000 and running costs are estimated at ₹ 6,000 for each of the first four years, increasing by ₹ 2,000 per year in the fifth and subsequent years. If the money is worth 10% per year, when the truck should be replaced. Assume that the truck will eventually be sold for scrap at a negligible price.

Answer: Replace at the end of the 9th year

5. The following mortality rates have been observed for a certain type of light bulbs.

Week (*t*)	1	2	3	4	5
% Failing at the end of week	10	25	50	80	100

There are 100 bulbs in use and it costs ₹ 2 to replace an individual bulb, which has burnt out. If all bulbs were replaced simultaneously, it would cost 50 paise per bulb. It is proposed to replace all the bulbs at fixed intervals, whether or not they have burnt out and continues to replace the bulbs as and when they fail. At what intervals all the bulbs should be replaced? **Answer:** Replace at the end of the 2nd week

6. The probabilities P_n of failures just before age *n* are shown below. If individual replacement costs ₹ 1.25 and group replacement costs ₹ 0.50 per item, find the optimal replacement policy. Assuming there are 1,000 items in use.

n	1	2	3	4	5	6	7	8	9	10	11
P_n	0.01	0.03	0.05	0.07	0.10	0.15	0.20	0.15	0.11	0.08	0.05

Answer: After every 5 weeks

7. A computer contains 10,000 resistors. When any resistor fails, it is replaced. The cost of replacing a resistor individually is ₹ 1 only. If all the resistors are replaced at a time, the cost of resistor would be reduced to 50 paise. The percent surviving by the end of month *t* as follows:

Month (*t*)	1	2	3	4	5	6
% Surviving by the end of the month	97	90	70	30	15	0

What is the optimum replacement plan? **Answer:** After every 3 months

8. If you wish to have return 10 % per annum on your investment, which of the following plan would you prefer.

	Plan A	Plan B
First cost (₹)	2,00,000	2,50,000
Scrap value for 20 years (₹)	1,50,000	1,80,000
Excess annual revenue over annual disbursement (₹)	25,000	30,000

Given:
  a. Annual series present worth factor at 10% for 20 years = 8.514
  b. Single payment present worth factor at 10% for 20 years = 0.2472.

Part X

Operations Research Models

32

Introduction to Operations Research Models and Linear Programing

32.1 DEFINITIONS AND EXPLANATIONS OF OPERATIONS RESEARCH

There are number of definitions or explanations of operations research in the management literature. The subject of operations research is introduced by listing topics like linear programing, queuing theory, simulation, inventory models, etc.

Some of the definitions of operations research are given below:

1. Mathematical solutions to the management problems
2. Operations research is applied for decision theory.
3. Operations research is the scientific method applied for making operations decisions.

Subject of operations research is also called optimization techniques because operations research models consist of minimizing or maximizing objective function under the given constraints. These models are also useful for making decisions. That is why these models are also called decision support system models. Operations research models consist of mathematical structure and these models are solved by using mathematical techniques. These models are useful for solving management problems. That is why these models are called management science. Operations research techniques were first applied in defense operations. Every problem is a new one. That is why this subject is called operations research.

32.2 OVERVIEW OF OPERATIONS RESEARCH MODELS OR VARIOUS OPERATIONS RESEARCH MODELS

Operations research models fall under the category of mathematical models. OR models consist of symbols, notations, objective function and constraints. Important OR models are:

1. Linear programing models: In this model, objective function and constraints are in linear form. The problem is to minimize or maximize the given objective function subject to the constraints.

    Important Linear programing models:

    a. Allocation models: Determining optimum product mix. That means determining the number of units of each product to be manufactured such that profit is maximized after satisfying the constraints. Allocation problems are solved by using Graphical method, Simplex method, two phase Simplex method, Dual Simplex method and Dynamic programing.

Examples: BPL Company producing number of products likes televisions, tape recorders, washing machines, driers, refrigerators, etc. under the resource constraints. Here the product mix means to determine number of units of each product to be manufactured so that profit is maximized.

Blending problem: Optimum mix of chemicals so that total cost of production is minimum.

   b. Transportation model: It is also a special class of linear programing problem in which the objective is to transport a single commodity from various sources to different destinations at a minimum total cost. Transportation problem is solved by Transportation algorithm.

   c. Assignment model: Assignment model is a special class of linear programing problems in which the objective is to assign the same number of jobs to the same number of machines at a minimum total cost. Assignment problem is solved by Hungarian method of assignment.

2. Integer linear programing models: In this type of linear programing problem, variables take only integers. There are four important methods for getting solutions to the integer linear programing problems.
   a. Branch and bound method
   b. Gomary cutting plane method
   c. Dynamic programing
   d. Genetic algorithms

3. Non-linear programing models: If the objective function or one or more constraints are in non-linear form then the problem is called non-linear programing problem. There are number of methods available for getting solution to the non-linear programing problem. Some important methods to solve non-linear programing problems are:

   a. Lagrangian method   e. Separable programing
   b. Kuhn-Tucker conditions   f. Search methods
   c. Quadratic programing   g. Dynamic programing
   d. Convex programing   h. Genetic algorithms

4. Inventory models: Determining the quantity to be ordered in each order so that the total cost of inventory is minimized. Inventory models are divided into probabilistic, deterministic models, models with price breaks and models with restrictions.

5. Game theory models: It is a kind of conflict in which two or more persons are playing the game. Game theory models fall under the category of making decisions under uncertainty.

6. Sequencing models: Determining the sequence or order in which jobs are performed so that total elapsed time is minimum.

7. Replacement models: Determining optimum replacement age of capital items whose maintenance cost increases over time and replacement of items that fail completely. Replacement models are divided into probabilistic and deterministic models.

8. Simulation models: These models also have a mathematical structure, but they cannot be solved purely by using techniques or tools of mathematics. A simulation model is to study the behavior of the system by using random numbers.

9. Network models: There are three types of network problems.

    a. Minimum span problems

    b. Shortest route problems

    c. Maximum flow problems

10. PERT and CPM: These models are useful for planning, scheduling and controlling of the projects.

11. Dynamic programing is a method of decomposing the optimization problem into a number of sub-problems each of lower dimensionality. Any problem with optimizing function and constraints can be divided into a number of sub-problems with the help of constraints, then that type of problem can be solved by dynamic programing approach. In dynamic programing problem, computations are carried out in stages by breaking the problem into sub-problems. Since sub-problems are interdependent, a procedure must be devised to link computations that generate a feasible solution for the entire problem.

    Applications of Dynamic programing:

    a. Shortest route problems

    b. Non-linear programing problems

    c. Integer non-linear programing problems

    d. Cargo loading problems

    e. Determining optimum number of scientists in a research project

    f. Reliability problems

    g. Capital budgeting problems

    h. Linear programing problems

12. Genetic algorithms: These models are useful for solving non-linear, traveling salesman problems, multi-objective optimization problems, etc.

32.3 TYPES OF OPERATIONS RESEARCH MODELS OR CLASSIFICATION OF OPERATIONS RESEARCH MODELS OR NATURE OF OPERATIONS RESEARCH MODELS

Operations research models are classified into:

1. Deterministic, probabilistic and uncertainty models

2. Static and dynamic models

3. Descriptive and mathematical models

4. Simulation models

1. Deterministic, probabilistic and uncertainty models

    a. Deterministic models: In the deterministic models, everything is known in advance and perfect information is available.

    Examples: Linear programing models, some of inventory models, some of replacement models, etc.

    b. Probabilistic models: In probabilistic models, only probabilistic information is available.

    Examples: Some of inventory models, some of replacement models, simulation models, etc.

    c. Uncertainty models: Here no information is available. Even probabilistic information is not available.

    Examples: Introducing new product, game theory models, etc.

2. Static and dynamic models
   a. Static models: In probabilistic mathematical models or deterministic mathematical models, if it is assumed that parameters remain unchanged over time then the models are called static models.
   b. Dynamic models: In probabilistic mathematical models or deterministic mathematical models, if it is assumed that the parameter changes with time then the model is called dynamic model.
3. Descriptive and mathematical models
   a. Descriptive model: A descriptive model explains the various operations in a non-mathematical language and tries to defined functional relationship and interactions between various operations. Descriptive model serves as preliminary to the development of mathematical models.
      Example: Cost increases demand decreases and temperature increases pressure increases.
   b. Mathematical models or analytical models: These models have a specific mathematical structure and thus can be solved by known mathematical techniques.
      Example: A general linear programing problem can be solved by using Simplex method.
4. Simulation models
   These models also have a mathematical structure, but they cannot be solved purely by using techniques or tools of mathematics. A simulation model is to study the behavior of the system by using random numbers.

32.4 DECISION MAKING

Management is essentially a decision making process. No activity can take place without taking decisions. Every aspect of management function is decided by decisions. Decision making is a continuous process. Decision making, therefore, involves choice of the best alternative among the two or more alternatives to achieve the objective. Operation Research techniques are helpful in making decisions.

32.5 STEPS IN DECISION MAKING

1. Identify the problem
2. Collecting needed information
3. Listing available alternatives
4. Assessing and weighing the risk associated with each alternative
5. Selecting the best alternative
6. Implementing the decision

Decision making, therefore, involves the choice of action among two or more available alternatives to achieve the objective.

Examples:
1. Purchasing a television
2. Make or Buy decision
3. Inventory decision
4. Replacement decisions
5. Product mix decision
6. Scheduling decision
7. Plant location decisions, etc.

Operations research techniques are used for making decisions. That is why operations research models are also called decision support system models.

32.6 DEFINITION OF LINEAR EQUATION

Linear equation of n variables is the first degree equation of n variables.

Example: $\quad a_1 x_1 + a_2 x_2 + \ldots + a_n x_n = b$

A linear equation can be represented in the form

$$ax = b$$

Where

a = Coefficient of variables in equation

x = Column vector of variables

b = Right hand side of equation

n = Number of variables

$$
\begin{bmatrix} a_1 & a_2 & \ldots & a_n \end{bmatrix}
\begin{bmatrix} x_1 \\ x_2 \\ \vdots \\ x_n \end{bmatrix} = b
$$

32.7 APPLICATIONS OF LINEAR PROGRAMING MODELS

1. Allocation models: Determining optimum product mix. That means determining the number of units of each product to be manufactured such that profit is maximized after satisfying the constraints. Allocation problems are solved by using Graphical method, Simplex method, Two phase Simplex method, Dual Simplex method and Dynamic programing.

   Examples:
   i. BPL company producing number of products like televisions, tape recorders, washing machines, driers, refrigerators, etc. under resource constraints. Here the product mix means to determine number of units of each product to be manufactured so that profit is maximized.
   ii. Blending problem: Optimum mix of chemicals so that total cost of production is minimum.

2. Transportation model: Transportation model is a special class of linear programing problem in which the objective is to transport a single commodity from various sources to different destinations at a minimum total cost. Transportation problem is solved by using Transportation algorithm.

3. Assignment model: Assignment model is a special class of linear programing problems in which the objective is to assign the same number of jobs to the same number of machines at a minimum total cost. Assignment problem is solved by using Hungarian method of assignment.

4. Integer linear programing models: In this type of linear programing problem, variables take only integers. There are three important methods for getting solutions to the integer linear programing problems.

   a. Branch and bound method

     b. Gomary cutting plane method
     c. Dynamic programing

32.8 METHODS OF SOLVING LINEAR PROGRAMING PROBLEM

1. Graphical method
2. Simplex method
3. Transportation algorithm
4. Assignment algorithm

32.9 GRAPHICAL METHOD

Graphical method is used to solve the linear programing problem, if the number of variables in linear program problem is equal to 2. The objective of drawing graph is to find solution space which satisfies the given constraints.

    Graphical method algorithm:

1. Graph the constraints and determine solution space. The objective of drawing graph is to find solution space.
2. Determine extreme points of solution space.
3. Determine solutions corresponding to extreme points. Solution of extreme point which maximizes or minimizes the objective function is the optimal solution.

32.9.1 Solved Problems

Application of linear programing to production planning problem

Problem 32.1: A firm manufactures two types of products A and B and sells them at a profit of ₹ 2 on type A and ₹ 3 on type B. Each product is processed on two machines G and H. Type A requires one minute of processing time on G and two minutes on H. Type B requires one minute on G and one minute on H. The machine G is available for not more than 6 hours 40 minutes while machine H is available for 10 hours during any working day. Solve the problem.

Solution:

    Let x_1 be the number of units of type A is to be manufactured

        x_2 be the number of units of type B is to be manufactured

        $z = 2x_1 + 3x_2$

    Subject to:

        $x_1 + x_2 \leq 400 \rightarrow$ (1) Machine G constraint

       $2x_1 + x_2 \leq 600 \rightarrow$ (2) Machine H constraint

          $x_1 \geq 0; x_2 \geq 0$

Number of variables in a given problem is equal to 2. So using graphical method can solve the problem. The objective of drawing graph is to find the solution space, which satisfies the given constraints 1 and 2.

Graphing the inequalities:

In the constraint equation 1

Let $x_1 = 0$; then $x_2 = 400$

Let $x_2 = 0$; then $x_1 = 400$

Joining these points gives us a line (Fig. 32.1) whose equation is $x_1 + x_2 = 400$. However, we are interested to find the solution which satisfies the constraints $x_1 + x_2 \leq 400$ and $x_1 \geq 0$; $x_2 \geq 0$. All the points on the line and below this line satisfies constraint $x_1 + x_2 \leq 400$. All the points in the shaded area satisfy $x_1 + x_2 \leq 400$ and $x_1 \geq 0$; $x_2 \geq 0$. The solution area is shown in Fig. 32.1.

Similarly, we are interested to find the solution space which satisfies the constraints $2x_1 + x_2 \leq 600$ and $x_1 \geq 0$; $x_2 \geq 0$. All the points on the line and below this line satisfy constraint $2x_1 + x_2 \leq 600$. All the points in the shaded area satisfy $2x_1 + x_2 \leq 600$ and $x_1 \geq 0$; $x_2 \geq 0$. The solution space is shown in Fig. 32.2.

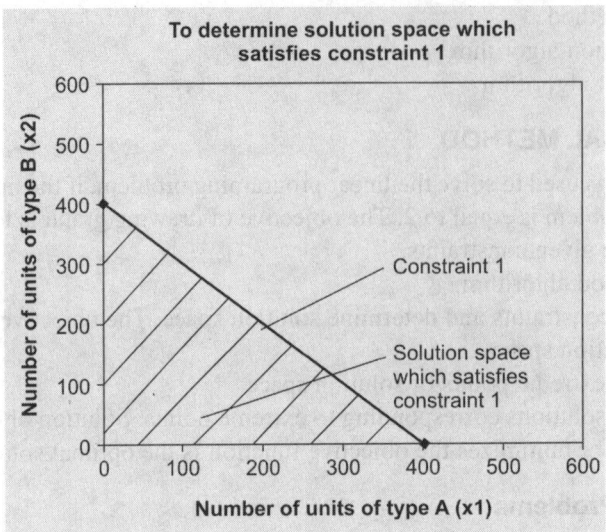

Fig. 32.1

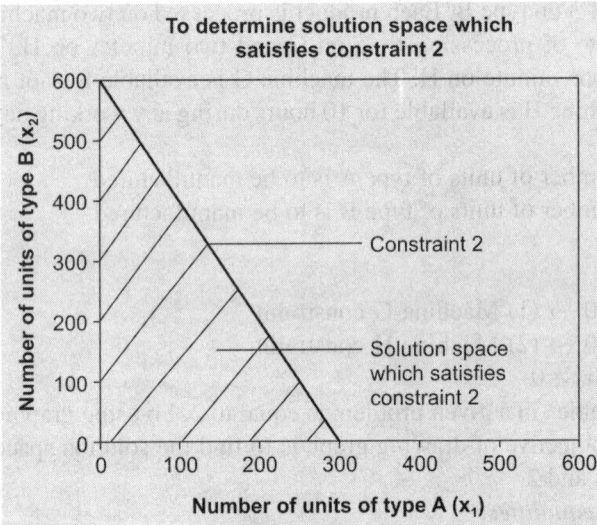

Fig. 32.2

But actual problem is to find the solution space which satisfies the constraints $x_1 + x_2 \leq 400$; $2x_1 + x_2 \leq 600$ and $x_1 \geq 0$; $x_2 \geq 0$. All the points in the shaded area satisfy $x_1 + x_2 \leq 400$; $2x_1 + x_2 \leq 600$ and $x_1 \geq 0$; $x_2 \geq 0$. The solution space is shown in Fig. 32.3. So there are number of solutions in the solution space.

Optimal solution: There are infinite number of solutions, which satisfy constraint 1 and constraint 2 and also non-negative constraints. According to the theory of linear programing, one of the solutions corresponding to the extreme point gives the best solution, which maximizes the profit. In Fig. 32.3, there are four extreme points for the solution space. They are A, B, C and D. One of these points gives the best solution.

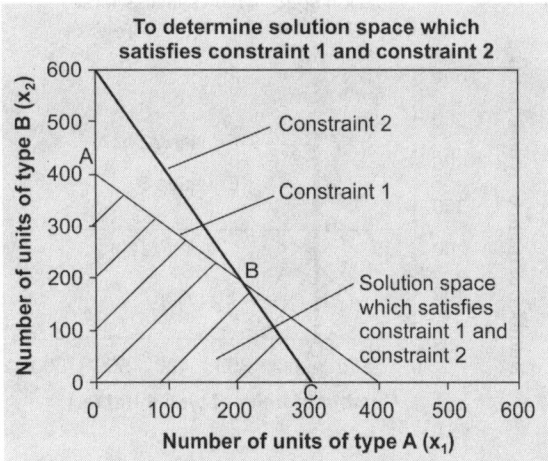

Fig. 32.3

Table 32.1 To determine optimal solution

Extreme point	Solution (x_1, x_2)	Value of objective function Maximize $z = 2x_1 + 3x_2$
A	(0, 400)	₹ 1,200
B	(200, 200)	₹ 1,000
C	(300, 0)	₹ 600
D	(0, 0)	₹ 0

Optimal solution:

$$x_1 = 0; \; x_2 = 400 \text{ and } z = ₹ \; 1,200$$

Application of linear programing problem to production planning and marketing problem

Problem 32.2: A company produces two types of hats. Labor time required to manufacture type 1 hat is two times more than type 2 hat. If all hats are of the second type only, the company can produce a total of 500 hats a day. The market limits daily sales of the first and second types to 150 and 250 hats. The profits per hat are ₹ 8 for type 1 and ₹ 5 for type 2. Formulate the problem as linear programing model in order to determine the number of hats to be produced of each type so as to maximize the profit. Also determine optimal solution.

Solution:

Let x_1 be the number of units of type 1 hat

x_2 be the number of units of type 2 hat

Maximize $z = 8x_1 + 5x_2$

Subject to:

$$2x_1 + x_2 \leq 500 \rightarrow (1) \text{ Labor time constraint}$$
$$x_1 \leq 150 \rightarrow (2) \text{ Marketing constraint}$$
$$x_2 \leq 250 \rightarrow (3) \text{ Marketing constraint}$$
$$x_1 \geq 0; x_2 \geq 0$$

Number of variables in a given problem is equal to 2. So using graphical method can solve the problem. The solution space is shown in Fig. 32.4.

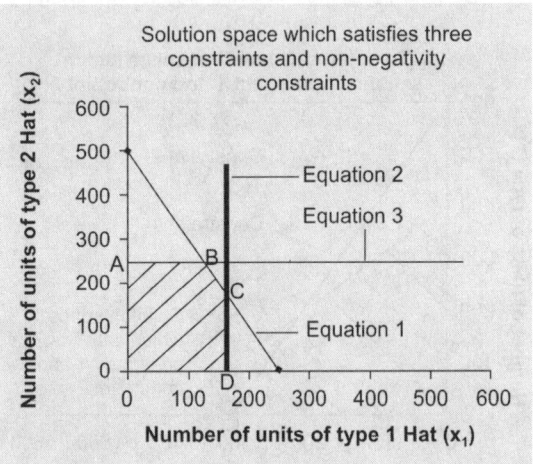

Fig. 32.4

Table 32.2 To determine optimal solution

Extreme point	Solution (x_1, x_2)	Value of objective function Maximize $z = 8x_1 + 5x_2$
A	(0, 250)	₹ 1,250
B	(125, 250)	₹ 2,250
C	(150, 200)	₹ 2,200
D	(150, 0)	₹ 1,200
E	(0, 0)	₹ 0

Optimal solution:

$x_1 = 125$; $x_2 = 250$ and $z = $ ₹ 2,250

Application of linear programing to blending problem

Problem 32.3: The manager of an oil refinery must decide on the optimal mix of two possible blending processes of which the inputs and outputs per production run are given in Table 32.3.

Table 32.3 Inputs and outputs per production run

Process	Input (units)		Output (units)	
	Crude A	Crude B	Gasoline X	Gasoline Y
1	5	3	5	8
2	4	5	4	4

The maximum amount available of crude A and crude B are 200 units and 150 units respectively. Market requirements show that at least 100 units of gasoline X and 80 units of gasoline Y must be produced. The profits per production run from process 1 and process 2 are ₹ 300 and ₹ 400 respectively. Solve linear programing problem.

Solution:

Let x_1 be the number of production runs of process 1

x_2 be the number of production runs of process 2

Maximize $z = 300x_1 + 400x_2$

Subject to:

$$5x_1 + 4x_2 \geq 200 \rightarrow (1) \text{ Crude A constraint}$$
$$3x_1 + 5x_2 \leq 150 \rightarrow (2) \text{ Crude B constraint}$$
$$5x_1 + 4x_2 \geq 100 \rightarrow (3) \text{ Gasoline X constraint}$$
$$8x_1 + 4x_2 \geq 80 \rightarrow (4) \text{ Gasoline Y constraint}$$
$$x_1 \geq 0; x_2 \geq 0$$

Number of variables in a given problem is equal to 2. So using graphical method can solve the problem. The solution space is shown in Fig. 32.5.

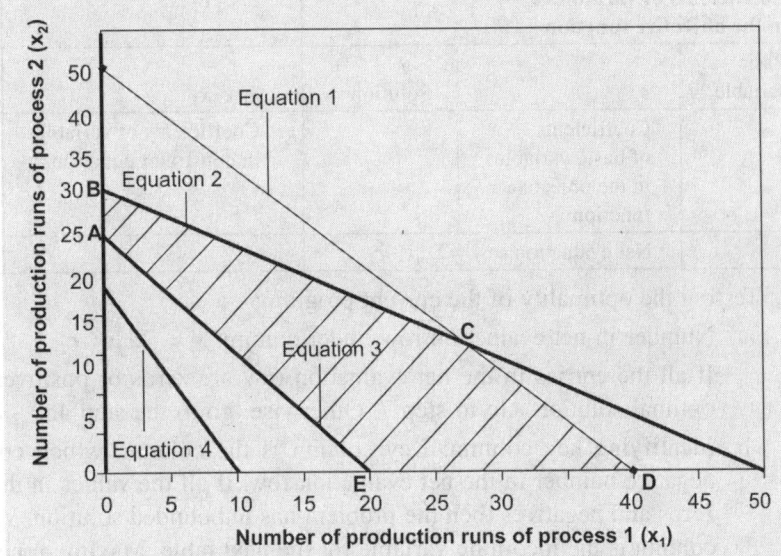

Fig. 32.5 Solution space

Table 32.4 To determine optimal solution

Extreme point	Solution (x_1, x_2)	Value of objective function Maximize $z = 300x_1 + 400x_2$
A	(0, 25)	₹ 10,000
B	(0, 30)	₹ 12,000
C	(30.76, 11.5)	₹ 13,846
D	(40, 0)	₹ 12,000
E	(20, 0)	₹ 6,000

Optimal solution:

$$x_1 = 30.76; x_2 = 11.5 \text{ and } z = ₹ \ 13,846$$

32.10 SIMPLEX METHOD

Simplex method is used to solve the linear programing problem, if the number of variables in linear programing problem is equal to 2 or more than 2 of the various methods of solving linear programing problem, the Simplex method is the most general and powerful. The Simplex method is based on the property that the optimum solution to a linear programing problem, if it exists can always be found in one of the basic feasible solutions. Thus in the Simplex method, the first step is always to find a basic feasible solution.

Simplex algorithm

Steps:

1. First convert the given linear programing problem into maximization problem.
2. Convert inequalities into equalities by adding slack variables.
3. $i = 1$

   Design of the Table 1

   (Other Simplex Tables will have similar interpretations.)

Table 1 in Simplex method

Coefficients of variables $c_i \rightarrow$ in the objective function			$c_1\ c_2\ c_3 \dots$	Replacement ratio
Basic variable	c_j	Solution	$x_1\ x_2\ x_3 \dots$	
	Coefficients of basic variables in the objective function		Coefficients of variables in constraint equations	
	Net evaluation row $= \Sigma c_j x_j \cdot c_i$			

4. Testing the optimality of the current program:
   a. Number in net evaluation row under column $x_j = \Sigma c_j x_j \cdot c_i$

      If all the entries in the net evaluation row are zeros or positives then solution is optimal solution. Go to step 7. Otherwise, go to the step 4b.

   b. Identifying key column: Key column is the column, which contains maximum negative number in the net evaluation row. If all the values in the key column are zeros and negatives then the problem has unbounded solution. Variable in the key column is the incoming variable for the next table. Maximum negative number in net evaluation row under two or more variables is same then there is a tie in terms of selection of the key column. Then take any one of the tied columns as key column.

   c. Replacement ratio: To get the number in the replacement ratio divide solution column with corresponding positive entry or zero in key column.

   d. Identifying key row: The row that contains minimum of (Positive, 0) replacement ratio is the key row. Variable in the key row is an outgoing variable. Minimum replacement ratio for two or more variables may be same. If this happens, there is a tie in terms of selection of the key row. Then take any one of the tied rows as key row.

   e. Identifying key number: Key number is a number at the intersection of key row and key column. Key number is represented in the rectangle.

5. Design of next table (Table $(i + 1)$):
   a. Replacement of key row: To replace key row divide all the entries in key row by key number.

   b. Replacement of non-key row: To replace non-key row first find fixed ratio for that non-key row.

   $$\text{Fixed ratio} = \text{FR} = \frac{\text{Number in the key column}}{\text{Key number}}$$

Number in Table $(i + 1)$ = Number in Table i – Corresponding number in key row in Table $i \times$ FR

6. Go to step 4 and repeat the steps until obtaining an optimum solution.
7. Optimal solution is reached.

32.10.1 Solved Problems

Problem 32.4: A farmer has 1,000 acres of land on which he can grow corn, wheat and soyabeans. Each acre of corn costs ₹ 100 for preparation requires 7 man-days of work and yields a profit of ₹ 30, an acre of wheat costs ₹ 120 to prepare for preparation, requires 10 man-days of work and yields a profit of ₹ 40, and an acre of soyabeans costs ₹ 70 to prepare and requires 8 man-days of work and yields a profit of ₹ 20. If the farmer has ₹ 1,00,000 for preparation and can count on 8,000 man-days of work, how many acres should be allotted to each crop to maximize profit?

Solution:

Let x_1 be the number of acres of corn

   x_2 be the number of acres of wheat

   x_3 be the number of acres of soyabeans

Converting the given data into linear programing problem

Maximize $z = 30x_1 + 40x_2 + 20x_3$

Subject to the constraints:

$$100x_1 + 120x_2 + 70x_3 \leq 1,00,000$$

or

$$10x_1 + 12x_2 + 7x_3 \leq 10,000$$
$$7x_1 + 10x_2 + 8x_3 \leq 8,000$$
$$x_1 + x_2 + x_3 \leq 1,000$$
$$x_1 \geq 0; x_2 \geq 0; x_3 \geq 0$$

Number of variables in the given problem is equal to 3. So the problem can be solved by using Simplex method.

Step 1: The given problem is a maximization problem.

Step 2: Here constraints are "$\leq$" type. So add slack variables to convert inequalities into equalities. Coefficients of slack variables in the objective function are zeros.

Maximize $z = 30x_1 + 40x_2 + 20x_3 + 0s_1 + 0s_2 + 0s_3$

Subject to the constraints:

$$10x_1 + 12x_2 + 7x_3 + s_1 = 10,000 \rightarrow (1)$$
$$7x_1 + 10x_2 + 8x_3 + s_2 = 8,000 \rightarrow (2)$$
$$x_1 + x_2 + x_3 + s_3 = 1,000 \rightarrow (3)$$
$$x_1 \geq 0; x_2 \geq 0; x_3 \geq 0; s_1 \geq 0; s_2 \geq 0; s_3 \geq 0$$

Number of variables in the problem is now equal to 6. Number of equations is 3. So number of basic variables in the solution is 3. The remaining variables have zero values.

Simplest basic feasible solution is,

Putting $x_1 = 0; x_2 = 0; x_3 = 0$

Then $s_1 = 10,000; s_2 = 8,000; s_3 = 1,000$ and $z = 0$. This program is given in Table 1.

Table 1 Initial simplex table

Coefficients of variables in the objective function $c_i \rightarrow$			30	40	20	0	0	0	Replacement ratio
Basic variable	c_j	Solution	x_1	x_2	x_3	s_1	s_2	s_3	
s_1	0	10,000	10	12	7	1	0	0	833.3
$\leftarrow s_2$	0	8,000	7	10	8	0	1	0	800 (key row)
s_3	0	1,000	1	1	1	0	0	1	1,000
Net evaluation row = $\Sigma c_j x_j - c_i$			−30	−40↑	−20	0	0	0	$z = 0$
				Key column					

Table 2 Revised simplex table from Table 1

Coefficients of variables in the objective function $c_i \rightarrow$			30	40	20	0	0	0	Replacement ratio
Basic variable	c_j	Solution	x_1	x_2	x_3	s_1	s_2	s_3	
$\leftarrow s_1$	0	400	1.6	0	−2.6	1	−1.2	0	250 (key row)
s_2	40	800	0.7	1	0.8	0	0.1	0	1,142
s_3	0	200	0.3	0	0.2	0	−0.1	1	666.6
Net evaluation row = $\Sigma c_j x_j - c_i$			−2↑	0	12	0	4	0	$z = 32,000$
			Key column						

Table 3 Revised simplex table from Table 2

Coefficients of variables in the objective function $c_i \rightarrow$			30	40	20	0	0	0
Basic variable	c_j	Solution	x_1	x_2	x_3	s_1	s_2	s_3
$\leftarrow x_1$	30	250	1	0	−1.625	0.625	−0.75	0
x_2	40	625	0	1	1.93	−0.4375	0.625	0
x_3	0	125	0	0	0.387	−0.1875	0.125	1
Net evaluation row = $\Sigma c_j x_j - c_i$			0	0	8.75	0.125	2.5	0

All the values in the net evaluation row are positives or zeros. So the Table 3 is an optimal solution.

Optimal solution from Table 3,

$$x_1 = 250;\ x_2 = 625;\ x_3 = 0;\ s_1 = 0;\ s_2 = 0;\ s_3 = 125;\ z = 32,500$$

Problem 32.5: You wish to export three products A, B and C. The amount available is ₹ 4,00,000. Product A costs ₹ 8,000 per unit and occupies 30 cubic feet (after packing). Product B costs ₹ 13,000 and occupies 60 cubic feet and product C costs ₹ 15,000 per unit and occupies 60 cubic feet. The profits per unit of A is ₹ 1,000, B is ₹ 1,500 and C is ₹ 2,000. The shipping company can accept a maximum of 30 packages and has a storage space of 1,500 cubic feet. How many of each product, if he bought and shipped to maximize the profit? The export potential for each product is unlimited.

Solution:

Let x_1 be the number of units of product A

x_2 be the number of units of product B

x_3 be the number of units of product C

Converting the given data into linear programing problem,

Maximize $z = 1,000x_1 + 1,500x_2 + 2,000x_3$

Subject to the constraints:

$$8,000x_1 + 13,000x_2 + 15,000x_3 \leq 4,00,000 \text{ or } 8x_1 + 13x_2 + 15x_3 \leq 400$$
$$30x_1 + 60x_2 + 60x_3 \leq 1,500 \text{ or } 3x_1 + 6x_2 + 6x_3 \leq 150$$
$$x_1 + x_2 + x_3 \leq 30$$

$x_1 \geq 0; x_2 \geq 0; x_3 \geq 0$

Number of variables in the given problem is equal to 3. So the problem can be solved by using Simplex method.

Step 1: The given problem is a maximization problem.

Step 2: Here constraints are "≤" type. So add slack variables to convert inequalities into equalities. Coefficients of slack variables in the objective function are zeros.

Maximize $z = 1,000x_1 + 1,500x_2 + 2,000x_3 + 0s_1 + 0s_2 + 0s_3$

$$8x_1 + 13x_2 + 15x_3 + s_1 = 400 \rightarrow (1)$$
$$3x_1 + 6x_2 + 6x_3 + s_2 = 150 \rightarrow (2)$$
$$x_1 + x_2 + x_3 + s_3 = 30 \rightarrow (3)$$

$x_1 \geq 0; x_2 \geq 0; x_3 \geq 0 ; s_1 \geq 0; s_2 \geq 0; s_3 \geq 0$

Number of variables in the problem is now equal to 6. Number of equations is 3. So number of basic variables in the solution is 3. The remaining variables have zero values.

Simplest basic feasible solution is,

Putting $x_1 = 0; x_2 = 0; x_3 = 0$

Then $s_1 = 400; s_2 = 150; s_3 = 30$ and $z = 0$. This program is given in Table 1.

Table 1 Initial simplex table

Coefficients of variables in the objective function $c_i \rightarrow$			1,000	1,500	2,000	0	0	0	Replacement ratio
Basic variable	c_j	Solution	x_1	x_2	x_3	s_1	s_2	s_3	
s_1	0	400	8	13	15	1	0	0	26.67
$\leftarrow s_2$	40	150	3	6	6	0	1	0	25 (key row)
s_3	0	30	1	1	1	0	0	1	30
Net evaluation row $= \Sigma c_j x_j - c_i$			−1,000	−1,500	−2,000↑	0	0	0	$z = 0$
					Key column				

Table 2 Revised simplex table from Table 1

Coefficients of variables in the objective function $c_i \rightarrow$			1,000	1,500	2,000	0	0	0
Basic variable	c_j	Solution	x_1	x_2	x_3	s_1	s_2	s_3
s_1	0	25	0.5	−2	0	1	−2.5	0
x_3	2,000	25	0.5	1	1	0	0.167	0
s_3	0	5	0.5	0	0	0	−0.167	1
Net evaluation row $= \Sigma c_j x_j - c_i$			0	500	0	0	334	0

All the values in the net evaluation row are positives or zeros. So the Table 2 is an optimal solution.

Optimal solution from Table 2,

$x_1 = 0$; $x_2 = 0$; $x_3 = 25$; $s_1 = 25$; $s_2 = 0$; $s_3 = 5$; $z = ₹\ 50,000$

32.11 TRANSPORTATION PROBLEM

Definition: Transportation problem is a special class of linear programing problem in which the objective is to transport a single commodity from various sources to different destinations at a minimum total cost. In the transportation problem, total capacity is equal to total requirement. In Transportation problem, variables take only integers. That is why Transportation problem is also called integer linear programing problem.

Transportation algorithm

Transportation algorithm consists of five steps:
1. Balancing the given transportation problem, i.e. making Capacity = Requirement
2. Obtaining initial basic feasible solution. There are three important methods for obtaining initial basic feasible solution. These methods are:
   i. North-West corner rule
   ii. Least cost method or inspection method
   iii. Vogel's approximation method
3. Testing the optimality of initial basic feasible solution. There are two important methods for testing the optimality of basic feasible solution. They are:
   i. Stepping stone method
   ii. Modified distribution method
4. If the solution is not optimal, revise the basic feasible solution.
5. Repeating the steps 3 and 4 until obtaining of optimal solution.

Obtaining initial basic feasible solution:

North-West corner rule: According to this rule, the first allocation is made to the cell occupying North-West corner cell. That is to the first cell (1, 1). If the origin capacity is exhausted first then move down the first column. If the destination requirement is exhausted, we move right in row 1. In this manner starting from first cell of the given transportation matrix, after satisfying the column requirements and row capacities, we move towards the last cell until all requirements and capacities are satisfied.

Inspection method: In the inspection method, allocations are made to the lowest cost cells.

Vogel's approximation method to obtain initial basic feasible solution:

Steps:
1. According to this method, a difference column and a difference row represent the difference between two cheapest routes from origin to destination.
2. Identify highest difference.
3. First assignment is made to the cheapest cell in the highest difference column or difference row.
4. Remove column or row or both which is exhausted.
5. Repeat the steps from 1 to 4 until all allocations are completed.

Stepping stone method for determining optimal solution: Stepping stone method consists of the following steps:
1. First step in Stepping stone method is determining the opportunity cost of empty cells. Opportunity cost of empty cells is determined by forming closed loops with positive or

occupied cells. Closed loops are formed with horizontal and vertical lines. Closed loops with empty cells are formed with four or six or eight cells such that all corners must have occupied cells. Different closed loops must be drawn for each empty cell. The opportunity cost associated with each empty cell is calculated by putting cost for empty cell + and alternatively –, + for the costs of remaining occupied cells. If the opportunity costs of all the empty cells are positives or zeros the solution is an optimal solution. Otherwise we have to revise the solution. Positive opportunity cost means, if we add one unit to this cell the cost of transportation increases equal to the opportunity cost. Negative opportunity cost means, if we add one unit to this cell the cost of transportation decreases equal to the opportunity cost.

2. The second step in Stepping stone method is revising the solution. Identify the cells which is having negative opportunity cost. Next identify the cell, which is having maximum negative opportunity cost value. Form the closed loop with the cell, which has maximum negative opportunity cost. Starting with empty cell alternatively put +, – for the remaining occupied cells of closed loop. Determine the minimum quantity in – sign cell of closed loop. Add minimum quantity to the quantity wherever + sign and subtract from the quantity wherever – sign.

3. Repeating the steps 1 and 2 until obtaining of optimal solution.

Modified distribution method for obtaining optimal solution (MODI): Modified distribution method consists of the following steps:

1. First step in Modified distribution method is determining the opportunity cost of empty cells.

   For the occupied cell $c_{ij} = (u_i + v_j)$

   Opportunity cost of empty cells $= c_{ij} - (u_i + v_j)$

   c_{ij} = Cost of transporting one unit from i to j

   u_i = Row number i

   v_j = Column number j

   Initially assume row number or column number for any row or column. Then calculate remaining row numbers and column numbers. Then determine opportunity cost of empty cells using the equation below.

   Opportunity cost of empty cells $= c_{ij} - (u_i + v_j)$

   If the opportunity costs of all the empty cells are positives or zeros then the solution is optimal solution. Otherwise, we have to revise the solution. Positive opportunity cost means, if we add one unit to this cell the cost of transportation increases equal to the opportunity cost. Negative opportunity cost means, if we add one unit to this cell the cost of transportation decreases equal to the opportunity cost.

2. The second step in modified distribution method is revising the solution. Identify the cells which is having negative opportunity cost. Next, identify the cell which is having maximum negative opportunity cost value. Form the closed loop with cell, which has maximum negative value. Starting with empty cell alternatively put +, – for the remaining occupied cells of closed loop. Determine the minimum quantity in – sign cell of closed loop. Add minimum quantity to the quantity wherever + sign and subtract from the quantity wherever – sign.

3. Repeating the steps 1 and 2 until obtaining optimal solution.

32.11.1 Solved Problems

Problem 32.6: A furniture company has plants in cities A, B and C which ship to four demand locations 1, 2, 3 and 4 with transporting costs (in hundred rupees) as shown in Table 32.5. Determine minimum transportation cost.

Table 32.5 Transporting costs (in hundred rupees)

Supply plants	Demand locations				Capacity
	1	2	3	4	
A	3	5	7	4	50
B	6	8	5	2	50
C	1	9	7	3	50
Requirement	20	60	30	40	

Solution:

Step 1: The first step is Balancing the given transportation problem, i.e. making Total capacity = Total requirement. In the given problem Total capacity = Total requirement = 150. So the given problem is a balanced transportation problem.

Step 2: The second step is obtaining initial basic feasible solution. There are three important methods for obtaining initial basic feasible solution. We obtain initial basic feasible solution by using Vogel's approximation method. Initial basic feasible solution using Vogel's approximation method is given in Table 1. Number of positive cells in the solution is equal to $m + n - 1$ that is $3 + 4 - 1 = 6$. So the initial solution is a basic feasible solution.

Initial total cost of transportation = $50 \times 5 + 10 \times 5 + 40 \times 2 + 20 \times 1 + 10 \times 9 + 20 \times 7$
= 630

Step 3: Third step is testing the optimality of initial basic feasible solution. There are two important methods for testing the optimality of initial basic feasible solution. Here Modified distribution method is used for obtaining optimal solution. Calculations of row numbers and column numbers for the initial basic feasible solution are shown in Table 1.

Table 1 Initial basic feasible solution by Vogel's approximation method

Supply plants	Demand locations				Capacity	Row number (u_i)
	1	2	3	4		
A	3	5 ⑤⓪	7	4	50	0
B	6	8	5 ⑩ +	2 ④⓪ −	50	2
C	1 ②⓪	9 ⑩	7 ②⓪ −	3 +	50	4
Requirements	20	60	30	40	150	
Column number (v_j)	−3	5	3	0		

Opportunity costs of empty cells are determined for the initial basic feasible solution and the results are given in Table 2.

Table 2 Opportunity costs of empty cells

Empty cell	A1	A3	A4	B1	B2	C4
Opportunity cost = $c_{ij} - (u_i + v_j)$	0	4	4	7	1	–1

Empty cell C4 has negative opportunity cost of ₹ –1. So we have to revise the solution by adding quantity to the cell C4. Form the closed loop with cell C4 (Table 1). Starting with empty cell C4 alternatively put +, – for the remaining occupied cells of closed loop. Determine the minimum quantity in – sign cell of closed loop. Minimum quantity in – sign cell of closed loop is 20. Add minimum quantity 20 to the quantity wherever + sign, i.e. to the cells B3 and C4 and subtract from the quantity wherever – sign, i.e. from the cells B4 and C3. The revised solution is given in Table 3.

Table 3 Revised solution

| Supply plants | Demand locations | | | | Capacity | Row |
	1	2	3	4		number (u_i)
A	3	5 . ⓼50	7	4	50	0
B	6	8	5 ㉚30	2 ⓴20	50	3
C	1 ⓴20	9 ⑩10	7	3 ⓴20	50	4
Requirements	20	60	30	40	150	
Column number (v_j)	–3	5	2	–1		

Number of positive cells in the solution is equal to $m + n - 1$ that is $3 + 4 - 1 = 6$. So the revised solution is a basic feasible solution. Calculation of row numbers and column numbers for the revised solution are shown in Table 3. Opportunity costs of empty cells are determined for the revised solution and the results are given in Table 4.

Table 4 Opportunity costs of empty cells for the revised solution

Empty cell	A1	A3	A4	B1	B2	C3
Opportunity cost = $c_{ij} - (u_i + v_j)$	6	5	5	6	0	1

Opportunity costs of all empty cells are positive. So the solution is an optimal solution.

Minimum cost of transportation = $50 \times 5 + 30 \times 5 + 20 \times 2 + 20 \times 1 + 10 \times 9 + 20 \times 3 = 610$

Problem 32.7: A company has factories at A, B, C which supply warehouses at D, E, F and G. Monthly factory capacities are 160, 150 and 190 units respectively. Monthly warehouse requirements are 80, 90, 110 and 160 units respectively. Unit shipping costs are given in Table 32.6. Determine the optimum distribution for this company to minimize shipping cost.

Table 32.6 Unit shipping costs

To → From ↓	D	E	F	G
A	42	48	38	37
B	40	49	52	51
C	39	38	40	43

Solution: Step 1: In the given problem, Total capacity is more than Total Requirement. So the given problem is unbalanced transportation problem. The problem is balanced by adding dummy column whose requirement is (500 – 440 = 60 units). Here dummy column is H.

Step 2: The second step is obtaining initial basic feasible solution. There are three important methods for obtaining initial basic feasible solution. We obtain initial basic feasible solution by using Vogel's approximation method. Number of positive cells in the solution is equal to 6. Number of positive cells in the solution is less than $m + n - 1$ that is $3 + 5 - 1 = 7$. So the initial solution is a degenerate basic feasible solution. In order to make the number of occupied cells or positive cells equal to $m + n - 1$, convert one empty cell into positive cell by adding very small quantity 'e' to any empty cell so that number of occupied cells is equal to $m + n - 1$. The value of e is almost equal to zero. But this cell CG is considered as positive cell for making calculations.

Table 1 Initial basic feasible solution by Vogel's approximation method

To → From →	D	E	F	G	H	Capacity	Row number (u_i)
A	42	48	38	37 (160)	0	160	–18
B	40 (80)	49	52 (10)	51	0 (60)	150	0
C	39	38 (90)	40 (100)	43 (e)	0	190	–12
Requirement	80	90	110	160	60	500	
Column number (v_j)	40	50	52	55	0		

Initial cost of transportation $= 160 \times 37 + 80 \times 40 + 10 \times 52 + 60 \times 0 + 38 \times 90 + 40 \times 100$
$= 17,060$

Table 2 Opportunity costs of empty cells

Empty cell	AD	AE	AF	AH	BE	BG	CD	CH
Opportunity cost = $c_{ij} - (u_i + v_j)$	20	16	4	18	–1	–4	11	12

Empty cells BE and BG have negative opportunity costs of ₹ –1 and –4 respectively. Here the opportunity of saving is more for the cell BG. So form the closed loop with cell BG (Table 1). Starting with empty cell BG alternatively put +, – for the remaining occupied cells of closed loop. Determine the minimum quantity in – sign cell of closed loop. Minimum quantity in – sign cell of closed loop is e units. The value of e is almost equal to zero. Solution will not change. So there is no need to make this change. So form the closed loop with cell BE. Starting with empty cell BE alternatively put +, – for the remaining occupied cells of closed loop. Determine the minimum quantity in – sign cell of closed loop. Minimum quantity in – sign cell of closed loop is 10 units. Add minimum quantity to the quantity wherever + sign, i.e. to the cells AE and CF and subtract from the quantity wherever – sign, i.e. from the cells BF and CE. The revised solution is given in Table 3. Number of positive cells in the revised solution is equal to 6. Number of positive cells in the solution is less than $m + n - 1$ that is $3 + 5 - 1 = 7$. So the revised solution is a degenerate basic feasible solution. In order to make the number of occupied cells or positive cells equal to $m + n - 1$, convert one or more empty cells into positive cells by adding very small quantity 'e' to the empty cells so that number of occupied cells equal to $m + n - 1$. The value of e is almost equal to zero. But this cell CG is considered as positive cell for making calculations.

Table 3 Revised solution

To → From →	D	E	F	G	H	Capacity	Row number (u_i)
A	42	48	38	37 (160)	0	160	–6
B	40 (80)	49 (10) –	52	51 +	0 (60)	150	11
C	39	38 (80) +	40 (110)	43 (e) –	0	190	0
Requirement	80	90	110	160	60	500	
Column number (v_j)	29	38	40	43	–11		

Table 4 Opportunity costs of empty cells for the revised solution

Empty cell	AD	AE	AF	AH	BF	BG	CD	CH
Opportunity cost = $c_{ij} - (u_i + v_j)$	19	16	4	17	1	–3	10	11

Empty cell BG has negative opportunity costs of ₹ –3. So form the closed loop with cell BG. Starting with empty cell BG alternatively put +, – for the remaining occupied cells of closed loop. Determine the minimum quantity in – sign cell of closed loop. Minimum quantity in – sign cell of closed loop is e units. The value of e is almost equal to zero. Solution will not change. So there is no need to make this change. So the solution is an optimal solution.

Minimum total cost of transportation = $160 \times 37 + 80 \times 40 + 10 \times 49 + 60 \times 0 + 38 \times 80 + 40 \times 110 = 17,050$.

32.12 THE ASSIGNMENT MODEL

Definition: The assignment model deals with a special class of linear programing problem in which the objective is to assign the number of origins to the same number of destinations at a minimum total cost. The assignment is to be made on one-to-one basis. That is each origin is associated with one and only one destination. Here the payoff matrix is a square matrix. There is only one assignment in each row and column. In assignment problem, variables take 0 or 1. That is why assignment problems are called 0-1 integer linear programing problems.

32.12.1 Differences between Transportation and Assignment Models

Differences between transportation and assignment models are given in Table 32.7.

Table 32.7 Differences between transportation and assignment models

Transportation model	Assignment model
1. Special class of linear programing problem in which the objective is to minimize the transportation cost.	1. Special class of linear programing problem in which the objective is to minimize the cost of assignment.
2. May or May not be a square matrix	2. Always square matrix.
3. There may be more than one assignment in each row or column.	3. There is only one assignment in each row or column.
4. Variables take only integer values.	4. Variables take only 0 or 1.
5. Transportation problem can be solved by using transportation algorithm.	5. Assignment problem can be solved by using Hungarian method or assignment algorithm.

Assignment algorithm or Hungarian method

Steps:
1. Obtain row opportunity matrix by subtracting lowest entry from all entries in that row.
2. Obtain total opportunity matrix by subtracting smallest entry of each column from the remaining entries in its column of row opportunity matrix.
3. Testing optimality of total opportunity matrix: Cover all the zeros with minimum number of horizontal and vertical lines. If the number of lines is equal to number of columns or rows then total opportunity matrix is the optimal solution. Otherwise, revise the total opportunity matrix.
4. Revising total opportunity matrix:
   a. The cells which have been covered by lines are called covered cells.
   b. The cells which have not been covered by lines are called uncovered cells.
   c. Determine smallest entry in the uncovered cell.
   d. Subtract this smallest entry from the uncovered cells.
   e. Add this smallest entry to those cells where horizontal and vertical lines cross each other.
   f. Keep the remaining covered cells unchanged.
5. Repeat steps 3 and 4 until obtaining of optimal solution.
6. Make assignment to optimal solution matrix.

32.12.2 Solved Problems

Problem 32.8: A company has four machines of which to do three jobs. Each job can be assigned to one and only one machine. The cost of each job on each machine is given in Table 32.8. Find optimal assignment.

Table 32.8 The cost of each job on each machine

Job/Machine	W	X	Y	Z
A	18	24	28	32
B	8	13	17	18
C	10	15	19	22

Solution: The given matrix is not a square matrix. Convert given matrix into square matrix by adding dummy row D. Assignment costs for dummy row are equal to zeros.

Table 1 Given matrix

Job/Machine	W	X	Y	Z
A	18	24	28	32
B	8	13	17	18
C	10	15	19	22

Table 2 Converting given matrix into square matrix

Job/Machine	W	X	Y	Z
A	18	24	28	32
B	8	13	17	18
C	10	15	19	22
D	0	0	0	0

Table 3 Row opportunity matrix

Job/Machine	W	X	Y	Z
A	0	6	10	14
B	0	5	9	10
C	0	5	9	12
D	0	0	0	0

Table 4 Total opportunity matrix

Job/Machine	W	X	Y	Z
A	0	6	10	14
B	0	5	9	10
C	0	5	9	12
D	0	0	0	0

Minimum number of lines to cover all zeros in total opportunity matrix is equal to 2. So solution is not an optimal solution. So we have to revise the solution.

Table 5 Revised solution 1

Job/Machine	W	X	Y	Z
A	0	1	5	9
B	0	0	4	5
C	0	0	4	7
D	5	0	0	0

Table 6 Revised solution 2 (Optimal solution)

Job/Machine	W	X	Y	Z
A	0	1	1	5
B	0	0	0	1
C	0	0	0	3
D	9	4	0	0

Minimum number of lines to cover all zeros in revised solution 2 matrix is equal to 4. So solution is an optimal solution. So we have to make assignment to the optimal solution.

Table 7 Revised solution 2 (Optimal solution) (optimal assignment)

Job/Machine	W	X	Y	Z
A	0	1	1	5
B	0	0	0	1
C	0	0	0	3
D	9	4	0	0

Optimal solution:

A to W = 18
B to X = 13
C to Y = 19
D to Z = 0
Total = 50

Maximization problem: The first step in maximization problem is converting into minimization problem by subtracting all the elements from the highest entry of the matrix. Then use assignment algorithm to solve the problem.

Problem 32.9: A company has 5 jobs to be done. Table 32.9 shows the returns in rupees on assigning i^{th} machine to the jth job. Assign the five jobs to the five machines so as to maximize the total profit.

Table 32.9 Returns in rupees

Machine/Job	A	B	C	D	E
1	5	11	10	12	4
2	2	4	9	3	5
3	3	12	5	14	6
4	6	14	4	11	7
5	7	9	8	12	5

Solution:

Table 1 Given matrix

Machine/Job	A	B	C	D	E
1	5	11	10	12	4
2	2	4	9	3	5
3	3	12	5	14	6
4	6	14	4	11	7
5	7	9	8	12	5

Given problem is a maximization problem. Convert maximization problem into minimization problem by subtracting all the entries from highest entry 14. The resultant matrix is given in Table 2.

Table 2 Minimization matrix

Machine/Job	A	B	C	D	E
1	9	3	4	2	10
2	12	10	5	11	9
3	11	2	9	0	8
4	8	0	10	3	7
5	7	5	6	2	9

Table 3 Row opportunity matrix

Machine/Job	A	B	C	D	E
1	7	1	2	0	8
2	7	5	0	6	4
3	11	2	9	0	8
4	8	0	10	3	7
5	5	3	4	0	7

Table 4 Total opportunity matrix

Machine/Job	A	B	C	D	E
1	2	1	2	0	4
2	2	5	0	6	0
3	6	2	9	0	4
4	3	0	10	3	3
5	0	3	4	0	3

Table 5 Revised matrix (optimal solution)

Machine/Job	A	B	C	D	E
1	0	1	(0)	0	2
2	2	7	0	8	(0)
3	4	2	7	(0)	2
4	1	(0)	8	3	1
5	(0)	5	4	2	3

Optimal solution:

    1 to C = 10
    2 to E = 5
    3 to D = 14
    4 to B = 14
    5 to A = 7
    Total   = 50

32.13 QUESTIONS AND PROBLEMS

1. What is operations research?
2. Give various definitions of operations research.
3. Explain various models in operations research.
4. Explain the nature of operations research models.

5. Distinguish the following models:
   a. Stochastic and deterministic model
   b. Static and dynamic models
6. What is decision making? What are the steps in making decision?
7. A plant manufactures washing machines and dryers. The major manufacturing departments are stamping department, motor transmission department and assemble department. The first two departments produce parts for both products, while the assembly lines are different for two products. The monthly department capacities are Stamping department: 1,000 washes or 1,000 dryers or any other linear combination. Motor transmission department: 1,600 washers or 7,000 dryers or any other linear combination.
   Washer assembly line: 9,000 washers only
   Dryer assembly line: 5,000 dryers only
   Profit per unit of washer and dryer are ₹ 270 and ₹ 300 respectively. Formulate and obtain the solution.
   Answer: Number of washers = x_1 = 0, Number of dryers = x_2 = 1,000, z = ₹ 3,00,000
8. Consider the following problem faced by a production planner in a soft drink plant. He has two bottling machines A and B. A is designed for 8 ounce bottles and B for 16 ounce bottles. However, each can be used on both types with some loss of efficiency. The following data is available.

Machine	8 ounce bottles	16 ounce bottles
A	100/minute	40/minute
B	60/minute	75/minute

The machine can be run 8 hours per day, 5 days per week. Profit on 8 ounce bottles is 15 paise and 16 ounce bottle is 25 paise. Weekly production on drink cannot exceed 3,00,000 ounces and the market can absorb 25,000 eight ounce bottles and 7,000 sixteen ounce bottles per week. Formulate linear programing problem to maximize profit. Use graphical method.

Hint: Maximize $z = 0.15x_1 + 0.25x_2$
Subject to:

$$\frac{1}{100}x_1 + \frac{1}{40}x_2 \leq 2,400$$
$$\frac{1}{60}x_1 + \frac{1}{75}x_2 \leq 2,400$$
$$8x_1 + 16x_2 \leq 3,00,000$$
$$x_1 \leq 25,000$$
$$x_2 \leq 7,000$$
$$x_1, x_2 \geq 0$$

Answer: x_1 = 25,000, x_2 = 6,250 and z = profit = ₹ 5,312

9. A company produces two types of leather belts, say type A and type B. Belt A is of superior quality and belt B is of lower quality. Profits on the two types of belts are 40 paise and 30 paise per belt respectively. Each of belts of type A requires twice as much as required by a belt of type B. If all the belts are type of B, the company can produce 1,000 belts per day. The supply of leather, however, is sufficient only for 800 belts per day. Belt A requires fancy buckles and only 400 fancy buckles are available for this per day. For belt of type B, only 700 buckles are available per day. How should the company

manufacture the two types of belts in order to have a maximum overall profit? Use graphical method.

Answer: $x_1 = 200$, $x_2 = 600$, profit $z = 260$

10. A pine apple firm is producing canned pineapple and canned juice. Requirement and availability of the resources are given below. The profits per unit of canned pineapple and one unit of canned juice are ₹ 2 and ₹ 5 respectively. Determine production schedule which maximizes the profit.

	Canned pineapple	Canned juice	Available resources
Labor (man-hours)	2	3	12
Equipment (machine-hours)	3	2	12
Raw material (units)	2	2	9

Answer: $x_1 = 0$, $x_2 = 4$ and $z = 20$

11. Describe steps of Simplex method.

12. A manufacture of leather belts makes three types of belts A, B and C which are processed on three machines M_1, M_2 and M_3. Belt A requires 2 hours on machine M_1, and 3 hours on machine M_3. Belt B requires 3 hours on machine M_1, 2 hours on machine M_2 and 2 hours on machine M_3 and belt C requires 5 hours on machine M_2 and 4 hours on machine M_3. There are 8 hours of time per day available on machine M_1, 10 hours of time per day available on machine M_2 and 15 hours of time per day available on machine M_3. The profits gained from belt A is ₹ 3 per unit, from belt B is ₹ 5 per unit and from belt C is ₹ 4 per unit. What should be the daily production of each type of belts so that the profit is maximum?

Hint:

Maximize $z = 3x_1 + 5x_2 + 4x_3$

Subject to the constraints:

$$2x_1 + 3x_2 \leq 8$$
$$2x_2 + 5x_3 \leq 10$$
$$3x_1 + 2x_2 + 2x_3 \leq 15$$
$$x_1 \geq 0; x_2 \geq 0; x_3 \geq 0$$

Answer: $x_1 = 0$; $x_2 = 2.67$; $x_3 = 2$ and $z = 21.33$

13. A company produces three products A, B and C. These products require three ores O_1, O_2 and O_3. The maximum quantities of ores O_1, O_2 and O_3 available are 22 tons, 14 tons and 14 tons respectively. For one ton of each of these product the ore requirements are:

	A	B	C
O_1	3	–	3
O_2	1	2	3
O_3	3	2	0

The company makes a profit of one, four and five thousands on each ton of the products A, B and C respectively. How many tons of each products A, B and C should company produce to maximize profit?

Hint:

Maximize $z = x_1 + 4x_2 + 5x_3$

Subject to the constraints:

$$3x_1 + 3x_3 \leq 22$$

$$1x_1 + 2x_2 + 3x_3 \leq 14$$
$$3x_1 + 2x_2 + 0x_3 \leq 14$$
$$x_1 \geq 0; x_2 \geq 0; x_3 \geq 0$$

Answer: $x_1 = 0$; $x_2 = 7$; $x_3 = 0$ and $z = 28$

14. Define transportation problem.
15. Explain the steps in transportation algorithm.
16. Explain the different methods for obtaining initial basic feasible solution.
17. Explain two important methods for testing the optimality of initial basic feasible solution.
18. Solve the following transportation problem.

Origin	Destination					Capacity
	D1	**D2**	**D3**	**D4**	**D5**	
O1	3	2	3	4	1	100
O2	4	1	2	4	2	125
O3	1	0	5	3	2	75
Requirement	100	60	40	75	25	

Answer: Minimum cost = 615

19. A company has three mines A, B and C and five factories F1, F2, F3, F4 and F5. The mines can supply 80, 100 and 140 tons of ore daily and the requirements of the factories are 40, 50, 70 and 80 respectively. The following Table gives the unit transportation cost of ore.

Mines	Factories				
	F1	**F2**	**F3**	**F4**	**F5**
A	4	2	3	2	6
B	5	4	5	2	1
C	6	5	4	7	3

Give a distribution plan to minimize the total cost of transportation.

Answer: Total minimum cost = 920

20. Solve the following transportation problem to maximize the total profit. Here entries are profits.

Origin	A	B	C	D	Capacity
S1	40	25	22	33	100
S2	44	35	30	30	30
S3	38	38	28	30	70
Requirement	40	20	60	30	

Hint: In general, transportation problem is minimizing algorithm. To convert maximization problem into minimization problem, subtract all the entries from highest entry. Remaining procedure is same as transportation algorithm. The resultant minimization algorithm is:

Origin	Destination				Capacity
	A	**B**	**C**	**D**	
S1	4	19	22	11	100
S2	0	9	14	14	30
S3	6	6	16	14	70
Requirement	40	20	60	30	

Answer: Total maximum profit = ₹ 5,130

21. Describe the steps in Hungarian method for solving assignment problem.
22. Four operators O1, O2, O3 and O4 are available to a manager who has to get four jobs J1, J2, J3 and J4 done by assigning one job to each operator. Given the times needed by different operators for different jobs in the matrix below. How should the manager assign the jobs so that the total time needed for all jobs is minimized?

Operator/Job	J1	J2	J3	J4
O1	12	10	10	8
O2	14	12	15	11
O3	6	10	16	4
O4	8	16	9	7

Answer: O1-J3, O2-J2, O3-J4, O4-J1 and $z = 34$

23. Find optimal solution to the assignment problem with the following cost matrix.

	I	II	III	IV
A	5	3	1	8
B	7	9	2	6
C	6	4	5	7
D	5	7	7	6

Answer: A-III, B-IV, C-II, D-I and $z = 16$

24. Five mechanics are available to work on six machines and the respective costs in rupees for each mechanic-machine combination are given in the matrix below. A sixth machine is available to replace one of the existing machines and the associated costs are given in Table below.

Machine→ Mechanic↓	1	2	3	4	5	6
A	19	15	–	16	13	22
B	13	–	15	–	21	14
C	15	17	19	20	12	18
D	20	22	16	18	17	–
E	–	16	14	19	18	15

Means cannot be assigned.
   a. Determine whether the new machine can be accepted.
   b. The optimal assignment and associated cost

Answer: Optimal Assignment: A-2, B-1, C-5, D-3, E-6, F-4
Optimal cost of assignment = 71
New machine can be accepted.

25. The efficiency of 5 machines on each of 5 jobs is given below. Determine an assignment schedule of the jobs to the machines such that total efficiency is maximum.

Machine/Job	1	2	3	4	5
I	62	78	50	101	82
II	70	85	60	75	55
III	88	96	118	85	71
IV	48	64	87	77	80
V	60	70	98	66	83

Answer: I-4, II-2, III-1, IV-5, V-3 and Total efficiency = 452

26. There are 5 jobs and 4 machines. The expected profits on each job on each machine are given blow. Determine an optimal assignment of the machine to the jobs so that total profit is maximum.

Machine/Job	1	2	3	4	5
I	62	78	50	101	82
II	71	84	61	73	59
III	87	92	111	71	81
IV	48	64	87	77	80

Answer: I-4, II-2, III-3, IV-5 and Total profit = 376

27. Solve the following problem for minimizing the cost of assignment.

Machine/Job	1	2	3	4
A	18	24	28	32
B	8	13	17	19
C	10	15	19	22

Answer: A-1, B-3, C-2 and Total cost = 50

Queuing Theory (Waiting Lines)

33.1 INTRODUCTION

Queue is formed, if the server is busy. Then remaining customers are waiting in the queue. Queuing theory is based on probability. Most of the basic work on queue or waiting line theory was carried by AK Erlang. A queuing system can be completely described by queuing system.

33.2 ELEMENTS OR PARAMETERS OF QUEUING SYSTEM

1. The input or arrival pattern
2. Inter arrival time
3. The service time
4. Capacity of the queue
5. Capacity of the source
6. Number of servers
7. The queuing discipline
8. Customers behavior
9. State of the system

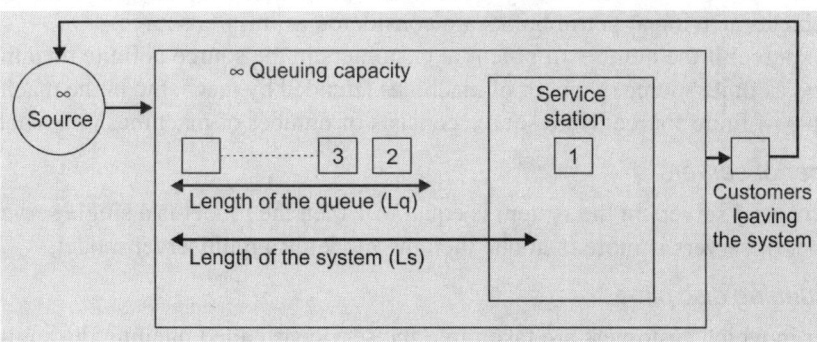

Fig. 33.1 Various parameters of queue (Single server model)

1. The input or arrival pattern

The input describes the way in which customers arrive from the source and join the queue. Generally, the customers arrive at random. At best, the arrival pattern can be described in terms of probabilities. The present chapter deals with the customers arrive in Poisson's distribution. Poisson's distribution is discrete random variable distribution.

Probability that n number of customers arrive in unit time $= p_n = \dfrac{\lambda^n e^{-\lambda}}{n!}$

λ = Mean arrival rate = Number of arrivals per unit time

$n = 0, 1, 2, 3, 4, ..., \infty$

Generally mean arrival rate is determined by conducting actual experiment or study.

2. Inter arrival time

Inter arrival time is also called time between the two successive arrivals. If the number of arrivals follows Poisson's distribution then inter arrival time follows negative exponential distribution.

$f(t)$ = Probability that inter arrival time is t units

$f(t) = \lambda e^{-t}$, where $0 < t \le \infty$

3. Service time

Service time per unit may be a constant or a random variable. Important probability service time distributions in practice are negative exponential distribution and Erlang distribution.

4. Capacity of queue

Capacity of the queue may be finite or infinite.

Infinite queue: If the number of persons allowed into the queue is more than 20 then the queue is considered as infinite queue. Queues at cinema theaters are the examples of infinite queue. Queue at barbershop is an example of finite queue.

Finite queue: Only limited number of customers is allowed into the queue, i.e. if the number of customers allowed into the queue is less than 20 then the queue is considered as finite queue.

5. Capacity of source

Capacity of source may be finite or infinite.

Infinite source: If the number of potential customers in the source is more than 20 then the capacity of source is infinite. Source from which people coming to theater and source from which vehicles arriving at petrol bunks are considered as infinite source.

Finite source: If the number of potential customers in the source is finite then the source is considered as finite source. Number of machines attended by mechanic in the machine shop is an example of finite source. Here source consists of number of machines in a machine shop.

6. Number of servers

If the number of servers in the system is equal to 1 then the model is a single server model. If the number of servers is more than one then the model is a multiserver model.

7. The queuing discipline

The order in which customers are taken into the service is called queuing discipline. The two important queuing disciplines are:

First come first serve: The person who is first in the queue will be taken into the service.

Shortest service time: The customer who has shortest service time is taken first into the service.

8. Customers behavior

Customers generally behave in four ways:

a. Balking: A customer may not enter in the queue when the queue is long and he has no time to wait.

b. Reneging: This occurs when a waiting customer leaves the queue due to impatience.

c. Jockeying: Customers may jockey or move from one queue to other queue when there is more than one queue.

d. Priorities: In certain applications customers are served before others regardless of their order of arrival. These customers have priority over others.

9. State of the system

a. Transient state: A system is said to be in transient state when its operating characteristics are dependent on time.

b. Steady state: A system is said to be in steady state when its operating characteristics are independent of time.

$$\text{Traffic intensity} = \frac{\text{Mean arrival rate}}{\text{Mean service rate}}$$

$$\rho = \frac{\lambda}{s\mu}$$

Where s = Number of servers

The unit of traffic intensity is Erlang. For steady state $\lambda < s\mu$.

33.3 KENDALL'S NOTATION FOR REPRESENTING QUEUING MODELS

Generally queuing model may be completely specified in the following form:

a/b/c/d/e/f

a = Probability law for the arrival

b = Probability law according to which customers are being served.

c = Number of servers

d = Capacity of system

e = Capacity of queue

f = Queue discipline

33.4 MODEL 1: SINGLE SERVER MODEL (M/M/1/∞/∞/FCFS)

Assumptions made in this model:

1. Poisson arrival process
2. Negative exponential service time distribution
3. Single server
4. ∞ source capacity
5. ∞ queuing capacity
6. First come first serve

To determine various parameters:

1. Probability that exactly n number of customers in the system = P_n

$$P_n = \left(\frac{\lambda}{\mu}\right)^n P_0$$

2. To determine probability that no customers in the system or probability that server is idle:

$$\sum_{n=0}^{\infty} P_n = 1$$

Probability that no customer in the system = Probability that server is idle = P_0

$$P_0 = 1 - \frac{\lambda}{\mu}$$

Probability that the server is busy $= 1 - P_0 = \frac{\lambda}{\mu}$

3. Probability that number of customers in the system $\geq N$

Probability that number of customers in the system $\geq N = P_{(n \geq N)}$

$$P_{n \geq N} = \left(\frac{\lambda}{\mu}\right)^N$$

4. Average or expected number of customers in the system (L_s):

$$L_s = \frac{\lambda}{\mu - \lambda}$$

5. Waiting time in the system (W_s): The average time spent by the customer in the system.

$$W_s = \frac{L_s}{\lambda}$$

$$W_s = \frac{1}{\mu - \lambda}$$

6. To determine average number of customers in the queue (L_q):

$$L_q = \sum_{n=1}^{\infty} (n-1) P_n$$

$$L_q = L_s - (1 - P_0)$$

$$L_q = \frac{\lambda^2}{\mu(\mu - \lambda)}$$

7. To determine waiting time in the queue (W_q):

$$W_q = \frac{L_q}{\lambda}$$

$$W_q = \frac{\lambda}{\mu(\mu - \lambda)}$$

8. Probability that waiting time in the queue $\geq t = \dfrac{\lambda}{\mu} e^{-(\mu - \lambda)t}$

9. Probability that waiting time in the system $\geq t = e^{-(\mu - \lambda)t}$

33.5 SOLVED PROBLEMS

Problem 33.1: In a railway yard goods train arrive at a rate of 30 trains per day. Assuming that inter arrival time follows an exponential distribution and the service time distribution is also exponential with an average of 36 minutes. Calculate the following:

i. The average number of trains in the queue

ii. Average number of trains in the system

iii. The probability that number of trains in the system exceeds 10.

Solution: Arrival rate = 30 trains/day

Service time = 36 minutes = $\dfrac{36}{60 \times 24} = 0.025$ day

Service rate $= \mu = \dfrac{1}{\text{Service time}} = \dfrac{1}{0.025} = 40$ trains/day

i. Average number of trains in the queue $= L_q$

$$L_q = \frac{\lambda^2}{\mu(\mu - \lambda)} = \frac{30^2}{40(40 - 30)} = 2.25 \text{ trains}$$

ii. Average number of trains in the system $= L_s$

$$L_s = \frac{\lambda}{(\mu - \lambda)} = \frac{30}{(40 - 30)} = 3 \text{ trains}$$

iii. The probability that the number of trains in the system exceeds 10.

$$P(n > 10) = \left(\frac{\lambda}{\mu}\right)^n = \left(\frac{30}{40}\right)^{10} = (0.75)^{10} = 0.0563$$

Problem 33.2: Customers arrive at a one-window drive in a bank according to Poisson's distribution with mean 10 per hour. Service time per customer is exponential with mean 5 minutes. The space in front of the window including that for serviced car can accommodate maximum of three cars. Other cars can wait outside the space.

  i. What is the probability that an arriving car can drive directly to the space in front of the window?
  ii. How long is an arriving customer expected to wait before starting the service?

Solution: Arrival rate $= \lambda = 10$ cars/hour

$$\text{Service time} = 5 \text{ minutes} = \frac{5}{60} = 0.083 \text{ hour}$$

$$\text{Service rate} = \mu = \frac{1}{\text{Service time}} = \frac{60}{5} = 12 \text{ cars/hour}$$

The probability that an arriving car can drive directly to the space in front of the window
$= P_0 + P_1 + P_2$

$$\frac{\lambda}{\mu} = 0.834$$

$$P_0 = 1 - \frac{\lambda}{\mu} = 1 - 0.834 = 0.166$$

$$P_n = \left(\frac{\lambda}{\mu}\right)^n P_0$$

$$P_1 = \left(\frac{\lambda}{\mu}\right)^1 P_0 = 0.834 \times 0.166 = 0.138$$

$$P_2 = \left(\frac{\lambda}{\mu}\right)^2 P_0 = 0.834^2 \times 0.166 = 0.115$$

$$P_0 + P_1 + P_2 = 0.166 + 0.138 + 0.115 = 0.419$$

ii. Waiting time in the queue $= W_q$

$$W_q = \frac{\lambda}{\mu(\mu - \lambda)} = \frac{10}{12(12 - 10)} = 0.416 \text{ hour}$$

Problem 33.3: Arrival rate of a telephone booth is according to Poisson's distribution with an average time of 9 minutes between two consecutive arrivals. The length of a telephone call is assumed to be exponentially distributed with mean of 3 minutes.

  i. Determine the probability that a person arriving at the booth will have to wait.
  ii. Average waiting length
  iii. What is the probability that an arrival will have to wait for more than 10 minutes before the phone is free.

iv. What is the probability that the customer will have to wait more than 10 minutes before the phone is available and to complete the call.

v. Find the fraction of a day that the phone will be in use.

vi. The telephone company will install a second booth when convinced that an arrival would expect to have to wait at least 4 minutes for the phone. Find the increase in flow of arrivals, which will justify a second booth.

Solution: Time between two consecutive arrivals = 9 minutes

Arrival rate $= \lambda = \dfrac{1}{9} = 0.11$ customer/minute

Service time = 3 minutes

Service rate $= \mu = \dfrac{1}{\text{Service time}} = \dfrac{1}{3} = 0.33$ customer/minute

$$\frac{\lambda}{\mu} = \frac{0.11}{0.33} = 0.33$$

i. Probability that the server is busy or probability that a person will have to wait$= 1 - P_0$

$$1 - P_0 = \frac{\lambda}{\mu} = \frac{0.11}{0.33} = 0.33$$

ii. Average number of customers in the queue $= L_q$

$$L_q = \frac{\lambda^2}{\mu(\mu - \lambda)} = \frac{0.11^2}{0.33(0.33 - 0.11)} = 0.166 \text{ customer}$$

iii. Probability that waiting time in the queue $\geq t = \dfrac{\lambda}{\mu} e^{-(\mu - \lambda)t}$

Probability that waiting time in the queue ≥ 10 minutes $= 0.33 e^{-(0.33 - 0.11)10} = 0.036$

iv. Probability that waiting time in the system $\geq t = e^{-(\mu - \lambda)t}$

Probability that waiting time in the system ≥ 10 minutes $= e^{-(0.33 - 0.11)10} = 0.11$

v. Fraction of a day that the phone will be in use = Probability that the server is busy
$$= 1 - P_0$$

Fraction of a day that the phone will be in use $= \dfrac{\lambda}{\mu} = 0.33$

vi. Required waiting time in the queue $= W_q = 4$ minutes

$$W_q = \frac{\lambda}{\mu(\mu - \lambda)} = \frac{\lambda}{0.33(0.33 - \lambda)} = 4$$

Hence, new arrival rate $= \lambda = 0.19$ arrivals/minute

Increase in arrival rate to justify second booth $= 0.19 - 0.11 = 0.08$ customer/minute.

Problem 33.4: A self-servicing store employs one cashier at its counter. Nine customers arrive at an average of 5 minutes while the cashier can service 10 customers in 5 minutes. Assuming Poisson's distribution for arrival rate. Find:

i. Average number of customers in system

ii. Average number of customers in queue

iii. Average time a customer spends in the system

iv. Average time a customer waits before being serviced

Solution: Arrival Rate $= \lambda = \dfrac{9}{5} = 1.8$ customers/minute

Service rate $= \mu = \dfrac{10}{5} = 2$ customers/minute

i. Average number of customers in the system $= L_s$

$$L_s = \frac{\lambda}{(\mu - \lambda)} = \frac{1.8}{(2 - 1.8)} = 9 \text{ customers}$$

ii. Average number of customers in the queue $= L_q$

$$L_q = \frac{\lambda^2}{\mu(\mu - \lambda)} = \frac{1.8^2}{2(2 - 1.8)} = 8.1 \text{ customers}$$

iii. Average time a customer spends in the system $= W_s$

$$W_s = \frac{1}{(\mu - \lambda)} = \frac{1}{(2 - 1.8)} = 5 \text{ minutes}$$

iv. Waiting time in the queue $= W_q$

$$W_q = \frac{\lambda}{\mu(\mu - \lambda)} = \frac{1.8}{2(2 - 1.8)} = 4.5 \text{ minutes}$$

Problem 33.5: A branch of Punjab National Bank has only one typist. Since the typing work varies in length, the mean service rate becomes 8 letters per hour. The letters arrive at a rate of 5 per hour during the entire 8 hours workday. If the typewriter is valued at ₹ 1.50 per hour, determine:

i. Equipment utilization
ii. The percent time that an arriving letter has to wait
iii. Average system time
iv. Average cost due to waiting on the part of typewriter (ldle time cost)

Solution: Arrival Rate $= \lambda = 5$ letters/hour

Service rate $= \mu = 8$ letters/hour

i. Probability that the server is busy or Equipment utilization $= 1 - P_0$

$$1 - P_0 = \frac{\lambda}{\mu} = \frac{5}{8} = 0.625$$

ii. The percentage time that an arriving letter has to wait = Probability that the server is busy $= 1 - P_0$

$$1 - P_0 = \frac{\lambda}{\mu} = \frac{5}{8} = 0.625 = 62.5\%$$

iii. Average time a customer spends in the system $= W_s$

$$W_s = \frac{1}{(\mu - \lambda)} = \frac{1}{(8 - 5)} = 0.33 \text{ hours} = 20 \text{ minutes}$$

iv. Idle time cost of typewriter per day $= 8 (1 - 0.625) 1.50 = ₹ 4.50/\text{day}$

Problem 33.6: On average 96 patients per 24 hours day requires the service of an emergency clinic. Also an average a patient requires 10 minutes of active attention. Assume that the facility can handle one emergency at a time. Suppose that it costs the clinic ₹ 100 per patient treated to obtain an average service time of 10 minutes and that each minute of decrease in the average time would cost ₹ 10 per patient to be treated. Determine present average size of queue. How much would have to be budgeted by the clinic to decrease the average size of the queue from 1.33 patients to 0.5 patients?

Solution: Arrival rate $= \lambda = \dfrac{96}{24} = 4$ patients/hour

Service time $= 10$ minutes $= 0.116$ hour

Service rate $= \mu = \dfrac{1}{\text{Service time}} = 6$ patients/hour

Average number of patients in the queue $= L_q$

$$L_q = \frac{\lambda^2}{\mu(\mu - \lambda)} = \frac{4^2}{6(6-4)} = 1.33 \text{ patients}$$

If the service time is 10 minutes then average length of queue is 1.33 patients/hour. To reduce average length of queue from 1.33 patients to 0.5 patient then service time has to be reduced.

$$L_q = \frac{\lambda^2}{\mu(\mu - \lambda)}$$

$$0.5 = \frac{4^2}{\mu(\mu - 4)}$$

$$0.5 \, \mu^2 - 2 \, \mu = 16$$

$$0.5 \, \mu^2 - 2 \, \mu - 16 = 0$$

$$\mu = \frac{-b \pm \sqrt{b^2 - 4ac}}{2a}$$

$$\mu = \frac{2 \pm \sqrt{2^2 - 4 \times 0.5 \times (-16)}}{2 \times 0.5}$$

$$\mu = 2 \pm \sqrt{36}$$

$$\mu = 8/\text{hour}$$

$$\text{Service time} = \frac{1}{8} \times 60 = 7.5 \text{ minutes}$$

To reduce average length of queue from 1.33 patients to 0.5 patient then service time has to be reduced from 10 minutes to 7.5 minutes.

Then cost per patient $= 100 + (10 - 7.5) \times 10 = 100 + 25 = ₹ 125/\text{patient}$

33.6 MODEL 2: M/M/1/∞/N/FCFS

In this model the capacity of the system is N.

To determine various parameters:

1. Probability that there are n customers in the system $= P_n = \left(\dfrac{\lambda}{\mu} \right)^n P_0$

2. To determine probability that server is idle or probability that 0 customer in the system (P_0)

$$P_n = \left(\frac{\lambda}{\mu} \right)^n P_0$$

$$P_0 = \frac{1 - \dfrac{\lambda}{\mu}}{1 - \left(\dfrac{\lambda}{\mu} \right)^{N+1}} = \frac{1 - \rho}{1 - \rho^{N+1}}$$

3. To determine average or expected number of customers in the system (L_s):

$$L_s = \sum_{n=0}^{N} n P_n$$

4. Waiting time in the system $= W_s$

$$W_s = \frac{L_s}{\lambda}$$

5. Average number of customers in the queue (L_q):
$$L_q = L_s - (1 - P_0)$$

6. Waiting time in the queue (W_q): Average time customer spends in the queue.

$$W_q = \frac{L_q}{\lambda}$$

33.7 SOLVED PROBLEMS

Problem 33.7: At a railway station only one train is handled at a time. The railway yard is sufficient only for two trains to wait while the other is given signal to leave the station. Trains arrive at the station at an average rate of 6 per hour and the railway station can handle them at an average of 12 per hour. Assuming Poisson's arrivals and exponential service distribution, find the steady state probabilities for the various number of trains in the system. Find also the average waiting time of a train coming into the yard.

Solution: Maximum number of trains allowed into system = $N = 3$

Arrival rare = λ = 6 trains/hour

Service rate = μ = 12 trains/hour

$$\frac{\lambda}{\mu} = \frac{6}{12} = 0.5$$

To find steady state probabilities:

Probability that 0 train in the system = P_0

Probability that 1 train in the system = P_1

Probability that 2 trains in the system = P_2

Probability that 3 trains in the system = P_3

$$P_0 = \frac{1 - \dfrac{\lambda}{\mu}}{1 - \left(\dfrac{\lambda}{\mu}\right)^{N+1}}$$

$$P_0 = \frac{1 - 0.5}{1 - (0.5)^4} = 0.533$$

$$P_1 = P_0 \left(\frac{\lambda}{\mu}\right)^1 = 0.533 \times 0.5 = 0.266$$

$$P_2 = P_0 \left(\frac{\lambda}{\mu}\right)^2 = 0.533 \times 0.5^2 = 0.133$$

$$P_3 = P_0 \left(\frac{\lambda}{\mu}\right)^3 = 0.533 \times 0.5^3 = 0.066$$

ii. Average or expected number of trains in the system (L_s):

$$L_s = \sum_{n=0}^{N} n P_n$$

$$L_s = \sum_{n=0}^{3} n P_n$$

$$= 0P_0 + 1P_1 + 2P_2 + 3P_3 = 0 + 1 \times 0.266 + 2 \times 0.133 + 3 \times 0.066 = 0.732 \text{ train}$$

iii. Average or expected number of trains in the queue (L_q):

$$L_q = L_s - (1 - P_0) = 0.732 - (1 - 0.533) = 0.265 \text{ train}$$

iv. Waiting time in the queue (W_q): Average time customer spends in the queue.

$$W_q = \frac{L_q}{\lambda} = \frac{0.265}{4} = 0.044 \text{ hour} = 2.65 \text{ minutes}$$

Problem 33.8: The capacity of yard is to admit of 9 trains. Calculate the following on the assumption that 30 trains/hour on average are received in the yard. Service rate is 40 trains/hour. Determine:

   i. The probability that the yard is empty.

   ii. Average queue length

Solution: Maximum number of trains allowed into system = N = 9 trains

Arrival rare = λ = 30 trains/hour

Service rate = μ = 40 trains/hour

λ/μ = 30/40 = 0.75

$$P_0 = \frac{1 - \dfrac{\lambda}{\mu}}{1 - \left(\dfrac{\lambda}{\mu}\right)^{N+1}}$$

$$P_0 = \frac{1 - 0.75}{1 - (0.75)^{10}} = 0.26$$

ii. Average or expected number of trains in the system (L_s):

$$L_s = \sum_{n=0}^{N} nP_n$$

$$L_s = \sum_{n=0}^{9} n\left(\frac{\lambda}{\mu}\right)^n P_0$$

$$L_s = P_0 \sum_{n=0}^{9} n\left(\frac{\lambda}{\mu}\right)^n$$

$$L_s = 0.26 \sum_{n=0}^{9} n(0.75)^n$$

$$L_s = \lambda = 2.79 \text{ trains}$$

iii. Average or expected number of trains in the queue (L_q):

$$L_q = L_s - (1 - P_0) = 2.79 - (1 - 0.26) = 2.02 \text{ trains}$$

Problem 33.9: A barber shop has space to accommodate on 10 customers. He can serve only one person at a time. If a customer comes to his shop and finds it full, he goes to the next shop. Customers randomly arrive at an average rate of 10 per hour and the barber service time is negative exponential with an average of 5 minutes per customer.

   i. Write recurrence relations for steady state queuing system.

   ii. Determine P_0 and P_n.

Solution: Maximum number of customers allowed into barber shop = N = 10 persons

Arrival rate = λ = 10 persons/hour

Service rate = μ = 12 persons/hour

Probability that 0 number of train in the system = P_0

$$\frac{\lambda}{\mu} = \frac{10}{12} = 0.833$$

$$P_0 = \frac{1 - \frac{\lambda}{\mu}}{1 - \left(\frac{\lambda}{\mu}\right)^{N+1}}$$

$$P_0 = \frac{1 - 0.833}{(1 - 0.833)^{11}} = 0.192$$

Probability that 10 number of trains in the system = P_{10}

$$P_{10} = P_0 \left(\frac{\lambda}{\mu}\right)^{10} = 0.192(0.832)^{10} = 0.030.$$

Problem 33.10: A stenographer has 5 persons for whom she performs stenographic work. Arrival rate is Poisson and service time is exponential. Average arrival rate is 4 per hour with an average service time of 10 minutes. Cost of waiting is ₹ 8 per hour, while the cost of servicing is ₹ 2.50 per customer. Calculate:

   i. The average waiting time of an arrival

  ii. The average length of the waiting line

 iii. The average time, which the arrival spends in the system.

  iv. The minimum cost service rate

Solution: Maximum number of customers to stenographer = λ = N = 5 persons

Arrival rate = λ = 4 persons/hour

Service rate = μ = 6 persons/hour

Probability that 0 number of customer in the system = P_0

$$\rho = \frac{\lambda}{\mu} = \frac{4}{6} = 0.666$$

$$P_0 = \frac{1 - \frac{\lambda}{\mu}}{1 - \left(\frac{\lambda}{\mu}\right)^{N+1}}$$

$$P_0 = \frac{1 - 0.666}{1 - (0.666)^6} = 0.365$$

Average or expected number of customers in the system (L_s):

$$L_s = \sum_{n=0}^{N} nP_n$$

$$L_s = \sum_{n=0}^{5} nP_0 \left(\frac{\lambda}{\mu}\right)^n$$

$$L_s = P_0 \sum_{n=0}^{5} n\left(\frac{\lambda}{\mu}\right)^n$$

$$L_s = 0.365 \sum_{n=0}^{5} n(0.666)^n$$

$$L_s = 0.365 \, (0.666) + 2 \times (0.666)^2 + 3 \times (0.666)^3 + 4 \times (0.666)^4 + 5 \times (0.666)^5$$
$$= 1.416 \text{ customers}$$

Average or expected number of customers in the queue (L_q):

$$L_q = L_s - (1 - P_0) = 1.416 - (1 - 0.365) = 0.781 \text{ customer}$$

Waiting time in the system = W_s

$$W_s = \frac{L_s}{\lambda} = \frac{1.461}{4} = 0.354 \text{ hour} = 21.24 \text{ minutes}$$

Waiting time in the queue = W_q

$$W_q = \frac{L_q}{\lambda} = \frac{0.78}{4} = 0.195 \text{ hour} = 11.7 \text{ minutes}$$

Cost/customer = Waiting time cost of customer + Server cost/hour

Cost/customer = $0.354 \times 8 + 2.50 = ₹ 5.33$

33.8 MODEL 3: MULTISERVER MODEL M/M/S/∞/∞/FCFS

The customers arrive in a Poisson fashion with mean arrival rate λ. This model is same as model 1 except number of servers in the system is equal to s. These s numbers of servers are arranged in parallel and customers can go to any of the free counters, where the service time at each counter is identical and follows the same negative exponential distribution. The mean service rate is μ.

1. To determine steady state probabilities P_n in multiserver model

Table 33.1 To determine steady state probabilities P_n in multiserver model

Steady state probabilities	Condition	Overall service rate	Number of idle servers	Number of persons waiting in the queue
$P_n = \dfrac{1}{n!} P_0 \left(\dfrac{\lambda}{\mu} \right)^n$	$n < s$	$n\mu$	$s-n$	Nil
$P_n = \dfrac{1}{s!s^{n-s}} P_0 \left(\dfrac{\lambda}{\mu} \right)^n$	$n \geq s$	$s\mu$	Nil	$n-s$

1. To determine probability that zero number of customer in the system

$$P_0 = \left(\sum_{n=0}^{s-1} \frac{(s\rho)^n}{n!} + \frac{(s\rho)^s}{s!(1-\rho)} \right)^{-1}$$

$$\rho = \frac{\lambda}{s\mu}$$

2. To determine average number of customers in the queue

$$L_q = \frac{\rho(s\rho)^s}{s!(1-\rho)^2} P_0$$

$$L_q = L_s - \frac{\lambda}{\mu}$$

3. To determine waiting time in the queue W_q

$$W_q = \frac{L_q}{\lambda}$$

4. To determine waiting time in the system (W_s)

$$W_s = \frac{L_s}{\lambda}$$

5. To determine average number of customers in the system (L_s)

$$L_s = L_q + \frac{\lambda}{\mu}$$

33.9 SOLVED PROBLEMS

Problem 33.11: A super market has two girls at the counters. The service time for each customer is exponential with mean of 4 minutes and people arrive in a Poisson fashion at the counter at the rate of 10 per hour.

   i. Calculate the probability that an arrival will have to wait for service.

  ii. Find the expected percentage of idle time for each girl.

 iii. Find waiting time in the queue.

Solution: Number of servers $= s = 2$

Arrival rate $= \lambda = 10$ customers/hour

Service time $= 4$ minutes $= 0.066$ hour

Service rate $= \mu = 1/$Service time $= 15/$hour

$$\rho = \frac{\lambda}{s\mu} = \frac{10}{15 \times 2} = 0.33$$

$$P_0 = \left(\sum_{n=0}^{s-1} \frac{(s\rho)^n}{n!} + \frac{(s\rho)^s}{s!(1-\rho)} \right)^{-1}$$

$$P_0 = \left(\sum_{n=0}^{2-1} \frac{(2 \times 0.33)^n}{n!} + \frac{(2 \times 0.33)^2}{2!(1-0.33)} \right)^{-1}$$

$$P_0 = (1 + 0.66 + 0.33)^{-1}$$

$$P_0 = 0.505$$

  i. Probability that an arrival has to wait for service $= 1 - (P_0 + P_1)$

$$P_n = \frac{1}{n!} P_0 \left(\frac{\lambda}{\mu} \right)^n$$

$$P_1 = \frac{1}{1!} \times 0.5 \left(\frac{10}{15} \right)^1 = 0.33$$

Probability that an arrival has to wait for service $= 1 - (P_0 + P_1) = 1 - (0.505 + 0.33) = 0.17$

ii. Expected percentage idle time of girl $= (2p_0 + 1p_1 + 0p_2)/2 = (2 \times 0.5 + 1 \times 0.3)/2 = 0.67$

Expected percentage idle time of girl $= 67\%$

iii. Length of the queue at each counter $= L_q$

$$L_q = \frac{\rho(s\rho)^s}{s!(1-\rho)^2} P_0$$

$$L_q = \frac{0.33(2 \times 0.33)^2}{2!(1-0.33)^2} \times 0.505 = 0.08 \text{ customer}$$

To determine waiting time in the queue (W_q)

$$W_q = \frac{L_q}{\lambda}$$

$$W_q = \frac{0.08}{10} = 0.008 \text{ hour} = 0.48 \text{ minute}$$

Problem 33.12: A bank has two tellers working on savings accounts. The first teller handles withdrawals only. The second teller handles deposits only. It has been found that the service time distribution for the deposits and withdrawals both are exponential with mean service time 3 minutes per customer. Depositors are found to arrive in a Poisson fashion throughout the day with mean arrival rate 16 per hour. Withdrawers also arrive in a Poisson fashion with mean arrival rate of 14 per hour. What would be the effect on the average waiting time for deposits and withdrawal, if each teller could handle both withdrawals and deposits? What would be the effect, if this could be only be accomplished by increasing service time to 3.5 minutes.
Solution:

Table 33.2 To determine waiting time in queue

Queue at withdrawal's counter	Queue at depositor's counter	Handling of withdrawal's and depositor's by both tellers
$\lambda = 14$ customers/hour Service time = 3 minutes $\mu = (60/3) = 20$/hour	$\lambda = 16$ customers/hour Service time = 3 minutes $\mu = (60/3) = 20$/hour	Number of servers = $s = 2$ Arrival rate = $\lambda = 30$ customers/hour Service time = $\mu = 3$ minutes/customer Service rate = $\mu = (60/3) = 20$/hour
$W_q = \dfrac{\lambda}{\mu(\mu - \lambda)}$	$W_q = \dfrac{\lambda}{\mu(\mu - \lambda)}$	$\rho = (\lambda/s\mu) = 30/(2 \times 15) = 0.75$
$= \dfrac{14}{20(20-14)}$	$= \dfrac{16}{20(20-16)}$	$P_0 = \left(\sum_{n=0}^{s-1} \dfrac{(s\rho)^n}{n!} + \dfrac{(s\rho)^s}{s!(1-\rho)} \right)^{-1}$
$= 0.116$ hour $= 7$ minutes	$= 0.2$ hour $= 12$ minutes	$P_0 = \left(\sum_{n=0}^{2-1} \dfrac{(2 \times 0.75)^n}{n!} + \dfrac{(2 \times 0.75)^2}{2!(1-0.75)} \right)^{-1}$
		$P_0 = 0.142$
		$L_q = \dfrac{\rho(s\rho)^s}{s!(1-\rho)^2} P_0$
		$L_q = \dfrac{0.75(2 \times 0.75)^2}{2!(1-0.75)^2} \times 0.142$
		$= 1.917$ customers
		$W_q = \dfrac{L_q}{\lambda} = \dfrac{1.917}{30} = 0.0643$ hour
		$= 3.86$ minutes

If each teller handles both deposits and withdrawals then waiting time in the queues is reduced to 3.86 minutes. So it is better to combine the duties.
Case 2:
Number of servers = $s = 2$
Arrival rate = $\lambda = 30$ customers/hour
Service time = 3.5 minutes/customer
Service rate = $\mu = 60/3.5 = 17.14$ customers/hour
$\rho = (\lambda/s\mu) = 30/(2 \times 17.14) = 0.875$

$$P_0 = \left(\sum_{n=0}^{s-1} \frac{(s\rho)^n}{n!} + \frac{(s\rho)^s}{s!(1-\rho)} \right)^{-1}$$

$$P_0 = \left(\sum_{n=0}^{2-1} \frac{(2 \times 0.875)^n}{n!} + \frac{(2 \times 0.875)^2}{2!(1-0.875)} \right)^{-1}$$

$$P_0 = 0.066$$

$$L_q = \frac{\rho(s\rho)^s}{s!(1-\rho)^2} P_0$$

$$L_q = \frac{0.875(2 \times 0.875)^2}{2!(1-0.875)^2} 0.066 = 5.7$$

To determine waiting time in the queue (W_q)

$$W_q = \frac{L_q}{\lambda}$$

$$W_q = \frac{5.7}{30} = 0.19 \text{ hour} = 11.4 \text{ minutes}$$

Problem 33.13: Four counters are being run on the frontier of a country to check the passports and necessary papers of the tourists. The tourists choose a counter at random. Arrival at the frontier is Poisson with parameter λ and the service rate is exponential with parameter $\lambda/2$. What is the steady state average queue at each counter?

Solution:

Number of servers $= s = 4$

Arrival rate $= \lambda$

Service rate $= \mu = \lambda/2 = 0.5$

$$\rho = (\lambda/s\mu) = (\lambda/(4 \times 0.5\lambda)) = 0.5$$

$$P_0 = \left(\sum_{n=0}^{s-1} \frac{(s\rho)^n}{n!} + \frac{(s\rho)^s}{s!(1-\rho)} \right)^{-1}$$

$$P_0 = \left(\sum_{n=0}^{4-1} \frac{(4 \times 0.5)^n}{n!} + \frac{(4 \times 0.5)^4}{4!(1-0.5)} \right)^{-1}$$

$$P_0 = \left(1 + \frac{2}{1!} + \frac{2^2}{2!} + \frac{2^3}{3!} + \frac{2^4}{4! \times 0.5} \right)^{-1} = 0.130$$

$$L_q = \frac{\rho(s\rho)^s}{s!(1-\rho)^2} P_0$$

$$L_q = \frac{0.5(4 \times 0.5)^4}{4!(1-0.5)^2} \times 0.130 = 0.1739$$

33.10 MACHINE SERVICING MODEL (M/M/R/k/k/FCFS)

In this queuing system, there are k machines, which are serviced by R repairmen. Whenever a machine breaks down, it will in a loss of production until it is repaired. This is equivalent to the finite calling source with maximum limit of k potential customers in the source.

To determine steady state probabilities:

$$P_n = kc_n \left(\frac{\lambda}{\mu} \right)^n P_0, \qquad 0 \le n \le R$$

$$P_n = \frac{kc_n n! \left(\frac{\lambda}{\mu}\right)^n}{R! R^{n-R}} P_0, \quad R+1 \le n \le k$$

$$P_0 = \left(\sum_{n=0}^{R} kc_n \left(\frac{\lambda}{\mu}\right)^n + \sum_{n=R+1}^{k} \frac{kc_n n! \left(\frac{\lambda}{\mu}\right)^n}{R! R^{n-R}} \right)^{-1}$$

Average number of breakdown machines in the system $= L_q = \sum_{n=R+1}^{k} (n-R) p_n$

$$W_q = \frac{L_q}{\lambda_{eff}}$$

$$L_s = L_q + \frac{\lambda_{eff}}{\mu}$$

$$W_s = \frac{L_s}{\lambda_{eff}}$$

$$\lambda_{eff} = \lambda \sum_{n=0}^{k} (k-n) P_n$$

33.11 SOLVED PROBLEMS

Problem 33.14: There are five machines each of which when running suffers breakdown at an average rate of 2 per hour. There are two servicemen and only one man can work on a machine at a time. Once a service man starts work on a machine, the time to complete the repair has an exponential distribution with mean of 5 minutes. Find the distribution of number of machines out of action at a given time. Find also the average time out of action machine has to spend waiting for the repair to start.

Solution:

k = Number of machines in the system = 5
R = Number of repairmen = 2
Arrival rate = 2 machines/hour
Service rate = 12 machines/hour

$$P_n = kc_n \left(\frac{\lambda}{\mu}\right)^n P_0, \quad 0 \le n \le R$$

$$P_n = \frac{kc_n n! (\lambda/\mu)^n}{R! R^{n-R}} P_0, \quad R+1 \le n \le k$$

$$P_n = kc_n \left(\frac{\lambda}{\mu}\right)^n P_0, \quad 0 \le n \le 2$$

$$P_n = \frac{kc_n n! \left(\frac{\lambda}{\mu}\right)^n}{R! R^{n-R}} P_0, \quad 3 \le n \le 5$$

$$P_0 = \left(\sum_{n=0}^{R} kc_n \left(\frac{\lambda}{\mu}\right)^n + \sum_{n=R+1}^{k} \frac{kc_n n! (\lambda/\mu)^n}{R! R^{n-R}} \right)^{-1}$$

$$P_0 = \left(\sum_{n=0}^{2} 5c_n (0.16)^n + \sum_{n=3}^{5} \frac{5c_n n! (0.16)^n}{2! 2^{n-2}} \right)^{-1}$$

$$P_0 = 0.457$$

$$P_n = kc_n \left(\frac{\lambda}{\mu} \right)^n P_0, \quad 0 \le n \le 2$$

$$P_1 = 5c_1 (0.16)^1 \, 0.457 = 0.38$$

$$P_2 = 5c_2 (0.16)^2 \, 0.457 = 0.129$$

$$P_n = \frac{kc_n n! \left(\frac{\lambda}{\mu} \right)^n}{R! R^{n-R}} P_0, \quad 3 \le n \le 5$$

$$P_3 = \frac{5c_3 3! (0.16)^3 \, 0.457}{2! (2)^{3-2}} = 0.031$$

$$P_4 = \frac{5c_4 4! (0.16)^4 \, 0.457}{2! (2)^{4-2}} = 0.003$$

$$P_5 = \frac{5c_5 5! (0.16)^5 \, 0.457}{2! (2)^{5-2}} = 0.0004$$

$$L_q = \sum_{n=3}^{5} (n-2) P_n$$

$$L_q = P_3 + 2P_4 + 3P_5 = 0.031 + 2 \times 0.003 + 3 \times 0004 = 0.044$$

$$W_q = \frac{L_q}{\lambda_{eff}} = \frac{0.044}{8.5} = 0.31 \text{ minute}$$

$$\lambda_{eff} = \lambda \sum_{n=0}^{k} (k-n) p_n = 2 \sum_{n=0}^{5} (5-n) P_n$$

$$\lambda_{eff} = 2(5P_0 + 4P_1 + 3P_2 + 2P_3 + P_4)$$

$$\lambda_{eff} = 2 (5 \times 0.457 + 4 \times 0.38 + 3 \times 0.126$$
$$+ 2 \times 0.031 + 0.003) = 8.5 \text{ machines/hour}$$

33.12 APPLICATIONS OF QUEUING MODELS

- Cinema theaters
- Railway stations
- Cafeteria
- Retail: Supermarkets, stores, banks
- Medical: Doctor's office, access to diagnostic procedures, specialist referrals
- Airports: Check-in, baggage collection, runway delays, waiting to land
- Traffic: Congestion
- Queues in manufacturing: Order backlogs, work-in-process inventories, distribution inventories
- Service: Mechanic or Job shops

33.13 QUESTIONS AND PROBLEMS

1. Explain various queuing models.
2. Describe various elements of queue.
3. Obtain steady state equation for the model (M/M/1/∞/∞/FCFS).
4. Obtain steady state equation for the model (M/M/1/∞/N/FCFS).
5. Explain multiserver model.
6. Explain machine repair queuing model.
7. A repair shop is attended by a single mechanic has an average of 4 customers per hour who bring small appliances for repair. The mechanic inspects them for defects and quite often can fix them right away or otherwise renders a diagnosis. This takes him 6 minutes on the average. Arrivals are Poisson and service time has the exponential distributions. You are required to find:
    i. The proportion of time during which the shop is empty.
    ii. The probability of finding at least one customer in the shop.
    iii. The average number of customers in the system
    iv. The average time including service time spent by customer
        Answer: i. $P_0 = 0.6$, ii. $1 - P_0 = 0.4$, iii. $L_s = 0.66$ and iv. $W_s = 10$ minutes
8. The tool room company's quality control department is manned by a single clerk who takes an average of 5 minutes in checking parts of each of the machine coming for inspection. The machines arrive once in every 8 minutes on the average. One hour of the machine is valued at ₹ 15 and clerk's time is valued at ₹ 4 per hour. What is the average hourly queuing system costs associated with the quality control department?
        Answer: $W_s = 0.22$, $P_0 = 0.375$, Average queuing cost /machine $= 15 \times W_s = ₹ 3.33$,
        Idle time cost of mechanic $= P_0 \times 4 = ₹ 1.5$/hour
9. The mean rate of arrival of planes at an airport during the peak period is 20/hour but the actual number of arrivals in an hour follows Poisson's distribution. The airport can land 60 planes per hour on an average in good weather or 30 per hour in bad weather. Find:
    i. How many trains would be flying over the field in the stack on an average in good weather and in bad weather?
    ii. How long a plane would be in stack and process of landing in good and bad weather?
    iii. How much stack and landing time should be allowed so that priority to land out of order would have to be requested only once in 20 times?
    Answer: i. $L_q = 0.166$, 0.33, ii. 1.5 minutes, 6 minutes and iii. 4.5 minutes, 18 minutes
10. Arrivals of machinist at a tool crib are considered to be Poisson distributed at an average rate of 6 per hour. The length of time the machinist must remain at the tool crib is exponentially distributed with average time being 0.05 hour.
    a. What is the probability that machinist arriving the tool crib will have to wait?
    b. What is the average number of machinists at the tool crib?
    c. The company will install a second booth when convinced that a machinist would expect to have spent at least 6 minutes waiting and being serviced at the tool crib. By how much the flow of machinists to the tool crib will be increased to justify the addition of a second tool crib.
        Answer: a. 0.3, b. 1.43 and c. 0.122
11. A ticket issuing office is being managed by single server. Customers arrive to purchase tickets according to Poisson's distribution with a mean arrival rate of 30 per hour. The

time required to serve a customer has exponentially distributed with mean of 90 seconds. Find the length of the system and waiting time in the system.

Answer: Length of the system = 3 and waiting time in the system = 6 minutes

12. A telephone exchange has two long distance operators. The telephone company finds that during the peak load long distance call arrives in Poisson fashion at an average rate of 15 per hour. The length of service on these calls is exponentially distributed with mean length of 5 minutes. ›

   i. What is the probability that a subscriber will have to wait for his long distance call during the peak hours of the day?

   ii. If the subscriber will wait and serviced in turn, what is the expected waiting time?

   Answer: i. $1 - P_0 - P_1 = 0.48$, ii. $W_q = 3.2$ minutes

13. A petrol bunk has two pumps. The service time follows the exponential distribution with a mean of 4 minutes and cars arrive for service in a Poisson process at the rate of 10 cars per hour. Find the probability that a customer has to wait for service. What proportion of time the pumps remain idle?

   Answer: 0.167, 67% for each pump

14. A car park contains 5 cars. The arrival rate is Poisson at a mean rate of 10 per hour. The service time has negative exponential distribution with mean of 5 hours. How many cars are in car park on average and what is the probability of a newly arriving customer finding the car park full and having to park his car elsewhere?

15. A company currently has two tool cribs, each having single clerk. One tool crib handles only tools for heavy machinery, while the second one handles all other tools. It is observed that for each tool crib the arrivals follow Poisson distribution with mean of 20 per hour and service time distribution is negative exponential with mean of 2 minutes. The tool manager feels that, if tool cribs are combined such a way that either clerk can handle any kind of tools as demand arises would be more efficient and the waiting problem could be reduced to some extent. It is believed that the mean arrival rate at the tool crib will be 40 per hour, while the service time will remain unchanged. Compare in status of queues in terms of average waiting time in the queue for each mechanic.

   Answer: Single server: $W_q = 4$ minutes, Multiserver: $W_q = 1.6$ minutes. Better to combine.

16. A mechanic services four machines. For each machine the mean time between service requirements is 10 hours and is assumed to be from an exponential distribution. The repair time tends to follow the same distribution with a mean of 2 hours. When a machine is down for repair the time lost has a value of ₹ 20 per hour. The mechanic costs ₹ 50 for day. Find:

   i. What is the expected number of machines in operation?

   ii. What is the expected down time cost for day?

   Answer: i. 3 machines, ii. ₹ 200/day

Inventory

34.1 INTRODUCTION

The stock, which is kept in the stores for the smooth running of an organization, is called inventory. Inventory may be 1 day stock or 15 days stock or 1month stock or 1 year stock.

The inventory or stock of the goods may be kept in any of the following forms:

1. Raw material inventory.

   Examples:

Factory	Raw material
Rice mill	Paddy
Oil refinery	Crude
Power plant	Coal
Cement plant	Lime stone
Automobile	Steel

2. Semi-finished goods inventory: Inventory is kept in semi-finished form. Examples: Castings of pistons, cylinders, etc.
3. Finished goods inventory: Inventory or stock is kept in finished goods form. Examples: Televisions, laptops, machine tools, cement bags, automobiles, etc.
4. Spare parts inventory: These spare parts are used when parts of a machine are worn out or failure of part in a machine.

Advantages of understocking:

1. Less space is required
2. Less capital investment
3. Less taxes
4. Less depreciation
5. Less damages

Disadvantages of understocking:

1. More possibility of company going out of stock
2. More number of orders

Advantages of overstocking:

1. Less possibility of company going out of stock
2. Less number of orders
3. Price discounts when company purchases large quantities.

Disadvantages of overstocking:

1. More space is required
2. More capital investment
3. More taxes
4. More depreciation
5. More damages

Inventory decisions:

There are two important decisions to be made regarding the inventory.

1. Ordering quantity or batch quantity: Determining quantity to be ordered per each order.
2. Ordering Time: Time between the orders

34.2 COSTS INVOLVED IN INVENTORY

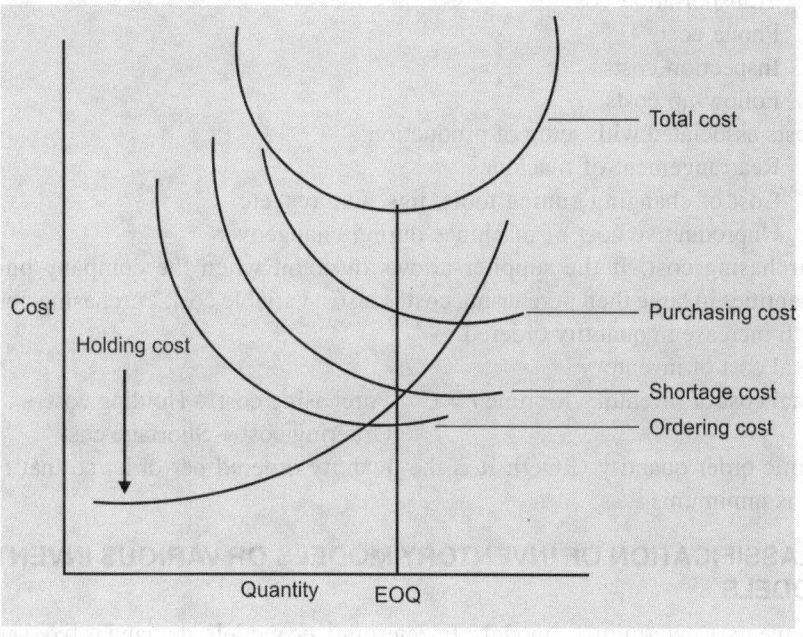

Fig. 34.1 Various cost curves

1. Holding cost or storage cost or inventory carrying cost: The cost associated with carrying or holding the goods in stores. Holding cost increases with the increase of cost. Holding cost per unit per unit time is denoted by c_1. Holding cost increases with increase of stock or inventory.

   Components of holding cost are:
   a. Interest on invested capital
   b. Cost of storage
   c. Taxes

    d. Insurance

    e. Depreciation cost, etc.

2. Shortage cost or stock out cost: The penalty cost that is incurred as a result of shortage of stock. Production is stopped when raw material or spare parts are not available. The loss incurred due to the shortage of stock or spare part is called shortage cost. Shortage cost per unit per unit time is denoted by c_2. Shortage cost decreases with increase in order size.

Shortage cost includes:

    a. Loss of production or sales

    b. Loss of profit

    c. Unproductive cost of men, machines and other capital resources

    d. Loss of goodwill

3. Setup cost or ordering cost: These include the fixed cost associated with obtaining goods by placing an order. Ordering cost per order or setup cost per setup is denoted by c_3. Ordering cost decreases with increase in order size.

Ordering cost includes:

    a. Cost of purchase

    b. Cost of stationery

    c. Transportation cost

    d. Phone cost

    e. Inspection costs

    f. Follow-up costs

Costs associated with setup of production:

    a. Rearrangement of machines

    b. Cost of changing cutting tools, jigs, fixtures, etc.

    c. Unproductive cost of machines during changeover

4. Purchasing cost: If the supplier allows discount when the company purchases the quantities in large then purchasing cost is also a variable cost. Purchasing cost decreases with increase in quantity ordered.

Total cost of inventory:

Total cost of inventory for time $t = C =$ Purchasing cost + Holding cost +
Ordering cost + Shortage cost

    Economic order quantity (EOQ): It is the quantity ordered per order so that total cost of inventory is minimum.

34.3 CLASSIFICATION OF INVENTORY MODELS OR VARIOUS INVENTORY MODELS

1. Elementary deterministic models: In deterministic models, demand is fixed and is known in advance.

Model 1a: Uniform demand, instantaneous replenishment without shortage cost

Model 1b: Uniform demand, instantaneous replenishment with shortage cost

Model 1c: Uniform demand, finite production rate without shortage cost

Model 1d: Uniform demand, finite production rate with shortage cost

2. Models with price breaks:

Model 2a: Model with two price breaks

Model 2b: Model with three price breaks

Model 2c: Model with number of price breaks

3. Models with restrictions
4. Probabilistic models:
   Model 4a: Discrete demand without setup cost
   Model 4b: Continuous demand without setup cost
5. Dynamic models:
   Model 5a: Fixed order quantity system
   Model 5b: Periodic review system
6. Inventory control:
   (a) ABC analysis
   (b) VED analysis
   (c) HML analysis
   (d) FSND analysis

34.4 MODEL 1a: UNIFORM DEMAND, INSTANTANEOUS REPLENISHMENT AND SHORTAGES ARE NOT ALLOWED (SIMPLE EOQ MODEL)

Assumptions made in simple EOQ model:

1. Uniform demand
2. Demand is known in advance.
3. Purchasing situation
4. Lead time is zero, i.e. time between placing an order and replenishment of stock is zero.
5. Shortages are not allowed, so shortage cost per unit is ∞.
6. Purchasing price per unit is constant.
7. Holding cost is calculated based on average time spent by the material in the production system.

Notations used:

R = Demand rate
c_1 = Holding cost per unit per unit time
c_2 = Shortage cost per unit per unit time
c_3 = Ordering cost per order
P = Purchasing cost per unit
t = Time between production or purchasing orders

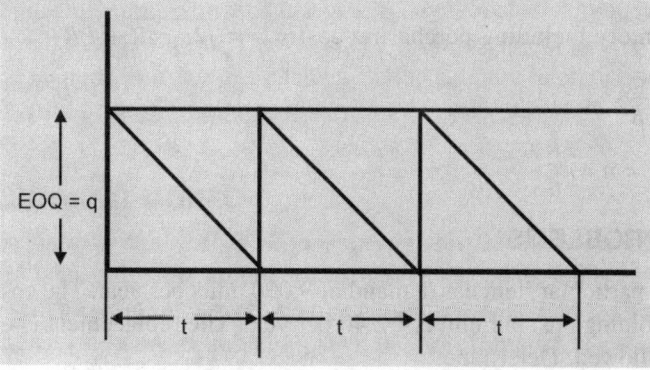

Fig. 34.2 Simple EOQ model 1a

In the simple EOQ model, there are two variable costs.
1. Holding cost
2. Ordering cost

Description of model:

Let q is the quantity ordered at the beginning of cycle t. These q units are consumed in time t at the rate of R units.

So $\qquad q = Rt$

$$t = \frac{q}{R}$$

Total inventory cost for time $t \quad = c =$ Holding cost + Ordering cost

$$= \left(\frac{q}{2}\right)c_1 t + c_3$$

Total inventory cost per unit time $= c = \dfrac{\dfrac{q}{2}c_1 t + c_3}{t}$

$$c = \left(\frac{q}{2}\right)c_1 + c_3 \frac{1}{t} \qquad \qquad ...(1)$$

Substituting $t = \dfrac{q}{R}$ in Equation 1

$$c = \left(\frac{q}{2}\right)c_1 + c_3 \frac{R}{q} \qquad \qquad ...(2)$$

$$\frac{dc}{dq} = \frac{c_1}{2} + c_3 R\left(\frac{-1}{q^2}\right) = 0$$

$$c_3 R\left(\frac{1}{q^2}\right) = \frac{c_1}{2}$$

$$EOQ = q = \sqrt{\frac{2c_3 R}{c_1}}$$

Substituting q in Equation 2

$$c = \frac{\sqrt{\dfrac{2c_3 R}{c_1}}}{2} c_1 + c_3 R\sqrt{\frac{c_1}{2c_3 R}}$$

Total cost of inventory $= c = \sqrt{2c_1 c_3 R}$

Total cost of inventory including purchasing cost $= c = \sqrt{2c_1 c_3 R} + PR$

Cycle time $= t = \dfrac{q}{R}$

Number of orders $= n = \dfrac{R}{q}$

34.5 SOLVED PROBLEMS

Problem 34.1: A particular item has demand of 9,000 units per year. The cost of procurement is ₹ 100 and the holding cost per unit is ₹ 2.40 per year. The replacement is instantaneous and no shortages are allowed. Determine:
   i. The economic lot size

ii. The number of orders per year

iii. The time between orders

iv. The total cost per year, if the cost of one unit is ₹ 1.

Solution:

R = Demand rate = 9,000 units/year

c_1 = Holding cost per unit per unit time = ₹ 2.40/unit/year

c_2 = Shortage cost per unit per unit time = ∞

c_3 = Ordering cost per order = ₹ 100

P = Purchasing cost per unit = ₹ 1

t = Time between production orders

i. $EOQ = q = \sqrt{\dfrac{2c_3 R}{c_1}}$

$EOQ = q = \sqrt{\dfrac{2 \times 100 \times 9,000}{2.40}}$

$EOQ = 866$ units/year

ii. $q = Rt$

Cycle time = Time between orders = $\dfrac{q}{R} = \dfrac{866}{9,000} = 0.0962$ year

iii. Number of orders = $\dfrac{R}{q} = \dfrac{9,000}{866} = 10.4$ orders/year

iv. Total cost of inventory including purchasing cost = $c = \sqrt{2c_1 c_3 R} + PR$

Total cost of inventory = $c = \sqrt{2 \times 2.4 \times 100 \times 9,000} + 9,000 \times 1$

$= ₹ 11,080$/year

Problem 34.2: A manufacturer has to supply 12,000 units of a product per year to the customer. The demand is fixed and the shortage cost is assumed to be infinite. The inventory holding cost is ₹ 0.20 per unit per month and the setup cost per run is ₹ 350. Determine:

i. The optimum run size

ii. Optimum scheduling period

iii. Minimum total variable yearly cost

Solution:

R = Demand rate = 12,000 units/year = 1,000 units/month

c_1 = Holding cost per unit per unit time = ₹ 0.2 /unit/month

c_2 = Shortage cost per unit per unit time = ∞

c_3 = Ordering cost per order = ₹ 350

i. $EOQ = q = \sqrt{\dfrac{2c_3 R}{c_1}}$

$EOQ = q = \sqrt{\dfrac{2 \times 350 \times 1,000}{0.2}}$

$= 1,870$ units/run

ii. Cycle time = $t = \dfrac{q}{R} = \dfrac{1,870}{1,000} = 1.87$ months

iii. Total cost of inventory = $c = \sqrt{2c_1 c_3 R}$

Purchasing cost is not given

Total cost of inventory excluding purchasing cost $= c = \sqrt{2 \times 0.2 \times 350 \times 1,000}$

$$= ₹ \ 374.1/\text{month} = ₹ \ 4,490/\text{year}$$

Problem 34.3: A stockist has to supply 400 units of product every Monday to his customers. He gets the product at ₹ 50 per unit from the manufacturer. The cost of ordering and transportation from the manufacturer is ₹ 75 per order. The cost of carrying inventory is 7.5% per unit per year of the cost of the product. Find:

    i. The economic lot size

    ii. The total optimal cost

Solution:

R = Demand rate = 400 units/week = 400 × 52 = 20,800 units/year

c_1 = Holding cost per unit per unit time = 0.075 × 50 = ₹ 3.75/unit/year

c_3 = Ordering cost per order = ₹ 75

p = Purchasing cost per unit = ₹ 50

    i. $EOQ = q = \sqrt{\dfrac{2c_3 R}{c_1}}$

$$EOQ = q = \sqrt{\dfrac{2 \times 75 \times 400 \times 52}{3.75}}$$

$$= 912 \text{ units/order}$$

    ii. Total cost of inventory including purchasing cost $= c = \sqrt{2c_1 c_3 R} + PR$

$$\text{Total cost of inventory} = c = \sqrt{2 \times 375 \times 75 \times 20,800} + 20,800 \times 50$$

$$= ₹ \ 10,43,420/\text{year}$$

$$= ₹ \ 20,065/\text{month}$$

34.6 MODEL 1b: UNIFORM DEMAND, INSTANTANEOUS REPLENISHMENT WITH SHORTAGES

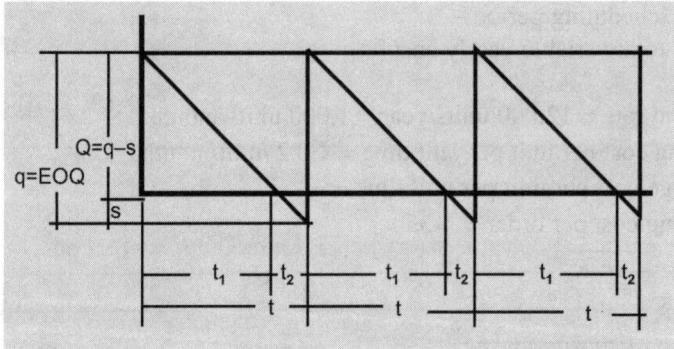

Fig. 34.3 Representation of model 1b

Description of model

Quantity ordered at the beginning of the time t_1 is q, i.e. economic ordered quantity. Out of which s units are utilized to meet previous shortages. Initial inventory at the beginning of time t_1 is $q\text{-}s$. These $q\text{-}s$ units are consumed at the rate of R units in period t_1. Stock at the end of period t_1 is zero. During the period t_2 shortages are allowed at the rate of R units. Shortage quantity at the end of period t_2 is s. At the end of period t_2, company places an order for quantity q. The cycle repeats.

Optimum shortage quantity $= s = q\dfrac{c_1}{(c_1 + c_2)}$

$$EOQ = q = \sqrt{\frac{2c_3 R}{c_1}}\sqrt{\frac{(c_1 + c_2)}{c_2}}$$

Total cost of inventory $= c = \sqrt{2c_1 c_3 R}\sqrt{\dfrac{(c_2)}{c_1 + c_2}}$

Maximum inventory $= Q = q - s = q\dfrac{c_2}{c_1 + c_2}$

Cycle time $= t = \dfrac{q}{R}$

Number of orders $= n = \dfrac{R}{q}$

34.7 SOLVED PROBLEMS

Problem 34.4: The demand of an item is uniform at a rate of 25 units per month. The fixed cost is ₹ 15 each time a production run is made. The production cost is ₹ 1 per item and the inventory carrying cost is ₹ 0.3 per item per month. If the shortage cost is ₹ 1.50 per item per month, determine how often to make a production run and what size it should be?

***Solution*:**

R = Demand rate = 25 units/month

c_1 = Holding cost per unit per unit time = ₹ 0.3/unit/month

c_2 = Shortage cost per unit per unit time = ₹ 1.5/item/month

c_3 = Ordering cost per order = ₹ 15

t = time between production runs or purchasing orders

$$EOQ = q = \sqrt{\frac{2c_3 R}{c_1}}\sqrt{\frac{(c_1 + c_2)}{c_2}}$$

$$q = EOQ = \sqrt{\frac{2 \times 15 \times 25}{0.3}}\sqrt{\frac{0.3 + 1.5}{1.5}}$$

$$= 54 \text{ items}$$

t = time between production or purchasing orders $= \dfrac{q}{R} = \dfrac{54}{25} = 2.16$ months

Problem 34.5: Given the data for an item of uniform demand, instantaneous delivery time and back order facility. Annual demand = 800 units; Cost of an item = ₹ 40; Ordering cost = ₹ 800/order; Inventory carrying cost = 40%/unit/year. Back order cost = ₹ 10/unit/year. Find out:

   i. Economic order quantity

  ii. Maximum number of backorders

 iii. Time between orders

 iv. Total annual cost

  v. Maximum inventory

Solution:

R = Demand rate = 800 units/year

c_1 = Holding cost per unit per unit time = ₹ 0.4×40 = ₹ 16/unit/year

c_2 = Shortage cost per unit per unit time = ₹ 10/unit/year

c_3 = Ordering cost per order = ₹ 800

p = purchasing cost per unit = ₹ 40

t = time between production or purchasing orders

i. $EOQ = q = \sqrt{\dfrac{2c_3R}{c_1}} \sqrt{\dfrac{(c_1 + c_2)}{c_2}}$

$= 456$ units

ii. Optimum shortage quantity = $s = q \dfrac{c_1}{c_1 + c_2}$

$s = 281$ units

iii. Cycle time = $t = \dfrac{q}{R}$ = 0.57 year

iv. Total cost of inventory = $c = \sqrt{2c_1c_3R} \sqrt{\dfrac{(c_2)}{c_1 + c_2}} + PR$

$= 2,807 + 40 \times 800$ = ₹ 34,806/year

v. Maximum inventory = $Q = q - s = q \dfrac{c_2}{c_1 + c_2} = 456 \dfrac{10}{16 + 10} = 175$ units

Problem 34.6: A contractor undertakes to supply diesel engines to a truck manufacturer at a rate of 25 per day. He finds that the cost of holding a completed engine in stock is ₹ 16 per month, and there is a clause in the contract penalizing him ₹ 10 per engine per day late for missing the scheduled delivery date. Production of engines is in batches and each time a new batch is started there is setup cost of ₹ 10,000. How frequently should batches be started and what should be the initial inventory level at the time each batch is completed.

Solution:

R = Demand rate = 25 units/day

c_1 = Holding cost per unit per unit time = 16/30 = ₹ 0.533/engine/day

c_2 = Shortage cost per unit per unit time = ₹ 10/engine/day

c_3 = Ordering cost per order = ₹ 10,000

t = time between production or purchasing orders

$$EOQ = \sqrt{\dfrac{2c_3R}{c_1}} \sqrt{\dfrac{(c_1 + c_2)}{c_2}}$$

$$q = EOQ = \sqrt{\dfrac{2 \times 10,000 \times 25}{0.533}} \sqrt{\dfrac{10 + 0.53}{10}}$$

$$q = 993 \text{ engines}$$

Maximum inventory = $Q = q - s = q \dfrac{c_2}{c_1 + c_2}$

$$= 993 \dfrac{10}{10 + 0.53}$$

$$= 943 \text{ engines}$$

34. 8 MODEL 1c: UNIFORM DEMAND, FINITE PRODUCTION RATE WITHOUT SHORTAGES

This model is an example of the company that is producing items at the rate of k units and at the same time they are consuming at the rate of R units.

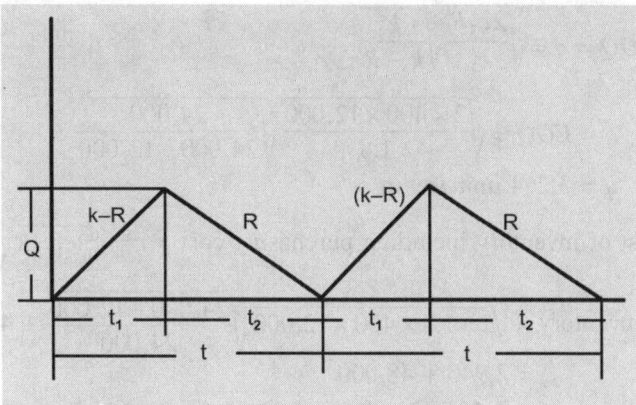

Fig. 34.4 Representation of model 1c

Description of model

At the beginning of the cycle inventory is zero. The inventory is building up at the rate of $(k - R)$ units during the period t_1. At the end of time t_1 maximum inventory is Q. Production is stopped at the end of period t_1. During the time t_2 inventory is reduced at the rate of R. There is no production in time t_2. At the end of time t_2 stock is zero. Again the cycle repeats.

$$EOQ = q = \sqrt{\frac{2c_3 R}{c_1}} \sqrt{\frac{k}{k - R}}$$

Total minimum cost of inventory $= c = \sqrt{2c_1 c_3 R} \sqrt{\frac{k - R}{k}}$

Total minimum cost of inventory including purchasing cost $= c = \sqrt{2c_1 c_3 R} \sqrt{\frac{k - R}{k}} + PR$

Maximum inventory $= Q = \dfrac{(k - R)q}{k}$

Production time $= t_1 = \dfrac{q}{k}$

Cycle time $= t = \dfrac{q}{R}$

34.9 SOLVED PROBLEMS

Problem 34.7: A company has a demand of 12,000 units per year for an item and it can produce 2,000 such items per month. The cost of one setup is ₹ 400 and the holding cost per unit per month is ₹ 0.15. Find the optimum lot size and the total cost per year. Assuming the cost of one unit as ₹ 4. Also find the maximum inventory, manufacturing time and total time.

Solution:

R = Demand rate = 12,000 units/year

k = Production rate = 24,000 units/year

c_1 = Holding cost per unit per unit time = 0.15×12 = ₹ 180/unit/year

c_2 = Shortage cost per unit per unit time = ∞

c_3 = Ordering cost per order = ₹ 400

t = Time between production or purchasing orders

P = Purchasing price = ₹ 4/unit

$$EOQ = q = \sqrt{\frac{2c_3R}{c_1}} \sqrt{\frac{k}{k-R}}$$

$$q = EOQ = \sqrt{\frac{2 \times 400 \times 12,000}{1.8}} \sqrt{\frac{24,000}{24,000 - 12,000}}$$

$$q = 3,264 \text{ units/setup}$$

Total minimum cost of inventory including purchasing cost = $c = \sqrt{2c_1c_3R} \sqrt{\frac{k-R}{k}} + PR$

Minimum cost of inventory = $\sqrt{2 \times 1.8 \times 400 \times 12,000} \sqrt{\dfrac{24,000 - 12,000}{24,000}} + 4 \times 12,000$

$$= 2,940 + 48,000$$

$$c = ₹ 50,940/\text{year}$$

Production time = $t_1 = \dfrac{q}{k} = \dfrac{3,264}{24,000} = 0.136$ year = 1.633 months

Maximum inventory = $Q = (k - R) t_1 = (24,000 - 12,000)\, 0.136 = 1,632$ units

Cycle time = $t = \dfrac{q}{R} = 3.266$ months

Problem 34.8: An item is produced at the rate of 50 items per day. The demand occurs at the rate of 25 items per day, if the setup cost is ₹ 100 per run and inventory cost is ₹ 0.01 per unit per day. Find the economic lot size for one run, cycle time, manufacturing time and maximum inventory.

Solution:

R = Demand rate = 25 units/day

k = Production rate = 50 units/day

c_1 = Holding cost per unit per unit time = ₹ 0.01/unit/day

c_2 = Shortages are not allowed

c_3 = Ordering cost per order = ₹ 100

t = Time between production or purchasing orders

$$EOQ = q = \sqrt{\frac{2c_3R}{c_1}} \sqrt{\frac{k}{k-R}}$$

$$q = EOQ = \sqrt{\frac{2 \times 100 \times 25}{0.01}} \sqrt{\frac{50}{50 - 25}}$$

$$q = EOQ = 1,000 \text{ units}$$

Production time = $t_1 = \dfrac{q}{k} = \dfrac{1,000}{50} = 20$ days

Cycle time = $t = \dfrac{q}{R} = \dfrac{1,000}{25} = 40$ days

Maximum inventory = $Q = (k - R) t_1 = (50 - 25)\, 20 = 500$ units.

34.10 MODEL 1d: FINITE PRODUCTION RATE, UNIFORM DEMAND WITH SHORTAGES

Description of model: Production starts at the beginning of period t_4. The shortages of previous cycle are reduced during the time t_4 at the rate of $(k - R)$ units. At the beginning of the period t_1 inventory is zero. The inventory is building up at the rate of $(k - R)$ units during the period t_1. At the end of time t_1 maximum inventory is Q. At the end of period t_1 production is stopped. During the time t_2 inventory is reduced at the rate of R. There is no production in time t_2. During time t_2 stock is consumed at the rate of R. At the end of time t_2 stock is zero. During the period t_3 shortages are going to build at the rate of R. At the end of period t_3 optimum shortage quantity is equal to s. Again the cycle repeats.

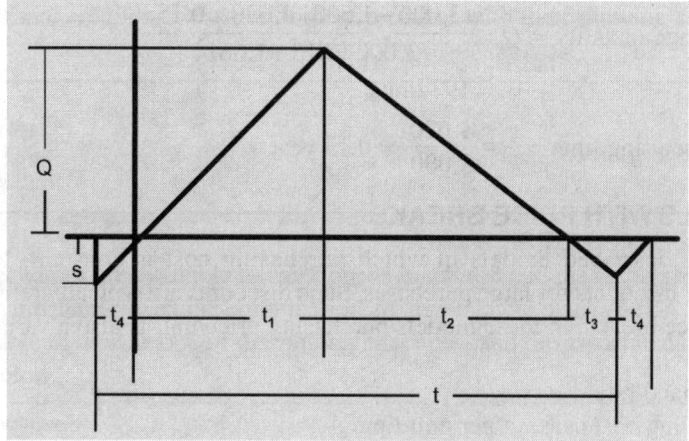

Fig. 34.5 Representation of model 1d

$$EOQ = q = \sqrt{\frac{2c_3 R}{c_1}} \sqrt{\frac{k}{k-R}} \sqrt{\frac{(c_1 + c_2)}{c_2}}$$

Total minimum cost of inventory $= c = \sqrt{2c_1 c_3 R} \sqrt{\frac{k-R}{k}} \sqrt{\frac{(c_2)}{c_1 + c_2}}$

Optimum shortage quantity $= Q = q - s = (k - R) \dfrac{q}{k} \dfrac{c_1}{(c_1 + c_2)}$

34.11 SOLVED PROBLEM

Problem 34.9: The demand for an item in a company is 18,000 units per year and the company can produce the item at a rate of 3,000 per month. The cost of one setup is ₹ 500 and the holding cost of one unit per month is ₹ 0.15 and the shortage cost of one unit is ₹ 20 per year. Determine the optimum manufacturing quantity and the number of shortages. Also, determine the manufacturing time and the time between setups.

Solution:

R = Demand rate = 18,000 units/year =1,500 units/month

k = Production rate = 3,000 units/month

c_1 = Holding cost per unit per unit time = ₹ 0.15/unit/month

c_2 = Shortage cost = ₹ 20/12 = ₹ 1.66/unit/month

c_3 = Ordering cost per order = ₹ 500

t = Time between production or purchasing orders

$$EOQ = q = \sqrt{\frac{2c_3R}{c_1}}\sqrt{\frac{k}{k-R}}\sqrt{\frac{(c_1+c_2)}{c_2}}$$

$$q = EOQ = \sqrt{\frac{2 \times 500 \times 1,500}{0.15}}\sqrt{\frac{3,000}{3,000-1,500}}\sqrt{\frac{0.15+1.66}{1.66}}$$

$$q = EOQ = 4,670 \text{ units}$$

Optimum shortage quantity = $Q = q - s = (k-R)\dfrac{q}{k}\dfrac{c_1}{(c_1+c_2)}$

Optimum shortage quantity = $Q = \dfrac{(3,000-1,500)\ 4,670 \times 0.15}{3,000\ (0.15+1.66)}$

$$= 193 \text{ units}$$

Manufacturing time $= \dfrac{q}{k} = \dfrac{4,670}{3,000} = 0.13 \text{ year}$

34.12 MODELS WITH PRICE BREAKS

So far we have described models in which purchasing cost is constant. However, many enterprises offer discounts for large purchases. Such discounts are typically referred as quantity discounts or price breaks. In these models purchasing price/unit is also a variable cost.

Notations used:

R = Demand rate

c_1 = Holding cost per unit per unit time

p = Purchasing cost per unit

t = Time between production runs or purchasing orders

I = Interest in percentage on one rupee capital invested /unit time

In this *EOQ* model, there are three variables.

1. Holding cost
2. Ordering cost
3. Purchasing cost

$$EOQ = q = \sqrt{\frac{2c_3R}{pI}}$$

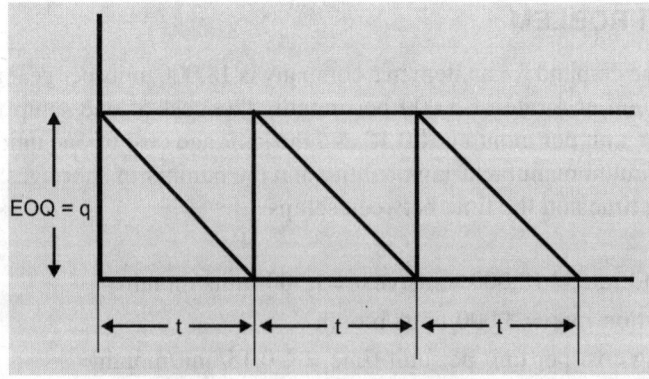

EOQ = q

Fig. 34.6 Representation of simple price breaks model

Total cost curves (*c*) for different prices are given in Fig. 34.7. So Economic order quantity is determined by using search procedure presented in sections 34.13 and 34.15.

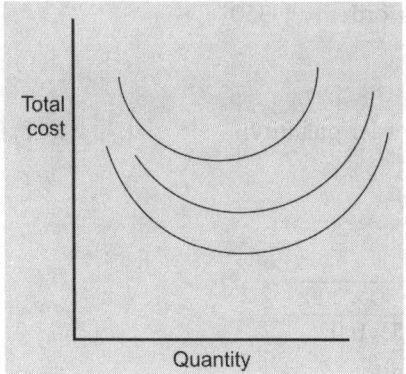

Fig. 34.7 Total cost curves for different purchasing prices per unit

34.13 MODEL WITH SINGLE PRICE BREAK

Price/unit	Quantity purchased
p_1	$1 \le q < b$
p_2	$q \ge b$

Steps to determine optimal order quantity:

Step 1: Determine

$$q_2 = \sqrt{\frac{2c_3 R}{p_2 I}}$$

If $q_2 \ge b$ then optimum order quantity is q_2.

If $q_2 < b$ determine

$$q_1 = \sqrt{\frac{2c_3 R}{p_1 I}}$$

Also determine

$$c(q_1) = p_1 R + \frac{c_3 R}{q_1} + \frac{c_3 I}{2} + \frac{p_1 q_1 I}{2}$$

$$c(b) = p_2 R + \frac{c_3 R}{b} + \frac{c_3 I}{2} + \frac{p_2 b I}{2}$$

Whichever be the minimum cost, i.e. optimal ordered quantity.

34.14 SOLVED PROBLEMS

Problem 34.10: Find the optimum order quantity for a product for which the price breaks are as follows:

Quantity	Unit cost (₹)
$0 \le q < 500$	10
$q \ge 500$	9.25

The monthly demand for a product is 200 units. The cost of storage is 2% of unit cost per month and the cost of ordering is ₹ 350.

Solution:

R = Demand rate = 200 units/month

I = 0.02/unit/month

c_3 = Ordering cost per order = ₹ 350

p_1 = ₹ 10/unit

p_2 = ₹ 9.25/unit

Steps to determine optimal order quantity:

Step 1: Determine

$$q_2 = \sqrt{\frac{2c_3R}{p_2I}}$$

$$q_2 = \sqrt{\frac{2 \times 350 \times 200}{9.25 \times 0.02}}$$

q_2 = 870 units

$q_2 \geq 500$

So optimum order quantity is q_2.

Optimum order quantity = q_2 = 870 units

Problem 34.11: Find the optimum order quantity for a product for which the price breaks are as follows:

Quantity	Unit cost (₹)
$0 \leq q < 500$	10
$q \geq 500$	9.25

The monthly demand for a product is 200 units. The cost of storage is 2% of unit cost per month and the cost of ordering is ₹ 100.

Solution:

R = Demand rate = 200 units/month

I = 0.02/unit/month

c_3 = Ordering cost per order = ₹ 100

p_1 = ₹ 10/unit

p_2 = ₹ 9.25/unit

Steps to determine optimal order quantity:

Step 1: Determine

$$q_2 = \sqrt{\frac{2c_3R}{p_2I}}$$

$$q_2 = \sqrt{\frac{2 \times 100 \times 200}{9.25 \times 0.02}}$$

q_2 = 465 units

$q_2 < 500$, so go to step 2.

Step 2

$$q_1 = \sqrt{\frac{2c_3R}{p_1I}}$$

$$q_1 = \sqrt{\frac{2 \times 100 \times 200}{10 \times 0.02}}$$

$q_1 = 447$ units

$$c(q_1) = p_1 R + \frac{c_3 R}{q_1} + \frac{c_3 I}{2} + \frac{p_1 q_1 I}{2}$$

$$c(447) = 10 \times 200 + \frac{100 \times 200}{447} + \frac{100 \times 0.02}{2} + \frac{10 \times 447 \times 0.02}{2}$$

$$= ₹ \ 2,090.42/\text{month}$$

$$c(b) = p_2 R + \frac{c_3 R}{b} + \frac{c_3 I}{2} + \frac{p_2 b I}{2}$$

$$c(500) = 9.25 \times 200 + \frac{100 \times 200}{500} + \frac{100 \times 0.02}{2} + \frac{9.25 \times 500 \times 0.02}{2}$$

$$= ₹ \ 1,937/\text{month}$$

$c(500) < c(447)$

So optimum order quantity is 500 units.

Problem 34.12: A company uses annually 24,000 units of raw materials which cost ₹ 1.25 per unit. Placing each order costs ₹ 22.50 and the carrying cost is 5.4% per year of the average inventory. Find the economic lot size and total inventory cost (including cost of material). Should the company accept the offer made by the supplier of discount of 5% on the cost price on a single order of 24,000 units?

Solution:

R = Demand rate = 24,000 units/year

I = 0.054/unit/year

c_3 = Ordering cost per order = ₹ 22.50

p = purchasing price per unit = ₹ 1.25

Case 1:

$$EOQ = q = \sqrt{\frac{2 c_3 R}{c_1}}$$

$$EOQ = q = \sqrt{\frac{2 \times 22.50 \times 24,000}{0.054 \times 1.25}}$$

$q = EOQ = 4,000$ units

Case 2: If the company purchases 24,000 units in single order then purchasing price per unit is $0.95 \times 1.25 = ₹ \ 1.187$

$$c(q) = p_1 R + \frac{c_3 R}{q_1} + \frac{c_3 I}{2} + \frac{p_1 q_1 I}{2}$$

$$c(4,000) = 1.25 \times 24,000 + \frac{22.5 \times 24,000}{4,000} + \frac{22.5 \times 0.054}{2} + \frac{1.25 \times 4,000 \times 0.054}{2}$$

$$= ₹ \ 30,270/\text{year}$$

$$c(b) = p_2 R + \frac{c_3 R}{b} + \frac{c_3 I}{2} + \frac{p_2 b I}{2}$$

$$c(24,000) = 1.18 \times 24,000 + \frac{22.5 \times 24,000}{24,000} + \frac{22.5 \times 0.054}{2} + \frac{1.18 \times 24,000 \times 0.054}{2}$$

$c(24,000) = ₹ \ 29,292/\text{year}$

The company can accept the single order of 24,000 units.

34.15 MODEL WITH NUMBER OF PRICE BREAKS

Price/unit	Quantity purchased
p_1	$1 \le q < b_1$
p_2	$b_1 \le q < b_2$
p_3	$q \ge b_2$

Steps to determine optimal order quantity:

Step 1: Determine

$$q_3 = \sqrt{\frac{2c_3R}{p_3I}}$$

If $q_3 \ge b_2$ then optimum order quantity is q_3.

If $q_3 < b_2$ then go to step 2.

Step 2 : Determine

$$q_2 = \sqrt{\frac{2c_3R}{p_2I}}$$

If $q_2 \ge b_1$ then determine

$$c(q_2) = p_2R + \frac{c_3R}{q_2} + \frac{c_3I}{2} + \frac{p_2q_2I}{2}$$

$$c(b_2) = p_2R + \frac{c_3R}{b_2} + \frac{c_3I}{2} + \frac{p_2b_2I}{2}$$

If $q_2 < b_1$ then go to step 3.

Step 3:

$$q_1 = \sqrt{\frac{2c_3R}{p_1I}}$$

And also determine

$$c(q_1) = p_1R + \frac{c_3R}{q_1} + \frac{c_3I}{2} + \frac{p_1q_1I}{2}$$

$$c(b_1) = p_2R + \frac{c_3R}{b_1} + \frac{c_3I}{2} + \frac{p_2b_1I}{2}$$

$$c(b_2) = p_2R + \frac{c_3R}{b_2} + \frac{c_3I}{2} + \frac{p_2b_2I}{2}$$

Whichever be the minimum cost, i.e. optimal order quantity.

34.16 SOLVED PROBLEMS

Problem 34.13: Find the optimal order quantity for which the price breaks are as follows:

Quantity	Unit cost (₹)
$0 \le q < 500$	10
$500 \le q < 750$	9.25
$q \ge 750$	8.75

The monthly demand for a product is 200 units and the cost of storage is 2% of the unit cost per month and the cost of ordering is ₹ 350.

Solution:

R = Demand rate = 200 units/month

I = 0.02/unit/month

c_3 = Ordering cost per order = ₹ 350

p_1 = ₹ 10/unit

p_2 = ₹ 9.25/unit

p_3 = ₹ 8.75/unit

Steps to determine optimal ordered quantity:

$$q_3 = \sqrt{\frac{2c_3R}{p_3I}}$$

$$q_3 = \sqrt{\frac{2 \times 350 \times 200}{8.75 \times 0.02}}$$

q_3 = 894 units

$q_3 \geq 750$ then optimum order quantity is q_3, i.e. 894.

Problem 34.14: The annual demand for a product is 64,000 units. The cost per order is ₹ 10 and the estimated cost of carrying one unit stock for a year is 20%. The normal price of the product is ₹ 10 per unit. However, the supplier offers a quantity discount of 2% on order of at least 1,000 units at a time and a discount of 5%, if the order is at least 5,000 units. Suggest the most economical purchase quantity per order.

***Solution*:**

R = Demand rate = 64,000 units/year

I = 0.2/unit/year

c_3 = Ordering cost per order = ₹ 10

p_1 = ₹ 10/unit

p_2 = ₹ 9.8/unit

p_3 = ₹ 9.5/unit

Quantity	Unit cost (₹)
$0 \leq q < 1,000$	10
$1,000 \leq q < 5,000$	9.8
$q \geq 5,000$	9.5

Steps to determine optimal order quantity:

$$q_3 = \sqrt{\frac{2c_3R}{p_3I}}$$

$$q_3 = \sqrt{\frac{2 \times 10 \times 64,000}{9.5 \times 0.2}}$$

q_3 = 820 units

$q_3 < 5,000$, go to step 2.

Step 2:

$$q_2 = \sqrt{\frac{2c_3R}{p_2I}}$$

$$q_2 = \sqrt{\frac{2 \times 10 \times 64,000}{9.8 \times 0.2}}$$

q_2 = 808 units

$q_2 < 1,000$, go to step 3.

Step 3:

$$q_2 = \sqrt{\frac{2c_3R}{p_1I}}$$

$$q_2 = \sqrt{\frac{2 \times 10 \times 64{,}000}{10 \times 0.2}} = 800 \text{ units}$$

$$c(q_1) = p_1R + \frac{c_3R}{q_1} + \frac{c_3I}{2} + \frac{p_1q_1I}{2}$$

$$c(800) = 10 \times 64{,}000 + \frac{10 \times 64{,}000}{800} + \frac{10 \times 0.2}{2} + \frac{10 \times 800 \times 0.2}{2} = ₹ \, 6{,}41{,}601$$

$$c(b_1) = p_2R + \frac{c_3R}{b_1} + \frac{c_3I}{2} + \frac{p_2b_1I}{2}$$

$$c(1{,}000) = 9.8 \times 64{,}000 + \frac{10 \times 64{,}000}{1{,}000} + \frac{10 \times 0.2}{2} + \frac{9.8 \times 1{,}000 \times 0.2}{2} = ₹ \, 6{,}28{,}821$$

$$c(b_2) = p_2R + \frac{c_3R}{b_2} + \frac{c_3I}{2} + \frac{p_2b_2I}{2}$$

$$c(5{,}000) = 9.5 \times 64{,}000 + \frac{10 \times 64{,}000}{5{,}000} + \frac{10 \times 0.2}{2} + \frac{9.5 \times 5{,}000 \times 0.2}{2} = ₹ \, 6{,}12{,}879$$

$$c(5{,}000) = ₹ \, 6{,}12{,}879$$

Optimal order quantity = 5,000 units

Problem 34.15: The annual demand for a product is 500 units. The cost of storage per unit per year is 10% of unit cost. The ordering cost is ₹ 180 for each order. The unit cost depends upon the quantity ordered. The range of quantity ordered and the unit cost price are as follows.

Range of quantity ordered	Unit cost (₹)
$0 \le q < 500$	25
$500 \le q < 1{,}500$	24.80
$1{,}500 \le q < 3{,}000$	24.60
$q \ge 3{,}000$	24.40

Find the optimal order quantity.
Solution:

R = Demand rate = 500 units/year
I = 0.1/unit/year
c_3 = Ordering cost per order = ₹ 180
p_1 = ₹ 25/unit
p_2 = ₹ 24.80/unit
p_3 = ₹ 24.60/unit
p_4 = ₹ 24.60/unit

Steps to determine optimal order quantity:
Step 1:

$$q_4 = \sqrt{\frac{2c_3R}{p_4I}}$$

$$q_4 = \sqrt{\frac{2 \times 180 \times 500}{0.1 \times 24.4}}$$

$q_4 = 271$ units

$q_4 < 3,000$, go to step 2.

Step 2:

$$q_3 = \sqrt{\frac{2c_3 R}{p_3 I}}$$

$$q_2 = \sqrt{\frac{2 \times 180 \times 500}{0.1 \times 24.6}}$$

$q_3 = 270.5$ units

$q_3 < 1,500$, go to step 3.

Step 3:

$$q_2 = \sqrt{\frac{2c_3 R}{p_2 I}}$$

$$q_2 = \sqrt{\frac{2 \times 180 \times 500}{0.1 \times 24.8}}$$

$q_2 = 269$ units

$q_2 < 500$, go to step 4.

Step 4:

$$q_1 = \sqrt{\frac{2c_3 R}{p_1 I}}$$

$$q_1 = \sqrt{\frac{2 \times 180 \times 500}{0.1 \times 25}}$$

$q_1 = 268$ units

$$c(268) = 25 \times 500 + \frac{180 \times 500}{268} + \frac{180 \times 0.1}{2} + \frac{268 \times 25 \times 0.1}{2}$$

$$= c(268) = ₹\ 13,179$$

$$c(500) = 24.8 \times 500 + \frac{180 \times 500}{500} + \frac{180 \times 0.1}{2} + \frac{500 \times 24.8 \times 0.1}{2}$$

$$= c(500) = ₹\ 13,209$$

$$c(1,500) = 24.6 \times 500 + \frac{180 \times 500}{1,500} + \frac{180 \times 0.1}{2} + \frac{1,500 \times 24.6 \times 0.1}{2}$$

$$= c(1,500) = ₹\ 14,214$$

$$c(3,000) = 24.4 \times 500 + \frac{180 \times 500}{3,000} + \frac{180 \times 0.1}{2} + \frac{3,000 \times 24.4 \times 0.1}{2}$$

$$= c(3,000) = ₹\ 15,899$$

Optimal order quantity = 268 units

34. 17 PROBABILISTIC MODELS OR STOCHASTIC MODELS

In all previous models, it is assumed that the demand is fixed and known. In probabilistic models, future demand is not known exactly, but probability distribution of demand is known.

There are two types of demands:

1. Discrete demand: Discrete demand occurs in discrete units. Demand occurs only in integers. Discrete probability demand distribution functions are assumed.

Examples: Demand for two wheelers, news papers, etc.

2. Continuous demand: Demand occurs in continuous units. Continuous probability demand distribution functions are assumed.

   Examples: Demand for cake, oil, etc.

34.18 MODEL 1: INSTANTANEOUS DEMAND (DISCRETE DEMAND) NO SETUP COST

In this model, it is assumed that demand occurs once and only once in a period, *t* and there is no setup cost.

Notations used:

q = Stock at the beginning of the cycle or optimum order quantity

c_1 = Holding cost/unit/unit time

c_2 = Shortage cost/unit/unit time

$p(d)$ = Probability that the demand is *d*

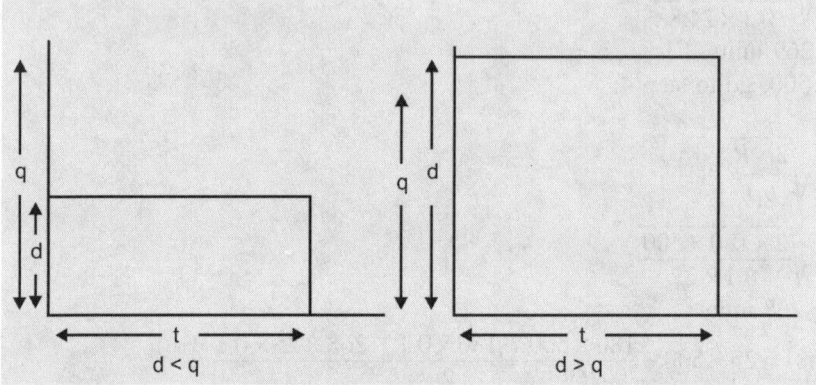

Fig. 34.8 Graphical representation of instantaneous demand in time *t*

There are three possibilities of demand in period *t*:

1. $d < q$ then there is holding cost for the $(q-d)$ units in period *t*.
2. $d = q$ then there is no holding or shortage cost for the period *t*.
3. $d > q$ then there is shortage cost for $(d-q)$ units in the period *t*.

q is optimum order quantity, if the following condition is satisfied.

$$\sum_{0}^{q-1} p(d) < \frac{c_2}{c_1 + c_2} < \sum_{0}^{q} p(d)$$

$$\sum_{0}^{q-1} p(d) = \text{Cumulative probability that demand} \le q - 1 \text{ units}$$

$$\sum_{0}^{q} p(d) = \text{Cumulative probability that demand} \le q \text{ units}$$

34.19 SOLVED PROBLEMS

Problem 34.16: A Newspaper boy has following probabilities of selling a magazine (Table 34.1).

Table 34.1 Probabilities of selling a magazine

The number of copies sold	10	11	12	13	14
Probability	0.10	0.15	0.20	0.25	0.30

Cost of copy is 30 paise and sale price is 50 paise. He cannot return unsold copies. Determine the number of copies to be ordered by the newspaper boy.

Solution:

Holding cost $= 30$ paise/unit/day $= ₹ 0.3$/unit/day

Shortage cost $= 20$ paise/unit/day $= ₹ 0.2$/unit/day

$$\frac{c_2}{c_1 + c_2} = \frac{20}{20 + 30} = 0.4$$

Table 34.2 To determine cumulative probability

The number of copies sold (d)	10	11	12	13	14
Probability $p(d)$	0.10	0.15	0.20	0.25	0.30
Cumulative probability $\Sigma p(d)$	0.10	0.25	0.45	0.70	1.00

q is optimum order quantity, if the following condition is satisfied.

$$\sum_0^{q-1} p(d) < \frac{c_2}{c_1 + c_2} < \sum_0^{q} p(d)$$

$$\sum_0^{12-1} p(d) < \frac{20}{20 + 30} < \sum_0^{12} p(d)$$

$25.25 < 0.4 < 0.45$

So optimal order quantity $= q = 12$ newspapers

Problem 34.17: The probability distribution of monthly sales of certain item is shown in the Table 34.3. The cost of carrying inventory is ₹ 30 per unit per month and the cost of unit shortage is ₹ 70 per month. Determine the optimum stock level which minimizes the total expected cost.

Table 34.3 The probability distribution of monthly sales of certain item

Monthly sales	0	1	2	3	4	5	6
Probability	0.01	0.06	0.25	0.35	0.20	0.03	0.1

Solution:

Holding cost $= c_1 = ₹ 30$/unit/month

Shortage cost $= c_2 = ₹ 70$/unit/month

$$\frac{c_2}{c_1 + c_2} = \frac{70}{70 + 30} = 0.7$$

Table 34.4 To determine cumulative probability

Monthly sales (d)	0	1	2	3	4	5	6
Probability $p(d)$	0.01	0.06	0.25	0.35	0.20	0.03	0.1
Cumulative probability $\Sigma p(d)$	0.01	0.07	0.32	0.67	0.87	0.9	1.0

q is optimum order quantity, if the following condition is satisfied.

$$\sum_0^{q-1} p(d) < \frac{c_2}{c_1 + c_2} < \sum_0^q p(d)$$

$$\sum_0^{4-1} p(d) < 0.7 < \sum_0^4 p(d)$$

$$0.67 < 0.7 < 0.87$$

So optimal order quantity $= q = 4$ units

Problem 34.18: Some of the spare parts of a ship cost ₹ 1,00,000 each. These spare parts can only be ordered together with the ship. If not ordered at the time when the ship is constructed, these parts cannot be available on need. Suppose that a loss of ₹ 1,00,00,000 is suffered for each spare part that is needed when none is available in the stock. Further, suppose that, probabilities that spare parts will be needed as replacement during the life term of the class of the ship is given in Table 34.5.

Table 34.5 Probabilities that spare parts will be needed as replacement during the life term

Spare parts	0	1	2	3	4	5 or more
Probability	0.9488	0.0400	0.0100	0.0010	0.0002	0.00

How many spare parts should be procured?

Solution:

Holding cost $=$ ₹ 1,00,000/unit

Shortage cost $=$ ₹ 1,00,00,000/unit

$$\frac{c_2}{c_1 + c_2} = \frac{1,00,00,000}{1,00,00,000 + 1,00,000} = 0.99009$$

Table 34.6 Determining cumulative probability

Spare parts (d)	0	1	2	3	4	5 or more
Probability $p(d)$	0.9488	0.0400	0.0100	0.0010	0.0002	0.00
Cumulative probability $\Sigma p(d)$	0.9488	0.9888	0.9988	0.9998	1.00	1.00

q is optimum order quantity, if the following condition is satisfied.

$$\sum_0^{q-1} p(d) < \frac{c_2}{c_1 + c_2} < \sum_0^q p(d)$$

$$\sum_0^{2-1} p(d) < 0.990009 < \sum_0^2 p(d)$$

$$0.9488 < 0.990009 < 0.9988$$

So optimal order quantity $= q = 2$ units

34.20 MODEL 2: INSTANTANEOUS DEMAND (CONTINUOUS DEMAND) NO SETUP COST

In this model, it is assumed that demand occurs once and only once in a period t and there is no setup cost. The demand function is assumed to be continuous.

Notations used:

    q = Stock at the beginning of the cycle or optimum order quantity

    c_1 = Holding cost/unit/unit time

    c_2 = Shortage cost/unit/unit time

    $f(x)$ = Probability that the demand is x

q is the optimal order quantity, if the following condition is satisfied.

$$\int_0^q f(x)\,d(x) = \frac{c_2}{c_1 + c_2}$$

34.21 SOLVED PROBLEMS

Problem 34.19: An ice cream company sells one of its types of ice cream by the weight. If the product is not sold on the day it is prepared, it can be sold at a loss of 50 paise per kg. But there is unlimited market for one day old ice cream. On the other hand, the company makes a profit of ₹ 3.20 on every kg of ice cream sold on the day it is prepared. Past daily distribution with

$$f(x) = 0.02 - 0.0002x,\ 0 \le x \ge 100$$

Determine optimum order quantity.

Solution:

    Holding cost = c_1 = ₹ 0.5/unit/day

    Shortage cost = c_2 = ₹ 3.20/unit/day

Probability distribution function of demand x

$$f(x) = 0.02 - 0.0002x,\ 0 \le x \ge 100$$

q be the optimal order quantity, if the following condition is satisfied.

$$\int_0^q f(x)\,d(x) = \frac{c_2}{c_1 + c_2}$$

$$\int_0^q (0.02 - 0.0002x)\,d(x) = \frac{3.2}{0.5 + 3.2}$$

$$\left[0.02x - \frac{0.0002x^2}{2} \right]_0^q = 0.864$$

Substituting values of q and zero

$$\left[0.02q - \frac{0.0002q^2}{2} \right] = 0.864$$

Solving the above equation,

Optimal order quantity = q = 63.2 kg

Problem 34.20: A baking company sells one of its types of cakes by weight. It makes a profit of 95 paise a pound on every pound of cake sold on the day it is baked. It disposes all cakes not sold on the day they are baked at a loss of 15 paise a pound. If the demand is known to be rectangular between 3,000 and 4,000 pounds, determine the optimum amount to be backed.

Solution:

    Holding cost = ₹ 0.15/pound/day

    Shortage cost = ₹ 0.95/pound/day

Probability distribution function

$$f(x) = \frac{1}{4,000 - 3,000}$$

$$f(x) = \frac{1}{1,000}$$

$$\int_0^q f(x)\, d(x) = \frac{c_2}{c_1 + c_2}$$

$$\int_{3000}^q \left(\frac{1}{1,000}\right) d(x) = \frac{0.95}{0.95 + 0.15} = 0.8636$$

$$\left(\frac{1}{1,000}\right) |x|_{3000}^q = 0.8636$$

$$q - 3,000 = 863.6$$

Optimal order quantity = q = 3,863.6 units

Problem 34.21: The probability density of demand of a certain item during a week to be
$$f(x) = 0.1, \qquad 0 \le x \le 10$$

The demand is assumed to occur with a uniform pattern over the week. The unit carrying cost of the item in inventory is ₹ 2 per week and unit shortage cost ₹ 8/week. Determine optimal order level of the inventory.

Solution:

Holding cost = c_1 = ₹ 2/unit/week
Shortage cost = c_2 = ₹ 8/unit/week

Probability distribution function

$$f(x) = 0.1, \qquad 0 \le x \le 10$$

$$\int_0^q f(x)\, d(x) = \frac{c_2}{c_1 + c_2}$$

$$\int_0^q 0.1\, d(x) = \frac{8}{10} = 0.8$$

$$0.1 |x|_0^q = 0.8$$

$$0.1\, q = 0.8$$

Optimal order quantity = q = 8 units

34.22 REPLENISHMENT SYSTEMS (INVENTORY CONTROL SYSTEMS)

A Replenishment system means to decide when to order and how much to order. Two main replenishment systems are:

1. Fixed order quantity system
2. Fixed order interval system

Notations used:

L = Lead time
R = Demand rate
S = Safety stock
EOQ = Economical order quantity

1. Fixed order quantity system: This system is also called fixed order quantity system or reorder system or *EOQ* ordering system. In this system, reorder quantity is always same and is equal to economic order quantity. The time intervals between two successive orders may be different. The various parameters of the system are shown in Fig. 34.9.

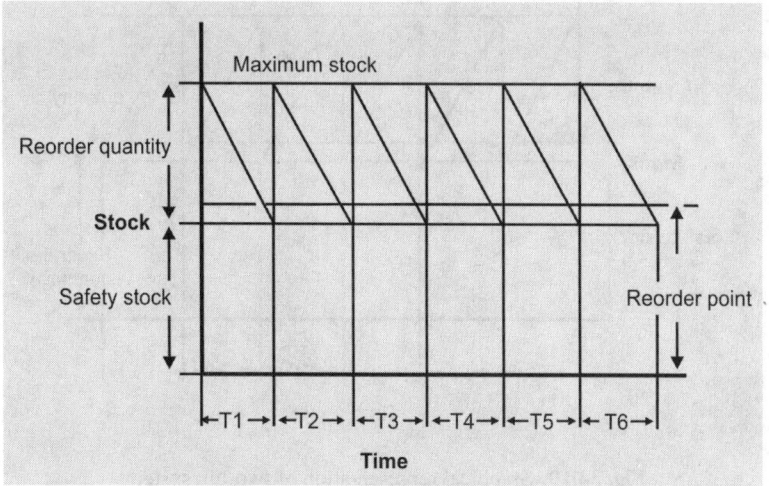

T1, T2, T3, T4, T5 are time intervals between two successive orders

Fig. 34.9 Graphical representation of fixed order quantity system

Safety stock is kept to take care of increase in demand during lead time as well as to meet the demand when the lead time is extended. All the items in stock are ordered according to their respective ordering cycles.

Maximum stock $= S + EOQ$

Minimum stock $= S$

Average stock $= S + EOQ/2$

Reorder level $= S + LR$

$$EOQ = q = \sqrt{\frac{2c_3 R}{c_1}}$$

Suitability of fixed order quantity system: Fixed order quantity system is generally suitable for *B* class and *C* class items. To operate fixed order quantity system efficiently, stock records must be kept up to date. If there is any error in recording of the stock, it may affect production. Adopting two-bin system or *S-s* policy can minimize error.

Two-bin system: Two-bin systems, which operate on the principle of reorder level physically segregate the entire stock of an item into two bins. The second bin contains stock equivalent to reorder level. The first bin contains order quantity minus lead time consumption. The empty of first bin indicates that stock has reached the reorder level and place order for the quantity equal to reorder quantity. Reorder quantity in two-bin system is fixed and is generally the economic order quantity. The basic parameters are shown in Fig. 34.10.

Reorder quantity in two-bin system is fixed and is generally the economic order quantity.

Second bin quantity $= S + L \times R$

First bin quantity $= EOQ - L \times R$

Maximum stock in two bins = $S + EOQ$

Average quantity = $S + EOQ/2$

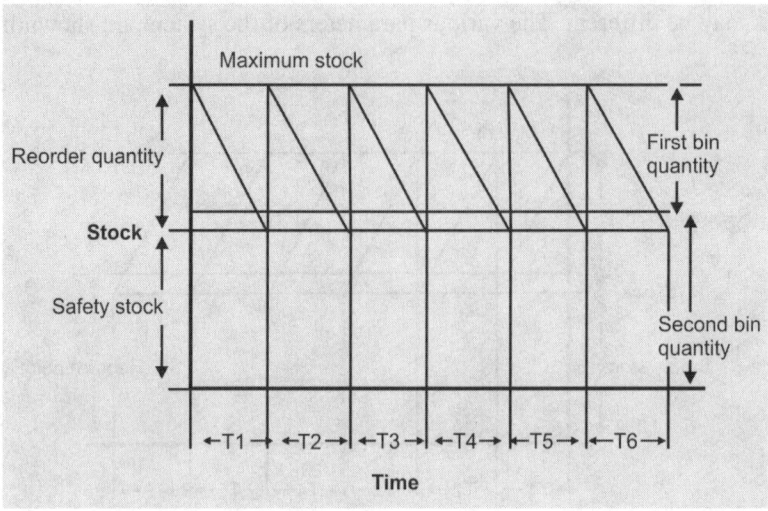

Fig. 34.10 Graphical representation of two-bin system

Two bins do not necessarily mean two separate bins. It may be two different locations. The system is most suitable for C class items that are having high volume but inexpensive and hence purchased in bulk.

Examples: Bolts, washers, chemicals, coal, iron, etc.

S-s policy: This system operates on the principle of reorder level; physically segregate the entire stock into two different locations. The second location "*s*" contains the stock equivalent to reorder level. The first location "*S*" contains the stock equivalent to reorder quantity minus lead time consumption. The empty of first location indicates that the stock has reached the reorder quantity. Reorder quantity in *S-s* policy is fixed and is generally the economic order quantity. The basic parameters to operate the system are shown in Fig. 34.11.

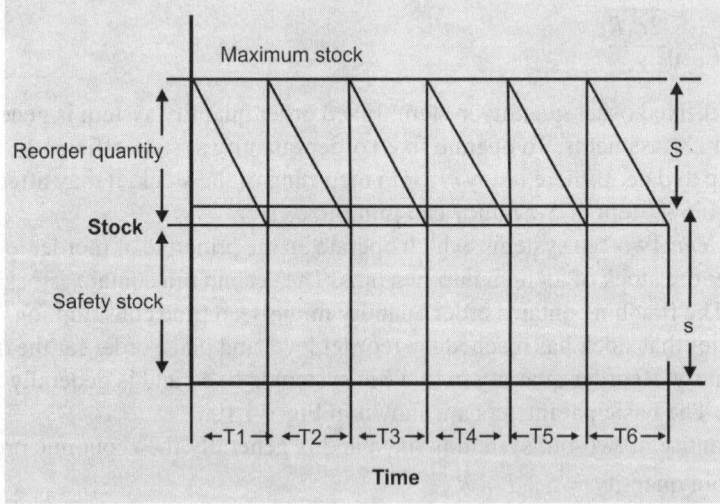

Fig. 34.11 Graphical representation of *S-s* policy

Reorder quantity in this system is equivalent to the economic order quantity.

The stock in "*s*" location = Safety stock + Lead time consumption = $S + L \times R$

The stock in "*S*" location = Maximum quantity – Reorder level = $EOQ - L \times R$

Maximum stock = Safety stock + EOQ

Average stock = Safety stock + $EOQ/2$

This system is most suitable for "*C*" class items.

Fixed interval system: Fixed interval system is also called periodic review system or periodic ordering system. In this system, the stock is ordered at fixed intervals as shown in Fig. 34.12.

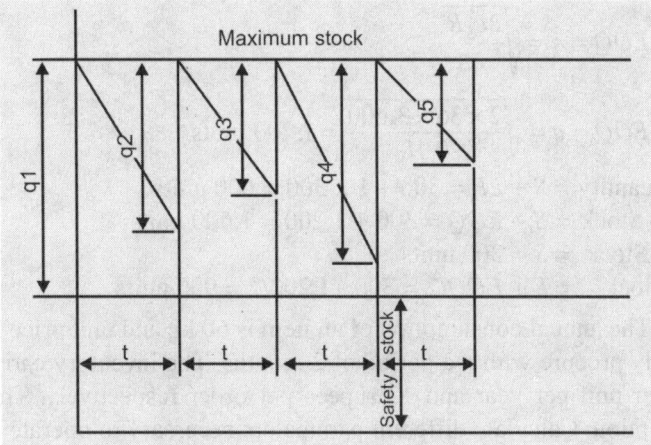

Fig. 34.12 Graphical representation of fixed interval system

In this system, the time between orders is fixed but reorder quantity at each time is different.

q_1, q_2, q_3, q_4, q_5, etc. are reordering quantities at the beginning of periods T_1, T_2, T_3, T_4 and T_5 respectively.

To determine optimum review period:

Let "*t*" is the optimum review period or time between orders.

$$EOQ = q = \sqrt{\frac{2c_3 R}{c_1}} = Rt$$

$$t = \sqrt{\frac{2c_3}{c_1 R}}$$

Advantages of fixed interval system: (1) The fixed interval system is used for high value items requiring strict control on stock levels. (2) Items which are required in different sizes (drills, reamers, taps, etc.) and are procured from a single source.

The main disadvantage of this method is that even for small quantities the orders should be placed.

34.23 SOLVED PROBLEMS

Problem 34.22: Monthly consumption of an item having unit price ₹ 1 has been estimated at 300 units. The inventory carrying cost and the procurement cost for the company have been computed at 18% per unit per year and ₹ 36 per order respectively. Stock records show that this item can normally be procured of one month. If the company adheres to the policy of one month safety stock for all A and B items, calculate: a. Reorder quantity b. Minimum stock c. Reorder level d. Maximum level e. Average inventory.

Solution:

Demand $= R = 3,600$ units/year

Unit cost of the item $= ₹ 1$

Holding cost $= c_1 = 0.18 \times 1 = ₹ 0.18$/unit/year

Lead time $= 1$ month

Safety stock $= S = 300$ units

Ordering cost $= c_3 = ₹ 36$/order

$$EOQ = q = \sqrt{\frac{2c_3 R}{c_1}}$$

$$EOQ = q = \sqrt{\frac{2 \times 36 \times 3,600}{0.18}} = 1,200 \text{ units}$$

a. Reorder quantity $= S + LR = 300 + 1 \times 300 = 600$ units
b. Maximum Stock $= S + EOQ = 300 + 1,200 = 1,500$ units
c. Minimum Stock $= S = 300$ units
d. Average Stock $= S + EOQ/2 = 300 + 1,200/2 = 900$ units

Problem 34.23: The annual consumption of an item is 60 kg and unit price is ₹ 8 per kg. The items can normally procure within a period of 2 months. The inventory carrying cost worked out to be 15% per unit per year and 12 rupees per order respectively. The safety stock is 2 months consumption. Calculate different parameters necessary to operate system.

Solution:

Demand $= R = 60$ kg/year

Unit cost of the item $= ₹ 8$/kg

Holding cost $= c_1 = 0.15 \times 8 = ₹ 1.2$/unit/year

Lead time $= 2$ months

Safety stock $= S = 2$ months consumption $= (60/12) \times 2 = 10$ kg

Ordering cost $= c_3 = ₹ 12$/order

$$EOQ = q = \sqrt{\frac{2c_3 R}{c_1}}$$

$$EOQ = q = \sqrt{\frac{2 \times 60 \times 12}{1.2}} = 35 \text{ kg}$$

Reorder level $= S + L \times R = 10 + 2 \times 5 = 20$ kg
Second bin quantity $= S + L \times R = 10 + 2 \times 5 = 20$ kg
First bin quantity $= EOQ - L \times R = 35 - 10 = 25$ kg
Maximum stock in two bins $= S + EOQ = 10 + 35 = 45$ kg
Average quantity $= S + EOQ/2 = 10 + 35/2 = 27.5$ kg

Problem 34.24: The Company under study has decided to buy its tool requirement in a limited number of orders in a year. Then estimated next year consumption of drills is ₹ 6,400. It takes one month to receive drills once an order is placed. The company management estimates that it costs ₹ 25 to place an order and the cost of inventory is 12.5% per year. How frequently should drills to be purchased, if the company adopts fixed interval system?

Solution:

$pR = ₹\ 6,400$

Safety stock $= S$

Lead time $= 1$ month

Ordering cost $= c_3 = 25/\text{order}$

Holding cost $= c_1 = Ip = 0.125p$

$$t = \sqrt{\frac{2c_3}{c_1 R}} = \sqrt{\frac{2c_3}{IpR}}$$

$$t = \sqrt{\frac{2 \times 25}{0.125 \times 6400}} = 0.25 \text{ year}$$

Optimum review period $= t = 0.25$ year $= 3$ months

34.24 EMPIRICAL FORMULAE FOR SAFETY STOCK

Safety stock = Maximum lead time × Maximum demand rate – Average lead time × Average demand rate

Reorder level (ROL) = Safety stock + Average lead time × Average demand rate

Determination of safety stock under normal distribution demand:

Let

D = Demand during lead time

X = Mean demand during lead time

σ = Standard deviation of x for 95% confidence level

Safety stock $= 1.645\ \sigma$

Reorder level $= L \times d + 1.645\ \sigma$

34.25 SOLVED PROBLEMS

Problem 34.25: From the past records it has been observed that,

Case 1: Lead time = 10 days

Annual demand rate = 600 items

Demand rate for the past two months = 70 items/month

Case 2: Lead times are 15, 17, 20, 16 days

Annual demand rate = 1,020 items

Case 3: Lead time distribution and demand distribution are shown in Tables 34.7 and 34.8 respectively.

Table 34.7 Lead time distribution

Lead time in days	0	1	2	3	4	5	6	7
Frequency	1	0	0	2	3	2	5	6

Table 34.8 Demand distribution

Demand in units	0	1	2	3	4	5	6
Frequency	2	0	4	5	5	1	2

Determine the safety stock and reorder level in each case.

Solution:

Case 1:

Safety stock = Maximum lead time × Maximum demand rate – Average lead time × Average demand rate

$$= 10 \times (70/30) - 10 \times (50/30) = 23.3 - 16.6 = 6.63 \text{ items}$$

Reorder level (ROL) = Safety stock + Average lead time × Average demand rate

$$= 6.63 + 1.66 \times 10 = 23.23 \text{ items}$$

Case 2:

Average lead time = $(15 + 17 + 20 + 16)/4 = 17$ days

Maximum lead time = 20 days

Safety stock = Maximum lead time × Maximum demand rate – Average lead time × Average demand rate

$$= 20 \times (1,020/360) - 17 (1,020/360) = 56.6 - 48.16 = 8.5 \text{ items}$$

Reorder level (ROL) = Safety stock + Average lead time × Average demand rate

$$= 8.5 + 48.16 = 56.6 \text{ items}$$

Case 3:

$$\text{Average lead time} = \frac{0 \times 1 + 1 \times 0 + 2 \times 0 + 3 \times 2 + 4 \times 3 + 5 \times 2 + 6 \times 5 + 7 \times 6}{1 + 0 + 0 + 2 + 3 + 2 + 5 + 6}$$

$$= 5.263 \text{ days}$$

$$\text{Average demand rate} = \frac{0 \times 2 + 1 \times 0 + 2 \times 4 + 3 \times 5 + 4 \times 5 + 5 \times 1 + 6 \times 2}{2 + 0 + 4 + 5 + 5 + 1 + 2}$$

$$= 3.15 \text{ items/day}$$

Safety stock = Maximum lead time × Maximum demand rate – Average lead time × Average demand rate

$$= 7 \times 6 - 5.263 \times 3.15 = 42 - 16.62 = 25.4 \text{ items}$$

Reorder level (ROL) = Safety stock + Average lead time × Average demand rate

$$= 25.4 + 16.62 = 42 \text{ items}$$

Problem 34.26: A company wants to provide 95% service level to its customers. Using the past history the following data is available. Daily demand follows Normal distribution with average daily demand of 20 units and standard deviation of 5 units. The lead time for procurement is 4 days. The cost of placing an order is ₹ 10 and the inventory carrying cost is ₹ 1 per unit per year. Shortages are not allowed. Determine safety stock.

Solution:

Demand = R = 20 units/day = $20 \times 365 = 7,300$ units/year

Standard deviation of demand = σ = 5 units

Lead time = 4 days

Ordering cost = c_3 = ₹ 10/order

Holding cost = c_1 = ₹ 1/unit/year

$$EOQ = q = \sqrt{\frac{2 c_3 R}{c_1}}$$

$$EOQ = q = \sqrt{\frac{2 \times 10 \times 7,300}{1}} = 381 \text{ units}$$

Safety stock = $1.645\sigma = 8$ units

Reorder level = $L \times d + 1.645\sigma = 20 \times 4 + 8 = 88$ units

34.26 SELECTIVE CONTROL

Every organization consumes several thousands of items, since all items are not of equal importance. Classifying the items in groups depending upon their importance is known as selective control.

Examples: Raw materials, spare parts, consumables, etc.

Important Selective control techniques:

1. ABC Analysis
2. VED analysis
3. FSND analysis

1. ABC analysis: Classification of items into three categories A, B and C in descending order of annual consumption value.

A items: 10% of the volume of items accounts 75% of consumption value.

B items: 15% of the volume of items accounts 15% of consumption value.

C items: 75% of the volume of items accounts 10% of total consumption value.

Above percentages are not fixed but varies from industry to industry.

Mechanics of ABC analysis or steps in conducting ABC analysis:

1. Determine consumption value of each item by multiplying consumption volume and unit cost of item.
2. Arranging the items in descending order as per consumption value.
3. Determining cumulative consumption value and consumption volume of items.
4. Determine % of cumulative consumption value and % cumulative consumption volume.
5. Classify the items based on % cumulative consumption value.

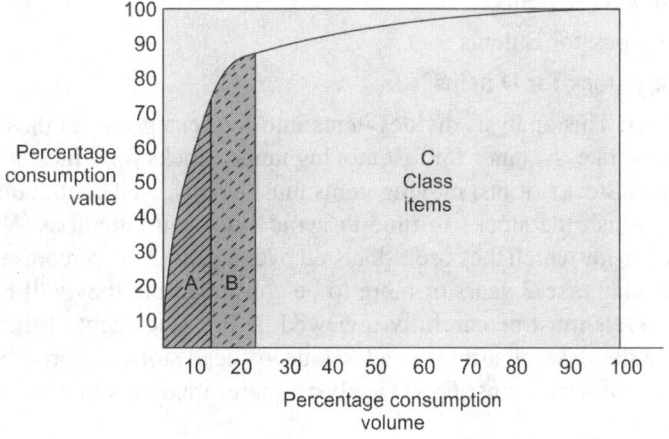

Fig. 34.13 Graphical representation of ABC analysis

Table 34.9 Controlling of A, B and C items

A items	B items	C items
1. Order frequently in small number	1. Medium number of orders	1. Once in a year
2. Review periodically	2. Review 2 times in a year.	2. Once in a year
3. Safety stock is less	3. Medium safety stock	3. Large safety stock
4. Strict inventory control is required	4. Medium control	4. Less control
5. More number of suppliers	5. Two or more suppliers	5. One supplier is enough.
6. Handled by top level management.	6. Handled by middle level management.	6. Handled by low level people.

Limitations of ABC analysis:

1. Criticality or importance of item is not considered.

2. Unit cost of the item is not taken into consideration for classification.

2. VED analysis: The VED analysis is based on criticality of items. The items are arranged in descending order of their criticality.

V stands for vital items

E stands for essential items

D stands for desirable items

Vital items: Vital items are not available when the whole system is in operative condition.

Examples: 1. Clutch wire of an automobile, 2. Cutting tools, 3. Belts, 4. Gears, 5. Coal in thermal plant, etc.

Essential items: These items which when demanded are not available reduce the efficiency of the system.

Examples: 1. Telephone, 2. Coolants and 3. Lubricants

For E items some risk can be taken.

Desirable items: Desirable items neither stop nor reduce the efficiency, but it will be good, if they are present in the system.

Example: Mirror and Seat cover in an Automobile.

The VED analysis is used in stock controlling of spare parts required for maintenance.

High safety stock for V items

Medium safety stock for E items

Less or no safety stock for D items

3. *FSND analysis*: This analysis divides items into four categories in the descending order of their consumption rate. F stands for fast moving items, stocks for which are consumed over a short span of time. Stocks of fast moving items must be observed continuously and all steps to be taken to replenish the stocks in time to avoid stock out situations. N denotes normal moving items' stocks for which they are exhausted over a year or so. S represents slow moving items. Such items may take 2 years or more to be consumed but they will be consumed any way. Their stock levels must be carefully reviewed before placement of their replenishment orders to minimize the risks of surplus stock. D means dead stock. Efforts should be made to find alternative uses of such items. *FSND* analysis, therefore, helps to control obsolescence.

34.27 SOLVED PROBLEM

Problem 34.27: From the details of Table 34.10, draw a plan of ABC selective control.

Table 34.10 Demand and unit cost of each item

Item No.	1	2	3	4	5	6	7	8	9	10	11	12
Units	7,000	24,000	1,500	600	38,000	40,000	60,000	3,000	300	29,000	11,500	4,100
Unit cost (₹)	5	3	10	22	1.5	0.5	0.2	3.5	8	0.4	7.1	6.2

Solution: Determine consumption value of each item by multiplying consumption volume and unit cost of item. Usage value of each item is given in Table 34.11.

Table 34.11 Determination of consumption value

Item No.	Units	Unit cost (₹)	Usage value
1	7,000	5	35,000
2	24,000	3	72,000
3	1,500	10	15,000
4	600	22	13,200
5	38,000	15	57,000
6	40,000	0.5	20,000
7	60,000	0.2	12,000
8	3,000	3.5	10,500
9	300	8	2,400
10	29,000	0.4	11,600
11	11,500	7.1	81,650
12	4,100	6.2	25,420

Arrange the items in descending order according to the consumption value or usage value (Table 34.12). Calculation to classify the items into A, B and C is shown in Table 34.12.

Table 34.12 Grouping the items into A, B and C items

Item No.	Units	Unit cost (₹)	Usage value (₹)	Cumulative usage value (₹)	%Cumulative usage value (₹)	Usage volume	Cumulative usage volume	%Cumulative usage volume
11	11,500	7.10	81,650	81,650	23	11,500	11,500	5.3
2	24,000	3	72,000	1,53,650	43.2	24,000	35,500	16.6
5	38,000	1.5	57,000	2,10,650	59.2	38,000	73,500	33
1	7,000	5	35,000	2,45,650	69	7,000	80,500	36.7
12	4,100	6.2	25,420	2,71,070	76.2	4,100	84,600	38.6
6	40,000	0.5	20,000	2,91,070	81.8	40,000	1,24,600	56.8
3	1,500	10	15,000	3,06,070	86.03	1,500	1,26,100	57.5
4	600	20	13,200	3,19,270	89.7	600	1,26,700	57.66
7	60,000	0.2	12,000	3,31,270	93.1	60,000	1,86,700	85.25
10	29,000	0.4	11,600	3,42,870	96.4	29,000	2,15,700	98.4
8	3,000	3.5	10,500	3,53,370	99.3	3,000	2,18,700	99.8
9	300	8	2,400	3,55,770	100	300	2,19,000	100

From the Table 34.12:

i. A class items: 11, 2, 5 and 1
ii. B class items: 6, 3 and 4
iii. C class items: 7, 10, 8 and 9

34.28 QUESTIONS AND PROBLEMS

1. What are the objectives of inventory control?
2. Define inventory.
3. What are the two important decisions to be made in inventory problems?
4. Describe the method of carrying out ABC analysis.
5. Explain the methods of determining safety stock.
6. What are the different costs associated with inventory control problems?
7. Derive EOQ in simple deterministic model.

8. With regard to inventory model explain the following:
   a. Lead time
   b. Probabilistic demand
   c. Safety stock

9. A company purchases lubricants at the rate of ₹ 42 per tin. The requirement is 1800 per year. The cost of placing an order is ₹ 16 and holding charges are 20 paise per rupee per year. Find economical order quantity and the time between the orders.
   Answer: EOQ = 83 units, Cycle time = 0.046 year

10. An aircraft company requires rivets at the rate of 5,000 kg per year. The cost of the rivet is ₹ 20 per kg and the ordering cost is ₹ 200. The holding cost is 10% per kg per year. How frequently should orders be placed? What quantity should be ordered for?
    Answer: EOQ = 1,000 units, Cycle time = 0.2 year

11. A company has to supply 10,000 auto parts per day. They can produce 25,000 parts per year. Holding cost is 20 paise per unit per year and the set up cost is ₹ 180 per run. How frequently should production be made?　　　　**Answer:** Cycle time = t = 10.5 days

12. The production rate of tomato ketchup is 600 bottles per day and the demand is 150 bottles per day. The cost per bottle is ₹ 6. The setup cost is ₹ 90 per run and the holing cost is 20% of cost per annum. Find the economic lot size and the number of production runs.
    Answer: EOQ = 3,309 units, number of production runs = 17 per year

13. The demand of an item is at the rate of 25 units per month. The fixed cost is ₹ 15 per run. The carrying cost is ₹ 0.30 per item per month. If the shortage cost is ₹ 1.50 per item per month, determine the economic lot size and period of a run.
    Answer: EOQ = 17 units, Time between production runs = 20 days

14. The annual demand of an automobile part is 6,000 units. The setup cost per production run is ₹ 500 and the storage cost per unit per year is ₹ 8. If the shortage cost is ₹ 20 per unit per year, find:
    a. Economic lot size
    b. Period of one run
    c. Number of shortages per run
    Answer: EOQ = 1,024 units, Time between production runs = 2 months, Shortage quantity = 293 units

15. The demand for a food commodity has the probability distribution is given below:

Demand	10	11	12	13	14
Probability	0.10	0.15	0.20	0.25	0.30

The cost is ₹ 30 each and the selling price is ₹ 50. Unsold items become useless. Determine how many items should be prepared.
    Answer: EOQ = 12 units

16. Find the EOQ, reorder level and the optimum annual inventory cost given that the annual demand is 1,000 units and holding cost is 0.15 per unit per year, ordering cost ₹ 10 and lead time is 2 years.
    Answer: EOQ = 365, Reorder level = 2,000 units, Annual inventory cost = ₹ 5,480

17. Find the EOQ for product for which the price breaks are as follows:

Price	Quantity
10	$q < 50$
9	$q \geq 50$

The monthly demand is 200 units. The ordering cost is ₹ 350 and the holding cost is 2% of the cost per month.

18. A manufacturer of engines is required to purchase 4,800 castings per year. These castings are subjected to quality discounts. The price schedule is as follows:

For less than 500 units ₹ 150 per unit

For 500 or more less than 750 units ₹ 140 per unit

For 750 or more units ₹ 132 per unit

Holding cost expressed as a percentage value of the units = 2%. Ordering cost is ₹ 750 per order. Find optimum purchase quantity per order.

Answer: *EOQ* = 1,656

19. A Subcontractor has to supply bushings to manufacturer who requires 83 bushings per day.

No shortages are to be allowed. Procurement cost is ₹ 90 per purchase. The cost per unit is a function of purchase quantity is as follows.

Purchase quantity	1–199	200–499	500 or more
Price per bushing (₹)	115	110	100

The holding cost is ₹ 0.45 per bushing per day. Calculate economical purchase quantity from the supplier.   **Answer:** *EOQ* = 1,656

20. The following information is known about a group of items. Classify the items as A, B and C items.

Item no.	501	502	503	504	505	506	507	508	509	510
Annual consumption in pieces	30,000	2,80,000	3,000	1,10,000	4,000	2,20,000	15,000	80,000	60,000	8,000
Unit price in paise	10	15	10	5	5	10	5	5	15	10

21. For the following data, determine approximately the *EOQ* for each of the three products 1, 2 and 3 when total value of average inventory levels of three products is less than ₹ 10,000.

Costs	Products		
	1	2	3
Holding cost/year%	20	20	20
Setup cost (₹)	50	40	60
Cost per unit (₹)	6	7	5
Yearly demand	10,000	12,000	7,500

22. The average inventory system is storing three items. The company cannot invest more than ₹ 14,000 on stock. Find the optimum sizes of inventory for each item using the following information. The system uses 20% in carrying the inventory cost for each item

Costs	Products		
	1	2	3
Holding cost/year%	20	20	20
Setup cost (₹)	50	70	100
Cost per unit (₹)	20	100	50
Yearly demand	1,000	500	2,000

35

Computer-aided Quality Control

35.1 INTRODUCTION

The two major parts of quality control are inspection and testing, which are traditionally performed manually with the help of gages, measuring devices and the testing apparatus. Measuring devices provide a quantitative value of the part features of interest. Gages determine whether the part feature falls within a certain acceptable range of values. Both techniques are widely used for post-process inspection of piece parts in manufacturing and measuring. The use of the computers for quality control of the product is called as the computer-aided quality control or CAQC. The two major parts of computer-aided quality control are computer-aided inspection (CAI) and computer-aided testing (CAT). CAI and CAT are performed by using the latest computer automation and sensor technology. CAI and CAT are the standalone systems and without them the full potential of CAQC cannot be achieved.

The main objectives of the CAQC are to improve the quality of the product, increase the productivity in the inspection process and reduce the lead times in manufacturing. The implementation of CAQC in the company results in the major change in the way the process of quality control is carried out in the company.

35.2 ADVANTAGES OF COMPUTER AIDED QUALITY CONTROL OR CAQC

Advantages of computer-aided quality control or CAQC are:
- 100% testing and inspection: In the traditional manual process, the testing and inspection is done by the sampling process out of the hundreds and thousands of products or parts manufactured by the company since it is not feasible to check each and every product. With CAI and CAT hundred percent inspection and testing can be accomplished without much difficulty. With 100% inspection the company does not have to depend on statistical quality control method in which it is assumed that anything less than 100% of quality is acceptable. With computer controlled inspection, it is not necessary for the quality control department to settle for less than perfection.
- Inspection integrated with manufacturing process: In the traditional process, there is separate quality control department where the manufactured product is taken for the inspection and testing. In CAQC, the inspection process is integrated with the manufacturing process and it is located along the production line. Thus as soon as the product is manufactured, it is

tested immediately by the computerized process without moving it to some other location. This helps in reducing the overall time required for manufacturing the product.

• Use of non-contact sensors: In the traditional process, the product or the part to be inspected is handled manually since it has to be positioned properly for inspection on the desk or suitable location. In CAQC, non-contact sensors are used for the inspection purpose and they inspect the product without coming in contact with the product. The non-contact sensors operated by the computer are kept along the production line and they can check the product very quickly in the fraction of seconds. In future with further advancements in the technology, the robots would be used to carry out the inspection process thus further automating and speeding the process.

• Computerized feedback control system: The data collected by the non-contact sensors is sent as the feedback to the computerized control systems. These systems would carry out the analysis of the data including statistical trend analysis. This helps in identifying the problem going on in the manufacturing line and find appropriate solution to it. For instance, the results from non-contact sensors may indicate that the parts manufactured are not within the acceptable tolerance limits. This would help the production or quality control personnel to find out the precise location of the problem and its exact cause. The corrective action taken quickly saves lots of time and money due to reduced wastages and also improves the quality of the product.

• Computer-aided quality control and CAD/CAM integration: Apart from inspection and testing, computers are used in a number of other areas of the quality control. All the applications of CAQC can be integrated with CAD/CAM to make the whole process of designing and manufacturing controlled by the computers converted into fully automated process.

The timing of the inspection procedure in relation to the manufacturing process is an important consideration in quality control. Two alternative situations can be distinguished:

Offline inspection: Offline inspection is performed away from the manufacturing process, and there is generally a time delay between processing and inspection.

Online inspection: The alternative to offline inspection is online inspection, in which the procedure is performed when the parts are made.

35.3 CHARACTERISTICS OF MEASURING INSTRUMENTS

This chapter focuses on automated inspection techniques. Some general characteristics of measuring instruments used in the inspection are outlined in Table 35.1. These characteristics are used to determine the correct choice of inspection equipment for a particular inspection procedure.

Table 35.1 Characteristics of measuring instruments

Characteristic	Comments
Accuracy and Precision	Accuracy is the degree to which the measured value agrees with the true value that has been pre-defined for the item. Precision is a measure of repeatability in a measurement process, such that precision reflects the consistency of the measurement results achieved.
Resolution and Sensitivity	This aspect of a measuring instrument is its capacity to distinguish very small differences in the quantity of interest. The indication of this characteristic is the smallest variation of the quantity that can be detected by the instrument.

(Contd...)

(contd...)

Characteristic	Comments
Speed of response	Measures the time required for the measuring device to indicate the quantity measured. Ideally, the time lag should be zero, but this is obviously impossible.
Wide operating range	This is the capability of a measuring instrument to measure the physical variable throughout the entire span of practical interest to the user.
High reliability	A measure of the absence of frequent malfunctions and failures of the measurement device.
Cost	The expense of purchasing and operating the measuring device, plus the expense of training on the measuring device.

35.4 INSPECTION TECHNIQUES

There are two types of inspection techniques. They are:
  a. Contact inspection techniques
  b. Non-contact inspection techniques

35.4.1 Contact Inspection Techniques

In contact inspection, physical contact is made between the object to be inspected, and the measurement device. Typically, contact is achieved using a mechanical probe or other device that touches the item, and allows the inspection procedure to occur. By its nature, contact inspection is concerned with some physical dimension of the part, and so contact methods are widely used in manufacturing and production industries. Contact metrology includes everything from simple manual tools to coordinate measurement machines (CMMs).

Principal contact inspection technologies include:
  • Conventional measuring and gaging instruments such as micrometers, calipers, protractors, and go/no-go gages.
  • Coordinate measuring machines (CMMs) and related techniques to measure mechanical dimensions.
  • Stylus type surface texture measuring machines to measure surface characteristics such as roughness and waviness.
  • Electrical contact probes for testing integrated circuits and printed circuit boards.
  • Electronic gages deploy transducers capable of converting linear displacement into a proportional electrical signal, which in turn is amplified and transformed into a suitable data format such as a digital read-out. Advantages of electronic gages include: Good sensitivity, accuracy, precision, repeatability, and speed of response; ability to sense very small dimensions; ease of operation; reduced human error; ability to display electrical signal in various formats; and capability to be interfaced with computer systems for data processing.

Contact inspection techniques are the most widely used inspection techniques. They are possessing considerable accuracy and reliability. In many cases, they represent the only methods available to accomplish inspection.

35.4.1.1 Coordinate measuring machines

In coordinate metrology, the actual shape and dimensions of an item are measured, and compared against desired shape and dimensions, as specified on a part drawing.

CMM consists of the following:
  • Mechanical structure: Mechanical structure provides motion of the probe in three cartesian axes.

- Probe head and probe: Probe head and probe to contact work part surface.
- Work table: Work table that passes underneath the probe, upon which the item to be inspected is placed.
- Displacement transducers: Displacement transducers to measure the coordinate values of each axis.
- Drive system: Drive system and control unit to move each axis.
- Computer system: Computer system that takes and records the probe results as they occur.

The overall CMM construction is illustrated in Fig. 35.1.

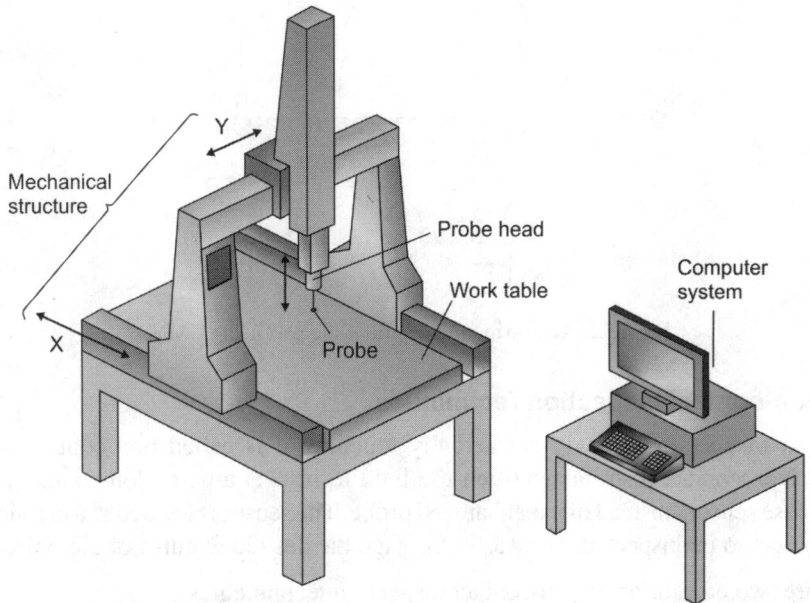

Fig. 35.1 Coordinate measuring machine

The tip of the probe in the CMM is usually a ruby ball. Ruby is a form of corundum (aluminum oxide) with high hardness for wear resistance, and low density for minimum inertia, thus making it ideal for probing applications. Probes can be single or multiple tip (see Fig. 35.2).

Fig. 35.2 Single-tip probe

Probe positioning may be accomplished using several methods, ranging from manual operation to direct computer control. Computer-controlled CMMs operate much like CNC machine tools, and these machines must be programed. After contact between probe and part surface, displacement transducers associated with the three linear axes, record the coordinate positions of the probe, and pass the results to the CMM controller. Compensation is made for the radius of the probe tip, and over-travel of the probe nib due to momentum is neglected. The probe returns to a neutral position when it leaves the part surface. A number of different physical configurations exist for the mechanical structure of the CMM. Measuring the length of the part using CMM is shown in Fig. 35.3.

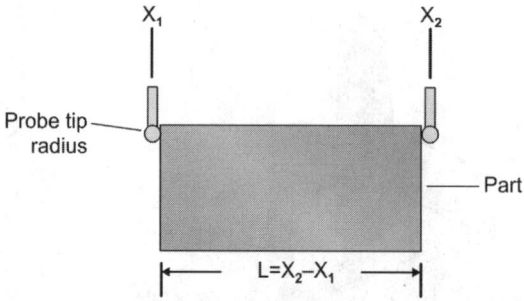

Fig. 35.3 Measuring the length of the part using CMM

35.4.2 Non-contact Inspection Techniques

Method that inspects a part without actually touching it is called non-contact inspection technique. Non-contact instruments often use light to inspect a part. Non-contact inspection techniques use sensors instead of a mechanized probe. The sensor is located at a certain distance from the object to be inspected, to measure or gage the desired features of the object.

There are two categories of non-contact inspection technologies:

- Optical inspection technologies: Optical inspection technologies use light to accomplish the measurement or gaging cycle. The most important technique is machine vision.

- Non-optical inspection technologies: Non-optical inspection technologies use other forms of energy than light to perform the inspection. Various energies utilized include: electrical fields, radiation, and ultrasonic.

35.4.2.1 Machine vision

Machine vision is the creation of an image and the collection of data derived from the image, and the subsequent processing and interpretation of the data by a computer from some useful application. Machine vision is also known as computer vision, and its principal application is in industrial inspection.

The operation of machine vision has three functions. They are:

- Image acquisition and digitization
- Image processing and analysis
- Interpretation

Figure 35.4 outlines basic functions of a machine vision system.

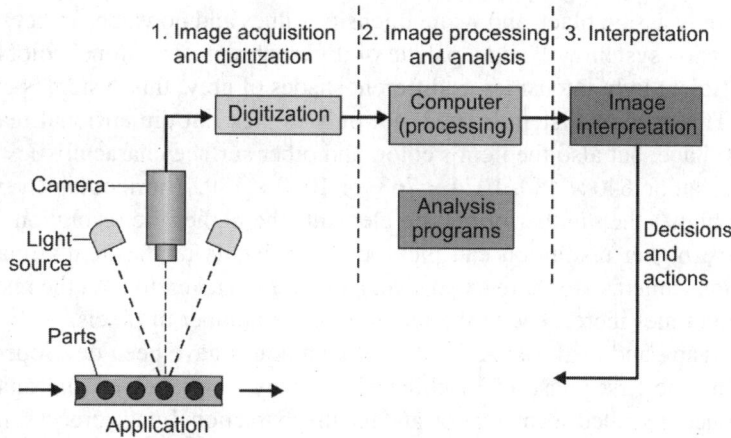

Fig. 35.4 Basic functions of a machine vision system

Image acquisition and digitization: Image acquisition and digitization is performed by a video camera to capture the image, and the use of a digitizing system to store the image data for subsequent analysis. The scene that the camera is focused upon must be well illuminated, if an image of sufficient quality is to be captured. Illumination must be well-placed and constant over the time required to capture the image; this usually means that special lighting must be deployed for a machine vision application, rather than relying upon ambient lighting. Then camera is focused upon the surface, and an image consisting of discrete pixel elements is captured in the viewing area; each pixel has a value proportional to the light intensity of that portion of the scene. The intensity value of each pixel is converted into its equivalent digital value by an analogue-to-digital converter. This operation in diagram format is depicted in Fig. 35.5.

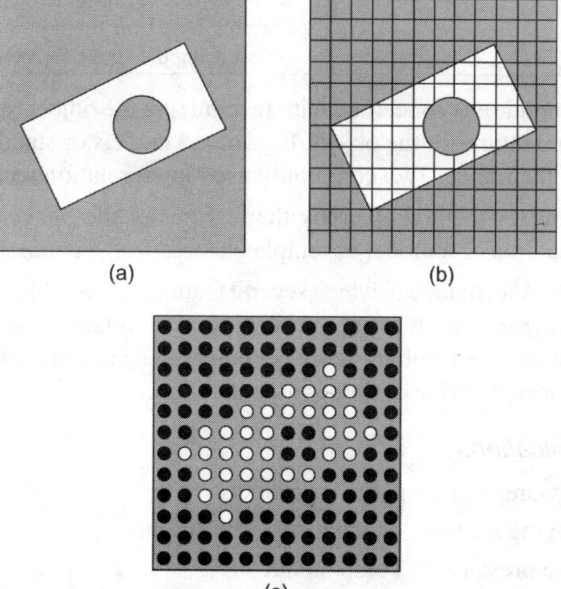

Fig. 35.5 Image acquisition and digitization: (a) Scene presentation; (b) 12 × 12 matrix super-imposed; and (c) Assignment of intensity values, in black or white

Figure 35.6 outlines the simplest type of machine vision, called binary vision (so called because it can only assign black and white intensity values and no values in between). A more sophisticated vision system will add a palette of different representational colors, in grey, that can capture different light intensities as different shades of grey; this system is called the gray scale system. This type of system is used, not only to pick-out dimensional features and the items size and shape, but also the item's color, and other surface characteristics.

Pixel arrays can be 640 × 480, 1024 × 768, or 1040 × 1392 (horizontal × vertical) picture elements. The higher the number of picture elements the higher the resolution of the camera achieved, where higher resolution can pick-out finer details of the item's image; however, higher resolution cameras are more expensive, and the time taken to read the resultant images is slower as read times increase with the increase in the number of pixels.

Image processing and analysis: A number of techniques have been developed so that data produced during the first phase of machine vision may be processed and analyzed. These general techniques are called segmentation and feature extraction. Image processing and analysis techniques under these general headings are outlined in Table 35.2.

Table 35.2 Image processing and analysis techniques

General category	Sub-category	Description
Segmentation	Thresholding	Involves the conversion of each pixel intensity level into a binary value (black or white); performed by comparing each pixel value to a defined threshold value.
Segmentation	Edge detection	Concerned with determining the location of the boundaries between an object and its surroundings in an image. Accomplished by identifying the contrast in light intensity that exists between adjacent pixels at the borders of the object.
Feature extraction		Methods that are designed to determine an object's features based on the area and boundaries of the object using the above segmentation techniques.

Interpretation: Interpretation is concerned with recognizing the object (object recognition) and recognizing the major features of the object. Predefined models or standard values are used to identify the object in the image. Two commonly used interpretation techniques are:

- Template matching: A method whereby the features of the image are compared against corresponding features of a model or template stored in the computer memory.

- Feature weighting: A technique in which several features are combined into a single measure by assigning a weight to each feature according to its relative importance in identifying the object, and where the resultant score is compared against an ideal object score stored in computer memory, to achieve proper identification.

Machine vision applications

- Dimensional measurement on parts or products
- Dimensional gauging on parts or products
- Verification of the presence of components
- Verification of hole location and number of holes
- Detection of surface flaws and defects

35.4.2.2 *3D Laser scanning*

3D Laser scanning or Laser digitizing utilizes a beam of laser light to capture the physical shape of the subject part, and produces a digital image, or point cloud. Laser digitizing is among the most versatile non-contact inspection services because it is highly portable, suitable for a wide range of part sizes, and capable of obtaining a large amount of data.

35.4.2.3 *Non-contact non-optical inspection techniques*

Other potential inspection techniques are non-contact non-optical inspection techniques. Non-contact non-optical inspection techniques are:

- Electrical field: An electrically active probe creates an electrical field which is affectedly the proximity of an object to the probe. In typical applications, the object to be inspected is placed at a set proximity to the probe, and the effect on the electrical field is measured. This procedure is repeated at different distances from the probe, and results are compared against each other to complete inspection procedure.
- Radiation: Uses X-ray radiation to accomplish non-contact inspection on metals and weld-fabricated products. The amount of radiation absorbed by the metal is measured and compared against standards. This allows metals that do not absorb sufficient amounts of radiation to be quickly spotted as flawed.
- Ultrasonic inspection: Uses very high frequency sound as an inspection mechanism. Methods can be either manually performed or performed automatically.
- Automated methods include emitting ultrasonic waves from a probe and reflecting them off the object to be inspected, to create a sound pattern. This sound pattern can be compared against the sound pattern produced by an ideal object for inspection purposes. If the produced sound pattern matches the standard pattern, the object passes the test; otherwise, it fails.

35.4.2.4 *Advantages of non-contact inspection techniques*

The advantages of non-contact inspection techniques over contact inspection techniques include:

- They avoid possible surface damage that can be caused upon contact.
- Inspection cycle times are faster as the contact probe must be re-positioned for each new part inspected, while the non-contact sensor remains stationary.
- Parts handling is lower with non-contact inspection than with contact inspection, as parts in the later methodology usually require special handling and adjustments so that inspection can occur.
- It allows for the possibility of 100% automated inspection.

35.5 INTEGRATION OF CAQC WITH CAD/CAM

The design department creates the product definition and the manufacturing department develops the manufacturing plan. It is important to add the QC to the CAD/CAM framework. All the applications of CAQC can be integrated with CAD/CAM to make the whole process of designing and manufacturing controlled by the computers converted into fully automated process. In the traditional process, there is a separate quality control department where the manufactured product is taken for the inspection and testing. In CAQC, the inspection process is integrated with the manufacturing process and it is located along the production line. Thus as soon as the product is manufactured, it is tested immediately by the computerized process without moving it to some other location. This helps in reducing the overall time required for manufacturing the product.

Although many important benefits result from the use of computer-aided quality control, additional benefits can be obtained by integrating CAQC with CAD/CAM. The quality control department must use the same CAD/CAM database to perform its function. Indeed, quality was defined as the degree to which a product or its components conform to the standards specified by the designer. These standards are all contained in the CAD/CAM database; available for QC to use. CAD/CAM database can be used to develop the NC programs to operate computer-controlled coordinate measuring machines. These programs can be generated automatically. These programs would then be downloaded to the CMM through a DNC link from the central computer to the controller unit for the CMM. The same sort of downloading process is possible for some of the non-contact inspection methods discussed earlier.

35. 6 QUESTIONS

1. Explain the concept of computer-aided quality control or CAQC.
2. Explain advantages of computer-aided quality control or CAQC.
3. What are the parameters of measuring instruments that are used to ensure correct device selection?
4. What are the two types of inspection techniques?
5. List the components of a coordinate measuring machine (CMM).
6. Describe the construction of a CMM probe.
7. What is machine vision?
8. List the three functions of machine vision.
9. How is image acquisition and digitization performed?
10. List some image processing and analysis techniques used in machine vision.
11. Name some common methods of interpretation in machine vision.
12. Explain various contact inspection methods.
13. Explain various non-contact inspection methods.
14. Explain optical non-contact inspection method.
15. Explain non-optical non-contact inspection method.
16. Explain integration of CAQC with CAD/CAM.

Index

A

Agile manufacturing 33, 46
ABC analysis 509
Actuators 170
Aggregate production planning 253
ALDEP 106
Arrival pattern 459
Assembly line balancing 125
Assignment model 449
Automated guided vehicles 34, 120, 160, 205
Automated manufacturing systems 43
Automatic tool changer 198
Average outgoing quality 347

B

Backward scheduling 282
Batch production system 41
Bathtub curve 69
Breakeven point 23

C

CAD software 66
Capacity planning 32, 261
Capital investment decision 425
Cellular manufacturing 43, 147
Centre of gravity method 79
Combination layout 94
Composite part 156
Compute-aided design 33, 60

Compute-aided engineering 167
Compute-aided manufacturing 33, 158
Compute-aided process planning 34, 163, 220
Compute-integrated warehousing 162
Computerized facility layouts 161
COMSOAL 34, 134, 162
Concurrent engineering 33, 55
Constructive solid geometry 65
Continuous demand 500
Continuous production system 40
Control charts 329
Controller 171
Conveyors 118
CORELAP 111, 161
CRAFT 105, 161
Cranes and hoists 118
Crashing of network 385
Critical path method 286, 372, 374
Critical ratio scheduling 286
Cutting parameters 213
Cycle time 116, 127

D

Decision making 433
Deming's 14 points 364
Deming cycle 364
Depreciation 20
Design for assembly 57
Design for manufacturability 57
Discrete demand 498
Drives 197

E

E-manufacturing 163
Elements of cost 14
End effecter 170
Enterprise resource planning 35, 275
Equipment failure rate 407
Extrude 63

F

Factor rating system 79
Finite element analysis 54
Fixed order quantity system 503
Fixed position layout 94
Flexible manufacturing system 34, 44, 161, 203
Flow patterns 90
Flow production system 40
FMS layout 208
FMS components 204
FMS computer control system 207
Forecast error 247
Forecasting 31, 229
Forward scheduling 281
FSND analysis 509

G

G-codes 201
Gantt chart 293
Generative CAPP system 223
Geometric models 62
Graphical method 301
Group replacement policy 422
Group technology 34, 43, 147
Group technology layout 95

H

Heuristic methods of project scheduling 397
Human resources 208
Hybrid approach 227

I

Index method 284
Individual replacement policy 422
Inspection 325
Intermittent production system 41
Interpolation 181
Inventory 478

J

Job shop and flow shop sequencing problems 292
Job shop production system 42
Johnson's algorithm 284, 293
Just in time production system 33, 47

K

Kanban system 48
Kendal's notation 461
Kilbridge-Wester method 142

L

Largest candidate rule 140
Lean production 33, 44
Line balancing 32
Line model 62
Line of balance 266
Linear equation 434
Linear programing 256, 434
Loading 42

M

M-codes 201
Machine control unit 183
Machine servicing model 473
Machining center 196
Maintenance 407
Maintenance management 32
Manipulators 169
Manual approach 220
Manufacturing operations 11
Manufacturing resource planning 163, 273
Mass production system 40
Master production schedule 32
Material handling 115, 204
Material requirements planning 32, 35, 162, 259
Mean time between failures 68
Multiplant locations 84
Multiserver model 470

N

Net present value 426
Normal distribution 351
Numerical control 34, 158, 178

O

OC curves 345
Operations management 30

Operations research 430
Opitz classification system 150
Optimum cutting speed 214
Overhead costs 18

 P

Part design 150
Part family 148
Part program 184
PERT 372, 381
Predictive maintenance 410
Preventive maintenance 410
Price breaks 490
Priority sequencing 288
Probabilistic models 497
Process capability 331
Process charts 96
Process layout 93
Process planning 31, 217
Process selection 31, 212
Product design 31, 52
Product layout 93
Product life cycle 50
Production concepts 1
Production control 262
Production flow analysis 152
Production line balancing 125
Production management 30
Production planning 32, 252
Production systems 4, 37
Productivity 5
Project scheduling under resource
    constraints 397
Project type production system 42

 Q

Qualitative forecasting techniques 230
Quantitative forecasting techniques 230
Quality 74
Quality control 32
Queuing theory 459

 R

Reactive maintenance 409
Relationship chart 102
Reliability centered maintenance 411
Reliability 74
Replacement models 417

Replenishment systems 502
Resource leveling 391
Resource aggregation 391
Resource allocation 391
Resource management 391
Retrieval CAPP system 221
Return on investment 79
Revolve 64
Robot 34, 121, 160, 168
Route sheet 153, 218
RPW technique 127

 S

S-s policy 504
Safety stock 503, 507
Sampling plans 343
Scheduling 280
Selective control 509
Selling price 15
Sensors 170
Shop floor control 32, 262
Simplex method 439
Single server model 461
Six sigma 35, 351, 356
Smoothing methods 240
Statistical quality control 325
Supply chain management 276
Surface model 62
Sweep 64

 T

Total productive maintenance 412
Total quality management 35, 366
Tracking signal 248
Transportation model 86
Transportation problem 444
Travel charts 97
Traveling salesman problem 304
Two-bin system 503

 U

Unit load 121
VED analysis 509
Volume model 62

 W

Workstation 126